KU-798-089

LADY MARGARET HALL LIBRARY
OXFORD

302639740Y

Chemical Analysis by Nuclear Methods

Chemical Analysis by Nuclear Methods

Editor Z. B. Alfassi
Ben-Gurion University of the Negev, Israel

JOHN WILEY & SONS
Chichester · New York · Brisbane · Toronto · Singapore

Baffins Lane, Chichester,
West Sussex PO 19 1UD, England
Telephone National Chichester (0243) 779777
International +44 243 77977

Other Wiley Editorial offices

Johen Wiley & Sons, Inc., 605 Third Avenue,
New York, NY 10158-0012, USA

Jacaranda Wiley Ltd, 33, Park Road, Milton,
Queensland 4064, Australia

John Wiley & Sons (Canada) Ltd, 22 Worcester Road,
Rexdale, Ontario M9W 1L1, Canada

John Wiley & Sons (SEA) Pte Ltd, 37 Jalan Pemimpin #05-04,
Block B, Union Industrial Building, Singapore 2057

Library of Congress Cataloging-in-Publication Data

Chemical analysis by nuclear methods / editor, Z. B. Alfassi.
p. cm.
Includes bibliographical references and index.
ISBN 0 471 93834 3
1. Radiochemical analysis. I. Alfassi, Zeev B.
QD605.C48 1994
543'.008—dc20 93-33546
CIP

British Library Cataloguing in Publication Data

A catalogue record for this book is available from the British Library

ISBN 0 471 93834 3

Typeset in 10/12pt Times by Thompson Press (India) Limited, New Delhi
Printed and bound in Great Britain by Biddles Ltd, Guildford, Surrey

Dedicated to
Professor Mo Hsiung Yang of the Institute of Nuclear Sciences of the National Tsing Hua University, Hsinchu, Taiwan.

For his true friendship and hospitality through the period this book was thought of and structured

CONTENTS

Part 1: Basic Background in Nuclear Physics and Chemistry

CHAPTER 1: INTERACTION OF RADIATION WITH MATTER
A. P. Kushelevsky

1 INTRODUCTION 3
1.1 General concepts and definitions 3
1.2 The macroscopic description 4
1.2.1 Radiation field intensity 4
1.2.2 The attenuation and stopping power coefficients 4
1.2.3 The absorbed dose 5
1.3 The microscopic description 5
1.3.1 The radiation field 5
1.3.2 Interaction cross-sections 5
1.4 Conversion between microscopic and macroscopic parameters 6
1.5 Types of radiation 6
2 CHARGED PARTICLE INTERACTIONS 7
2.1 The continuous slowing down approximation (CSDA) 7
2.1.1 Stopping power of heavy charged particles 7
2.1.2 The range of heavy charged particles in matter 8
3 HIGH ENERGY ELECTRONS AND BETA-RAYS 9
3.1 Electrons 9
3.1.1 The stopping power of electrons 9
3.1.2 Bremsstrahlung 10
3.1.3 Cherenkov radiation 10
3.2 Beta-rays 10
4 UNCHARGED RADIATIONS 11
4.1 High energy X-rays and gamma-rays 11
4.1.1 Absorption and scatter coefficients 13
4.1.2 Buildup factors 13
4.2 Photon reactions 13
4.2.1 Elastic scattering 13
4.2.2 Compton scattering 14

4.2.3 The photoelectric effect . . . 14
4.2.4 Pair production . . . 16
4.2.5 The photonuclear effect . . . 19
5.1 Neutron interactions . . . 19
5.1.1 Neutron elastic scattering . . . 20
5.1.2 Inelastic scattering . . . 21
5.2 Neutron capture . . . 21
5.2.1 The (n, γ) reaction . . . 21
5.2.2 (n, nucleon) reactions . . . 21
5.2.3 (n, fission) reactions . . . 22

CHAPTER 2: NUCLEAR DETECTION METHODS AND INSTRUMENTATION
A. Tsechanski

1 INTRODUCTION . . . 23
2 NUCLEAR DETECTORS . . . 23
2.1 Gas-filled detectors . . . 23
2.2 Scintillation counters . . . 28
2.2.1 Introduction . . . 28
2.2.2 Organic scintillators . . . 30
2.2.3 Inorganic scintillators . . . 32
2.2.4 Photomultiplier tubes . . . 33
2.3 Semiconductor detectors . . . 35
2.3.1 Introduction . . . 35
2.3.2 Basic principles . . . 36
2.3.3 Physical properties of pure Ge and Si . . . 36
2.3.4 Physical properties of real (doped) Ge and Si . . . 38
2.3.5 Basic properties of p–n junctions . . . 41
2.3.6 Charged particle detectors . . . 43
2.3.7 Germanium detectors . . . 46
3 GAMMA-RAY SPECTROSCOPY . . . 51
3.1 Instrumentation for nuclear spectroscopy . . . 51
3.2 Energy measurements . . . 53
REFERENCES . . . 56

CHAPTER 3: RADIATION SOURCES
Z. B. Alfassi and K. Strijckmans

1 NEUTRON SOURCES . . . 57
Z. B. Alfassi

1.1 Research nuclear reactors . . . 57
1.1.1 Sample introduction . . . 59
1.1.2 Irradiation with neutrons of limited energy range . . . 59

1.1.3 Irradiation with fast neutrons 61
1.2 Neutrons from accelerators and neutron generators 61
1.3 Radioactive neutron sources 62
2 CHARGED PARTICLE ACCELERATORS 64
K. Strijckmans
2.1 Introduction 64
2.1.1 A charge in an electric field 64
2.1.2 A charge in a magnetic field 65
2.1.3 Relativity 65
2.2 DC accelerators 66
2.3 Cyclic accelerators 68
2.3.1 The linear accelerator 69
2.3.2 The cyclotron 69
2.3.3 The synchrocyclotron 71
2.3.4 The isochronous cyclotron 72
2.3.5 The synchrotron 72
2.3.6 The betatron 73
REFERENCES 74

CHAPTER 4: NUCLEAR REACTION CHEMICAL ANALYSIS: PROMPT AND DELAYED MEASUREMENTS
M. D. Glascock

1 INTRODUCTION 75
2 NUCLEAR REACTIONS 77
2.1 Scattering reactions 78
2.2 Compound nucleus formation 78
2.3 Reaction cross-sections 79
2.4 Excitation functions 80
3 IRRADIATION SOURCES 81
3.1 Nuclear reactors 81
3.2 Neutron generators 82
3.3 Cyclotrons and accelerators 83
3.4 Radioisotopic sources 83
4 DELAYED ANALYSIS METHODS 83
4.1 Delayed gamma-ray neutron activation analysis 84
4.1.1 NAA with epithermal neutrons 85
4.1.2 Fast neutron activation analysis (FNAA) 85
4.1.3 Sensitivities and applications for DGNAA 86
4.2 Delayed gamma-rays following charged particle activation . . 86
4.3 Photon activation analysis 90
4.4 Delayed neutrons following neutron activation 91
5 PROMPT METHODS OF ANALYSIS 91

5.1 Prompt gamma-ray neutron activation analysis 91
5.2 Particle-induced gamma-ray emission 94
5.3 Particle-induced prompt particle emission 94
5.4 Neutron depth profiling 95
6 ADVANTAGES AND DISADVANTAGES OF PROMPT vs. DELAYED ANALYSIS 95
7 CONCLUSIONS 96
REFERENCES 97

CHAPTER 5: RADIATION PROTECTION
T. Schlesinger

1 INTRODUCTION 101
2 TYPES AND ROUTES OF EXPOSURE 102
3 PHYSIOLOGICAL AND RADIOLOGICAL CONCEPTS, PARAMETERS, QUANTITIES AND UNITS USED IN RADIATION PROTECTION 102
3.1 Types of ionizing radiation 102
3.2 Penetration range of ionizing radiations into matter 102
3.2.1 Range of alpha-radiation 102
3.2.2 Range of beta-radiation 103
3.2.3 Range of gamma-radiation and X-rays 103
3.3 Special unit of energy 103
3.4 Units of activity and specific activity 103
3.5 Radiological quantities and units 103
3.5.1 Exposure 103
3.5.2 Dose 104
3.5.3 Dose equivalent 104
3.5.4 Effective dose equivalent 105
3.6 The K-factor 105
3.7 Physical, biological and effective half-lives (or half-times) . . . 106
3.8 Shielding 106
3.8.1 Half value thickness 106
4 RELATION BETWEEN EXPOSURE, ACTIVITY, DISTANCE AND TIME 107
5 BIOLOGICAL EFFECTS OF IONIZING RADIATION 108
5.1 Somatic effects 108
5.1.1 Acute effects (also called deterministic or non-stochastic effects 108
5.1.2 Late (delayed) effects 109
5.2 Genetic damage 110
6 RESTRICTIONS ON EXPOSURE TO IONIZING RADIATION—INTERNATIONAL STANDARDS 111
6.1 Annual dose limits for workers 112

6.2 Dose limits for members of the public 113
6.3 Dose limits for pregnant women 113
6.4 Medical exposures 113
6.5 Emergency exposures 113
6.6 Internal contamination—limits of intake 114
7 METHODS OF RADIATION PROTECTION 115
7.1 External radiation—time, distance and shielding 115
7.2 Internal contamination 115
8 MONITORING AND CONTROL 116
9 NATURAL BACKGROUND RADIATION 116
REFERENCES 117

Part 2: Elemental Analysis with Neutron Sources

CHAPTER 6: DELAYED INSTRUMENTAL NEUTRON ACTIVATION ANALYSIS
S. Landsberger

1 INTRODUCTION 121
2 THEORY OF NEUTRON ACTIVATION ANALYSIS 122
2.1 Neutron-induced reactions 123
2.2 Reaction rate 124
3 NEUTRON ACTIVATION ANALYSIS CALCULATIONS . . . 125
4 IRRADIATION AND COUNTING PROCEDURES 127
5 INTERFERENCES IN NAA 130
6 ACCURACY AND PRECISION 133
7 SENSITIVITY 137
8 QUALITY CONTROL AND REFERENCE MATERIALS 139
9 FUTURE DEVELOPMENTS 140
REFERENCES 140

CHAPTER 7: RADIOCHEMICAL NEUTRON ACTIVATION ANALYSIS
Z. B. Alfassi

1 INTRODUCTION 143
1.1 The different steps in RNAA 147
2 SAMPLE DISSOLUTION 148
2.1 Geological samples 148
2.2 Metal samples 149
2.3 Biological samples 149
3 DETERMINATION OF CHEMICAL YIELD 155
4 RADIOCHEMICAL SEPARATIONS 157
REFERENCES 160

CHAPTER 8: PROMPT ACTIVATION ANALYSIS
C. Chung

1 INTRODUCTION 163
2 PRINCIPLES 164
2.1 Nuclear reactions 164
2.2 Prompt gamma-rays 166
2.3 Prompt gamma detection 168
3 QUANTITATIVE DETERMINATION 169
3.1 Spectroscopic analysis 170
3.2 Counting statistics 173
3.3 Detection limit 174
4 PGAA FACILITIES 176
4.1 General layout 176
4.1.1 Neutron sources 176
4.1.2 Neutron collimator and shield 177
4.1.3 Sample handling device 178
4.1.4 Gamma-ray spectrometer 179
4.2 Typical PGAA facility 179
4.3 Applications 184
4.4 Comparison with other nuclear methods 185
REFERENCES 186

CHAPTER 9: CHEMICAL ANALYSIS BY THE THERMALIZATION, SCATTERING AND ABSORPTION OF NEUTRONS
C. M. Bartle

1 INTRODUCTION 189
1.1 Analysis by the thermalization of neutrons 190
1.1.1 Measurement of the moisture content of soil and materials in field locations 191
1.1.2 Water and hydrogen measurement in industrial materials 193
1.2 Analysis by the scattering of neutrons 194
1.2.1 The kinematics of scattering 194
1.2.2 Analysis based on neutron transmission 195
1.2.3 Analysis based on the simultaneous transmission of neutrons and gamma-rays (Neugat) 197
1.2.4 Measurements of chemical composition using the Neugat method 200
1.3 Analysis by the absorption of neutrons 203
1.3.1 Polyenergetic neutron transmission 204
1.3.2 Monoenergetic neutron transmission 207
REFERENCES 207

Part 3: Elemental Analysis with Particle Accelerators

CHAPTER 10: CHARGED PARTICLE ACTIVATION ANALYSIS
K. Strijckmans

1 INTRODUCTION 215
2 THEORY 216
2.1 Interaction of charged particles with matter 216
2.1.1 Stopping power of elements 216
2.1.2 Stopping power of mixtures and compounds 216
2.1.3 Range 217
2.1.4 Calculation of stopping power and range data for elemental matter, mixtures and compounds 218
2.2 Nuclear reaction 219
2.2.1 Q-value and threshold energy 219
2.2.2 Coulomb barrier 220
2.2.3 Nuclear reaction cross-section 221
2.2.4 Beam intensity reduction 222
2.2.5 Radionuclides formed by nuclear reaction 223
2.3 Interferences 224
2.3.1 Nuclear interference 224
2.3.2 Spectral interference 224
2.3.3 Matrix interference 224
2.4 Standardization 225
2.4.1 Introduction 225
2.4.2 Approximative standardization methods 227
2.4.3 The average stopping power method 229
2.4.4 The internal standardization method 230
2.4.5 The two reactions method 232
2.4.6 Standard addition method 233
3 EXPERIMENTAL 233
3.1 Irradiation 233
3.1.1 Accelerator 233
3.1.2 Target 234
3.1.3 Direct beam intensity monitoring 234
3.2 Quantitative beam intensity monitoring 237
3.3 Chemical etch 238
3.4 Energy reduction calculation 239
3.5 Recoil nuclides 241
3.6 Radiochemical separation 242
3.7 Standards 245
3.8 Activity measurement 246

3.8.1 Gamma spectrometry 246
3.8.2 Positron counting 247
3.8.3 Decay curve analysis 249
3.8.4 Decay correction 249
3.9 Quantitation 249
3.9.1 No interferences 249
3.9.2 Correction for spectral interference 250
3.9.3 Correction for nuclear interference 250
REFERENCES 251

CHAPTER 11: ION BACKSCATTERING AND ELASTIC RECOIL DETECTION
E. Rauhala

1 INTRODUCTION 253
2 THEORETICAL PRINCIPLES 255
2.1 Mass perception—energetics of the elastic collision 256
2.2 Depth analysis—energy loss of particles in matter 256
2.3 Quantitative analysis—scattering probability and cross-section 258
2.3.1 Rutherford scattering 259
2.3.2 Non-Rutherford scattering 259
3 SPECTRUM EVALUATION 260
3.1 Mass analysis and depth scale 261
3.1.1 Backscattering spectrum 261
3.1.2 Elastic recoil spectrum 262
3.2 Quantitative analysis 263
4 EXPERIMENTAL TECHNIQUES 265
4.1 Ion backscattering 265
4.1.1 Ion channeling 265
4.1.2 Special arrangements 267
4.2 Elastic recoil 268
4.2.1 Conventional set-ups 268
4.2.2 Special arrangements 269
5 DATA ANALYSIS 271
6 ANALYTICAL CHARACTERISTICS 274
6.1 Ion backscattering 274
6.2 Elastic recoil detection 276
7 APPLICATIONS TO ELEMENTAL ANALYSIS 276
7.1 Ion backscattering 276
7.2 Elastic recoil detection 281
REFERENCES 284

CHAPTER 12: NUCLEAR REACTION ANALYSIS OR MORE GENERALLY CHARGED PARTICLE (ACTIVATION) ANALYSIS
F. Sellschop

1 INTRODUCTION–THE THIRD GENERATION OF NUCLEAR ANALYSIS 293
2 PRINCIPLES 294
3 APPLICATIONS OF ELEMENTAL ANALYSIS 299
4 BEYOND PURE ELEMENTAL ANALYSIS 300
4.1 Ion channelling 300
4.2 Time differential perturbed angular distributions (TDPAD): microscopic CP(A)A 311
4.3 Muon spin rotation and muonium 315
4.4 The special role of μ beams in CP(A)A: $\bar{\mu}$ SR 320
5 THE IMMEDIATE FUTURE 321
6 ACKNOWLEDGEMENTS 321
REFERENCES 322

CHAPTER 13: PARTICLE-INDUCED X-RAY EMISSION
U. A. S. Tapper, W. J. Przybyłowicz and H. J. Annegarn

1 INTRODUCTION 323
2 X-RAY PRODUCTION 328
2.1 Ionization cross-sections 328
2.2 Characteristic radiation 329
2.3 Continuous background 332
2.4 Detection limits 335
2.5 Intensity of fluorescence radiation 336
2.5.1 Basic formalism 336
2.5.2 Thin target approximation 338
2.5.3 Thick targets 339
3 X-RAY SPECTROSCOPY 340
3.1 X-ray detectors 340
3.2 Spectrum fitting 343
3.3 Calibration factor 345
3.4 Sample chamber 346
3.5 Beam charge measurements 347
4 APPLICATIONS 348
4.1 Biology and medicine 348
4.2 Aerosol studies 350
4.3 Geoscience 353
5 COMPARISON WITH OTHER TECHNIQUES 355
REFERENCES 356

CHAPTER 14: USE OF MICROPROBES
U. Lindh

1 INTRODUCTION . . . 361
2 NUCLEAR MICROSCOPY OF CELLS AND TISSUES . . . 362
3 MINI-REVIEW OF BIOMEDICAL NUCLEAR MICROSCOPY. . . . 363
4 SLIM-UP—AN EXAMPLE OF A NUCLEAR MICROSCOPE. . . . 366
5 ALZHEIMER'S DISEASE—A CONTROVERSY OF TRACE ELEMENT INVOLVEMENT . . . 368
6 APPLICATIONS IN DERMATOLOGY . . . 370
7 APPLICATIONS TO CARDIOVASCULAR DISEASE (ATHEROSCLEROSIS) . . . 373
8 APPLICATION TO AIDS RESEARCH . . . 378
9 APPLICATIONS TO THE CHRONIC FATIGUE SYNDROME. . . . 380
10 NUCLEAR MICROSCOPY AS A MEANS TO STUDY PROTECTIVE EFFECTS . . . 383
11 CONCLUDING REMARKS . . . 385
REFERENCES . . . 385

Part 4: Use of Radioactive (alpha, beta and gamma) Sources

CHAPTER 15: X-RAY FLUORESCENCE ANALYSIS WITH RADIOACTIVE SOURCES
T. Biran-Izak and M. Mantel

1 INTRODUCTION . . . 391
2 GENERAL . . . 391
3 X-RAY FLUORESCENCE SYSTEM . . . 393
3.1 Excitation . . . 393
3.1.1 Sources . . . 393
3.1.2 Configuration . . . 397
3.2 Detection . . . 399
3.2.1 Detectors . . . 399
3.2.2 Collimation and shielding. . . . 399
4 MATRIX EFFECT . . . 400
5 SAMPLE PREPARATION . . . 402
6 APPLICATIONS . . . 403
6.1 Industry . . . 403
6.2 Geology . . . 408
6.3 Environment . . . 410

6.4 Archaeology 411
6.5 Biology 412
REFERENCES 414

CHAPTER 16: SCATTERING OF ALPHA-, BETA- AND GAMMA-RADIATION FOR CHEMICAL ANALYSIS
E. M. A. Hussein

1 INTRODUCTION 417
2 ALPHA-PARTICLES 418
2.1 Scattering modalities 418
2.2 Angular distribution 420
2.3 Kinematics 421
2.4 Applications 422
3 BETA-PARTICLES 423
3.1 Scattering modalities 424
3.2 Applications 425
4 GAMMA-RAYS 426
4.1 Compton scattering 427
4.2 Rayleigh scattering 427
4.3 Applications 428
REFERENCES 430

CHAPTER 17: MÖSSBAUER SPECTROSCOPY IN CHEMICAL ANALYSIS
E. Kuzman, S. Nagy and A. Vértes

1 PRINCIPLES OF MÖSSBAUER SPECTROSCOPY 433
1.1 Mössbauer parameters 434
1.2 Dependence of the Mössbauer parameters on external physical parameters 434
1.3 Measurement of Mössbauer spectra 442
2 ANALYTICAL INFORMATION FROM MÖSSBAUER SPECTRA. 446
2.1 The fingerprint method 446
2.2 Pattern analysis 447
2.3 Spectrum decomposition 451
2.4 Quantitative analysis 454
3 EXAMPLES FOR ANALYTICAL APPLICATIONS 455
3.1 Corrosion 456
3.2 Phase analysis in alloys 459
3.3 Elemental analysis in alloys 467
REFERENCES 473

CHAPTER 18: CHEMICAL ANALYSIS BY POSITRON ANNIHILATION
Y. C. Jean

1 POSITRON ANNIHILATION AND POSITRONIUM CHEMISTRY 477
1.1 Positrons 477
1.2 Positronium 478
1.3 Quenching of Ps 479
1.4 Inhibition 481
2 EXPERIMENTAL TECHNIQUES FOR POSITRON ANNIHILATION 482
2.1 Positron sources 482
2.1.1 Fast positrons 482
2.1.2 Slow positrons 483
2.2 Positron annihilation lifetime spectrosocopy (PAL) 485
2.3 Angular correlation of positron annihilation radiation (ACAR) 487
2.4 Doppler broadening spectroscopy (DBS) 489
3 MICROSTRUCTURAL ANALYSIS OF FREE VOLUME HOLES IN POLYMERS 490
3.1 Hole size, distribution, and fraction 490
3.2 Anisotropic structure of free volume holes 493
4 SURFACE-STATE ANALYSIS OF POROUS MEDIA 497
4.1 In situ characteristics for Ps 497
4.2 Determination of total surface area 498
4.3 Determination of chemical state in catalysts 499
5 POSITRON ANNIHILATION-INDUCED AUGER ELECTRON SPECTROSCOPY 500
6 POSITRON IONIZATION MASS SPECTROMETRY 501
7 OTHER POSITRON SPECTROSCOPIES SIMULATING ELECTRON SPECTROSCOPY 502
7.1 Positron microscopy 502
7.2 Positron energy loss spectroscopy 505
7.3 Low energy positron diffraction 505
7.4 Positron polarimetry 505
8 CONCLUSIONS 506
9 ACKNOWLEDGMENT 506
REFERENCES 506

Part 5: Use of Radiotracers

CHAPTER 19: ISOTOPE DILUTION ANALYSIS
K. Masumoto

1 INTRODUCTION 511
2 IDA BASED ON SPECIFIC ACTIVITY MEASUREMENT . . . 512
2.1 General types of IDA 512
2.1.1 DIDA 512
2.1.2 RIDA 512
2.1.3 Conditions for IDA 512
2.1.4 Application of IDA 513
2.2 Substoichiometric IDA 513
2.2.1 Sub-IDA 513
2.2.2 Substoichiometry after activation 514
2.2.3 Conditions for sub-IDA 516
2.2.4 Application of sub-IDA 516
2.3 Sub- and super-equivalence (SSE) method 517
2.3.1 Principle of SSE 517
2.3.2 Application of SSE 519
3 IDA BASED ON ISOTOPE RATIO MEASUREMENT 519
3.1 Isotope dilution mass spectrometry 519
3.2 Isotope dilution alpha spectrometry 520
3.3 Stable isotope dilution activation analysis 520
4 SPECIAL FEATURES OF ISOTOPE DILUTION 522
4.1 Isotope effect 522
4.2 Isotopic equilibrium 522
REFERENCES 522

CHAPTER 20: RADIOIMMUNOASSAYS AND RELATED RADIOACTIVE METHODS FOR THE QUANTITATION OF HORMONES AND OTHER SUBSTANCES IN BIOLOGICAL FLUIDS
J. Levy, S. Glick and Y. Sharoni

1 RADIOIMMUNOASSAY 525
1.1 Principles 527
1.2 Reagents and antibody 528
1.3 Reagents and antigen labeling 530
1.4 Separation methods 532
1.5 Calculations and presentation of results 533
1.6 Validation 535
1.7 Assay sensitivity 536
1.8 Assay specificity 536

1.9 Problems 537
1.9.1 Effects of experimental conditions on the immune reaction 537
1.9.2 Effect of degradation of labeled antigen and/or antibody 538
1.9.3 Heterologous hormone standards 539
1.9.4 Immunologically related but different hormones 539
1.9.5 Heterogeneity of the test compound 540
1.9.6 Circulating endogenous antibody 540
2 IMMUNORADIOMETRIC ASSAY 541
3 RADIORECEPTOR ASSAY 543
REFERENCES 544

INDEX. 547

Part 1

Basic Background in Nuclear Physics and Chemistry

Chapter 1

INTERACTION OF RADIATION WITH MATTER

A. P. Kushelevsky
Department of Nuclear Engineering, Ben-Gurion University of the Negev, Beer-Sheva, Israel

1 INTRODUCTION

This chapter discusses how radiation interacts with matter, a topic required for optimizing elemental analysis by nuclear methods in terms of sensitivity, accuracy and precision.

It opens by discussing a number of general concepts and definitions used in studying the interaction of all types of radiation with matter. These are given in Table 1, and are discussed in detail in the following sections.

1.1 General concepts and definitions

Two approaches are used in describing the interaction of radiation with matter. The first, a macroscopic approach, uses parameters which are familiar and relatively easy to measure, describing the interacting radiation field in terms of its energy content, and matter in terms of its linear dimensions, mass, density and chemical composition.

The second, a microscopic approach, focuses on the discrete nature of the interactions, e.g. on the interactions of the individual particles in the radiation beam with the individual atoms, nuclei and electrons in matter.

Each of these approaches has advantages. In shielding calculations, for example, the macroscopic approach is the more convenient one to use because output data is required in macroscopic units, e.g. thickness and weight, of the shielding material.

Chemical Analysis by Nuclear Methods Edited by Z. B. Alfassi

TABLE 1.
Quantities and units used to describe the interaction of radiation and matter

Name	Symbol	Units
Particle fluence	Φ	cm^{-2}
Particle flux density	ϕ	$cm^{-2} s^{-1}$
Energy fluence	F	erg cm^{-2}
Energy flux density	I	erg $cm^{-2} s^{-1}$
Attenuation coefficient	μ	cm^{-1}
Mass attenuation coefficient	μ/ρ	$cm^{-2} g^{-1}$
Stopping power	S	erg cm^{-1}
Mass stopping power	S/ρ	erg $cm^2 g^{-1}$
Absorbed dose	D	erg g^{-1}

On the other hand, in activation analysis, where it is necessary to calculate the number of nuclei transformed in various nuclear reactions, the microscopic approach is the more convinent, since most of the relevant input data is given in microscopic terms, e.g. atomic cross-sections.

1.2 The macroscopic description

1.2.1 Radiation field intensity

In the macroscopic approach, the radiation intensity at a point in the field is defined by the local energy fluence F and energy flux density I at that point. To define these quantities, we consider a very small sphere with cross-section ΔS. The energy fluence F is the quotient of ΔQ by ΔS, where ΔQ is the total energy, exclusive of rest energy, crossing the sphere, and the flux density I is the quotient of the energy fluence by Δt, where Δt and ΔS are small enough to ensure uniformity. In the case of a uniform radiation beam incident perpendicularly on a plane surface, I is simply the energy transferred through a unit area of the surface per second.

1.2.2 The attenuation and stopping power coefficients

Two parameters—the attenuation coefficient μ, for non-charged radiation, and the stopping power S for charged particle radiation—characterize the ability of the medium to interact with the radiation field.

The former is defined as the fractional loss of intensity per unit path length, or $(\Delta I/I)\Delta x$, where Δx is a very small distance in the attenuating medium. The latter describes the rate at which the radiation beam loses its energy per unit path length in the absorbing medium, i.e. $\Delta E/\Delta x$ in units of (energy length^{-1}).

1.2.3 The absorbed dose

Another very important macroscopic parameter is the absorbed dose D, defined as the quotient of ΔE_d by Δm, where ΔE_d is the energy imparted by ionizing radiation to matter in a volume element of mass Δm:

$$D = \Delta E_d/\Delta m \tag{1}$$

This parameter is used extensively in radiation dosimetry, and is measured either in rads, defined as 100 erg/g, or in grays, defined as 1 J/kg.

1.3 The microscopic description

1.3.1 The radiation field

The microscopic approach, which takes into account the discrete structure of the radiation field, uses the particle fluence Φ and particle flux ϕ to describe the field, where Φ is defined as the quotient $\Delta N/\Delta S$; ϕ as the quotient $\Delta\Phi/\Delta t$, and ΔN is the number of particles entering a very small sphere with a cross-sectional area ΔS.

1.3.2 Interaction cross-sections

The ability of the radiation field to interact with the field in microscopic terms is described by the interaction cross-section of the medium σ, defined as the probability of an interaction occurring between a beam of unit fluence, e.g. a single particle, and a single interaction site in the path of the beam. Geometrically, it describes the apparent reaction area exhibited by individual interacting sites, and hence σ has the dimensions of area.

Because of the central role that the cross-section plays in radiation studies, considerable effort has been made to develop theories to predict its value theoretically. Unfortunately, this has met with little success because σ has been found experimentally to be a complicated function of the atomic properties of the medium (atomic number and atomic weight), of the interacting particles (type of particle and its energy) and of the type of interaction considered.

The approach taken in practice is to use experimental data which have been organized in tabular form and classified according to the type of interacting particle, energy, nature of the target nuclei and specific nuclear interaction. In many of these tables the data is given in barns, a unit of area equal to 10^{-24} cm^2, which is of the order of magnitude of the geometric cross-section of typical nuclei.

1.4 Conversion between microscopic and macroscopic parameters

The formal similarity between the definitions of particle fluence and energy fluence lead to simple equations for conversion between the microscopic and macroscopic parameters. For example, the intensity of a monoenergetic beam given in particle fluence units may be converted into energy fluence units using the equation:

$$F = E\Phi \tag{2}$$

where E is the energy of the particles.

For a multi-energetic beam, conversion is effected by dividing the energy spectrum of the beam into small energy intervals, multiplying each energy interval by the corresponding particle fluence and summing over the various energy intervals, i.e.:

$$F = \Sigma\,(\Phi(E)\cdot E) \tag{3}$$

The equation which converts microscopic cross-sections into macroscopic attenuation coefficients is $\mu = \sigma n$. This results from the following consideration: Let a uniform beam of monoenergetic particles of energy E and flux density ϕ bombard a thin slab of matter, δx cm thick, containing n atoms/ cm^3, which occupy the volume so sparsely that the front atoms do not shield the atoms behind them. Since each atomic target has a cross-sectional area of σ, the projected area of all the atoms on to the front face of the slab per unit area is $\sigma\, n\, \delta x$ cm^2 and the total number of interactions per second $= \phi\, \sigma n \cdot \delta x$.

Assuming that each particle carries E units of energy, the total energy of the beam entering the slab per cm^2 per second is ϕE; the energy ΔE removed from it per second due to the interactions is $\sigma\, n\, \phi\, E\cdot \delta x$, and the fraction of energy removed from the beam per unit thickness is $(\Delta E/\delta x)\,\phi\, E = \sigma n$. This, by definition, is exactly equal to the attenuation coefficient (see Section 1.2.2). Thus:

$$\mu = \sigma\, n \tag{4}$$

When dealing with neutron reactions, the product σn is often designated Σ and is called the macroscopic cross-section.

Equation (4) may also be calculated by substituting $(\rho/A) \times N$ for n, where ρ is the density in g/cm^3, A is the atomic weight of the attenuating medium and N is Avogadro's number $= 0.602 \times 10^{24}$, giving

$$\mu = \sigma\,(\rho/A)\cdot N \tag{5}$$

1.5 Types of radiation

In the study of the interaction of radiation with matter, it is convenient to divide the various types of radiation into two groups: charged particle radiations—high energy protons, alpha-particles, fast electrons and beta-particles and highly charged large fission fragments—and uncharged radiations—X-rays, gamma-rays and neutrons.

These two groups of radiation differ in a number of ways, but most dramatically in the depth by which each penetrates matter. For example, 5 MeV alpha-particles penetrate only a few cm in air, whereas γ-rays of the same energy penetrate many meters without appreciable attenuation.

As we shall see in the following sections, most of the energy of all types of radiation is ultimately transferred to matter by means of charged particle interactions, e.g. via secondary electron reactions. It is therefore appropriate to begin our detailed discussions with charged particles.

2 CHARGED PARTICLE INTERACTIONS

2.1 The continuous slowing down approximation (CSDA)

Two important experimental results may be deduced by observing the tracks of heavy charged particles in cloud and bubble chambers, which appear as thick straight cores of ionization with relatively short tracks branching off at random distances from the main core. First, high energy charged particles lose their energy in matter gradually by electrical interactions with bound atomic electrons. Second, the average energy lost per interaction is very small.

The continuous slowing down model (CSDA) based on these observations successfully explains the experimental track structure, and also permits us to develop the following equation to calculate heavy particle stopping powers (in MeV/cm):

$$-\mathrm{d}E/\mathrm{d}x = 4z^2e^4n[\ln(2mV^2/I) - \ln(1-b^2) - b^2]/(mV^2) \qquad (6)$$

where z = the charge of the particle, V = the velocity of the particle, c = velocity of light, 3×10^8 m/s, m = the rest mass of the electron, e = the electron charge, n = The number of electrons in the medium per cm^3, and I = the mean excitation and ionization potential of the atoms of the medium.

Using this equation, the stopping powers of different particles with non-relativistic velocities, may be easily compared in a given medium if their relative charges and masses are known.

2.1.1 Stopping power of heavy charged particles

Stopping powers of high energy protons in soft tissue as a function of energy are given in Table 2. As predicted by equation (6), at low energies or low velocities the slopping powers are relatively large compared with the much lower values at high energies, and small changes in the proton energy at very low energies lead to

TABLE 2.
Stopping powers and ranges of high energy protons in soft tissue

Energy, MeV	dE/dx, MeV/cm	Range, cm	Energy, MeV	dE/dx, MeV/cm	Range, cm
1	268.04	0.00	25	21.38	0.643
2	161.09	0.005	50	12.21	2.262
3	118.04	0.012	100	7.14	7.85
4	94.24	0.022	200	4.39	26.50
5	78.97	0.034	300	3.44	52.56
6	68.30	0.047	400	2.965	84.06
7	60.31	0.063			
8	54.15	0.08			
9	49.22	0.10			
10	45.19	0.12			
12	38.96	0.169			
14	34.35	0.224			
16	30.80	0.285			
18	27.97	0.353			
20	25.66	0.428			

relatively large changes in stopping power compared with the much smaller changes in the stopping power for small changes in the proton energy at high energies.

2.1.2 The range of heavy charged particles in matter

The average distance a charged particle travels in the medium until it loses its excess kinetic energy, or its range R, may be calculated by integrating equation (6) from E_{initial} to 0, i.e.

$$R = \int_{E_{\text{initial}}}^{0} (-\,dE/dx)^{-1}\,dE \tag{7}$$

Calculated ranges of high energy protons in soft tissue as a function of energy are shown in Table 2. Note the non-linear relationship between range and energy, with the ranges of high energy protons considerably greater than would be expected from simple proportionality of range with energy.

If only approximate values are required, empirical range energy formulae may be used. The following equations are examples of useful formulae developed for alpha-particles with energies up to 8 MeV in air at STP conditions (0 °C and 760 mm Hg):

$$R\,(\text{cm}) = 0.56\,E\,(\text{MeV}) \quad \text{for} \quad E < 4\,\text{MeV} \tag{8}$$

and

$$R\,(\text{cm}) = 1.24\,E\,(\text{MeV}) - 2.62 \quad \text{for} \quad 4 < E < 8\,\text{MeV}$$

3 HIGH ENERGY ELECTRONS AND BETA-RAYS

High energy electrons produced by electron accelerators and beta-rays emitted by radioisotopes interact with matter very similarly to heavy charged particles. Like heavy charged particles, they leave behind a train of excited and ionized atoms in passing through matter, losing very small amounts of energy per interaction as they interact with bound electrons via long range electrical forces. As with heavy charged particles, therefore, the CSDA model may be used to describe their interaction with matter.

Nevertheless, two important reasons exist for considering these particles separately from heavy charged particles. First, they are easily scattered by the electrons at rest in the medium, a fact which causes them to be sharply deflected from their original directions and results in tortuous tracks differing sharply from the straight paths traced by heavy charged particles. Second, unlike heavy charged particles, they lose significant amounts of energy in the form of electromagnetic radiation (bremsstrahlung) emitted as they are decelerated by the electric fields surrounding the atoms with which they interact.

3.1 Electrons

3.1.1 The stopping power of electrons

Because of the differences in the behavior of high energy electrons and heavy charged particles mentioned above, and due to relativistic and quantum mechanical exchange phenomena, the stopping powers of high energy electrons are smaller than those of heavy charged particles. This is seen by comparing the stopping powers of electrons (Table 3) with those of high energy protons in soft tissue (Table 2).

TABLE 3.
Stopping power and range of electrons in tissue

Energy, MeV	Stopping power, $MeV \cdot cm^2/g$	Range, g/cm^2	Bremsstrahlung, %
0.1	4.145	0.0142	0.07
0.25	2.550	0.0632	0.14
0.5	2.052	0.1754	0.26
0.75	1.927	0.3019	0.37
1.0	1.890	0.4332	0.48
2.0	1.922	0.9604	0.87
5.0	2.125	2.442	1.98
10.0	2.374	4.661	3.84

TABLE 4.
Radiation (bremsstrahlung) yields as percentages for various materials

Energy, MeV	Hydrogen	Water	Tissue	Air	Al	Iron	Lead
0.1	0.01	0.07	0.07	0.08	0.76	0.37	1.67
0.25	0.03	0.15	0.14	0.08	0.33	0.77	3.34
0.5	0.05	0.26	0.26	0.30	0.59	1.34	5.48
0.75	0.07	0.38	0.37	0.43	0.84	1.84	7.19
1.0	0.09	0.44	0.48	0.55	1.06	2.31	8.67
2.0	0.18	0.89	0.87	1.08	1.88	3.94	13.3
5.0	0.46	2.08	1.98	2.23	8.2	4.08	22.92
10.0	0.97	4.15	3.84	4.28	7.72	7.08	33.63

3.1.2 Bremsstrahlung

As mentioned above, fast electrons radiate a fraction of their energy as bremsstrahlung radiation, which is produced as the electron is decelerated by target nuclei and emitted at a rate proportional to the square of the deceleration. The fraction of energy converted into bremsstrahlung is given approximately by:

$$(\mathrm{d}T/\mathrm{d}x)_{\mathrm{bremss}}/(\mathrm{d}T/\mathrm{d}x)_{\mathrm{coll}} = ZE/800 \quad (9)$$

where Z is the atomic number of the absorber and E is the energy of the fast electrons in MeV. Thus, bremsstrahlung losses are about 10% of the collision losses for 1 MeV electrons incident on a gold target. Losses for other materials and energies are shown in Table 4.

3.1.3 Cherenkov radiation

Electrons whose velocity exceeds the velocity of light in the medium lose energy by Čerenkov radiation. In water, for example, the linear rate of loss of energy by Čerenkov radiation is about one thousandth of the loss by excitation and ionization Although small, even compared with bremsstrahlung radiation, Čerenkov radiation is important for detection purposes.

3.2 Beta-rays

Beta-particles are emitted with a continuous energy distribution ranging from zero to a maximum value, determined by the nuclear structure of the emitting radioisotope. Because they are not monoenergetic, their behavior as a group cannot be described by a single equation. Nevertheless, based on the experimental measurements, approximate equations can be written to calculate their range in matter. Such an equation is:

$$R = 412\exp(1.265 - 0.0954/\ln E) \tag{10}$$

where E is the maximum energy of the beta-particles in MeV and R, the range, is measured in mass units (mg/cm^2). Note that, provided that the ranges are quoted in mass units, this equation may be applied to all types of materials, as it depends only on E.

4 UNCHARGED RADIATIONS

High energy photons, e.g. X-rays and gamma-rays of nuclear origin, and neutrons are two important types of uncharged radiation used in elemental analysis. High energy photons are used to activate samples in a number of analytical applications, e.g. X-ray fluorescence, but their main importance is in counting the sample following activation.

Neutrons, on the other hand, are rarely used to count samples but are used extensively to activate samples. Because of the great differences in their reaction cross-sections, it is customary to distinguish between fast and thermal neutrons.

4.1 High energy X-rays and gamma-rays

Photons interact with matter by absorption, a process in which the photon gives up all its energy to the medium in a single event, and by scattering, whereby the direction of the photon is altered with or without energy loss. These processes are stochastic and result in exponential attenuation:

$$I(x) = I(0)\exp(-\mu x) \tag{11}$$

where $I(x)$ is the intensity of the beam at a distance x within the absorbing medium, μ is the linear attenuation coefficient of the medium, and $I(0)$ is the incident intensity of the beam. Exponential attenuation (Fig. 1) arises if the decrease ΔI in the beam's intensity, as it penetrates the attenuating medium by a small distance Δx, is proportional both to the intensity of the beam I and to Δx, that is:

$$-\Delta I = \mu/I \cdot \Delta x \tag{12}$$

As mentioned above, μ has the dimensions of (length^{-1}) and expresses the probability of the appropriate interaction of the beam per unit distance in the medium. Alternatively, equation (12) may be expressed in mass units (mass $\cdot$ area^{-2}) given by the product of the density ρ and the thickness Δx in the medium.

$$-\Delta I = (\mu/\rho) \cdot I \cdot (\rho/\Delta x) \tag{13}$$

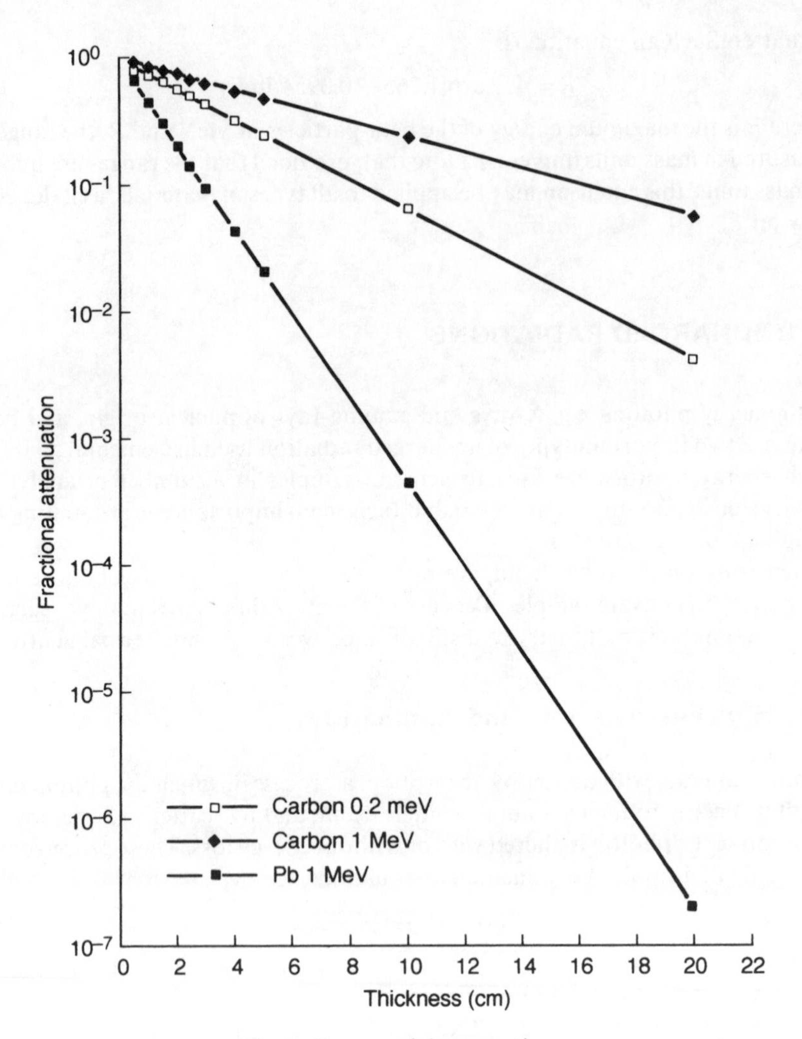

Fig. 1. Exponential attenuation.

The ratio μ/ρ is called the mass attenuation coefficient, and is particularly useful when comparing the intrinsic attenuation properties of different materials.

In the following sections we discuss the functional dependence of the attenuation coefficients of the main photon interaction mechanisms—elastic scattering, photoelectric absorption, Compton scattering, pair production and photonuclear absorption—on photon energy and on the nuclear properties of the absorber.

4.1.1 Absorption and scatter coefficients

Absorption (μ_a) and scatter (μ_s) coefficients are related to the attenuation coefficient and express the probabilities of absorption and scattering from the beam, respectively, per unit path length in the medium. They, too, are functions of photon energy and of the type of absorbing material, and are related to each other and to the attenuation coefficient by the equation:

$$\mu_t = \mu_a + \mu_s \quad (14)$$

which states mathematically that the photons interact with the medium either by absorption or scattering but by no other mechanism, so that the total probability of interaction is simply the sum of the probabilities that the photon is either absorbed or scattered.

4.1.2 Buildup factors

Were it not for scattering, calculation of the depth of penetration of X-rays in matter would be simple. Scattering, however, changes the directions of the photons, producing secondary photons with a continuous energy spectrum. For wide beams produced, for example, by uncollimated point sources, adequate correction is obtained by multiplying the exponential attenuation equation (11) by a buildup factor $B(\mu x)$, which is a function of the attenuation coefficient and the depth of penetration. Appropriate buildup factors for different energies and absorbing materials may be found in books on radiation shielding.

4.2 Photon reactions

4.2.1 Elastic scattering

High energy photons scattered by electrons in the interacting medium without losing any of their kinetic energy are said to be scattered elastically. Although energy is not transferred to the medium in this interaction, and therefore elemental analysis based on radiation changes in the medium is not possible, elastic scattering is nevertheless important in elemental analysis for two reasons. First, X-ray diffraction analysis, which is used extensively to obtain crystal structures, is based on elastic scattering. Second, scattered radiation may interfere with nuclear analysis and therefore must be allowed for in the design of analytical procedures.

X-ray scattering is particularly important at X-ray energies $< 10\,\mathrm{keV}$. The intensity I_s of elastically scattered radiation by free electrons in terms of the initial intensity I_0 and scattering angle θ is

$$I_s = I_o[e^4(1 + \cos^2\theta)][2r^2m^2c^4] \quad (15)$$

where e, m and c are as defined in Section 2.1.

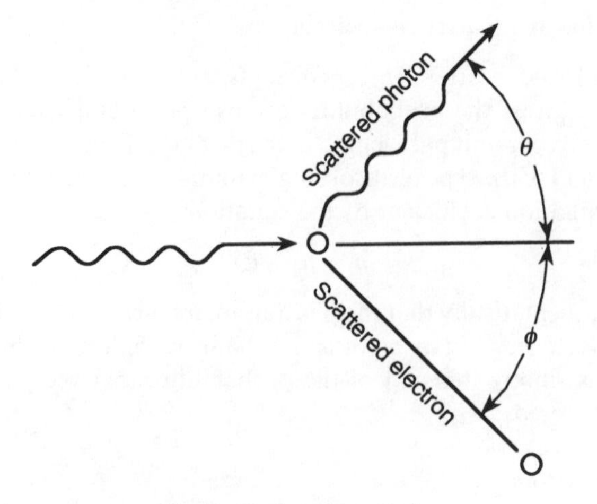

Fig. 2. Compton scattering.

4.2.2 Compton scattering

At high energies (> 100 keV) elastic scattering becomes highly improbable and the dominant scattering reaction is inelastic Compton scattering. In this interaction, the photon transfers part of its energy to a target electron which is dissipated locally, and is scattered in a new direction with lower energy E' and longer wavelength, as shown schematically in Fig. 2. The fraction of energy lost by the photon is related to the scattering angle by the equation:

$$E'/E = 0.51[1 - \cos\alpha + 0.51/E] \quad (16)$$

where E and E' are in MeV and α is the scattering angle.

Assuming conservation of momentum and energy in the collision, and using the quantum relationship between the photon wavelength and its energy, namely $E = hc/\lambda$, an expression for $\Delta\lambda$, the change of wavelength following scattering, as a function of the scattering angle α may be written as:

$$\Delta\lambda = h/mc\,(1 - \cos\alpha) \quad (17)$$

Figure 3 gives $\Delta\lambda$ as a function of α for various photon energies.

Compton scattering cross-sections per free electron, as a function of energy and angle, may be calculated using the Klein and Nishima equation, and Compton attenuation coefficients (Fig. 4) can then be obtained by multiplying the electron cross-sections by the number of electrons per unit volume, since each electron scatters independently of the others.

4.2.3 The photoelectric effect

A reaction in which a photon is absorbed by an atom and an electron is ejected is called photoelectric absorption. For it to occur, the energy of the photon must be

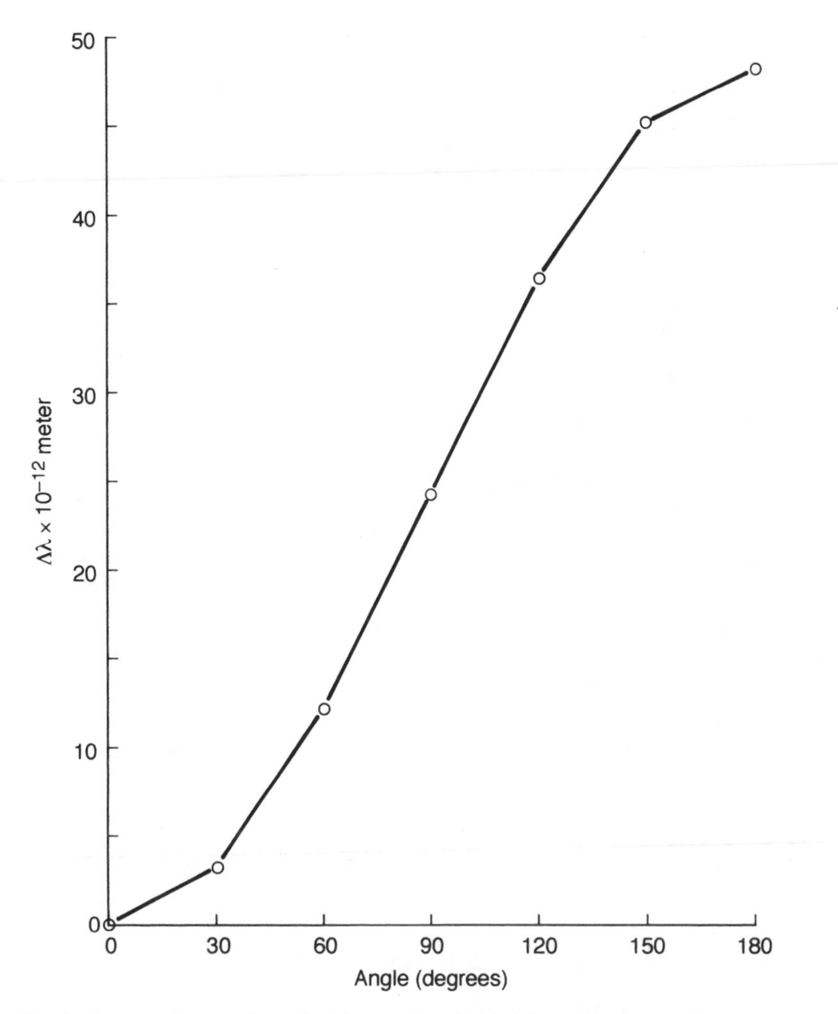

Fig. 3. Increase in wavelength, $\Delta\lambda$, as a function of Compton scattering angle.

equal to or greater than the binding energy of the emitted electron E_b, which is ejected with kinetic energy E_{kin} given by

$$E_{kin} = hf - E_b \tag{18}$$

where h is Planck's constant and f is the photon frequency.

Following the interaction, the atom is left in an unstable state, and in returning to a stable electronic configuration emits a series of secondary (Auger) electrons or fluorescence X-rays which in high Z materials may exceed 60 keV.

The cross-section of photoelectric absorption is a function of photon energy and atomic number of the absorbing atom. For high atomic number materials the

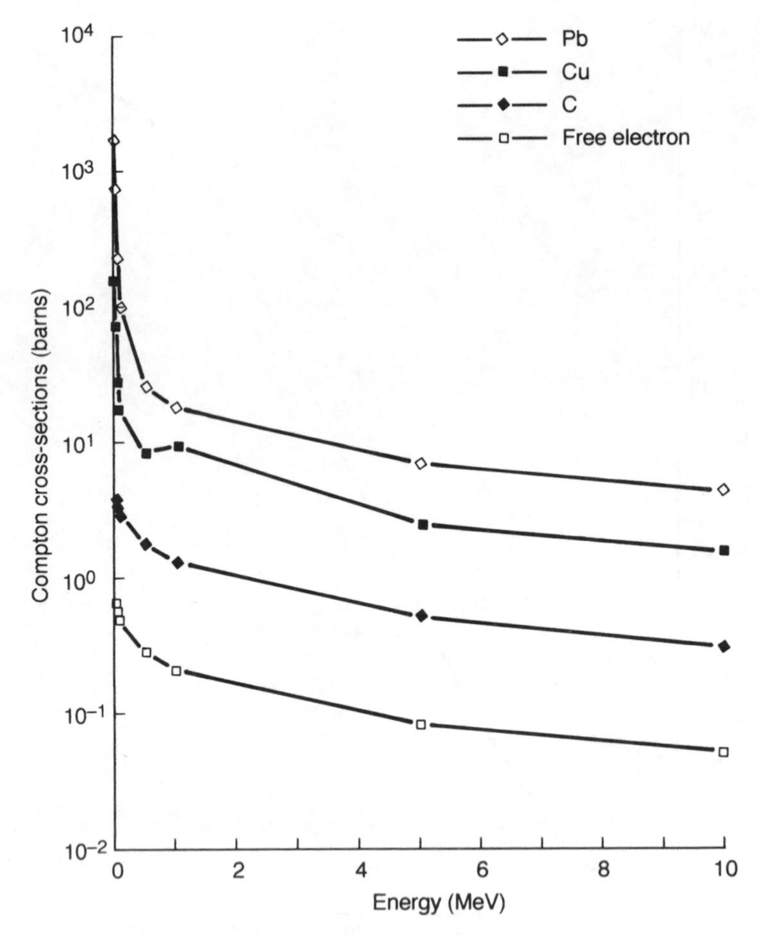

Fig. 4. Compton scattering cross-sections as a function of energy in barns.

cross-section may be approximated as:

$$\sigma_{\text{photoelectric}} = \text{constant}(Z^n/E^3) \tag{19}$$

where n varies between 3 for low energy X-rays and 5 for high energy X-rays.

Figure 5 shows photoelectric cross-sections for a number of materials. Note the strong dependence of the photoelectric effect on atomic number, explaining why lead, with $Z = 82$, is a preferred material for shielding photons with energies up to 1 MeV.

4.2.4 Pair production

When a photon passes by the intense electric fields close to the nucleus and, to a lesser extent, orbital electrons, it is sometimes converted into an elec-

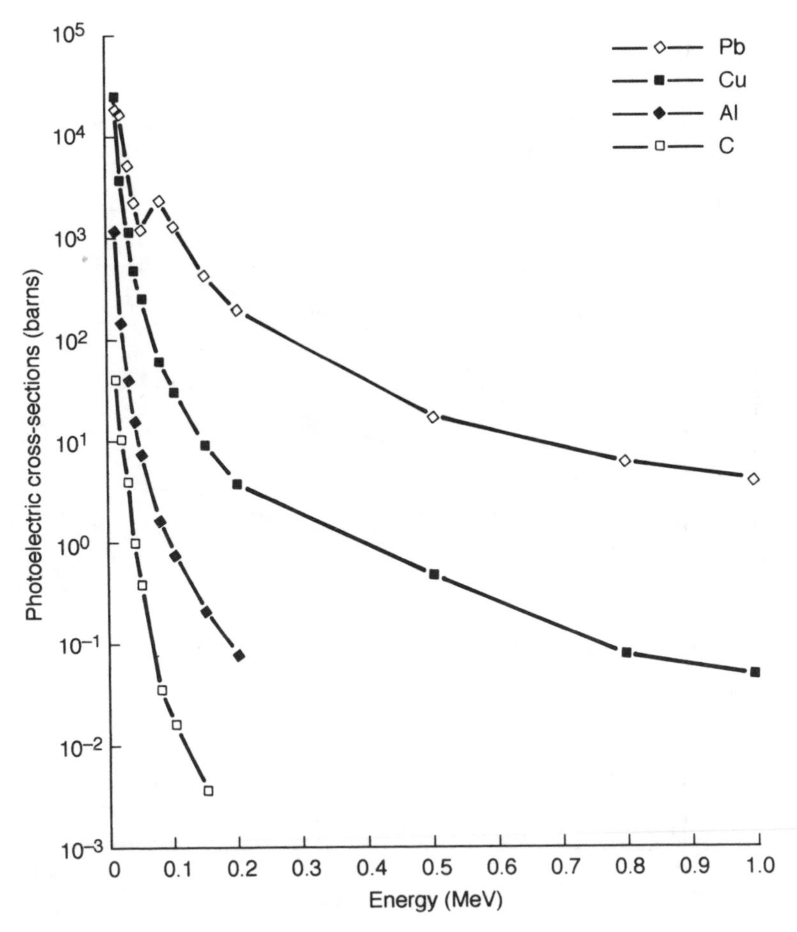

Fig. 5. Photoelectric cross-sections as a function of energy in barns.

tron–positron pair (a positron is a positively charged electron). This reaction, called pair production, is a beautiful illustration of the relativistic equivalence of energy and matter, whereby a photon having energy but no mass is converted into two electrons provided the energy of the photon is above the threshold value of 1.02 MeV, the energy equivalent of the rest masses of the particle pair. Energy of the photon in excess of the threshold value is shared between the electron and positron as kinetic energy, with a small fraction of it being transferred to the atomic nucleus.

The cross-section for pair production is approximately proportional to $(Z^2 + Z)$ and therefore is particularly significant for high atomic number absorbers. Its cross-sections increase slowly from 1.02 to about 5 MeV, and above this it rises at a rate approximately proportional to the logarithm of the photon energy

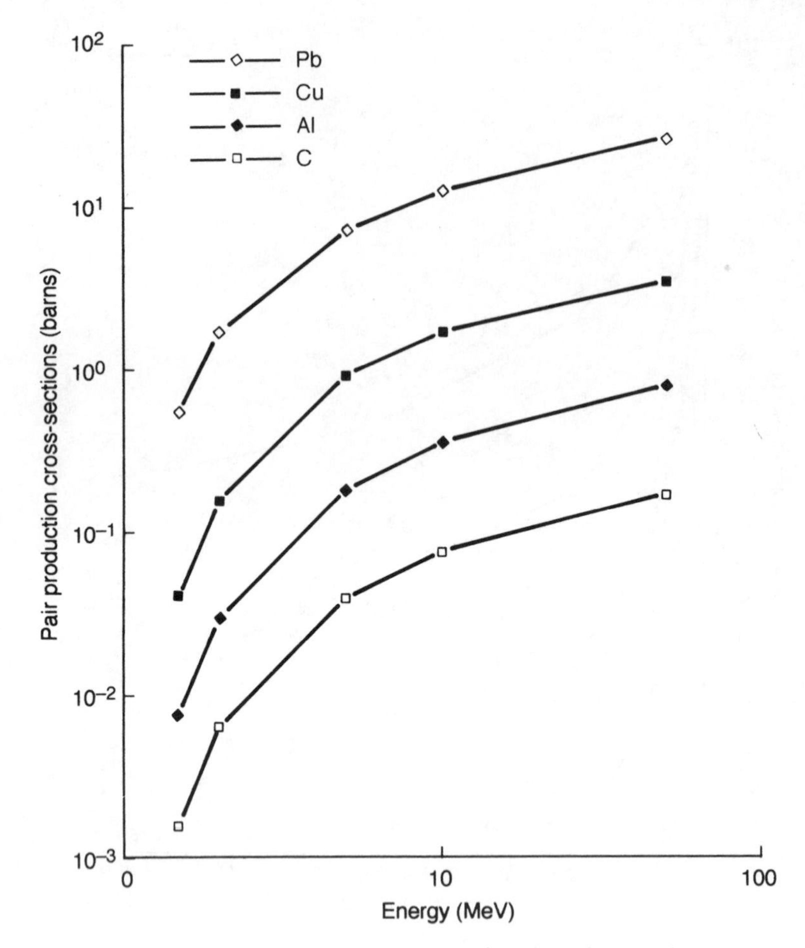

Fig. 6. Pair production cross-sections as a function of energy in barns.

(Fig. 6). Since the probabilities of both the photoelectric and the Compton reaction decrease at high energies, pair production is the dominant absorption process at very high energies.

Following its creation, the positron–electron pair loses its excess kinetic energy to the absorbing medium by excitation and ionization. As the positron comes to rest it combines with an electron in the absorbing medium and the two undergo mutual annihilation, producing in this process two photons of energy 0.51 MeV.

Since annihilation occurs when the total momentum of the annihilating positron and electron is very close to zero, the two annihilation photons emerge at approximately 180° to one other to preserve the zero momentum of the system. The very small observed deviations from 180° are a function of the local

TABLE 5.
Threshold values of photonuclear reactions

Element	Reaction	Threshold (MeV)
Antimony	(γ, n)	8.9
Gold		8.1
Lead		8.4
Calcium	(γ, p)	10.7
Magnesium		12.1
Yttrium		22.3
Iron	(γ, np)	20.9
Zirconium	$(\gamma, 2n)$	19.2

environment of the annihilating pair and thus provide the basis for positron spectroscopy, a useful although rather specialized analytical technique.

4.2.5 The photonuclear effect

Photons with very high energies may be absorbed by a nucleus with the emission of one or more nucleons. Like photoelectric absorption, this reaction, called the photonuclear reaction, occurs only if the energy of the photon is greater than the binding energy of the ejected particles.

At energies of the order of 8–13 MeV, neutron ejection is favored, since neutrons have no coulombic potential barrier to overcome in escaping from the nucleus. At higher photon energies charged particle photonuclear emission becomes possible (e.g. protons, deuterons and alpha-particles), with photo-disintegration of the nucleus taking place at very high energies.

Two isotopes with exceptionally low thresholds are deuterium and beryllium (1.666 MeV for ^{9}Be and 2.23 MeV for deuterium). They are used to produce laboratory sources of monoenergetic neutrons.

Following the absorption of the photon and emission of a nucleon, the nucleus is left in an excited state, and in returning to its ground state emits very high energy X-rays, leaving the target nucleus radioactive.

Table 5 gives threshold values of different photonuclear reactions of some common elements.

5.1 Neutron interactions

In discussing neutron reactions, the basis of a number of important nuclear methods of elemental analysis, it is convenient to divide them into two categories: scattering reactions and capture reactions. The former do not change the identity of the target but alter the neutron spectrum, slowing down fast neutrons and modifying their distribution in space and time. In contrast, capture reactions,

TABLE 6.
Neutron reactions

Reaction name	Reaction	Cross-section
Elastic scattering	(n, n)	σ_s
Inelastic scattering	(n, n′)	σ_n
Neutron capture	(n, γ)	σ_c
	(n, p)	$\sigma_{(n,p)}$
	(n, α)	$\sigma_{(n,\alpha)}$
Fission	(n, f)	σ_f

which add an extra neutron to the target nuclei, alter the nuclear identity of the targets by increasing their atomic weight by one unit, or by initiating secondary reactions which can change the atomic numbers and atomic weights of the targets.

It is generally believed that capture reactions involve the formation of a highly excited compound nucleus, which then rapidly returns to a more stable state by emission of a gamma-ray photon, by expulsion of a particle or a combination of particles including protons, neutrons and alpha-particles, or by breaking up of the compound target into two large fragments.

Six general types of neutron reaction are listed in Table 6 and are discussed separately in the following sections.

5.1.1 Neutron elastic scattering

Neutron elastic scattering is defined as a collision in which the sum of the kinetic energies of the neutron and target nucleus particles remains constant. For head-on collisions the respective energies are given by the following equations:

$$E = \alpha^2 E_0 = [(M - m)/(M + m)]^2 \cdot E_0 \tag{20}$$

$$E' = E_0(1 - \alpha^2) \tag{21}$$

where E_0' is the initial energy of the neutron, E and E' are the energies of the scattered neutron and the target, respectively, and m and M are the masses of the incident neutron and of the target nucleus, respectively.

These equations, show that in colliding with hydrogen the neutron can lose all its kinetic energy in a single collision, whereas in colliding with heavier nuclei only a fraction of the initial energy is lost. It is not surprising, therefore, that hydrogenous materials are used for slowing down neutrons.

A more complete treatment of elastic collision, assuming isotropic scattering, shows that the fraction f of the incident neutron's energy transferred to a target nucleus of mass M averaged over all scattering angles is:

$$f = 2M/(M + 1)^2 \tag{22}$$

For hydrogen ($M = 1$), the average energy transferred in an elastic collision is 50% of the initial kinetic energy of the neutron, while for carbon ($M = 12$) it is only 14.2%.

Using this equation, the average depth a neutron penetrates into matter until thermalized, also called the slowing down length, may be calculated. Once thermalized, the neutron diffuses within the material by thermal motion until it is absorbed. The average distance it travels until it is absorbed is called the thermal diffusion length. Values of these two 'lengths' for a large number of materials may be found in special compilations.

5.1.2 Inelastic scattering

In inelastic scattering, the nuclei are excited from their ground state to higher energy levels, and in returning to their ground states they emit characteristic high energy photons.

The thresholds of inelastic scattering with elements of moderate and high mass number are between 0.1 and 1 MeV. For elements with low mass numbers there is a general tendency for the threshold energies to increase. The threshold energy for oxygen is about 6 MeV, and in hydrogen inelastic scattering does not occur at all.

5.2 Neutron capture

5.2.1 The (n, γ) reaction

Reactions in which the neutron is absorbed by the target nucleus are classified as neutron capture reactions. If there is no immediate particle emission, the atomic weight of the target nucleus increases by one unit and the excess energy, consisting of the binding energy of the absorbed neutron and its kinetic energy, is emitted as a prompt gamma photon of the order of 6–8 MeV. This process may be written symbolically as:

$$_{Z}X^{A} + {}_{0}n^{1} \longrightarrow {}_{Z}X^{A+1} + \gamma \tag{23}$$

Many target nuclei remain unstable after prompt gamma emission, emitting delayed beta- and gamma-radiation. In these cases they are said to have become activated or radioactive. Capture cross-sections span many orders of magnitude, as may be seen in Table 7 for thermal neutrons.

5.2.2 (n, nucleon) reactions

Reactions in which one or more nucleons are expelled following absorption constitute another type of capture reaction. If a charged particle, e.g. a proton or alpha-particle, is emitted, the target is transformed into a different chemical

TABLE 7.
Thermal neutron cross-sections

Element		Cross-section (barn)
Nitrogen	N^{15}	2.4×10^{-5}
Oxygen	O^{16}	1.8×10^{-4}
Iron	Fe^{56}	2.7
Silver	Ag^{107}	35
Gold	Au^{197}	98.8
Dysprosium	Dy^{164}	2800

element; if only neutrons are emitted only the isotopic nature of the target changes. These absorption reactions usually require high energy neutrons to initiate them. Generally speaking their cross-sections are considerably smaller than those of (n,γ) reactions; nevertheless the neutron reactions which transform the target atoms chemically are very important in radioanalytical work, provided the target atoms become radioactive, since powerful chemical techniques may be used to separate the activated atoms from the bulk of the sample, allowing high resolution spectroscopy to be carried out with minimum interference.

5.2.3 (n, fission) reactions

In the case of certain nuclei, such as those of uranium and thorium, neutron capture can cause the nuclei to break up into large fragments and to emit large amounts of energy and additional free neutrons. These neutrons may be used to sustain a chain reaction in nuclear reactors, and serve as a source of neutrons for neutron activation.

The energy required to cause fission depends on the nuclear properties of the target; ^{235}U, for example, has a large cross-section for thermal neutrons (549 barn), whereas ^{238}U undergoes fission only if irradiated by fast neutrons (0.29 barn).

Chapter 2

NUCLEAR DETECTION METHODS AND INSTRUMENTATION

A. Tsechanski
Ben-Gurion University of the Negev, Department of Nuclear Engineering, P.O. Box 653, Beer-Sheva 84105, Israel,
and
Soroka Medical Center, Oncology Department, P.O. Box 151, Beer-Sheva 84101, Israel

1 INTRODUCTION

In the following pages we shall present a concise description of nuclear detectors such as gas-filled detectors, scintillation counters and solid-state detectors, along with a short excursion into gamma-ray spectroscopy. Those interested in comprehensive coverage of the subject are referred to the book by G. F. Knoll [1].

2 NUCLEAR DETECTORS

2.1 Gas-filled detectors

In general, detection of nuclear radiation is based on registration of the different kinds of interactions of radiation with the atoms of the detector material. In most cases, such interaction releases a large number of secondary electrons, which are collected and shaped into a voltage or current pulse for subsequent analysis by electronic means.

One of the first and best known types of nuclear detector is based on using a gas as the sensitive medium, and it is called a gas-filled detector. The simplest detector

Chemical Analysis by Nuclear Methods Edited by Z. B. Alfassi

of this type is the ionization chamber, which can simply be regarded as a gas-filled parallel-plate capacitor. When an energetic charged particle passes through a gas inside the ionization chamber, it ionizes a large number of gas molecules or atoms along its track. As a result, electrons and positive ions are produced. The electric field between the plates is strong enough to keep the ions from recombining with the electrons, so directing them in opposite directions. The negative free electrons drift toward the positive plate, and the positive ions are attracted to the negative one. When this charge is collected on the plates, an electric pulse is produced. The magnitude of this pulse can be easily estimated. The average energy needed to produce an electron–ion pair in air is about 35 eV, thus a 1 MeV charged particle produces about 3×10^4 electron–ion pairs. For a square 10×10 cm chamber with a plate separation of 1 cm, the capacitance is 8.85×10^{-12} F and therefore the resulting voltage pulse is 1.6 C per ion $\times\ 3 \times 10^4/(8.5 \times 10^{-12}\,\mathrm{F}) \approx 0.5$ mV. This is too small a signal amplitude, which must be considerably amplified before it can be analyzed by standard electronics.

If, after production of ionization by a charged particle, both the electrons and the ions are completely collected, the final voltage signal will be proportional to the total ionization and, therefore, to the charged particle energy. Usually, however, this requires too long a time (milliseconds). This is an exceedingly long time for handling reasonably high counting rates, and thus the ion chamber is of no use in counting individual particles. It does find wide use as a radiation monitor, especially in intense radiation fields. In this case, the radiation intensity is recorded as a current resulting from interaction of many nuclear particles during the response time of the chamber.

The gas in an ionization chamber should be free from electronegative impurities. From this point of view, the use of air in ion chamber is disadvantageous, since air contains oxygen, an extremely electronegative gas. Therefore ion chambers employing air have bad saturation characteristics and require rather high operating voltages. This is particularly so when chambers are filled to high pressures to obtain high sensitivities. To solve this problem some filling gas other than air is used. The most common choice is nitrogen or an argon–nitrogen for gamma sensitivity or hydrogen and hydrogen–argon mixture for neutron sensitivity. The gas pressure should be chosen to give an approximate range of the charged particle, and is typically between 0.1 and 40 atm.

The main disadvantages of the ion chambers are the very small output signal and the poor energy resolution compared with proportional counters and semiconductor detectors. To use a gas-filled detector, such as an ionization chamber, for detecting individual nuclear particles, the electric field in the ionization chamber has to be increased significantly. In such a case, the electrons drifting to the anode can acquire between two successive collisions kinetic energy sufficient to cause the further (secondary) ionization of gas molecules. The new electrons produced by the secondary ionization are in turn accelerated towards the anode and can produce further ionization. This rapid amplification by production of secondary ionizations is called a Townsend avalanche. Provided that the electric field is not too high, the total charge produced is proportional to the initial ionization of the primary particle and therefore to its energy. As a result of development of the gas discharge avalanche, the charge collected by the electrodes may be up to 10^4–10^5 times that of the initial ionization. Since the pulse height obtained is proportional to the ionization capability of the primary particle, the device is known as a *proportional counter*.

The most common geometry of the proportional counter is a cylinder with a very thin wire anode coaxial to a cylindrical cathode. The high-field region is concentrated very close to the anode wire. In this intense-field narrow region around the anode gas multiplication takes place, resulting in almost total independence of the gas multiplication factor on the initial position of ionization. In general, the voltage induced on the collecting electrode by a charged particle is proportional to the fraction of voltage through which the charged particle falls before collection. Since about half of the total number of ion pairs is formed within one mean free path of the surface of the anode, the positive ions fall through a much greater potential before collection than the electrons. The signal induced on the anode is therefore due mainly to the positive ions rather than the electrons. Since most of the positive ions are formed in the region of high electrostatic field near the central wire, the rise time of the signal is very rapid (of the order of microseconds), and, therefore, the proportional counter can be operated at high counting rates of the order of 10^6/s.

Because of the gas amplification, the signal in the proportional counter is much larger than in an ion chamber. Therefore the proportional counter is particularly useful for the detection of low energy radiations and particularly photons (down to about 0.1 keV). Very thin entry windows (down to 1 μm thick) of plastic or beryllium should be employed in these counters. The gas filling of a proportional counter is selected from the noble gases neon, argon, krypton and xenon. In addition, up to 10% of a quench gas (methane is most commonly used) is added to stabilize the operation at high values of gas multiplication.

Proportional counters are widely used for neutron detection too. For this purpose the filling gas in BF_3 or the counter is lined with ^{10}B. For proton recoil fast neutron spectrometry, counters with polyethylene lined windows are used. Proportional counters are used in neutron flux mapping and monitoring in nuclear reactors, where the high temperature and neutron flux would make the use of other types of detectors problematic.

If the electric field in a proportional counter is increased to even larger values, a new factor begins to play an increasingly important role. This factor is a considerable excitation of atoms and molecules of the gas. As a result of very fast de-excitation of these molecules, ultraviolet photons are emitted that, in turn, produce photoelectrons, mostly from the counter wall. These photons can migrate far from the place of the original avalanche, causing secondary avalanches by means of photoelectrons, and so the gas in the entire tube is involved in the discharge. As a result, the amplification is very large, up to 10^{12}. Since in every registration effect the entire active volume (gas) of the tube is participating, the amount of charge collected on the electrodes is independent of the initial ionization, and virtually a single primary ion pair can fire the discharge. As a result, there is no information at all on the type and energy of the original radiation. This region of operation is called the *Geiger–Müller region*, and gas counters employing this principle are usually known as Geiger–Müller counters (or simply Geiger counters).

Just after beginning the gas discharge in the Geiger counter, the electrons, being considerably more mobile than positive ions, are effectively collected by the anode, so the output signal consists mainly of the electrons from the many avalanches. The electron collection time is of the order of a few microseconds. During this short time interval, the positive ions, having a lower mobility, barely move away from the avalanche region around the anode. They are left in the counter as a positive charge sheath, which is continuously building up and reduces the electric field intensity in the counter to a level which eventually terminates the multiplication process. The discharge would then be ended after the positive ion sheath arrives at the cathode. However, when the positive ions arrive at the cathode, they cause photon emission, which in turn can cause the gas discharge in the counter to start again. To prevent this from occurring, some method of quenching the discharge must be used.

In general, quenching can be implemented by external electronic means or by adding a second type of gas (quenching gas) to the tube. In electronic quenching, some reduction in the counter operating voltage is brought about immediately after the counter pulse has appeared. The reduction must be enough to bring the counter voltage below the starting point of the second discharge. After allowing time for all the positive ions to reach the cathode and become neutralized (up to 1 ms), the voltage is restored to its original working value.

In the so-called self-quenching counters, the quenching action is accomplished by the addition to the main (counting) gas of a small amount of a second gas with complex molecules which have lower ionization potential than the simple counting gas molecules. In addition, the quenching gas, when ionized, should have a higher probability to lose its energy by dissociation rather than by emission of UV photons. A typical filling for a Geiger counter is 90% argon and 10% ethanol. As the positive charge sheath, consisting mostly of argon ions, begins to move towards the cathode, collisions will occur with the molecules of the quenching gas, resulting in transferring of an electron from the ethanol molecule to an argon ion. Thus, only ions of the quenching gas reach the cathode and are neutralized there. However, the energy of the ions that formerly went into emission of an UV photon can now be absorbed in the dissociation of the ethanol molecule (a process not possible for a simple argon atom). The quenching gas is thus gradually used up, and the Geiger tube life expectation will be about 10^8 to 10^9 counts. Another widespread category of Geiger tubes is the so-called halogen counters. Their counting gas is composed of noble gases (Ne, Ar, or Kr) with a small quantity (0.1–1%) of halogen vapour (Cl_2, Br_2, or I_2) as a quenching agent. As in the previous case, the diatomic molecules of the halogen gas are dissociated in the quenching process, but the subsequent spontaneous regeneration of the undissociated halogen molecules eliminates, in principle, the need to replace the Geiger tube. However, contamination of the quench gas by reaction products and changes on the anode surface ultimately limit the lifetime of the halogen Geiger counters to about 10^{10} counts, as compared to 10^8 typical life expectation counts of organic vapour-quenched counters.

In the period immediately following the Geiger discharge, the positive space charge around the anode effectively reduces the electric field below the ignition point. If during this period of time another charged particle occurs inside the counter, it will not be observed, because gas multiplication is prevented. This *dead time* when the tube is unable to respond to radiation is of the order of 50–200 μs . After the dead time there is a progressive recovery, when the Geiger counter can respond again but the resulting pulse does not attain its regular magnitude. Registration of such a partial pulse depends on the sensitivity adjustment of the counting electronics. The usual procedure is to arrange the electronic circuit to have a defined dead time so that the loss of counts due to finite dead time may be accurately estimated.

The counting efficiency of the Geiger counter for any charged particle that enters its active volume is close to 100%, since even one charged particle inside the tube is enough to trigger full-scale Geiger discharge. As to the efficiency for counting of gamma-rays this depends on the probability of interaction of the incident gamma-ray in the solid wall of the counter and on the probability that the secondary electron released in this interaction reaches the active volume before the end of its track. Since the probability of gamma interaction within the sensitive layer is small even for high-Z cathode materials, the gamma-ray counting efficiency is generally low and at 1 MeV will be $\approx 1\%$. Most Geiger counters have a cylindrical stainless steel cathode and may have mica window thicknesses down to 1.5 mg/cm^2; gas pressure of the order of 100 Torr is usual. A large variety of geometrical configurations of Geiger counters is available for different applications.

The various regions of operation of gas-filled detectors are presented in Fig. 1. The amplitude of the observed pulse is given as a function of the applied voltage within the detector for two particles depositing two different amounts of energy within the gas. At very low voltages the field is insufficient to prevent the recombination of the primary electrons and ions, and the collected charge is less than the original one. As the voltage is increased, recombination is suppressed and the region of ion saturation is achieved. This is the *ionization chamber region*, where the pulse output is proportional to the primary ionization and thus to the energy of the radiation, but is independent of the applied voltage. As the voltage is increased still further, gas multiplication begins in the detector, and the observed pulse amplitude will increase substantially with applied voltage. This is the *proportional region*, in which the pulse amplitude rises proportionally to the

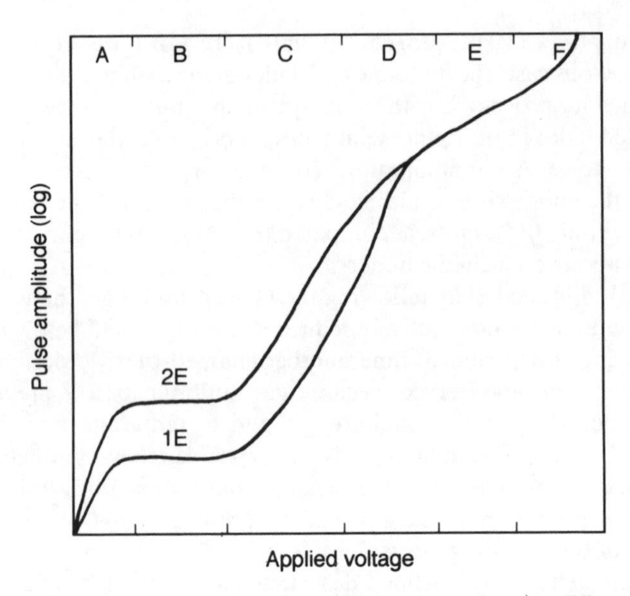

Fig. 1. Pulse amplitude as a function of applied voltage for the different regions of operation of gas-filled detectors. The plots are given for charged particles of two different energies. A, pulse amplitude reduced by recombination; B, ionization chamber region; C, proportional counter region; D, limited proportionality region; E, Geiger counter region; F, continuous discharge region.

voltage, but the output pulse remains proportional to the energy of the radiation, thus simplifying identification of the primary radiation and its energy. Further increase of the applied voltage results in increasing concentration of the positive ions (space charge) around the anode, which can significantly diminish the electric field within the detector. As a result, some non-linearities in gas multiplication will begin to be observed. This is the region of *limited proportionality*, in which the pulse amplitude still increases with energy of the primary radiation but not in a linear fashion. Finally, as the applied voltage is made sufficiently high, the gas multiplication is determined by the space charge created by the positive ions. Under these conditions, the gas discharge is self-limiting and all output pulses are of the same amplitude, regardless of the kind of the primary radiation and its energy. This is the *Geiger-Müller region* of operation characteristic for Geiger counters.

2.2 Scintillation counters

2.2.1 Introduction

The main disadvantage of gas-filled detectors is their low efficiency for several nuclear radiations and particularly for gamma-rays. This arises from the fact that

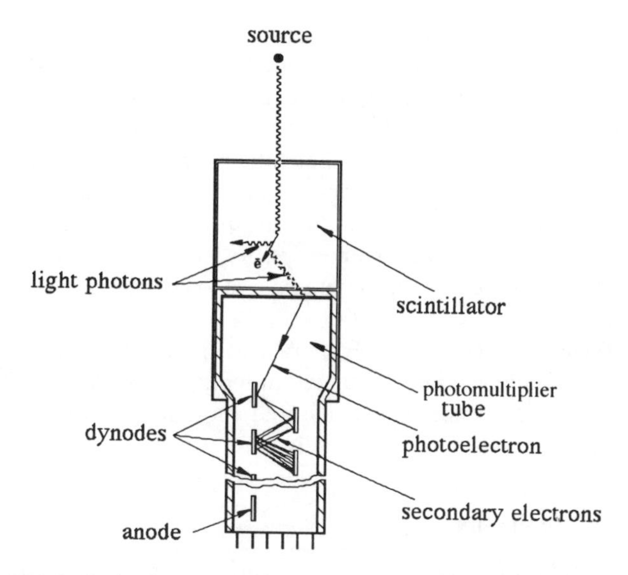

Fig. 2. Basic elements and processes in a scintillation detector.

the detection medium, gas, is low in density. From this point of view solid detectors are more advantageous, since their higher densities result in acceptable absorption probabilities for a detector of reasonable size. Of the detectors with a solid detection medium, the scintillation counter is one of the earliest and most widely used detectors of nuclear radiation. A schematic representation of the basic elements and processes in a scintillation detector is given in Fig. 2. The detector consists of a detection medium (scintillator) and a photomultiplier tube. The incident particle enters the scintillator and interacts repeatedly with the atoms of the scintillator, raising them to excited states. De-excitation follows very rapidly by emission of photons of visible or near-visible light. The emitted photons strike a photocathode, releasing a maximum of one photoelectron per photon. Finally, these photoelectrons are multiplied in the photomultiplier tube (PMT), resulting in an output signal pulse of up to several volts.

In general, for use as a scintillator for nuclear radiation the material should possess the following properties:

(a) it should convert the energy of charged particles into emitted light linearly and with a high scintillation efficiency;
(b) the scintillator should be transparent to its own emitted light;
(c) the decay time of the luminescence should be short;
(d) the spectral distribution of the emitted light should match as closely as possible the response of the photocathode of the PMT used.

Many materials are known to luminescence under the influence of nuclear radiation. They can be in gaseous, liquid or solid form. Scintillating materials are usually divided into organic and inorganic materials. The most widely used inorganic scintillators have generally higher light output and linearity but slower time response than organic scintillators. The high photoelectric cross-section and high density of inorganic scintillators make them especially valuable for gamma detection, whereas organic scintillators are preferable for charged particle and neutron spectroscopy. Organic scintillators in liquid or plastic form may be easily fabricated in large and irregular geometries. As to the size of inorganic scintillators, this is strictly limited by the maximum size to which a single crystal may be grown.

A general review of the basic theory, practice and application of scintillation detectors can be found in the exhaustive monograph by Birks [2]. Since the mechanisms of luminescence in organic and inorganic scintillators are quite different they will be considered separately.

2.2.2 Organic scintillators

In organic scintillators (whether in the gaseous, liquid or plastic state), the molecules are relatively loosely bound and, as a result, retain their individual electronic structure and luminescence features. A charged particle passing into the scintillator interacts with many molecules, losing a few eV at each interaction. Generally, a scintillator molecule either absorbs energy by excitation of the electrons to higher excited states, or atoms in the molecule can vibrate against one another. As a result of the interaction many vibrational states also become excited. However, since any excited vibrational state is not in thermal equilibrium with its neighbors, it quickly decays (during ≈ 1 ps) to the lowest vibrational state of the electronic excited state. De-excitation of these molecules (during a few nanoseconds) to one of the vibrational states of the ground-level electronic state produces the principal scintillation light. The vibrational states of the electronic ground state again decay quickly to the vibrational ground state.

Of the pure organic materials, only anthracene and stilbene (aromatic hydrocarbons with linked benzene ring structure) have achieved widespread popularity. Anthracene features the greatest light output per unit energy (scintillation efficiency). Stilbene has a lower scintillation efficiency but it exercises pulse shape discrimination property to distinguish between scintillation induced by heavy charged particles (e.g. protons) and by electrons. Both materials are relatively fragile and may be fabricated as single crystals having a side of no more than several centimeters.

If a small concentration of impurity is added to a bulk scintillator (solvent) there is, in some cases, high probability that the excitation energy absorbed mainly by the solvent molecules be transfered ultimately to the impurity molecules. If these molecules de-excite by radiationless transition (mainly by collisions) the process reduces the scintillation efficiency and the effect its known as 'quenching'. The prominent example is dissolved oxygen which degrades substantially the light output of many scintillators. On the other hand, if the added impurity is an efficient scintillator it may increase effectively the overall fluorescent efficiency. These 'binary' organic scintillators are widely used in liquid and plastic form and represent a separate type of organic scintillator called liquid and plastic solutions. The primary excitation takes place mainly in the host material or solvent which may be toluene, xylene or benzene for liquid scintillators and polyvinyltoluene or polystyrene for plastics. The excitation is effectively transferred to solute molecules such as *p*-terphenyl, PPO or PBD. Sometimes a third additive is present in a concentration of a few percent or less and serves as a 'waveshifter'. Its function is to improve matching between the fluorescence spectra and the PMT spectral response and to increase the transparency of the scintillator to its own light. The most commonly used waveshifter is POPOP.

Liquid scintillators are often used as large-volume detectors with dimensions of up to several meters. In such applications the liquid scintillator may be the only choice from a price standpoint. But probably the most common application of liquid scintillators is the assay of radioactive material which can be dissolved or dispersed in the scintillator solution. In this case the scintillator acts as an almost 100% efficient detector for the emitted charged particles. This so-called liquid scintillation counting method is especially valuable for counting of low energy beta-rays of tritium, ($E_\beta^{max} = 20\,\text{keV}$) and carbon-14 ($E_\beta^{max} = 200\,\text{keV}$). As to the plastic scintillators, being relatively inexpensive, they can be used in large-volume solid scintillator detectors and in a large variety of forms (flat sheets, rods, cylinders, etc.).

In general, organic scintillators are used for detection and spectrometry of charged particles (electrons, protons, alpha-particles, etc.). As such they are widely used for the detection and spectrometry of gamma-rays through the Compton electrons and of fast neutrons through recoiled protons. To expand the detection possibilities of both liquid and plastic organic scintillators, it is possible to load them with some additional materials. Thus, for instance, boron, lithium or gadolinium may be added to give a thermal neutron detector. In the case of boron and lithium, the (n, α) reaction results in direct detection of alpha-particles in the organic scintillator. In the case of gadolinium, very effective capture for thermal neutrons results in beta and gamma activity which is detected in the organic scintillator.

For most organic scintillators, the variation of the light output with electron energy is slightly non-linear for low energies up to about 0.1 MeV and linear at higher energies. Heavy charged particles such as protons or alpha-particles are considerably less efficient in producing light in organic scintillators than electrons. This subject is extensively treated by Birks [2].

As to the time response of organic scintillators, it should be pointed out that this category of scintillators is the fastest type. In a simple approach, the time profile of the light pulse can be described by a single exponent $I = I_0 e^{-t/\tau}$, where τ

is the time constant describing the decay of the optical levels. In the fastest organic scintillators τ is 1.5–2.5 nanoseconds. This exponent represents the prompt fluorescence responsible for most of the observed scintillation light. In some scintillators, however, the slower components are also observed with a characteristic decay time of several hundred nanoseconds. It should be pointed out that the relative amounts of light in the fast and slow components often depend on the nature of the exciting particle. This dependence can be used to discriminate between different kinds of particle. This is so called pulse shape discrimination (PSD), and is used in several applications; the most common use is for discrimination between gamma-rays (i.e. via Compton electrons) and fast neutrons (i.e. via recoil protons).

2.2.3 Inorganic scintillators

In general, inorganic scintillators are single crystal insulated materials. A monocrystalline structure is necessary because multiple scattering and absorption of the emitted light at the many faces of a polycrystal would make such a scintillator useless. The energy band structure of a crystalline scintillator includes the lower band, called the valence band, which is generally full and represents the energy states of the electrons strictly bound at crystal sites. The upper band, called the conduction band, contains energy states that are empty. Between the valence and conductions bands there exists the forbidden energy band, in which electrons can never be found. In organic scintillators, the width of the forbidden band is about 4 eV. Incoming radiation can excite an electron and elevate it from the full valence band across the forbidden band into the conduction band. In the pure crystal, the excited electron loses its energy by emission of a photon and returns into the valence band. Although pure inorganic crystals can be used as scintillators, they are inefficient because of the high self-absorption of the emitted light by the crystal itself. To reduce self-absorption of the light and to increase the probability of visible photon emission, a small amount of a suitable impurity, called an activator, is added to the crystal. Such activators create additional energy states within the forbidden band and the photon emission takes place between the activator states. Since the photon energy is less than the forbidden band-width of the host material, self-absorption cannot occur, and the scintillation process results in emission of photons of lower energy lying in the visible range. Another advantage of the change in photon energy from the ultraviolet (pure crystal) to the visible (activated crystal) results in better overlap between the emission spectra of inorganic scintillators and the photosensitivity of most photocathodes.

The most notable property of inorganic scintillators is their high efficiency for detection of gamma-rays. This follows from the higher density and higher atomic number Z of inorganic scintillators, resulting in much better efficiency of detection by the photoelectric effect that in any organic material. Of the inorganic scintillators, the most common material is sodium iodide to which 0.1% mole fraction of thallium has been added as an activator. Such NaI(Tl) crystals have a high light yield. They may be fabricated in irregular geometries and in large sizes, up to 30 cm. Sodium iodide is deliquescent, and must therefore be encapsulated in an air-tight container for normal use. The dominant decay time of the scintillation of NaI(Tl) is long (≈ 230 ns) compared with that of organic scintillators and it is therefore not the material of choice in applications involving high counting rates and fast timing. The response of the NaI(Tl) crystal to gamma-rays (via the nuclear photoeffect) is virtually linear over the energy range of practical interest. This feature, in addition to its excellent light yield, makes NaI(Tl) the generally accepted standard scintillator for gamma-ray spectrometry.

Of other inorganic scintillators, CsI(Tl) and CsI(Na) have become popular for use as gamma detectors. Both materials have a higher effective atomic number than NaI(Tl) and as a result their gamma-ray absorption coefficient is higher. Cesium iodide crystals are less brittle than NaI(Tl) and are not as hygroscopic. CsI crystals can be cut into thin sheets of thickness only sufficient to stop heavy charged particles. Because of its low sensitivity to gamma-ray and electrons, this configuration is ideally suited to the detection of heavy charged particles.

Of the recently developed inorganic scintillation materials, bismuth germanate, Bi_4GeO_{12}, cannot be omitted. Commonly abbreviated as BGO, it was first introduced commercially in 1979. Its prominent feature is a high stopping power (Z of Bi = 83, density = 7.13 g/cm^3), resulting in the highest probability of any scintillator for the photoelectric absorption of gamma-rays. The light yield is relatively low, only $\approx 21\%$ of that of NaI(Tl). This factor limits low energy applications of BGO and impairs its energy resolution. BGO is therefore the material of choice when the need for high gamma-ray counting efficiency outweighs consideration of the energy resolution. The principal decay time of BGO, 300 ns, is close to that of NaI(Tl); however, the slower decaying components responsible for the afterglow are more than 100-fold weaker than those in sodium iodide. This feature, together with the high gamma cross-section of BGO and its non-hygroscopicity, have led to BGO finding widespread application as a scintillation scanner in computerized tomography and subsequently in positron emission tomography. In both cases, scintillators must accurately follow rapid changes in photon intensity.

The properties of some commonly used scintillators are presented in Table 1.

2.2.4 Photomultiplier tubes

A schematic diagram of a PMT is presented in Fig. 2. It consists of a photocathode, a series of multiplying and focusing electrodes called *dynodes* and finally, a collector electrode, or anode. A small number of photoelectrons released from the photocathode by the scintillation light is accelerated (and focussed) through 100–200 V onto the first dynode. The dynodes are made of alloys having a high

TABLE 1.
Properties of some common scintillators

Name	Type	Density (g/cm^3)	Index of refraction	Light output (%) relative to anthracene for γ-rays	Decay constant of main component (ns)	Wavelength of max. emission (nm)	Principal application
Anthracene	Organic crystal	1.25	1.62	100	30	447	α, β, γ, fast neutrons
Stilbene	Organic crystal	1.16	1.626	50	4.5	410	γ, fast neutrons PSD
PILOT U	Organic plastic	1.032	1.58	67	1.36	391	Ultra fast timing
NE 213	Organic liquid	0.874	1.508	78	3.7	425	γ, fast neutrons PSD
NaI(Tl)	Inorganic crystal	3.67	1.85	230	230	415	γ
CsI(Tl)	Inorganic crystal	4.51	1.80	95	1000	540	γ, α
BGO	Inorganic crystal	7.13	2.15	48	300	505	γ

probability of secondary electron emission. As a result, an increased number of secondary electrons (a typical value is 4 to 6 for every primary incident electron) is produced by the first dynode, and these, in turn, are accelerated onto the second dynode, where further multiplication takes place. The process of multiplication is repeated on each dynode. If a PMT has N multiplication stages (dynodes) and the multiplication factor for a single stage is δ, then the overall gain K for the PMT is given by

$$K = \alpha\, \delta^N \tag{1}$$

where α is the fraction of all photoelectrons collected by the multiplier structure (α typically ≈ 1). A common number of stages is between 9 and 14. With typical value of $\delta = 5$, this will result in an overall gain of between 10^6 and 10^9 at the anode. This high gain is sufficient to produce a pulse of several volts on the anode without any external amplification. The multiplication factor δ varies (and generally increases) with the interdynode voltage, therefore the overall gain K is a very sensitive function of the high voltage applied to a PMT.

One of the important elements of a PMT is the photocathode, which is responsible for the conversion of incident light photons into photoelectrons. The photocathode can be deposited as either an opaque or a semi-transparent layer on the inside of the glass envelope.

The coupling of a scintillator to a PMTcan be achieved in many ways. Some detectors include the PMT integrally connected to the scintillation crystal and enclosed in a light-tight assembly. Others incorporate the scintillator in one container with the PMT mounted in such a way that it can be removed by the user and installed again. In any case, transparent optical grease should be used between the scintillator and the PMT to provide a smooth change in index of refraction and minimize internal reflection. Scintillators are either mounted directly on the PMT face plate or optically coupled by means of a light pipe typically made of quartz or glass. Light pipes are used to enhance uniform light collection or when the PMTmust be located far away from the scintillator. Both the scintillator and the light pipe must be painted with reflective material to enhance light collection; magnesium oxide is the most commonly used reflective material.

2.3 Semiconductor detectors

2.3.1 Introduction

Historically, semiconductor detectors have been developed as solid-state ionization chambers. Highly insulating diamond crystals were first used as a sensitive material (instead of gas) to obtain a high field, low current, solid-state device for detection of nuclear radiation. However, such crystals were quickly abandoned because of inadequate charge collection characteristics of the diamond crystals resulting from the charge trapping at defect centres. Only after the successful development of the technology for production of highly purified single crystals of silicon (Si) and germanium (Ge) for transistor electronics did it become possible to develop Si- and Ge-based solid-state detectors by forming reverse-biased junctions on these materials. The most prominent feature of the semiconductor detectors is their excellent energy resolution. In addition, they exercise fast timing capabilities, and their effective thickness, and thus their efficiency, can be varied in accordance with the requirements of the experiment.

Detailed information on the subject of semiconductor detectors is available in the current literature [1, 3, 4]

2.3.2 Basic principles

At first sight, it seems that many solids could be used as the sensitive material in solid-state detectors, but this is not so for a number of reasons. First of all we have to take into account the noise considerations. Even in the absence of radiation the conductivity of the active volume is not negligible, so that inevitably a leakage current will exist and obscure the small useful signal. From this point of view, insulators (with resistivity of up to $10^{16}\,\Omega$-cm) seem to be the best choice. However, insulators suffer from such grave shortcomings as charge carrier trapping and formation of a space charge. Under these circumstances, the electrons and holes formed in a insulator would quickly be trapped or recombine and so would induce only a negligible voltage pulse on the detector electrodes. Therefore semiconductors are the more appropriate choice for solid-state detectors, despite the fact that their resistivity under normal conditions is considerably less than that prescribed by the signal-to-noise considerations ($\approx 10^{10}\,\Omega$-cm). Of all the semiconductor materials, germanium and silicon are used almost exclusively in modern solid-state detectors, since only for these materials is an adequate purification technology available. Using this most advanced purification technology, silicon and germanium having resistivity of no more than $\approx 50\,000\,\Omega$-cm and $50\,\Omega$-cm, respectively, can be obtained. Both values are too small to enable a straightforward application of a homogeneous block of Si or Ge as a semiconductor detector. Therefore some special measures should be taken to increase sharply the resistivity of the semiconductors. These measures include cooling of the semiconductor to liquid nitrogen temperature (77 K), use of a reverse-biased P–n junction, compensation by the lithium drift method, and other means that will be elaborated later.

2.3.3 Physical properties of pure Ge and Si

Both Ge and Si have a crystal structure consisting of a regular repetition of a tetrahedron unit cell with an atom at each vertex. Each atom in such a structure contributes four electrons, so that the atom is tetravalent. The connection between neighboring atoms stems from the fact that each of the four valence electrons of a Ge or Si atom is shared by one of its four adjacent neighbors. This is the so-called electron-pair, or covalent, bond. Since all four valence electrons take part in binding of one atom to another, no free carriers of electricity are available and the crystal behaves as an insulator, at least at a very low temperature (close to 0 K). In such a pure crystal an electron is allowed to exist only in the valence band, where it is immobile, or in the conduction band, where it is free to move in an applied electric field. Therefore, in this situtation, the valence band remains full, the conduction band stays empty, and the crystals are insulators at absolute zero. As the temperature is raised, an increasing number of the valence electrons can acquire energy greater than the forbidden band width E_g and thus transfer into the conduction band. Once in the conduction band these electrons are free to move under the influence of an applied electric field. The elevation of an electron to the conduction band leaves a vacancy in the valence band. This vacancy is known as a positive 'hole' which, under the influence of an electric field, will move in a direction opposite to that of the electron. The field-induced motion of both electrons and holes contributes to the overall conductivity of the semiconductor. The production of equal amounts of conduction electrons and holes by thermal agitation is known as the intrinsic mechanism of conductivity. In an intrinsic semiconductor the number of holes p is equal to the number of free electrons n. Whereas thermal excitation produces new hole–electron pairs, other hole–electron pairs disappear as a result of recombination. It is found that the intrinsic concentration $n_i (n = p = n_i)$ varies with temperature T as

$$n_i = AT^{3/2} e^{-E_{g0}/2kT} \quad (2)$$

where E_{g0} is the forbidden band width at 0 K in eV, k is the Boltzmann constant in eV/K, and A is a constant ($A = 1.8 \times 10^{16}\ cm^{-3}$ for Si and $8.5 \times 10^{15}\ cm^{-3}$ for Ge). From equation (2) it follows that the intrinsic hole or electron concentrations at 300 K are $1.5 \times 10^{10}\ cm^{-3}$ in Si and $2.4 \times 10^{13}\ cm^{-3}$ in Ge. Another important conclusion is that the equilibrium intrinsic concentration, being an exponential function of temperature, will decrease sharply as the material is cooled down.

If an electric field is applied to the semiconductor material, the charge carriers (free electrons and holes) start to drift in the direction dictated by the field. Because of the inevitable inelastic collisions with ions of the crystalline lattice, a steady-state condition is reached in which an ultimate value of drift velocity is attained. This drift velocity is proportional both to the applied field and to the mobility of the charge carriers [5].

The main constants describing the most important physical characteristics of Ge and Si are given in Table 2[1, 6].

In gas-filled detectors, the mobility of free electrons is typically 1000 times greater than for ions, but in semiconductor materials the mobility of the holes is comparable with that of the electrons (see Table 2). As a result, the holes

TABLE 2.
Some basic properties of intrinsic Ge and Si

Property	Ge	Si
Atomic number	32	14
Atomic weight	72.60	28.09
Density (300 K), g/cm^3	5.32	2.33
Atoms/cm^3	4.41×10^{22}	4.96×10^{22}
Dielectric constant (relative)	16	12
Forbidden energy gap, eV, at 0 K	0.746	1.165
Forbidden energy gap, eV, at 300 K	0.665	1.115
Intrinsic carrier density, n_i, cm^{-3}, at 300 K	2.4×10^{13}	1.5×10^{10}
Intrinsic resistivity, Ω-cm, at 300 K	45	2.3×10^5
Electron mobility, cm^2/V-s, at 300 K	3900	1350
Hole mobility, cm^2/V-s, at 300 K	1900	480
Electron mobility, cm^2/V-s, at 77 K	3.6×10^4	2.1×10^4
Hole mobility, cm^2/V-s, at 77 K	4.2×10^4	1.1×10^4
Carrier saturation velocity, cm/s, at 300 K	5.9×10^6	8.2×10^6
Carrier saturation velocity, cm/s, at 77 K	9.6×10^6	10^7
Energy per electron–hole pair, eV, at 300 K	(not applicable)	3.62
Energy per electron–hole pair, eV, at 77 K	2.96	3.76

contribute a significant part of the overall conductivity and so improve considerably the timing characteristics of semiconductor detectors. Morever, as can be seen from Table 2, the mobilities of the charge carriers increase by an order of magnitude at the temperature of liquid nitrogen (77 K). Thus the timing characteristics of semiconductors can be improved even more by cooling down to this temperature. As a result, semiconductors detectors can be expected to be among the fastest of all radiation detectors.

From equation (2) it follows that the resistivity of semiconductor materials increases very rapidly with cooling of the material. Calculations based on the data presented in Table 2 show that the intrinsic resistivity of absolutely pure (i.e. without any impurities) Ge and Si increases at 77 K to such an extent that both materials turn out to be excellent insulators at this temperature. It may appear at first sight that either material could be successfully used at low temperatures to form the basis of a homogeneous solid-state detector.

However, in practice any semiconductor material inevitably contains some small amounts of impurities which cannot be removed by even the most advanced contemporary purification technologies. Moreover, the electrical properties of real semiconductor materials are largely dictated by these very small amounts of residual impurities. The intrinsic semiconductor is a very useful theoretical concept but in practice it is virtually unattainable.

2.3.4 Physical properties of real (doped) Ge and Si

Let us consider the effect of the small concentration of impurity which is present in the semiconductor material, either as a residual amount after the best purification possible or as a small amount intentionally added to the material to change its properties. If to tetravalent Si or Ge there is added a small percentage of trivalent or pentavalent impurity atoms, a doped, impure, or *extrinsic* semiconductor is formed. The impurity atoms will displace some of the host atoms in the crystal lattice. If the dopant atom has five valence electrons, four of the five electrons will occupy covalent bonds; the fifth will be bound only very loosely to the original impurity atom and, as a result, will be available as a carrier of current. The energy required to detach the fifth electron from the atom is very small, being of the order of 0.01 eV for Ge and 0.05 eV for Si. Suitable pentavalent impurities are P, As, and Sb. Since such impurities donate excess negative charge carriers (electrons) they are referred to as donor, or n-type, impurities, and the semiconductor material containing them is called an n-type semiconductor. Donor impurities introduce a set of discrete donor states just slightly below the conduction band. Since the energy spacing between the donor levels and the bottom of the conduction band is very small (0.01 eV for Ge and 0.05 eV for Si), almost all of the donor impurities are ionized at room temperature [see equation (2)], and therefore almost all the excess electrons of the donor atoms are raised into the conduction band. Since in most practical cases the impurity atoms are present in concentrations of the order of a few parts per million, the concentration of impurity atoms N_D is large compared with the concentration of intrinsic electrons n_i. Therefore, in an n-type material, the free electron concentration becomes completely dominated by the contribution from the donor atoms ($n \approx N_D$). In an n-type semiconductor material, not only does the number of electrons increase but the number of holes decreases, since the large number of electrons present increases the rate of recombination of electrons with holes.

If, on the other hand, a trivalent impurity such as In, B, or Ga is added to an intrinsic semiconductor, only three of four covalent bonds can formed, leaving a vacancy (hole) in the fourth one. These impurities are known as acceptor, or p-type, impurities. They introduce discrete acceptor states just above the valence band, and the material itself is referred to as a p-type semiconductor because the principal charge carriers are the positively charged holes. Just as in the case of the donor impurity, only a very small amount of an acceptor must be added to change appreciably the conductivity of the semiconductor material. The ionized acceptor sites represent fixed negative charges, which balance the excess hole charges and do not take part in current conduction in the semiconductor. Usually the concentration N_A of acceptor impurities is made large enough to be compared with the intrinsic concentration of holes p_i. In such a case the number of holes is totally determined by the concentration of aceptors ($P \approx N_A$). As mentioned above, n-type impurities decreases the number of holes. Similarly, doping with p-type impurities decrease considerably the concentration of free electrons. The mass action law controls the number of the free charge carriers in the semiconductor. It states [5] that, under thermal equilibrium, the product of the free negative (n) and positive (p) concentrations is a constant, independent of the amount of donor and acceptor impurity doping:

$$np = n_i^2 \tag{3}$$

where n_i is the intrinsic concentration given by equation (2). The important result following from equation (3) is that the doping of an intrinsic semiconductor not only increases the conductivity, but also produces a material in which the electric carriers are either predominantly holes or predominantly electrons. In an n-type semiconductor, the electrons are called the majority carriers and the holes are called minority carriers. In a p-type material, the holes are the majority carriers and the electrons are the minority carriers.

In practice, a semiconductor material will always contain both types of impurities and their effects will at least partly cancel one another out, since the electrons produced by the donor impurities will recombine with the holes generated by the acceptor impurities. The ultimate character of the semiconductor material (n-type or p-type) will depend on whether the number of electrons generated by the donors exceeds the number of holes generated by the acceptors or vice versa.

The action of nuclear radiation on a semiconductor material ultimately results in the creation of electron–hole pairs. The average energy ε necessary to produce an electron–hole pair is found experimentally to be independent of the type and the energy of the incident radiation. The values of ε are: 3.62 eV in Si at 300 K; 3.76 eV in Si and 2.96 eV in Ge at 77 K (Table 2). The smallness of the average energy ε compared with the average energy necessary to create an electron–ion pair in a gas (typically about 30 eV) represents an outstanding feature of semiconductor detectors and confers the superior energy resolution of these detectors. This follows from the increased (by ≈ 10 times) number of charge carriers produced in the semiconductor as compared with the gas detector for the same energy deposited in the detector. The relative statistical fluctuation in the number of charge carriers per event becomes smaller as the total number is increased.

Once, as a result of the action of nuclear radiation, electrons and holes are formed in a semiconductor, these may recombine with one another, be trapped by impurities or imperfection centers, or be collected at the electrodes of the detector, producing a useful signal. The charge carrier lifetime is measured as being 10^{-2}–10^{-4} s in the purest semiconductor materials available. On the other hand, the mean charge carrier lifetime due to direct electron–hole recombination is estimated to be as high as ≈ 1 s. Therefore direct recombination is rather improbable, and the main factor determining the lifetimes of the charge carriers is the trapping effect. The trapped charge carrier can be released back to the relevant band; however, the time delay (detrapping time) is often too long and, as a result, the charge carrier is lost to the charge collection process or is collected with significantly reduced efficiency. On the other hand, if the mean detrapping time is substantially shorter than the charge collection time, then the trap has no effect on the charge collection process. In modern semiconductor detectors, the maximum permissible concentration of trapping centers is of the order of 10^{10} cm^{-3}, corresponding to approximately 1 for every 10^{12} atoms of the host material [6]. This results in high charge collection efficiency of the order of 0.999. The corresponding collection time is of the order of 10^{-7} s and carrier lifetime is of the order of 10^{-4}–10^{-5} s.

In summary, the physical properties required for a semiconductor detector material are as follows: (a) high specific electric resistivity to give low leakage current; (b) freedom from recombination and trapping of charge carriers to give efficient (preferably 100%) charge collection; (c) small width of the forbidden band to give a large number of electron–hole pairs; (d) high mobilities of both charge carriers to give a fast time response; (e) high atomic number Z to give a high photoelectric cross-section for gamma- and X-ray spectrometry.

2.3.5 Basic properties of p–n junctions

Virtually all semiconductor detectors exploit the remarkable properties of the junction created between two semiconductor materials with different types of conductivity. Such a p–n junction is formed if donor impurities are introduced into one side and acceptors into the opposite side of a single crystal of a semiconductor. When n- and p-type semiconductors are brought into contact, the electrons from the n-type material begin to diffuse across the junction into the p-type material and the holes from the p-type material into the n-type material. This occurs because there is a density gradient of the free charge carriers across the junction. Free electrons diffusing from the n-type into the p-type material combine there with the holes and disappear as a result of annihilation. Each free electron that disappears from the n-side of the junction leaves behind an immobile positively charged donor site. Similarly, the holes diffusing in the opposite direction annihilate with the free electrons in the n-type material, leaving behind immobile negatively charged acceptor sites. The accumulated space charge from the immobile donor and acceptor sites creates an electric field which eventually halts further diffusion. Since the region of the junction is depleted of mobile charge carriers, it is called the depletion region, or the space charge region. If the concentrations of the donors on the n-side of the junction and the acceptors on the p-side are approximately the same, the depletion region extends equally into both sides across the junction. However, this is not the case in semiconductor detector practice. Here an intentional strong assymmetry in the doping levels on each side of the junction is produced. In such a case, the depletion layer extends almost entirely into the low-level doping side of the junction.

The configuration just described is virtually a junction diode. The width of the depletion region of such a diode is of the order of 0.5 μm and the potential difference across the junction (the contact potential) amounts to a few tenths of a volt. Electric field existing across the junction sweeps the free charge carriers out of the depletion region. As a result, the carrier density concentration remaining in the depletion region is approximately eight orders of magnitude lower than typical carrier densities in the bulk material [1]. Such a low concentration of free charge carriers in the depletion region means that the specific resistivity of this region is very high. As mentioned, the high specific resistivity of the semiconductor is one of the main properties required from the material for semiconductor detectors. It should be pointed out that the p–n junction diode can be used for the detection of nuclear radiation even with no external voltage applied. This is because a nonzero electric field exists across the junction. If a charged particle enters the depletion region and creates electron–hole pairs, these will be swept out of the depletion region by the electric field towards the respective electrodes, and their flow forms an useful electronic pulse. However, the thickness of the depletion region in the unbiased junction diode is very small ($< 10^{-4}$ cm). As a result, the active volume of the detector is too small to be of any practical interest. For the same reason, the capacitance of an unbiased junction is high, resulting in very bad signal-to-noise properties of the detector. Therefore unbiased junction detectors are not used as practical detectors of nuclear radiation.

The thickness of the depletion region of the junction diode can be increased considerably by applying a reverse bias to the junction. The reverse bias causes both the holes in the p-side of the junction and the electrons in the n-side to move away from the junction. Consequently, the region of uncompensated negative charge is spread to the left of the junction and that of uncompensated positive charge is spread to the right. Since practical detectors are operated with a very large reverse bias voltage (up to 3000 V) compared with the contact potential, the thickness of the depletion region is increased considerably, reaching more than 0.1 cm for Si diodes. As mentioned, the depletion region width is determined not only by the biasing voltage but also by the free carrier densities. Since the net charge must be zero, then:

$$N_A W_p = N_D W_n \quad (4)$$

where N_A and N_D are the acceptor and donor concentrations in the corresponding sides of the p–n junction, and W_p and W_n are the depths to which the space charge extends into the p- and n-sides of the junction. In semiconductor detector practice, the n-side of the junction is usually doped to a much greater extent than the p-side. Consequently, the depletion layer extends much further into the p-side than the n-side. In such a case, the width d of the depletion layer is given by [1]:

$$d \cong W_p = (2\varepsilon V \mu \rho_d)^{1/2} \quad (5)$$

where ε is the dielectric constant of the semiconductor, V is the reverse bias and ρ_d is the resistivity of the doped semiconductor. From equation (5) it follows that the largest depletion layer is obtained when a semiconductor material of highest resistivity is chosen and the largest possible reverse bias is applied to the junction. For the material to be of highest resistivity it has to have the highest purity possible. The largest reverse bias is limited by the maximum electric field occurring at the junction, and must be no more than 2×10^4 V/cm under typical conditions. On the other hand, the reverse bias should not be increased above the point at which the detector becomes totally depleted.

The pulse magnitude observed in a semiconductor detector after accomplishment of the charge collection process is inversely proportional to the sum of the capacitance of the p–n junction and the input circuit. The signal-to-noise ratio of the system is improved if the capacitance of the junction decreases. From this point of view, use of the largest possible reverse bias is preferable.

2.3.6 Charged particle detectors

In most cases, detectors for charged particle spectrometry are reverse biased diodes made with highly purified single crystal silicon. Silicon is chosen for this purpose because its energy gap is large enough to ensure that the resistivity is sufficiently high even at room temperature. In addition, the low atomic number of Si is of no importance for charged particle detectors.

For the detection of charged particles, a depletion layer must be located as close as possible to the surface of the detector, since the regions outside the depletion layer act as a dead layer for incoming charged particles. There are a number of technologies for manufacturing silicon detectors with surface junction construction. For example, the diffusion method of fabrication starts with diffusion of a donor impurity such as phosphorus at a high temperature (≈ 800 °C) through the front face of a homogeneous crystal of p-type silicon. A p–n junction is formed at a distance of 0.1–1 μm from the surface of the crystal. Since the superficial n-type layer is heavily doped compared with the original p-type, the depletion layer extends primarily into the p-side of the junction. Therefore the n-side of the junction represent a dead layer. The thickness of the dead layer should be determined experimentally to make possible accurate measurement of the energy of the incoming charged particles.

To avoid a large dead layer and the high temperature involved in the diffusion process, another type of surface junction detector was developed—the surface barrier detector—which successfully replaced the diffusion detector in many applications. Typically, a surface barrier detector is a diode made of an etched n-type silicon wafer with a very thin (40 μg/cm^2) layer of gold deposited by evaporation on the Si wafer. The gold layer forms a p-type rectifying contact on the diode. The other surface of the wafer is coated with a layer of 40 μg/cm^2 of evaporated aluminum to produce good ohmic contact. When a reverse bias is applied to the contacts (i.e. a negative voltage on the gold electrode), electrons from the vicinity of the gold electrode are swept away, creating a depletion region that forms a sensitive medium for detection of charged particles.

The entrance window (gold layer) of surface barrier detectors is so thin ($\approx 0.02\ \mu m$) that it is transparent to the photons of visible light. To avoid a high noise level generated by normal light, surface barrier detectors should be used in a light-proof environment, usually provided by the vacuum chamber required for most charged particle applications.

Surface junction detectors are available in a wide range of active areas from 7 to $2000\ mm^2$ and sensitive depths from 10 to $5000\ \mu m$, and are therefore satisfactory for the detection of many charged particles in wide energy range.

From equation (5) it follows that the width of the depletion layer is proportional to the applied voltage. According to the magnitude of the applied reverse bias voltage, the detector can be partially depleted, fully (totally) depleted or over-depleted. In the first case, the depth of the depletion region W is less than the physical thickness L of the detector. The partially depleted detectors are therefore sensitive to charged particles incident only on the junction surface of the detector.

Thin, totally depleted silicon detectors are widely used as transmission detectors for identification of charged particles on the basis of their rate of energy loss dE/dx. These are so called dE/dx detectors. For this application, a very thin detector is chosen so that the particle emerges with minimum energy loss and can be further detected by another detector (telescope arrangement). The dE/dx detector should have layers as thin as possible on both the entrance and the exit side of the detector. Planar totally depleted silicon surface barrier detectors are currently available in thickness down to $10\ \mu m$.

As follows from equation (5), the width of the depletion layer in the surface junction detector is determined by the resistivity (purity) of the starting silicon wafer and by the bias voltage applicable without breakdown. The currently available purity of silicon is such that depletion thicknesses are limited to no more than a few millimeters. Greater thickness in silicon can be obtained only with intrinsic material or through the use of material compensated by the lithium-drifted process.

Generally, all real semiconductor materials contain impurities of both n- and p-type. Since the electrons originating from donor impurities will recombine with holes from acceptor atoms, it is obvious that a partial mutual compensation of impurities takes place. In such a case, the net concentration ($N_D - N_A$), where N_D and N_A are the concentrations of donor and acceptor atoms, determines the type of semiconductor. If $N_D > N_A$, it is an n-type material, and vice versa. If, on the other hand, $N_D = N_A$, then the semiconductor behaves like an intrinsic material. Such compensated materials are designated with the letter 'i'.

The method of impurity compensation based on the process of lithium ion drifting was developed by E. M. Pell [7] in 1960. The method can be used effectively for compensating p-type silicon and germanium crystals after the crystals have been grown. The lithium drift method is based on the fact that lithium ions, being of small radius, are fast interstitial diffusants in both Ge and Si and act as donor impurities with very low ionization potential (0.033 eV in Si and

0.0043 eV in Ge). The mobility and therefore the coefficient of diffusion of Li ions is some 10^7 times greater than that of ordinary diffusants like boron and phosphorus.

It should be pointed out that, in both Si and Ge, the material with highest available purity is of p-type. The process of lithium compensation starts with diffusing lithium, which acts as a donor, into the front surface of a p-type semiconductor. To increase lithium mobility, the temperature of the crystal is elevated to 500 °C. As a result of diffusion, an n-type region is created near the exposed surface of the crystal. The resulting p–n junction is then back-biased and the positive lithium ions are slowly pulled further by the electric field into the p-region, where their concentration increases and approaches the concentration of the p-type impurities in the original material. Simultaneously, the lithium donor concentration decreases in the n-region. The lithium ions, having high mobility, redistribute themselves under the influence of the local fields in such a way that the total space charge in the drifted region tends towards zero at every point. Otherwise the local field would essentially reverse direction and would sweep the lithium ions back in the opposite direction. An equilibrium point is thus established in which the lithium ions go on to drift into the p-region, spreading themselves over the region in such a way that the number of lithium donors compensates exactly the acceptors of the original material. It has been shown by Pell [7] that any deviation from exact compensation by the lithium ions during drifting is unstable, and any instantaneous imbalance in lithium concentration is quickly eliminated by itself, thus restoring exact compensation. Using the lithium drifting process, thick compensated regions of up to 5–10 mm in Si and 10–15 mm in Ge can be achieved. In this case, the drift process can take hundreds of hours. More detailed information about the lithium drift process can be found in Ref. 4.

Lithium ions have high mobility at room temperature, particularly in germanium. Therefore, to preserve the lithium concentration required for compensation, the Ge crystal must be cooled to liquid nitrogen temperature (77 K) immediately after the desired compensation is attained. In silicon, however, the lithium ion mobility is lower, so that the lithium-drifted silicon detector can be temporarily stored at room temperature without cooling.

As mentioned, the problem of insufficient thickness of the depletion zone in semiconductor detectors was solved with the development of the lithium drifting process for obtaining compensated material. Such silicon-based detectors are called *lithium-drifted silicon detectors*, or Si(Li) detectors (pronounced ‘silly’). Junctions of the Si(Li) detectors are known as p–i–n junctions and their properties are different from the ordinary n–p junctions discussed earlier. Because, theoretically, no net charge exists in the compensated i-region, the resulting electric field is constant across the i-region, dropping sharply to zero at its boundaries. Therefore the active volume of the detector is determined by the dimensions of the i-region.

The increased thickness of the compensated (active) zone of Si(Li) detectors (5–10 mm) makes them suitable for the detection and spectroscopy of beta-particles and for the detection of low-energy X- and gamma-rays. For the same reason, the bulk generated leakage current of Si(Li) detectors is a significant contributor to the noise. As a result, the detectors in these applications must be cooled to liquid nitrogen temperature. For this purpose they are mounted in a vacuum-tight cryostat incorporating a built-in sorption pump. A thin beryllium or Mylar window enables entry of the low energy radiation into the detector. The cryostat is mounted on a liquid nitrogen dewar with a capacity of about 30 liters. To diminish the contribution of electronic noise in the energy resolution of Si(Li) detectors, they are supplied with a low-noise preamplifier incorporating a cooled first stage. Being low energy gamma-ray spectrometers, Si(Li) detectors are extensively used for the observation of the Mössbauer effect, where energies below 40 keV are of main interest. As X-ray spectrometers they are used in non-destructive testing of ores, alloys, etc.

2.3.7 Germanium detectors

Germanium detectors are mainly used in gamma-ray spectroscopy because the atomic number of Ge ($Z = 32$) is much higher than that of Si ($Z = 14$). The photoelectric probability, which depends upon Z^4 to Z^5, is thus about 60 times greater in Ge than in Si. In general, much greater sensitive thicknesses are required for the detection of gamma-rays than for the charged particles. Before the mid-1970s the required depletion depths could be achieved only by counter-doping p-type germanium crystals with lithium in the lithium ion drifting process described above in connection with Si(Li) detectors. These lithium compensated Ge-detectors are known as Ge(Li) (pronounced 'jelly') detectors.

To increase overall detector volume, and thus detection efficiency, Ge(Li) detectors are constructed in cylindrical or coaxial geometry. In *true coaxial geometry*, lithium diffuses on the outside of the p-type germanium crystal and drifts towards a central core under a strong electrical field, thus producing a cylindrical shell of high-resistivity (intrinsic) material. A central core of insensitive p-type germanium extends along the entire length of the cylindrical crystal. In a *closed-end coaxial configuration*, lithium is also allowed to drift from the front face of the Ge crystal. As a result, the sensitive volume of the detector is increased even more. Coaxial Ge(Li) detectors can also be constructed in *well* configuration, in which the insensitive central core of the germanium cylinder is removed. Radio-active samples can then be placed within the well, thus substantially increasing the counting efficiency. For lower gamma energies (up to several hundred keV), Ge(Li) detectors may be fabricated in a *planar* configuration, in which the electrical contacts are formed on the flat surfaces of a germanium disk a few centimeters thick.

In order to reduce the thermal charge carrier generation (electronic noise) to an acceptable level, Ge(Li) detectors must be cooled to liquid nitrogen temperatures. The cooling of Ge(Li) detectors prevents them from lithium precipitation and thus from destruction. This is necessary because of the high mobility of lithium ions in germanium at room temperature. For this reason, Ge(Li) detectors must be kept at liquid nitrogen temperature at all times (both in operation and for storage). The germanium crystal and the input stage of the preamplifier are installed in a vacuum chamber which is inserted in liquid nitrogen container (Dewar), just as for Si(Li) detectors (Fig. 3). At these temperatures, reverse leakage currents are very low, in the range 10^{-9}–10^{-12} A [8].

In the mid 1970s, advances in germanium purification technology made available high purity germanium with an impurity level as low as 10^{10} atoms/cm^3. Detectors produced from this ultrapure Ge material are called *high purity germanium* detectors, usually abbreviated as HPGe detectors. The outstanding feature of these detectors is that they do not have to be kept at liquid nitrogen

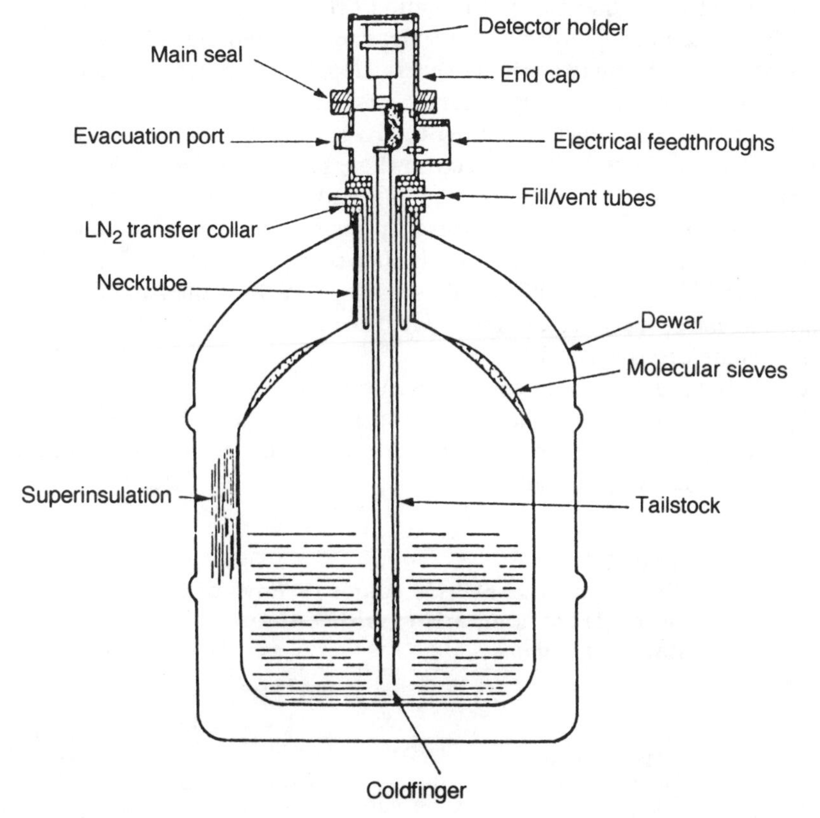

Fig. 3. Ge detector attached to a dipstick cryostat inserted in a liquid nitrogen container (From Ref. 8).

temperature at all times. HPGe detectors must be cooled only in operation when a high voltage is applied. Between uses these detectors can be kept at room temperature.

Since HPGe and Ge(Li) detectors are virtually identical from the point of view of construction and operation, their main performance characteristics are practically the same. On the other hand, HPGe detectors are obviously much more convenient in operation than Ge(Li) detectors. This is the reason for almost complete replacement of Ge(Li) detectors by HPGe detectors in contemporary gamma-ray spectroscopy.

One of the main measures of the performance of nuclear detectors is their resolution. In general, resolution is determined as the ability of a detector to separate two peaks that are close together in energy. Therefore, the narrower the peak, the better resolution of the detector. The resolution is measured as the full width of the peak measured at half the maximum amplitude (abbreviated FWHM). Germanium detectors feature very high energy resolution capabilities as compared with other types of radiation detectors (proportional counters or scintillation counters). Basically, the better energy resolution of Ge detectors can be attributed to the small amount of energy required to generate a pair of charge carriers (electron–hole) in germanium. As a result, a greatly increased number of charge carriers is produced, forming the large output signal relative to that of other detectors. At ≈ 3 eV per electron–hole pair (see Table 2), the number of charge carriers produced in Ge is about one and two orders of magnitude higher than in gas and scintillation detectors respectively. The energy resolution of Ge detectors expressed as FWHM in keV is often specified at 5.9 keV (^{55}Fe), 122 keV (^{57}Co) and 1332 keV (^{60}Co). The representative FWHM values and the resultant improvement in energy resolution of Ge detectors over other types of radiation detectors are presented in Table 3 [8].

The striking difference in energy resolution between a Ge detector and a 3 inch × 3 inch NaI(Tl) scintillation detector is illustrated in Fig. 4. The germanium detector used is a coaxial Ge(Li) detector with 7.5% efficiency. Since the absolute efficiency of a standard 3 in. × 3 in. NaI(Tl) crystal is well-known

TABLE 3.
Representative FWHM values (in keV) of Ge detectors in comparison with various detector types

Counter	Energy (keV)		
	5.9	122	1332
Proportional counter	1.2	–	–
3 in. × 3in. NaI(TI)	–	12.0	60
Si(Li)	0.16	–	–
Planar Ge	0.18	0.5	–
Coaxial Ge	–	0.8	1.8

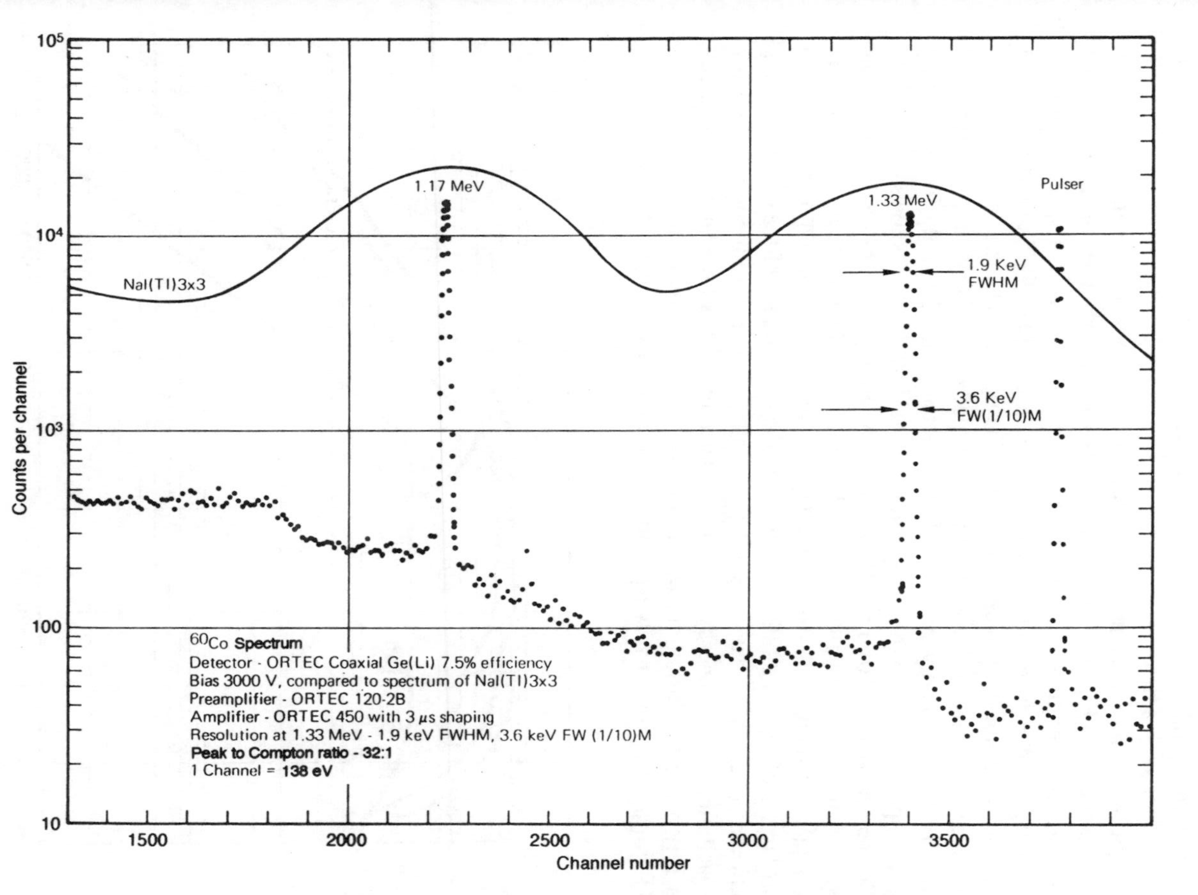

Fig. 4. Co-60 spectrum for NaI(Tl) and Ge(Li) detectors (From Ref. 6).

(1.2×10^{-3}), the relative detection efficiency can easily be obtained. Coaxial Ge detectors of greater than 40% relative efficiency have been fabricated from germanium crystals up to ≈ 65 mm in diameter [8]. About 1 kg of germanium is required for such a detector.

Typical absolute efficiency curves for various types of germanium detectors are presented in Fig. 5. Curve 1 represents a 10% relative efficiency coaxial high purity germanium detector. It is used in general gamma spectroscopy for the energy range from 50 keV to more than 10 MeV. Curve 2 represents a $200\,mm^2 \times 10\,mm$ thick planar detector. The detector element is made of high purity germanium. Its cryostat window is made from beryllium for improved sensivity to low energy photons. The detector is used mainly in low energy photon spectrometry in the energy range 3–200 keV. Curve 3 shows the absolute efficiency curve for a $10\,cm^2 \times 15\,cm$ thick low energy germanium detector. This type of detector is fabricated with an ultrathin boron implanted p-contact on the front face and on the cylindrical wall. The n-contact, a lithium diffused spot on the rear face, is of less than full area, and thus the capacitance of the detector is less than that of a planar device. Since preamplifier noise is a function of detector capacitance, the low energy germanium detector affords lower noise and consequently better resolution at low and moderate energies than any other detector [8]. For applications involving moderate gamma-ray energies (5–500 keV), this

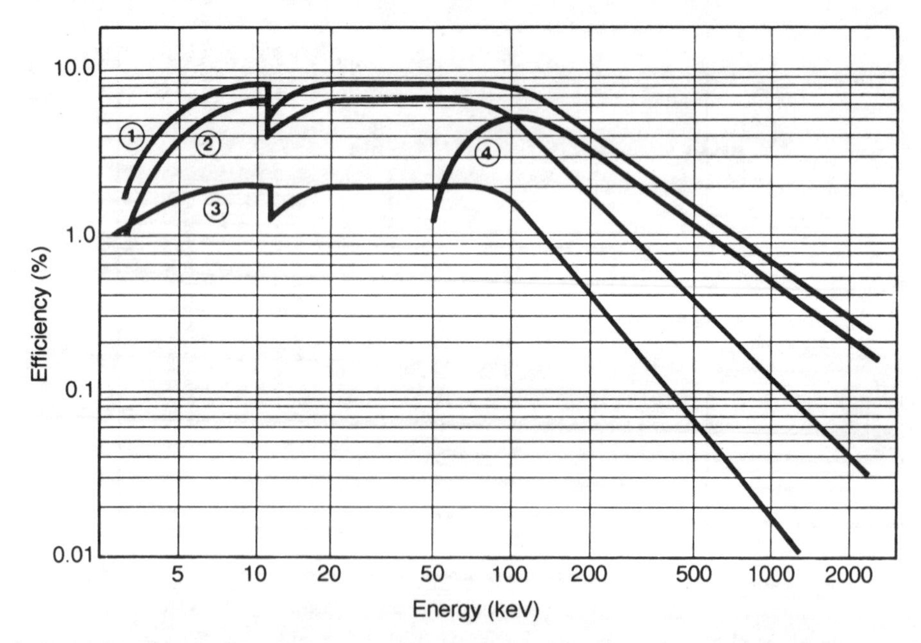

Fig. 5. Typical absolute efficiency curves for various Ge-detectors with 2.5 cm source to end-cap spacing (From Ref. 8).

type of detector may well out-perform an expensive large volume coaxial detector. Curve 4 gives the absolute efficiency data for a 15% relative efficiency reverse-electrode detector. The main design feature of the detector is the reversal of the electrodes from the configuration in a conventional coaxial detector. The p-type electrode (ion-implanted boron) is on the outside and the n-type contact (diffused lithium) is on the inside of the Ge crystal. The ion-implanted outside contact is extremely thin by comparison with a lithium-diffused contact. This, in conjunction with a thin cryostat window, results in an energy response down to 5 keV, giving this detector a dynamic range of 2000:1 (5 keV to 10 MeV). Another feature of this detector is that it is ≈ 10 times more resistant to radiation damage than conventional coaxial Ge detectors [8].

3 GAMMA-RAY SPECTROSCOPY

3.1 Instrumentation for nuclear spectroscopy

Nuclear detectors provide a variety of information on detected radiation (energy, type of radiation, timing data, etc.) in the form of electronic signals. To obtain this information, the signal must be processed by the appropriate nuclear electronics system. However, the subject of nuclear electronics is outside the scope of this chapter. Interested readers are referred to reference books [9, 10]. In what follows we present a short, qualitative description of the process of obtaining possibly the most important characteristic of nuclear radiation, its energy distribution (energy spectrum).

A schematic diagram of the basic electronic equipment for measurement of the energies of gamma-rays is presented in Fig. 6. A photon detector can be represented as a capacitor in which charge carriers are generated as a result of interaction of a nuclear particle within the detector. The high voltage provides the detector bias to collect the electric charge. As a result, a small (pulse) current flows, producing the pulse voltage drops across the bias resistor R. The voltage pulse produced enters into the preamplifier isolated from the high voltage by the capacitor C. In general, the preamplifier serves for the initial amplification of the detector output signal. The preamplifier is usually located as close to the detector as possible to reduce excessive cable capacitance, which lowers the pulse height as well as degrading the rise time of the pulse. Also, such positioning amplifies the signal before it enters a long cable leading to the main amplifier. Since the capacitance of most solid state detectors depends upon the high voltage bias, a charge-sensitive preamplifier should be used. The voltage pulse produced at the output of the preamplifier will then be proportional to the collected charge and independent of detector capacitance.

The main amplifier serves to shape the pulse for further processing as well as amplifying the pulse from the preamplifier to make it compatible with the input range of the spectrum analyzer. In energy spectroscopy the pulse height information is of primary importance. Therefore very strict proportionality between the input and output amplitudes of the main amplifier must be maintained. Generally, the pulse at the output of a preamplifier has a fast rise time, ranging from a few nanoseconds up to a few microseconds, and a slow decay time of about 50 microseconds. The amplitude of this pulse is proportional to the energy of the radiation to be measured. Therefore the long decay time of the preamplifier output pulse may cause distortion of the energy information of the second pulse, if it rides on the tail of the first pulse. This is the so-called pile-up effect. Another shortcoming of tailed pulses is their bad signal-to-noise ratio [10]; in fact, it would be more advantageous to have a Gaussian or triangular pulse shape. For this purpose, *RC* clipping can be used, in which the tailed pulse is differentiated to remove the slowly varying tail. Subsequent integration reduces the noise. The resulting pulse is much shorter and has a near-Gaussian shape, yielding improved signal-to-noise characteristics and count rate capabilities.

Specially shaped pulses from the output of the main amplifier are directed into the pulse height analysis system. This may consist of either a single channel analyzer and counter or a multichannel analyzer (Fig. 6).

The single channel analyzer (SCA) is an electronic device which analyzes and sorts incoming pulses according to their amplitudes. It has both lower level and upper level discriminators, which can be set to define a certain amplitude range in the main amplifier output. The span between the lower level and upper level discriminator thresholds is called the SCA window. Only pulses with amplitudes falling within the SCA window result in a standard logic output pulse from the SCA. Pulses with amplitudes below the lower level setting or greater than the upper level setting are rejected and do not produce output pulses. With the detector output proportional to energy, the SCA can be used to obtain energy spectra of the incident radiation. For this purpose, the lower level and upper level of the SCA are set to give a fixed narrow window, and full sweeping of the window is performed across the pulse height range under consideration. The number of pulses counted by scaler (counter) at each step and normalized per unit time can then be presented as a histogram of the energy spectrum. However, full data collection and processing by means of a single channel analyzer is a tedious and time-consuming process. Automatic and fast acquisition of energy spectra data is achieved by means of the multichannel analyzer (MCA). A MCA can be considered as a succession of single channel analyzers with the same window width settings. All windows are arranged sequentially in order of increasing energy. Thus, by representing the number of counts of each SCA versus the mean energy of the corresponding window, a histogram of a number of counts versus energy, i.e. an energy spectrum, will be obtained.

In an actual MCA, the succession of single channel analyzers is replaced by a special device called an analog-to-digital converter (ADC). This is the core of a MCA. The ADC measures and digitizes the height of each pulse to be analyzed

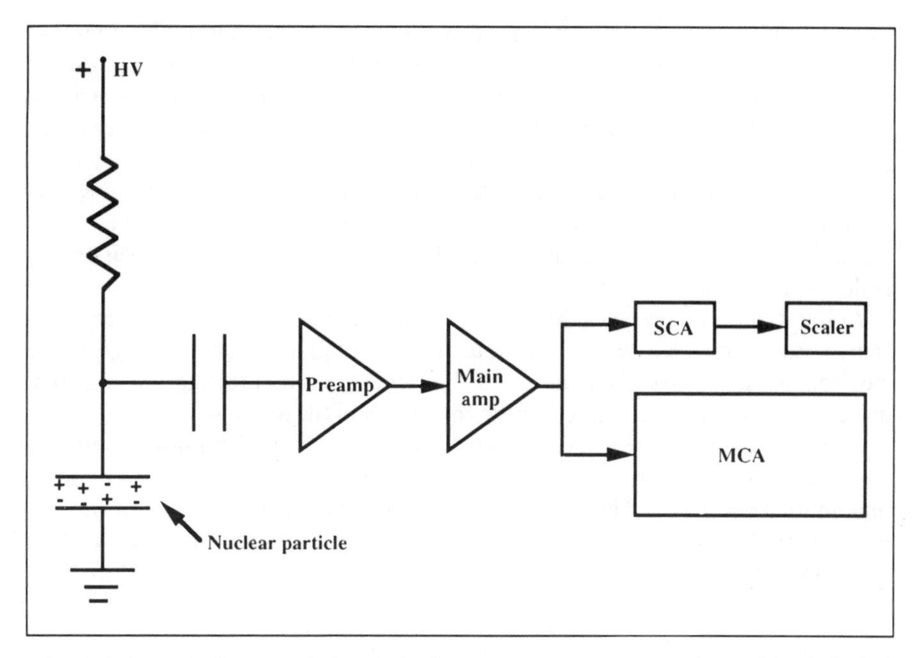

Fig. 6. Schematic diagram of electronics for energy measurements of a nuclear radiation.

and determines to which pulse height window this pulse corresponds. The MCA then takes this window number and increments a memory channel whose address corresponds to the digitized value. In this way all incoming pulses are sorted according to pulse heights represented by channels. The content of each channel is stored in the memory. The total number of channels into which the voltage range of input pulses (0–10 V) is digitized is called the conversion gain. This gain determines the resolution of the MCA and ranges from 64 up to 16K in contemporary MCAs.

An input pulse is delayed while checking is performed to see if it is within the selected SCA range. If it is, the input pulse is passed to the ADC. The ADC digitizes its amplitude, generating a number proportional to the pulse voltage. This number represents an address in the memory to which one count is added. The display reads the memory many times per second, producing a point plot of memory content versus memory location, which is equivalent to the number of pulses versus voltage, i.e. the energy spectrum.

3.2 Energy measurements

There are three principal kinds of interaction of photons with matter. These interactions are the photoelectric effect, Compton scattering and pair production. All three processes are strongly dependent upon the energy of the photon and the atomic number Z of the material. In the photelectric process, a photon is absorbed by an atom and, as a result of the absorption, a photoelectron is emitted. Its energy is equal to the photon energy minus the binding energy. The photoelectric cross-section increases very strongly with Z of the material and decreases with photon energy. It is the photoelectron that produces the electron–hole pairs in the sensitive medium of the detector, resulting in the output pulse. Electrons have a short range in the crystal and therefore all of their energy is deposited in the detector.

At higher photon energies the dominant interaction is Compton scattering. In the Compton process, the photon interacts with what might be termed a free electron, losing a part of its energy to the electron and scattering to a different direction. Since in Compton scattering the energy of the original photon is shared between two particles (the recoil electron and the scattered photon), there is a continuous distribution of pulse amplitudes of Compton electrons up to some maximum pulse height. This maximum pulse height produces the Compton edge, which is the maximum energy that can be imparted to an electron in head-on collision. It is equal to [1]

$$E_{\mathrm{e}}^{\max} = \frac{E_0}{1 + m_0 c^2 / 2E_0} \tag{6}$$

where $E_{\mathrm{e}}^{\max}$ is the maximum electron energy, E_0 is the incident photon energy, m_0 is the electron rest mass and c is the velocity of light.

Pair production is the third important process of photon interaction with matter. The photon enters the detector and creates an electron–positron pair in the field of a nucleus. From the law of conservation of mass and energy it follows that the initial photon must have an energy of at least 1.022 MeV ($= 2m_0c^2$) because this is the exact energy necessary to create both an electron and a positron.

All three processes determine the response of a solid-state detector to monoenergetic gamma-rays. The relative importance of these processes for detector response lies in the proportion of the incident gamma-ray energy that is deposited in the detector sensitive medium. A typically idealized response of a solid-state detector to monoenergetic gamma-rays is presented in Fig. 7.

The full-energy peak, or photopeak, results from the gamma photon losing all its energy by photoelectric absorption, or Compton scattering followed by photoelectric absorption in the detector. Compton scattering causes only a fraction of the energy of the initial photon to be deposited in the crystal because the scattered photon can escape from the crystal without further interaction. This process contributes to the energy response of the detector the so-called Compton continuum, ranging from zero up to a maximum determined by the Compton edge [equation (6)]. If the energy of an incident photon is above 1.022 MeV, pair

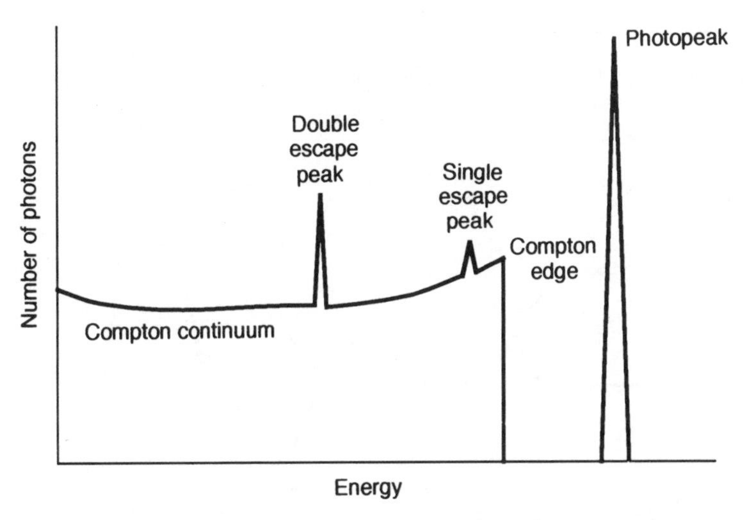

Fig. 7. Idealized detector response to monoenergetic gamma photons.

production can occur. The positron produced slows down in the crystal to an energy close to that of the atomic electron and annihilates the electron, producing two 0.511 MeV photons. Each of these photons can escape from the detector totally or deposit only part of its energy by Compton scattering. If one of these photons escapes totally, the total energy absorbed is 0.511 MeV less than the original incident photon energy. If both annihilation photons escape totally, the total deposited energy is 1.022 MeV less than the original photon energy. Therefore, in the measured, spectrum, in addition to the full-energy peak, there will be seen two additional peaks: a single-escape peak at energy $(E_\gamma - 0.511)$ MeV and a double-escape peak at energy $(E_\gamma - 2 \times 0.511)$ MeV (Fig. 7). It should be emphasized that the single- and double-escape peaks appear only if the incident photon energy is above 1.022 MeV.

The detector response presented in Fig. 7 is an idealized one, since it does not take into account the broadening of all peaks by the final resolution of the detector and the effect of multiple Compton scattering.

Figure 8 represents a real spectrum of the monoenergetic gamma photons from a ^{137}Cs source taken by a 2 in. × 2 in. NaI(Tl) scintillator. The photopeak at 0.662 MeV is clearly separated from the Compton edge (0.478 MeV). Multiple Compton scattering fills the gap between the Compton edge and the photopeak, forming continuity in the spectrum from zero to the photopeak energy. The final resolution of the detector broadens the photopeak and rounds off the Compton edge and backscatter peak at 0.184 MeV. Backscatter occurs when photons from the source make Compton interactions in the materials surrounding the detector,

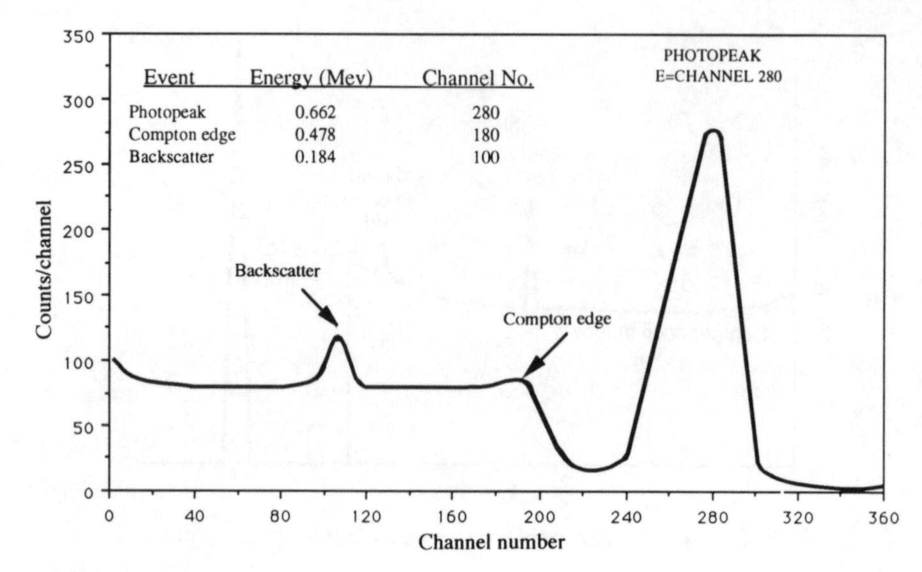

Fig. 8. Real spectrum of the monoenergetic gamma photons from a Cs-137 in a 2 in × 2 in NaI(Tl) scintillator.

e.g. Pb shielding. Backscattered photons are then detected through photoelectric absorption when they enter the detector. The energy of the backscatter peak is $E_0/(1 + 2E_0/m_0c^2)$.

REFERENCES

1. G. F. Knoll, *Radiation Detection and Measurements*, Wiley, New York, 2nd Ed. (1989).
2. J. B. Birks, *The Theory and Practice of Scintillation Counting*, Pergamon Press, Oxford (1964).
3. G. Dearnaley and D. C. Northrop, *Semiconductor Counters for Nuclear Radiation*, Wiley, New York, 2nd Ed. (1966).
4. G. Bertolini and A. Coche (Eds.), *Semiconductor Detectors*, Elsevier North-Holland, Amsterdam (1968).
5. J. Millman and C. C. Halkias, *Integrated Electronics*, McGraw-Hill Kogakusha, Tokyo (1972).
6. *Instruments For Research*, Catalogue 1002, 1970–1971 ORTEC, Oak Ridge, TN.
7. E. M. Pell, *J. Appl. Phys.* **31**, 291 (1960).
8. Canberra Catalogue, 7th Ed., Canberra Industries Inc., Meriden, CT.
9. P. W. Nicholsen, *Nuclear Electronics*, Wiley, London (1974).
10. W. R. Leo, *Techniques for Nuclear and Particle Physics Experiments*, Springer-Verlag, Berlin (1987).

Chapter 3

RADIATION SOURCES

Zeev B. Alfassi
Department of Nuclear Engineering, Ben-Gurion University of the Negev, Beer-Sheva, Israel

and

Karel Strijckmans
Laboratory of Analytical Chemistry, Institute for Nuclear Sciences, Universiteit Gent, Gent, Belgium

1 NEUTRON SOURCES

Zeev B. Alfassi

Three main sources of neutrons are available:

(1) research nuclear reactors (nuclear reactors used as neutron sources);
(2) ion and electron accelerators, including neutron generators;
(3) radioactive sources.

Research nuclear reactors have the highest neutron fluxes, but are limited as to on-site applicability and by their price and availability. Consequently, nuclear reactors will be used only for neutron activation analysis of very minute amounts or for sensitive neutron radiography. When on-site irradiation is important, neutron generators or radioactive sources are used.

1.1 Research nuclear reactors

Research nuclear reactors are usually large devices in which fissionable material, almost exclusively ^{235}U, is fissioned into two nuclides with simultaneous

Chemical Analysis by Nuclear Methods Edited by Z. B. Alfassi

emission of neutrons, which induce further fissions in a chain reaction. The fission-produced neutrons are very energetic. The cross-section for neutron-induced fission of fissionable nuclides increases with decreasing energy of the neutrons and, in order to increase the neutron activity, moderators which slow down the neutrons are incorporated in the reactor. To reflect back some of the neutrons which leak from the reactor core, reflectors are used. The fission process releases large amounts of energy, mainly due to stopping of the two recoiling fissioned particles, and the system is cooled by a coolant (either liquid or gas). The nuclear reactors are classified according to their fuel, moderator, coolant, reflector and configuration.

Almost all research nuclears reactors (neutron sources) are heterogeneous reactors in which the fuel is in the form of rods. The fuel is enriched ^{235}U (natural uranium has only 0.7% of ^{235}U—the fissile material). Most research reactors contain 93–99% ^{235}U. Many of the reactors have rods which are U–Al alloys; however, some of the newer designs (mainly those converted to 20% ^{235}U) are of the uranium silicide type. TRIGA reactors operate with uranium–zirconium hydride fuel, which owing to its large negative temperature coefficient of reactivity, allows the operation of the reactor in pulses.

In the light water reactor (LWR), ordinary water (H_2O) is used both as moderator and as coolant. The reflector is mainly graphite, but there are also Be or H_2O reflected reactors. The construction is of either pool type or tank-in-pool type. Because of the relatively high cross-section for capture of thermal neutrons by H atoms, the flux of neutrons in light water reactors always contain a large fraction of fast and epithermal neutrons. The available power is in the range 10–5000 kW with neutron fluxes of 5×10^{14}–1.5×10^{18} n m^{-2} s^{-1}.

Neutrons are usually divided into three groups according to their energy: (1) thermal neutrons with most prevalent energy of $kT = 0.025$ eV at room temperature—this group is usually extended up to the Cd (thermal neutron absorber) cut-off, which is 0.5 eV; (2) neutrons with energies between 0.5 eV and 1 MeV are called epithermal neutrons; and (3) neutrons with energies above 1 MeV are called fast neutrons. Many reactors are unique in design; however, some commercial types are fairly common, the American TRIGA and the Canadian Slowpoke. The TRIGA reactor is a popular multi-purpose research reactor. About 50 of this type are operating, with power levels of 18 kW to 3 MW and fluxes of 7×10^{15}–3×10^{17} n m^{-2} s^{-1}. The most common have power levels of 250 kW and 1 MW. They are of the pool type and graphite reflected, with uranium–zirconium hydride fuel having ^{235}U enrichment of 10–70%. The Slowpoke reactor is a low power (20 kW) reactor designed specifically as a teaching aid, with the additional purposes of activation analysis and production of small amounts of radioisotopes. The system is designed to operate remotely. It can be provided with up to five irradiation sites in the core with a flux of 10^{16} n m^{-2} s^{-1} and five further tubes outside the reflector with half that flux.

Heavy water research reactors are of the tank type, usually with enriched uranium fuel, heavy water as moderator and coolant, and heavy water and graphite reflected. Because of the low cross-section for thermal neutron absorption by D and O, such reactors are characterized by a well thermalized neutron flux (very small epithermal and fast neutron fluxes except inside the core). Owing to the lower moderation power of D compared with H, the physical size of heavy water reactors is larger and hence they have a large available irradiation volume. Their power is usually between 10 and 26 MW, (with fluxes of up to 2×10^{18}n $m^{-2} s^{-1}$).

1.1.1 Sample introduction

The way of introducing a sample into the neutron flux depends on the physical structure of the reactor. It is essential that the introduction of the sample should not affect the operation of the reactor. The irradiation site maybe within the reactor core or outside in the moderator/reflector region. If the reactor is an open pool type with access from above, vertical tubes or ropes can be installed to lower the samples, either inside the core or close to the side. The samples are enclosed in sealed ampules so that they are not in contact with the water surrounding the core. The ampules are usually made of aluminum because of its corrosion resistance and short lived activation products. For short irradiations, polyethylene capsules can also be used.

The manual loading of samples is neither quick nor reproducible in time when short irradiation is performed. In order to have quick and reproducible sample introduction, mechanical systems are used. Two types of mechanical system are used: chain driven racks and pneumatic devices. The latter type is the more common because of its shorter loading and unloading time, lower maintenance and fewer failures in operation. In the pneumatic device, the sample is pushed along a tube by pressurized gas (air or nitrogen). The transfer time depends on the distance transferred. In many systems the transit time is a second or less. The pneumatic device can be automated very easily.

In the case of closed tank reactors, the irradiation is done either by means of a pneumatic device or with neutron beams.

1.1.2 Irradiation with neutrons of limited energy range

Usually neutron irradiation is performed with the whole spectrum of neutrons; however, for special purposes it is preferable to use neutrons of limited energy range. In order to irradiate almost only with thermal neutrons, very well moderated reactors, usually D_2O moderated, have to be used. In order to irradiate with only epithermal and fast neutrons (with exclusion of thermal neutrons), thermal neutron absorbers have to be used between the neutron flux and the sample [1]. Two thermal neutron absorbers are used for this purpose, cadmium and boron, which have high cross-sections for reaction with thermal neutrons. The absorber can be used in connection with the sample, as a covering sheet wrapping the sample or as a capsule in which the sample is enclosed, or as a permanent lining of one of the irradiation tubes. Cadmium is used as commercially available sheets of different thicknesses. Boron is not machinable in the elemental form and various compounds have been used. Rosenberg [2] mixed B_2O_3 with the sample. The best machinable refractory compound of boron is boron nitride, BN [3], and consequently many epithermal and fast neutron irradiations are done in BN capsules. However, Ehmann Brückner and McKown [4] suggested that BN should not be used because of the relatively high cross-section (1.81 barn) for the ^{14}N (p, n) ^{14}C reaction, which will lead to high amounts of the long lived radionuclide ^{14}C. They preferred the use of boron carbide; however, boron carbide is hard and not machinable and a capsule can be made only by the hot pressing process [5]. Stuart and Ryan [6] prepared boron shields by forming a mixture of boron carbide powder and paraffin wax. The mixture was heated to 70 °C and cast into cylindrical forms. Parry used a similar method with B instead of boron carbide [7].

The difference between the two thermal neutron absorbers, boron and cadmium, lies in their excitation functions (the dependence of the cross-section on the energy), as given in Fig. 1, and in the products of the neutron absorption. Figure 1 shows that cadmium approximates to a perfect sharp filter for the thermal region,

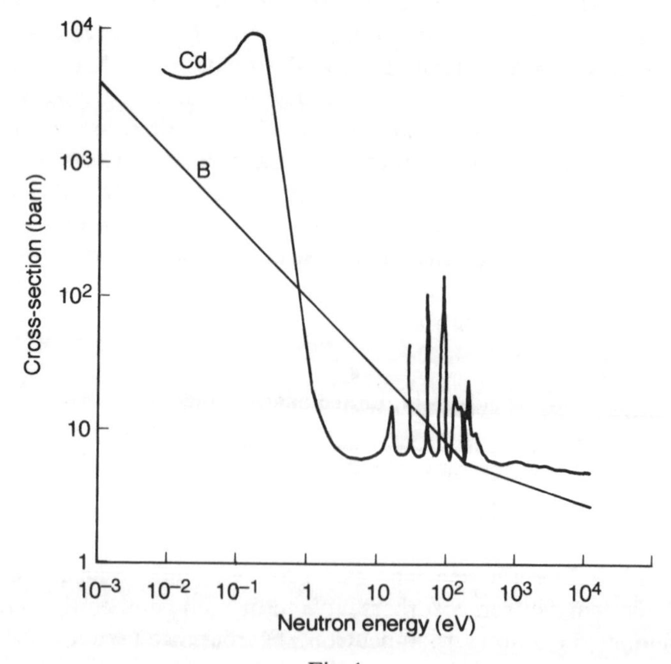

Fig. 1.

apart from a few resonances in the epithermal region. Boron behaves as almost a perfect $1/V$ absorber (linear dependence of σ on $E^{1/2}$) with no sharp energy cut-off.

The absorption of a neutron by boron lead to a stable nuclide by the reaction, ^{10}B (n, α) ^{7}Li, whereas thermal neutron absorption by cadmium leads to formation of short lived and long lived radionuclides. Consequently, the unloading and unpacking of cadmium capsules (or wrappings) pose radiation safety problems. The absorption of neutrons by cadmium occurs by the radiative capture (n, γ) reaction. The high energy γ-rays are not absorbed by the sample and hence the samples are not heated too much. On the other hand, the α-particles formed in neutron absorption by boron, ^{10}B (n, α) ^{7}Li, have very short range and are absorbed in the sample. When biological samples are irradiated, this heating accelerates the decomposition of the organic material, producing high pressure in air-tight sealed sample containers, which can lead to explosion or may cause volatilization of part of the sample. Stroube, Cunningham and Lutz [8] found that, in the 20 MW reactor at NIST, thermal heating of the boron nitride vessel limited the irradiation period for freeze-dried foods to 4 seconds. In order to lengthen the irradiation time the BN capsule was irradiated in a cadmium-lined irradiation position. Most of the thermal neutrons were absorbed in the Cd lining. Glascock, Tian and Ehmann [9] found that the irradiation time for a BN vessel in a flux of 10^{18} $n\,m^{-2}\,s^{-1}$ should not exceed 10 s in order to prevent melting and destruction of the polyethylene capsules. Chisela, Gawlik and Brätter [10] studied the temperature in a sintered boron carbide capsule in an air-cooled irradiation facility and found that the capsule reached steady-state temperatures of 163 and 194 °C for 4 MW and 5 MW reactors, respectively.

1.1.3 Irradiation with fast neutrons

A pure flux of fast neutrons can be obtained by irradiation of samples within shields of ^{235}U or LiD. The ^{235}U reacts with the thermal neutrons and produce more fission spectrum fast neutrons. LiD transform the thermal neutrons to 14 MeV neutrons by the sequence of reactions ^{6}Li (n,α) T, D (T, n) ^{4}He. Conversion rates between 2×10^{-4} and 9.6×10^{-4} were found for transformation of thermal neutrons to 14 MeV neutrons by the use of LiD.

1.2 Neutrons from accelerators and neutron generators

Charged particle accelerators can produce neutrons fluxes by (d, n), (p, n) or (α, n) reactions. Electron accelerating machines can produce neutron fluxes via (γ, n) reactions, where the gamma flux is obtained by stopping of electrons in a high-Z material. The main reactions for producing neutron fluxes are D (d, n) ^{3}He,

T (d, n) ^{4}He and ^{9}Be (d, n) ^{10}B. The first two reactions are exothermic and require very little acceleration of the deuteron beams. For this reason they are used in neutron generators, which are very small accelerators with acceleration to 150–500 keV. The disadvantage is that the target material is a gas (D_2 or T_2) adsorbed on a metal. If the target is heated too much the gas will be desorbed from the metal target. This limits the current of the bombarding deuteron beam and consequently the neutron flux. If a higher flux of neutrons is required, a target of Be is used, but the deuterons should have an energy of a few MeV. The most commonly applied reaction [11] is T (d, n) ^{4}He, because of the low acceleration needed and the higher cross-section of this reaction compared with D (d, n) ^{3}He. Different high voltage generators are used in various neutron generators: Cockcroft–Walton, insulating core transformer (ICT), Van de Graaff, and electrostatic rotor machines. A deuteron beam is produced by one of the various ion sources and accelerated to 1 to a few hundred keV. The beam may also be a mixture of 50% of tritium and 50% of deuterium. Typical beam intensities are a few mA, but intensities as high as 150 mA have been used. Neutrons of 14 MeV are produced from the interaction of the beam with a large surface tritiated target. The target consists of a few hundred μm of titanium, zirconium or palladium evaporated on a backing disk ≈ 30 mm in diameter and saturated with tritium gas. The disk is water cooled and made of heat-conducting material, silver or copper. Rotating targets are sometimes used to ensure heat dissipation.

Commercial compact sealed-tube neutron generators are also available. The most usual neutron generators have fluxes up to 5×10^{11} n s^{-1}, but special generators with an output of 5×10^{13} n s^{-1} have also been constructed [11].

Whereas D–T reaction with low energy projectiles from low voltage machines produces neutrons with energy of 14.1–14.9 MeV (according to the angle), neutrons with various energies can be obtained by the use of high energy accelerators. The higher the energy of the projectiles, the higher is the energy of the neutrons. The target material is metallic beryllium. Because of its better heat conduction, higher currents of deuteron beams can be used, leading to higher fluxes of neutrons.

1.3 Radioactive neutron sources

Radionuclide neutron sources are composed either of a radionuclide emitting alpha-particles or gamma-rays, together with an appropriate surrounding material, or a radionuclide decaying by spontaneous fission. The gamma emitter in photoneutron sources is surrounded by beryllium or deuterium (as D_2O) and neutrons are emitted due to the (γ, n) reaction

$$\gamma + {}^9\text{Be} \rightarrow 2\alpha + \text{n} - 1.67\ \text{MeV} \qquad \gamma + \text{D} \rightarrow \text{H} + \text{n} - 2.23\ \text{MeV}$$

Very few radionuclides with reasonable half-life emit gamma-rays with such high energies. For that reason, together with the disadvantage of the long range of γ-rays, the use of that kind of neutron source is very limited. The most commonly used photoneutron source is ^{124}Sb–Be, which emits neutrons of 26 ± 1.5 eV. The ^{124}Sb is produced by reactor irradiation of natural antimony; ^{124}Sb emits several gamma-rays including 1.692 MeV photons (with intensity of 48%), and has a half-life of 60.9 days. The neutron source has two parts, the core (a sphere or cylinder) made of irradiated antimony metal, and a shell of beryllium metal about 2 cm thick. The practical yield is $\approx 10^7$ n s^{-1} per 1 Ci of ^{124}Sb.

Most neutron sources use α interaction with ^{9}Be. The spectra of these neutrons extends up to 10–12 MeV. The common α emitters used are Ra, ^{210}Po, ^{239}Pu and ^{241}Am. The main properties of the sources are summarized in Table 1, and the interaction is $^9\text{Be} + \alpha \rightarrow {}^{12}\text{C} + \text{n} + 5.91$ MeV (^{12}C is in the 4.43 MeV excited state). The Po–Be and Am–Be sources are made by mixing fine beryllium powder with polonium metal or americium oxide. The Pu–Be source is an intermetallic compound, Pu–Be_{13}. The mixture or the intermetallic compound is doubly encapsulated, first in an inner capsule of tantalum and then in an outer capsule of stainless steel. In 80% of the (α, n) interactions the ^{12}C nucleus is left in the 4.43 MeV excited state, which decays with emission of 4.43 MeV gamma photons. However, the γ dose is lower than that associated with the previously used neutron source Ra–Be. An even lower γ dose is associated with neutron sources involving spontaneous fission. The main source used for spontaneous fission is ^{252}Cf.

$$^{252}\text{Cf} \xrightarrow[t_{1/2} = 2.65\text{ y}]{\text{spontaneous fission}} \text{two fission products} + 3.8\,\text{n} + 200\text{ MeV}$$

The half-life for spontaneous fission is 85.5 years and for alpha emission is 2.3 years; the effective half life is 2.65 years.

$$\frac{1}{t_{\text{eff}}} = \frac{1}{t_\alpha} + \frac{1}{t_{\text{sf}}}$$

The ^{252}Cf sources are very compact. The neutron spectrum is very similar to that of neutron-induced fission, with a mean energy of 2.348 MeV. The neutron yield is

TABLE 1.
Properties of neutron sources for ^{9}Be (α, n) ^{12}C reaction

	Alpha emitter		
	^{210}Po	^{239}Pu	^{241}Am
Neutron yield (n s^{-1} Ci^{-1})	2.5×10^6	1.7×10^6	2.2×10^6
Half-life	138.4 d	24360 y	458 y
Approximate size (cm^3 Ci^{-1})	0.1	12	3
Heating (mW Ci^{-1})	32	31	33
γ dose (rad h^{-1} Ci^{-1})	0.11 (4.43 MeV)	0.08 (4.43 MeV)	10 (60 eV), 0.1 (4.43 MeV)

$2.31 \times 10^6 \, n \, s^{-1} \, \mu g^{-1}$. In order to increase the flux of neutrons, it is cheaper to surround the ^{252}Cf with ^{235}U than to use larger amounts of ^{252}Cf. These devices are called neutron multipliers. One mg of ^{252}Cf combined with 1.4 g of ^{235}U (93.4%) and polyethylene as moderator is equivalent to a neutron source containing 33 mg of ^{252}Cf. This device has thermal and fast neutron fluxes of 4×10^8 and $6 \times 10^8 \, n \, cm^{-2} \, s^{-1}$, respectively [12].

2 CHARGED PARTICLE ACCELERATORS

Karel Strijckmans

2.1 Introduction

Accelerators make use of an electric field to accelerate ions (charged particles). Non-linear accelerators also apply a magnetic field to 'guide' the ions. Accelerated ions also gain mass.

2.1.1 A charge in an electric field

The electric field strength ($\boldsymbol{E}$) between two electrodes at distance d and with an electric potential difference of U (Fig. 2) is given by

$$\boldsymbol{E} = U/d \tag{1}$$

and its unit is V/m. When a charge Q is placed in an electric field $\boldsymbol{E}$, it experiences a force

$$\boldsymbol{F} = Q\boldsymbol{E} \tag{2}$$

If the charge Q is displaced over a distance d in a homogenous electric field, it gains a kinetic energy

$$T = Fd = QU \tag{3}$$

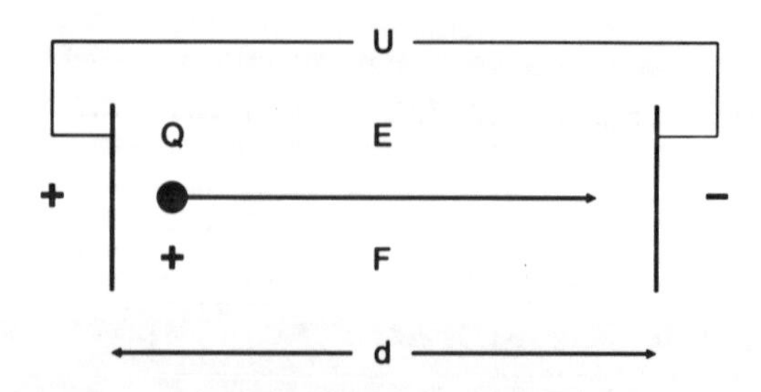

Fig. 2. A charge in an electric field.

For $Q = e \approx 1.6 \times 10^{-19}$ C (i.e. the elementary charge, the charge of an electron or proton) and $U = 1$ V, the kinetic energy is equal to $\approx 1.6 \times 10^{-19}$ J, more practically defined as 1 eV (electronvolt). Thus, a proton accelerated by an electric potential difference of 1 V requires a kinetic energy of 1 eV.

2.1.2 A charge in a magnetic field

A charge Q moving with velocity **v** in a magnetic field with magnetic induction (magnetic flux density) $\boldsymbol{B}$ experiences the Lorentz force (Fig. 3) given by

$$\boldsymbol{F}_{\mathrm{L}} = Q(\boldsymbol{v} \times \boldsymbol{B}) \tag{4}$$

As it is caused to deviate from its original direction, it experiences a centrifugal force. Thus a charge with mass m moving perpendicular to a uniform magnetic field will describe a circle in a plane perpendicular to the magnetic field. As the centrifugal force (F_{C}), is equal to the Lorentz (centripetal) force (F_{L})

$$mv^2/r = QvB \tag{5}$$

the angular velocity is

$$\omega = v/r = (Q/m)B \tag{6}$$

and as the kinetic energy is

$$T = mv^2/2 \tag{7}$$

from equations (6) and (7) the orbit radius is

$$r = v/\omega = (2Tm)^{1/2}/(QB) \tag{8}$$

2.1.3 Relativity

Ions being accelerated gain not only kinetic energy but also mass:

$$\begin{aligned} m &= m_0/(1 - v^2/c_0^2)^{1/2} \\ &\approx m_0[1 + T/(m_0 c_0^2)] \end{aligned} \tag{9}$$

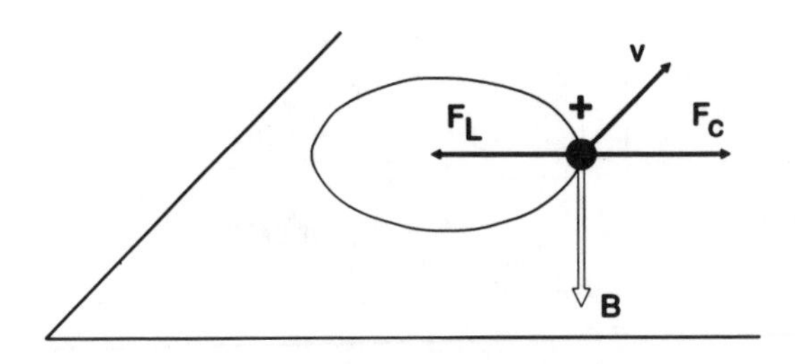

Fig. 3. A charge in a magnetic field.

with m_0 = rest mass and c_0 = velocity of light in a vacuum. As $m_0c_0^2$ equals 938 MeV for a proton and 511 keV for an electron, a significant ($>1\%$) relativistic mass increase is observed for >10 MeV protons and >5 keV electrons.

2.2 DC accelerators

DC accelerators apply a DC (direct current) electric potential difference to accelerate ions. Between the ion source (high voltage end) and the target (grounded), several acceleration tubes or electrodes are connected to a voltage divider (Fig. 4). These tubes both accelerate and focus the ions. The latter effect can be understood by observing the electric field lines in the gap between two tubes (Fig. 5). Ions traveling along the symmetry axis (i.e. focussed ions) will experience only an accelerating force along this axis. Ions traveling away from the axis (i.e. defocussed ions) will experience additionally a focussing force during the first half of the gap transit and a defocussing force during the second. As the ions move more slowly when entering than when leaving the gap, the focussing effect is stronger than the defocussing effect. High voltage supplies are the Cockcroft–Walton voltage multiplier and the electrostatic Van de Graaff generator.

The Cockcroft–Walton voltage multiplier charges capacitors in parallel and discharges them (over the voltage divider in Fig. 4) in series. The principle is shown in Fig. 6. An AC (alternating current) source charges one capacitor

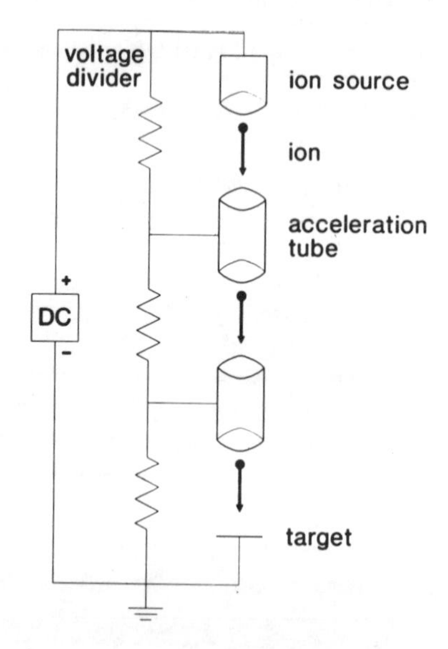

Fig. 4. Principle of a DC accelerator.

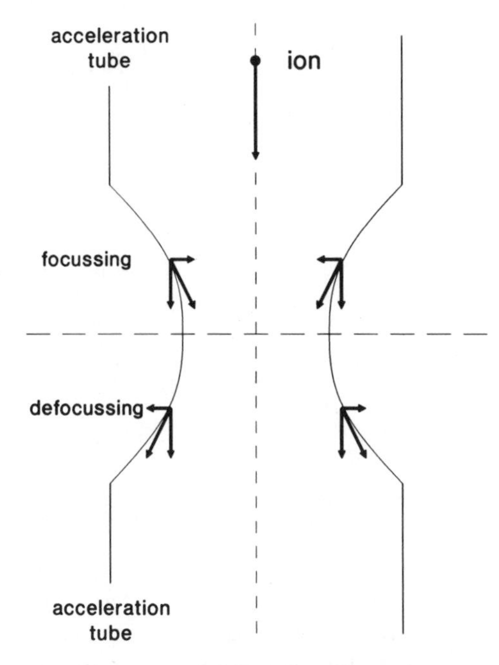

Fig. 5. Focussing in a DC accelerator.

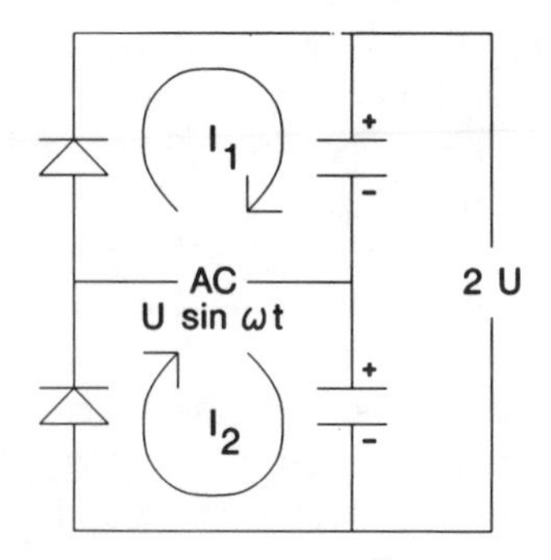

Fig. 6. Principle of a Cockcroft–Walton voltage multiplier: current during the first (I_1) and the second (I_2) half-cycle.

through its diode during the first half-cycle and the other capacitor during the second half-cycle. The DC output (i.e. that across the two capacitors) is twice the maximum electric potential difference of the AC source.

The electrostatic Van de Graaff generator can produce much higher electric potential differences. The generator, shown schematically in Fig. 7, consists of a

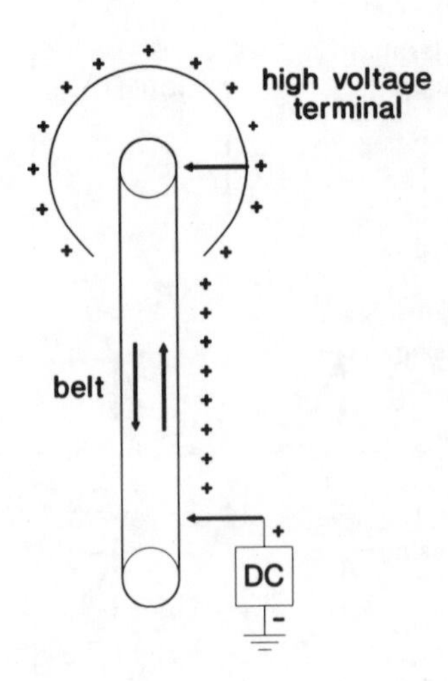

Fig. 7. Principle of an electrostatic Van de Graaff generator.

rounded high voltage terminal of conducting material and a moving belt of non-conducting material to carry charges to the terminal. An array of sharp points, connected to a positive high voltage source, sprays positive charges on to the belt (by transfer of electrons from the belt to the spray points). The belt moves the static charges to the interior of the insulated terminal. Inside the terminal another array of sharp points removes the positive charges and distributes them to the surface of the terminal. On one hand, the charge (and thus the electric potential) of the terminal continuously increases. On the other hand, charges return to ground by resistive leakage (through the voltage divider in Fig. 4) and by corona discharge. The latter increases markedly with the electric potential of the terminal. The electric potential of the terminal is determined (and can be adjusted) by the balance of the charging current and the corona discharge current.

The energy attainable by a Van de Graaff accelerator can be increased by the 'tandem' principle, i.e. the same electric potential difference is used twice to accelerate ions. Negative ions are first accelerated from a grounded ion source towards the positive terminal of the generator. Then the negative ions are stripped of electrons by passage through a metal foil or gas and positive ions are obtained. These are further accelerated from the terminal to the ground potential. In this way 20 MeV protons can be obtained with a 10 MV generator.

2.3 Cyclic accelerators

With the exception of the tandem Van de Graaff, the DC accelerators use their electric potential difference only once to accelerate the ions to their final energy.

Cyclic accelerators make use of the same (low) electric potential difference to accelerate ions over and over again, and consequently the final energy obtained is higher than for DC accelerators.

2.3.1 The linear accelerator

The linear accelerator or linac consists of a set of acceleration tubes alternately connected to the output of an AC source, as shown in Fig. 8. Suppose a (positive) ion is emitted when the first tube is at negative potential. The electric field between the grounded ion source and the first tube accelerates the ion. Inside the tube, the ion continues at constant velocity. For a tube of appropriate length, the ion leaves the tube one half-cycle later, when the second tube is at negative potential. The ion is accelerated again. This process continues and the final energy is the sum of all energy gains when traversing the gaps. For each acceleration the ion goes faster and faster, so the tube length should be longer and longer. Indeed, the time for reversal of the polarity on the tubes is constant. For relativistic velocity ($v \rightarrow c_0$) the ion gains mass rather than velocity and the tube length becomes constant ($c_0 T/2$, where T is the period of the AC source).

2.3.2 The cyclotron

The cyclotron consists of an electromagnet providing a uniform magnetic field (Fig. 9). The accelerated ions describe a circular orbit in a plane perpendicular to the magnetic induction, the radius being determined by their energy [equation (8)]. Between the poles of the electromagnet, two hollow acceleration electrodes are introduced, i.e. the two halves of a cylindrical box cut in two along the axis and slightly separated from each other. Because of the resemblance of this semicircular structure to the capital letter D, the electrodes have been called 'dees'. The dees are connected to an AC source. Positive ions are released in the centre of the cyclotron (Fig. 10). An ion is accelerated to the negative dee by the electric field in the gap between the two dees. Once the ion enters the (hollow) dee it experiences only the magnetic field, and the path of the ion is consequently half

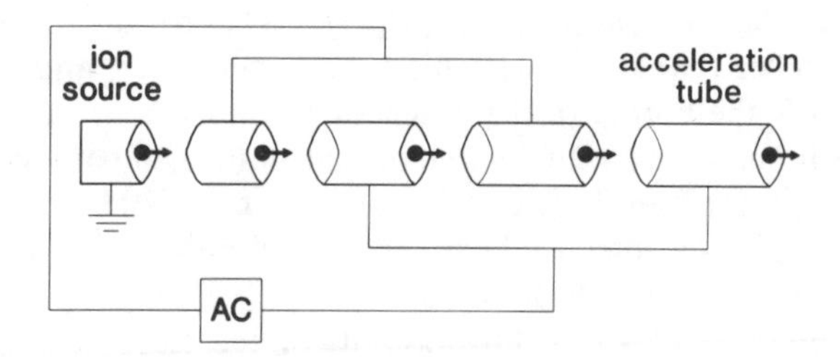

Fig. 8. Principle of a linear accelerator.

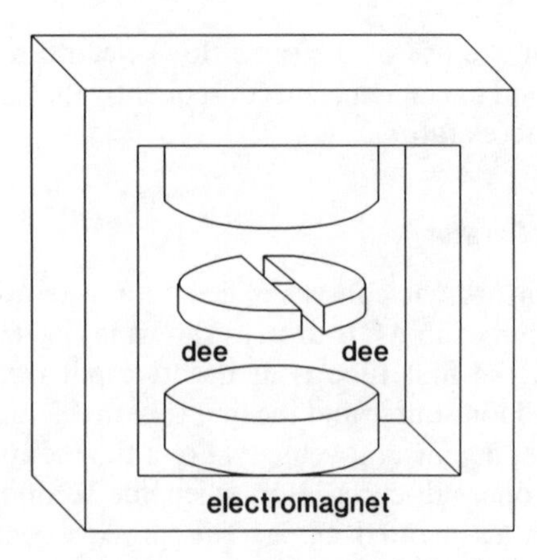

Fig. 9. Principle of a cyclotron: three-dimensional view.

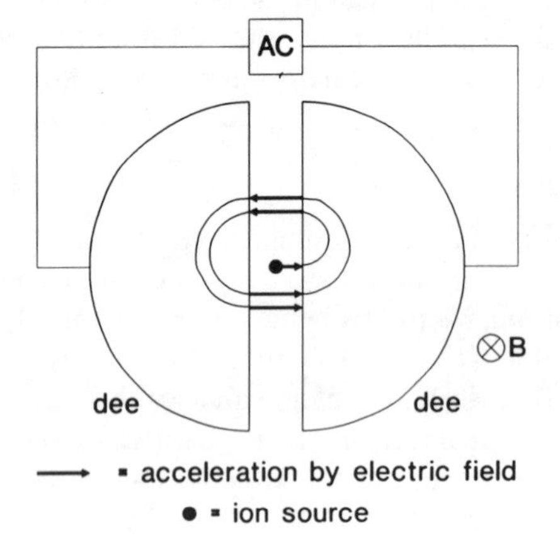

Fig. 10. Principle of a cyclotron: two-dimensional view in the mid-plane.

a circle. When the ion leaves the dee, the polarity on the dees is reversed, so that the ion is accelerated again to the opposite (negative) dee. The same process is repeated in the other dee, but the orbit radius is higher because the kinetic energy of the ion is higher [equation (8)]. This process continues, and the ion spirals outward towards the border of the magnetic field. The ion acquires an energy equal to the sum of all individual accelerations in the gap between the dees. To obtain 10 MeV protons, an electric potential difference of 10 MV was required for a DC accelerator. A cyclotron makes use, e.g., of only 10 kV, accelerating the proton in 1000 steps of 10 keV up to 10 MeV.

This situation only holds true if the polarity on the dees is changed every time the accelerated ion appears again at the gap between the dees. As was shown in equation (6), the angular velocity of the ion is a function only of the charge to mass ratio of the ion and the magnetic induction. It is not dependent on the orbit radius (or the velocity or kinetic energy) of the ion. Thus, the resonance condition of a cyclotron is that the angular velocity ($2\pi f$, with f = frequency of the AC source) of the electric field is set equal to that of the ion [equation (6)].

From equations (6) and (7) one obtains the final kinetic energy for a radius $r = R$, i.e. the border of the magnetic field

$$T/m = (Q/m)^2 B^2 R^2/2 \quad (10)$$

The kinetic energy per unit mass of the ion (T/m in MeV/u) is dependent on the charge to mass ratio of the ion Q/m, the magnetic induction B and the size of the magnet R. It is not dependent on the electric potential difference U applied on the dees (the lower the value of U the more rotations the ion makes before arriving at the radius R).

The magnetic field should not be uniform: the magnetic induction should decrease as a function of the radius in order to obtain axial stability. In such a magnetic field, ions moving away from the midplane between the magnet poles experience Lorentz forces, bringing them back to the midplane. Axial (as well as radial) stability is indispensable for ions traveling a long distance before reaching their final energy.

The maximum energy attainable by a cyclotron is limited by (1) the relativistic mass increase of the ions accelerated, and (2) the decrease in magnetic induction as a function of the radius. The angular velocity of the ions being accelerated [equation (6)] is no longer independent of the orbit radius, but decreases with (1) increasing ion mass and (2) decreasing magnetic induction. Consequently, both phenomena lead to decreasing angular velocity of the ion as a function of the radius. If the angular velocity of the electric field is kept constant, the ion comes increasingly late to the gap between the dees. Finally the ion will appear at the gap when the potential on the opposite dee is positive, and the ion will be decelerated (instead of accelerated). At this stage the energy has reached its physical limit. There are two possible ways to overcome this limit, applied, respectively, in the synchrocyclotron and the isochronous cyclotron.

2.3.3 The synchrocyclotron

The relativistic mass increase can be compensated by decreasing the angular velocity of the electric field *during* the acceleration process. Thus, the angular velocity (or *frequency*) of the electric field can be *modulated*, so that the angular velocities of both the ion and the electric field, although varying, are always equal (or *synchronous*) during the acceleration process. In this way a frequency-modulated or synchrocyclotron is obtained, which is not subject to the maximum energy limitation of a cyclotron. However, the intensity of the beam is much lower (typically by a factor of 1000). Indeed, each group of ions is accelerated to its maximum energy by tuning the frequency of the electric field, then the frequency of the electric field is reset to its starting value, and a new group of ions is accelerated to its maximum energy. This results in a pulsed beam having an average intensity much lower than that for a cyclotron.

2.3.4 The isochronous cyclotron

The relativistic mass increase is related to the energy (or velocity) of the ion [equation (9)], which is related to the orbit radius of the ions in the magnetic field through equation (8). The relativistic mass increase can be compensated by magnetic induction that increases as a function of the radius of the magnetic field. As the angular velocity of both the ion and the electric field remain constant (or *isochronous*) during the acceleration process, an isochronous cyclotron is obtained. An isochronous cyclotron is not subject to the maximum energy limitation of a cyclotron, nor to the maximum intensity limitation of a synchrocyclotron.

However, the increase in magnetic induction as a function of the radius results in axial instability (see Section 2.3.2). This is overcompensated by additional, axially focussing forces originating from the particular shape of the magnet poles of an isochronous cyclotron. Removal of radial sectors (or, better, spiral sectors) from the magnet poles (Fig. 11) affords hills and valleys in sectors with a strong or a weak magnetic field respectively. Ions moving in such an azimuthally varying field (azimuthally: along the orbit of the ion) experience axially focussing (Lorentz) forces. These focussing forces are much larger than the axially defocussing forces that exist in a magnetic field for which the magnetic induction increases as a function of the radius. An isochronous cyclotron is also called an AVF (azimuthally varying field) or sector-focussed cyclotron.

2.3.5 The synchrotron

The maximum energy attainable with a (synchro- or isochronous) cyclotron [equation (10)] for a given ion is determined by the practical limits of magnetic

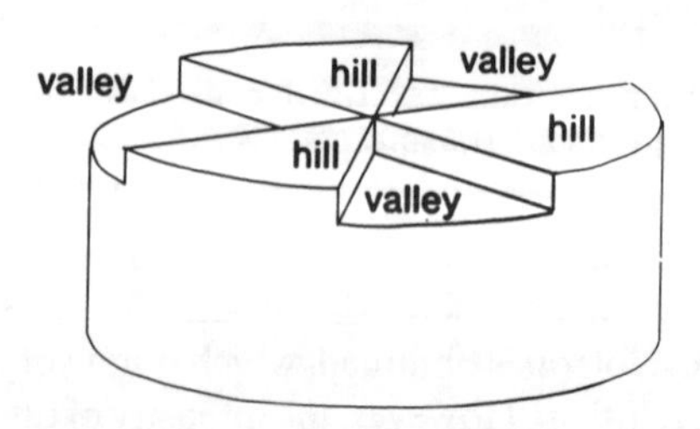

Fig. 11. Magnet pole shaping in an isochronous cyclotron.

induction and the size of the magnet (mass of iron). The latter limitation is overcome in a synchrotron. Ions do not spiral outward, which needs a largely sized magnetic field. Ions are kept at the same large radius during the acceleration process by increasing the magnetic induction along with the kinetic energy [equation (8)]. Consequently, there is no need for a magnetic field in the centre and the magnet becomes a narrow ring of large radius, resulting in substantial economies in the mass of iron (Fig. 12). Large dees are replaced by an acceleration tube. The frequency of the electric field should also increase along with the magnetic induction during the acceleration process. This can also be understood, as the path length between two accelerations is kept constant, in contrast to a (synchro- or isochronous) cyclotron, where the orbit radius of the ion (and thus the path length between two accelerations) becomes larger for higher energy ions. The beam intensity is much lower than for a synchrocyclotron. Indeed, each group of ions is accelerated to its maximum energy by tuning both the frequency of the electric field and the magnetic induction; both then have to be reset to their initial values before another group of ions can be accelerated. Since tuning the magnetic field from zero is practically impossible, another accelerator (DC or AC) is used to inject 'low' energy ions into the synchrotron. Table 2 summarizes the similarities and differences for a cyclotron, a synchrocyclotron, an isochronous cyclotron, and a synchrotron with respect to the time dependence of the magnetic induction, frequency of the electric field and radius of the ion orbit during the acceleration process and the attainable energy and beam intensity.

2.3.6. The betatron

Accelerators have been described for the acceleration of positively charged 'heavy' ions such as protons. Acceleration of electrons is also possible, except for a

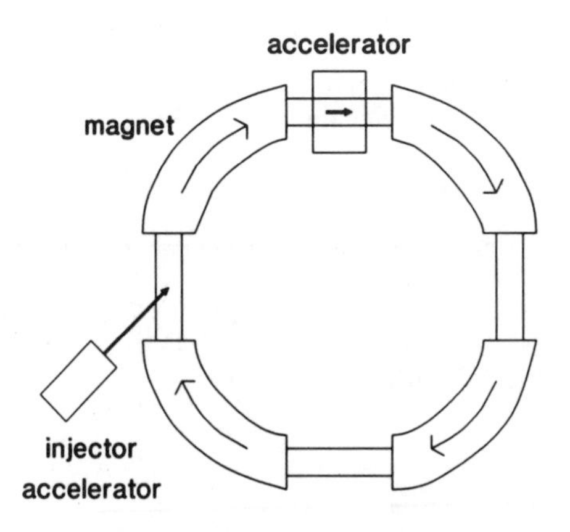

Fig. 12. Principle of a synchrotron.

TABLE 2.
Similarities and differences for a cyclotron, a synchrocyclotron, an isochronous cyclotron, and a synchrotron with respect to the time dependence of the magnetic induction, frequency of the electric field, and orbit radius of the ion during the acceleration process and the attainable energy and beam intensity

	Cyclotron	Synchro cyclotron	Isochronous cyclotron	Synchrotron
Magnetic induction	constant	constant	constant	increases
Frequency of the electric field	constant	decreases	constant	increases
Radius of the ion orbit	increases	increases	increases	constant
Maximum energy	low	high	high	very high
Maximum beam intensity	high	low	high	very low

cyclotron, because electrons are subject to relativistic mass increase at a 2000-fold lower energy than protons. A betatron is an electron accelerator, with some similarities to a synchrotron: the electron orbit is constant and the magnetic field is time dependent. However, acceleration of the electrons occurs just as in a transformer, the copper conductor of the output being replaced by the electron beam.

REFERENCES

1. Z. B. Alfassi, in *Activation Analysis*, Vol. 2, edited by Z. B. Alfassi, CRC Press, Boca Raton (1990), p.3.
2. R. J. Rosenberg, *J. Radioanal. Chem.* **62,** 145 (1981).
3. H. Soreq and H. C. Griffin, *J. Radioanal. Chem.* **79,** 135 (1983).
4. W. D. Ehmann, J. Brückner and M. D. McKown, *J. Radioanal. Chem.* **57,** 491 (1980).
5. F. Chisela, D. Gawlik and P. Brätter, *J. Radioanal. Nucl. Chem.* **112,** 293 (1987).
6. D. C. Stuart and D. E. Ryan, *Can. J. Chem.* **59,** 1470 (1981).
7. S. J. Parry, *J. Radioanal. Nucl. Chem.* **81,** 143 (1984).
8. W. B. Stroube, Jr., W. C. Cunningham and G. J. Lutz, *J. Radioanal. Nucl. Chem.* **112,** 341 (1987).
9. M. D. Glascock, W. Z. Tian and W. D. Ehmann, *J. Radioanal. Nucl. Chem.* **92,** 379 (1985).
10. F. Chisela, D. Gawlik and P. Brätter, *J. Radioanal. Nucl. Chem.* **98,** 133 (1986).
11. J. Czikai, *Handbook of Fast Neutron Generators*, CRC Press, Boca Raton (1987).
12. P. H. Permar, D. G. Karraker and S. C. Aiken, ^{252}Cf status and prospects, presented at the 4th International Transplutonium Element Symposium, Baden-Baden, Germany, 1975. Report DP-MS-75-34 (1975).

Chapter 4

NUCLEAR REACTION CHEMICAL ANALYSIS: PROMPT AND DELAYED MEASUREMENTS

Michael D. Glascock
Research Reactor Center, University of Missouri, Columbia, MO 65211, USA

1 INTRODUCTION

All nuclear techniques for chemical analysis depend on two major processes. First, an alteration in the composition or energy configuration of a target nucleus initiated by an interaction between an incoming projectile (i.e., neutron, charged particle, or gamma-ray) and the target nucleus. Second, measurement of the types, energies, and quantities of emitted radiation (neutrons, charged particles, or gamma-rays) following the interaction.

The dimensions of the nucleus (radius $10^{-13} - 10^{-12}$ cm) are small relative those of an atom ($\approx 10^{-8}$ cm) and offer such a small target for incident particles that the likelihood of an interaction is very low. When an interaction occurs, several events are possible: (a) deflection of the incoming projectile with no transfer of energy to the target nucleus, (b) temporary excitation of the nucleus followed by an instantaneous ($\approx 10^{-14}$ s) de-excitation through the emission of subatomic particles or gamma-rays, or (c) temporary excitation followed by an instantaneous de-excitation and transformation of the target nucleus into another nuclide with a different number of subatomic particles. In many instances, the transformed nucleus will undergo delayed de-excitations by emitting nuclear particles or gamma-rays, and the rate of the delayed de-excitation has a characteristic half-life. A diagram illustrating the possible interaction of a projectile with a target nucleus is shown in Fig. 1.

Chemical Analysis by Nuclear Methods Edited by Z. B. Alfassi

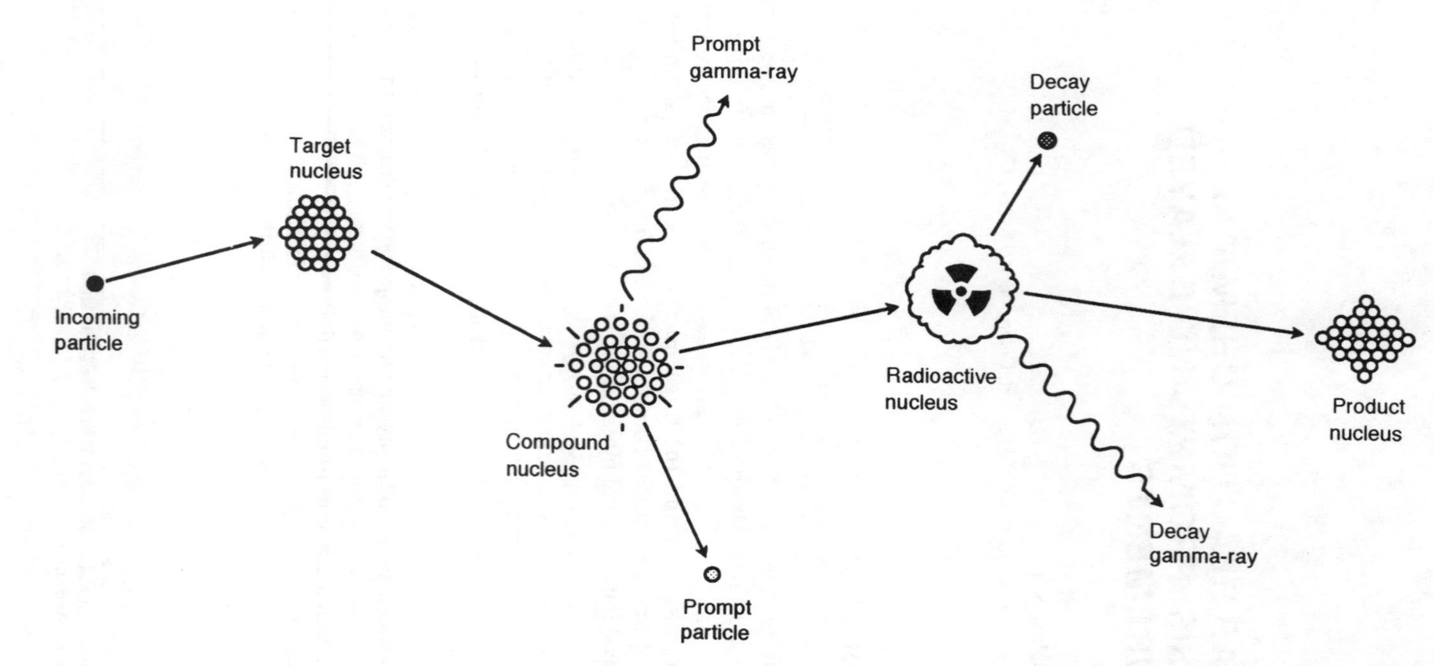

Fig. 1. A nuclear projectile interacting with a target nucleus.

As a result of irradiation, two distinct phases exist during which emission of particles and gamma-rays can take place. The first phase occurs during the nuclear interaction and the second phase occurs afterward. Both phases offer the potential for use in chemical analysis. In principle, therefore, all nuclear methods of analysis fall into two types, depending upon the phase used for analysis: (1) *prompt analysis*, where measurements of radiation occur during the interaction, or (2) *delayed analysis*, where measurements occur following decay. This chapter presents a description of the basic nuclear physics and chemistry of both prompt and delayed analysis.

2 NUCLEAR REACTIONS

Since nuclear reactions play such a central role in the various methods of nuclear analysis, it is necessary to approach the subject by providing a brief discussion of nuclear reactions. Nuclear reactions can be represented by the expression

$$\mathrm{a} + \mathrm{A} \rightarrow [\mathrm{X}] \rightarrow \mathrm{b} + \mathrm{B} + Q \quad \text{or} \quad \mathrm{A(a, b)B}$$

where a is the incoming particle or radiation, A is the target nuclide, [X] is a compound nucleus in an intermediate state of high excitation, B is the product nuclide, b is the exiting particle or radiation, and Q is the change in energy for the system. Both B and b are products of the reaction. If b is a particle, the only difference is that b is lighter than B.

Mass–energy, charge, nucleon number, and momentum are all conserved quantities for reactions of interest in nuclear analysis. The value of Q, above, accounts for the amount of energy released or absorbed by the reaction. When Q is positive, the reaction is exoergic. When Q is negative, the reaction is endoergic. Most reactions are of the latter type and require the incoming particle to have a certain amount of energy for the reaction to occur.

Since some incident particle energy is used in conserving momentum for the particle–nucleus system, and energy is also required to overcome the Coulomb repulsion forces between an incident charged particle and the nucleus, the incident particle must have a kinetic energy in excess of the Q-value. The minimum amount of energy necessary to induce a reaction is knwon as the threshold energy. For an exoergic reaction involving neutrons, the threshold energy is zero. Thus, the reaction can occur when the colliding neutron has a kinetic energy of nearly zero (e.g., for thermal neutrons). For projectiles with less than 30 MeV of energy, most nuclear reactions can be classified into one of two broad types: scattering reactions or compound nucleus formation.

2.1 Scattering reactions

Scattering reactions are those in which the target nucleus deflects incoming particles. The result is generally a change in direction of the incident particle but no change in identity of the particle or target nucleus. Scattering reactions are further subdivided into two types—that is, either elastic or inelastic. Elastic scattering results in no difference in the total kinetic energy for the projectile–target system before and after the interaction. On the other hand, inelastic scattering reactions are processes that involve a transfer of energy from the incoming particle to the target nucleus, resulting in some excitation of the target nucleus. The excited nucleus undergoes prompt de-excitation by emitting gamma-rays until it returns to its original energy configuration. Since both types of scattering reactions occur almost instantaneously, their applications for nuclear analysis must occur in the prompt mode.

Elastic scattering reactions are described simply by the expression:

$$\mathrm{A(a, a)A}$$

A method that uses the elastic scattering of charged particles to determine stoichiometry, structure and impurity concentrations for material surfaces is Rutherford backscattering spectrometry (RBS). Reviews by Chu, Mayer and Nicolet [1] and by Feldman and Mayer [2] describe the fundamentals of RBS as applied for the analysis of surfaces and thin films. Yet another method uses the forward scattering of charged particles [3] and is called elastic recoil detection analysis (ERDA).

The notation used to characterize inelastic scattering reactions is

$$\mathrm{A(a, a'\gamma)A^*}$$

where a′ refers to a scattered particle that has the same identity as the incident particle, and A* represents the target nucleus in an excited state. The scattered particle changes both its direction and kinetic energy and the excited nucleus undergoes rapid de-excitation by emitting prompt gamma-rays.

2.2 Compound nucleus formation

Compound nucleus formation reactions are more frequently called direct reactions [4]. These reactions occur through a two-step process that involves both formation and evaporation. During the formation step, the projectile merges with the target nucleus such that the total energy (kinetic plus binding) becomes randomly distributed among all nucleons. Once a state of statistical equilibrium is achieved, the original particle is no longer distinct from the other nucleons. The resulting compound nucleus is often unstable with respect to particle emission. If adequate energy is present, a single nucleon or a collection of nucleons will acquire adequate energy to escape. The most probable result gives only part of the excitation energy to the exiting particle(s); thus it is probable that the residual nucleus continues to be in a state of excitation, and there may be subsequent emissions of particles and/or gamma-rays. This breakup process for the compound nucleus is known as nucleon evaporation, because it resembles the way in which molecules in a drop of hot liquid escape into the vapor phase.

Nuclear fission is another compound nucleus reaction [5] involving only the heaviest elements ($Z \geq 90$). It involves absorption of an incoming particle (typically a low energy neutron), resulting in a splitting of the compound nucleus into two large fragments with an accompanying release of several free neutrons. The large escaping fragments have a kinetic energy of ≈ 200 MeV. These fragments are also unstable and eventually de-excite by emitting particles and/or gamma-rays.

In many instances, a product nuclide created by compound nucleus formation is moderately unstable, and achieves a more stable configuration through a delayed emission of particles or gamma-rays. These latter de-excitations (or decays) proceed at rates according to the characteristic half-life of each product nuclide.

Neutron capture (n, γ) reactions are among the various reactions that proceed by compound nucleus formation. In most cases involving low energy neutrons, the compound nucleus has insufficient energy for particle emission; thus only the emission of gamma-rays is possible. At higher neutron energies the compound nucleus will undergo particle emission.

In general, reactions involving compound nucleus formation are more useful for nuclear analysis than other types of reactions because they offer greater selectivity and sensitivity for elemental analysis. The application of nuclear reactions involving compound nucleus formation to 'activate' target elements is commonly known as activation analysis. The chemical analysis techniques that employ measurement of prompt and delayed radiation following activation are known as prompt activation analysis and delayed activation analysis, respectively.

2.3 Reaction cross-sections

The interaction rate for a given nuclear reaction depends on the number of incident particles, the number of target nuclei available for interaction, and the probability of occurrence for that interaction. This probability, called the cross-section (σ) for the reaction, is usually expressed in terms of an area per incident particle. Cross-sections are different for each type of incident particle, target nuclide, and reaction channel.

In general, cross-sections for low energy neutron-induced reactions are a few orders of magnitude greater than those for reactions involving high energy neutrons, charged particles or photons. Neutrons at low energies favor the radiative capture reaction (n, γ). Higher energy neutrons and charged particles favor more complex reactions involving nuclear transmutation. For example, high energy particles prefer transmutation reactions such as (n,p) and (p,n), stripping reactions like (d, p) and (d, n), or pickup reactions like (d, ^{3}He) and (d, α).

An equation that expresses the rate R of interactions for a particular nuclear reaction in a target per unit of time is given by

$$R = inx\sigma \tag{1}$$

where i = the number of incident particles per second, n = the number of target nuclei per unit volume (cm^{-3}), x = the thickness of the target (cm), and σ = the reaction cross-section per incident particle (cm^2).

Cross-section has the dimensions of area and expresses the probability that an individual particle will cause the specified reaction. Since most cross-sections are of the order 10^{-24} cm^2, it has become convenient to express cross-sections in units of barns, where

$$1 \text{ barn} = 10^{-24} \text{ cm}^2$$

Equation (1) is based on the following assumptions:

(1) the product nuclide has a half-life much longer than the length of irradiation. It does not decay appreciably during the irradiation period;

(2) the number of target nuclei is not depleted significantly during the irradiation. For moderate length irradiations this assumption is usually true. For high cross-sections or long irradiations target burnup is a concern;

(3) the cross-section is constant. This assumption is valid when the target is thin and the incident particle energy remains unchanged.

The details of target burnup and irradiation of thick targets are available in texts by Kruger [6] and De Soete, Gijbels and Hoste [7].

In a nuclear reactor, the target is placed in a uniform flux of neutrons, and equation (1) can be rewritten as follows:

$$R = N\phi\sigma \tag{2}$$

where N is the total number of target nuclei, ϕ is the flux density of neutrons (cm^{-2} s^{-1}), and σ is the reaction cross-section (cm^2).

2.4 Excitation functions

For each reaction channel, the cross-section varies with the energy of the incident particle. A detailed variation of the reaction cross-section with bombarding energy is known as the excitation function of the incident particle for the target nuclide. Extensive tabulations of the excitation functions for various reactions are available in Brune and Schmidt [8] and Mughabghab, Divadeenam and Holden [9]. Reactions involving charged particles, high energy neutrons, and photons require kinetic energies in excess of the threshold energy before the reaction can occur.

3 IRRADIATION SOURCES

Application of a particular reaction for analytical purposes rests to a considerable extent on the availability of irradiation facilities. The selection of an irradiation source implies that a suitable nuclear reaction has been identified. When choosing an irradiation source, three factors must be considered:

(1) the type of particle or radiation;
(2) the energy of the particle or radiation;
(3) the magnitude or intensity of the beam or flux.

Some possible sources of nuclear particles for inducing nuclear reactions are research reactors, accelerators, and radioisotopic sources. Chapter 3 in this volume gives a full description of the various radiation sources. The main properties of the sources are presented here.

3.1 Nuclear reactors

Neutrons are the most widely employed particles for nuclear analysis, owing to their greater range of penetration into target materials, large reaction cross-sections and high available fluxes. The neutron fluxes of present day research reactors range from 10^{12} to 10^{15} n cm^{-2} s^{-1}.

Nuclear reactors produce neutrons as the by-products of neutron-induced fission of uranium. In general, fission-produced neutrons have energies that range from 0 to 15 MeV with average energies of ≈ 2 MeV. Through only a few elastic collisions with moderator nuclei, the most energetic fission neutrons rapidly lose energy. This slowing down creates a broad distribution of neutron energies with three main components (thermal, epithermal, and fast). The thermal component consists of low energy neutrons at thermal equilibrium with the moderator atoms. At room temperature, the energy spectrum for thermal neutrons approximates a Maxwell–Boltzmann distribution with an average energy of 0.025 eV and a most probable velocity of 2200 m/s. The product of the number of neutrons per unit volume and the most probable velocity is defined as the thermal neutron flux. An upper energy limit for thermal neutrons is established by a cadmium foil of 1 mm thickness. The foil functions as a filter through which only neutrons with energies greater than 0.5 eV will pass.

Above the 0.5 eV threshold, the epithermal neutrons represent those neutrons that have only partially slowed down by interacting with the moderator. Their distribution follows a $1/E$ slope, beginning at the cadmium threshold energy and ranging up to about 1 MeV. The epithermal neutron fluxes in research reactors are on the order of 2% of the thermal flux.

Both thermal and epithermal neutrons can induce radiative capture (n, γ) reactions. Thus, borrowing from equation (2), a non-rigorous but commonly used expression for the total reaction rate per target atom for (n, γ) reactions in a reactor is given by:

$$R = N(\varphi_{th}\sigma_{th} + \varphi_{epi} I) \tag{3}$$

where φ_{th} = thermal neutron flux (n cm^{-2} s^{-1}); φ_{epi} = epithermal neutron flux (n cm^{-2} s^{-1}); σ_{th} = average thermal neutron cross-section (b); I = effective resonance integral (b).

The cross-section distribution functions for many nuclides in the epithermal region are characterized by a number of resonance peaks. To calculate reaction rates for this region, the effective resonance integral is given by the expression:

$$I = \int_{0.5\ \mathrm{eV}}^{\infty} \sigma(E)\,\mathrm{d}E/E \tag{4}$$

At energies above 1 MeV, the reactor neutron spectrum consists of the primary fission neutrons. These high energy neutrons still have most of their original energy from the fission reaction. At high neutron energies, the cross-sections for (n, γ) reactions are very small, and nuclear reactions that result in the ejection of one or more particles—(n, p), (n, α) and (n, 2n) reactions—are dominant. These reactions can occur only above a minimum threshold energy E_T. Because the energy distribution is complex, the average cross-section for a reaction in a ^{235}U fission neutron spectrum is defined by:

$$\bar{\sigma}_f = \frac{\int_{E_T}^{\infty} \sigma(E)\,\varphi(E)\,\mathrm{d}E}{\int_{E_T}^{\infty} \varphi(E)\,\mathrm{d}E} \tag{5}$$

Using this average cross-section, the reaction rate in a fission neutron spectrum is given by the expression:

$$R = \bar{\sigma}_f\,\varphi_f \tag{6}$$

where φ_f represents the average fission neutron flux.

3.2 Neutron generators

Neutron generators are small charged particle accelerators designed to produce neutrons following a suitable nuclear reaction. A common design is the Cockcroft–Walton accelerator, which bombards a tritium target with energetic deuterons to produce neutrons with an average energy of 14 MeV. A typical neutron generator for research emits about 2×10^{11} n/s. However, in a realistic irradiation geometry, neutron fluxes on the samples are only $\approx 2 \times 10^{9}$ n cm^{-2} s^{-1}.

3.3 Cyclotrons and accelerators

Cyclotrons and Van de Graaff accelerators produce energetic beams of protons, deuterons, tritons, alphas, and ^{3}He particles. Secondarily, these charged particles can produce high energy neutrons through a variety of nuclear reactions. Analytical applications with high energy neutrons and energetic charged particles are dominated by reactions involving particle evaporation.

Two other types of accelerators are the linear accelerator (linac) and the synchrotron, used mainly to accelerate electrons. Although electrons themselves have few analytical applications, the bremsstrahlung radiation (photons) emitted when high energy electrons interact with a tantalum or tungsten target can be used for photon activation analysis (PAA).

3.4 Radioisotopic sources

Two types of isotopic neutron source are possible. One type uses an alpha- or photon-emitting radionuclide to induce neutron emission from a target material consisting of a low-Z nuclide with a low binding energy for neutrons (e.g., beryllium). The emitted neutron then becomes the incident particle used for analysis of another target material. Some examples of these are the alpha emitters ^{239}Pu and ^{241}Am and the high energy gamma emitters ^{88}Y and ^{124}Sb. Radioisotopic sources can be made small and portable. However, their main disadvantage is a low neutron output, on the order of 10^5–10^7 n s^{-1} per Ci of activity.

A second type of isotopic neutron source is the spontaneous fission source (i.e., ^{252}Cf, with a half-life of 2.64 years). The neutron yield depends on the amount of ^{252}Cf in the source. A 1 g source of ^{252}Cf will emit 2.3×10^{12} n s^{-1} into 4π space. The energy range of neutrons emitted by the spontaneous fission of ^{252}Cf is 1 to 3 MeV. The rather short half-life is also sometimes a disadvantage.

4 DELAYED ANALYSIS METHODS

Most nuclear analytical methods involve the detection of radiation emitted from the radioactive product after irradiation. All radioactive products undergo decay by either α, β, γ, or delayed neutron emission processes. Since the product nuclides have characteristic half-lives, the irradiation and decay parameters can be chosen to enhance the production of specific products. In general, the radioactive products are easier to measure because backgrounds are lower, spectra are less complex, and more elements are observed. After irradiation, it is also possible to enhance sensitivity for a particular radionuclide by performing chemical separations to concentrate the product and remove interfering activities.

4.1 Delayed gamma-ray neutron activation analysis

The most familiar nuclear analytical technique is delayed gamma-ray neutron activation analysis (DGNAA). In DGNAA, the gamma-rays emitted during the decay of radioactive nuclides are measured following irradiation of the analytical specimen. The method affords a simultaneous multi-element analysis of the unknown specimen.

In DGNAA, the count rate A for decay gamma-rays during the measurement interval depends on the disintegration rate D_0 at the end of the irradiation. The distintegration rate is proportional to the amount of target element in the irradiated sample. Using the reaction rate R from equation (3), the respective activation and decay equations for a radionuclide with a half-life $t_{1/2}$ are:

$$D_0 = (N_0\, WF/M)\, R(1 - e^{-\lambda T_i}) \tag{7}$$

$$A = \varepsilon \Gamma D_0 e^{-\lambda T_d} \tag{8}$$

where N_0 = Avogadro's number, W = the weight of the element in the target (g), F = the fractional abundance of the target isotope, M = the atomic weight of the element, λ = the decay constant ($\ln 2/t_{1/2}$) for the radioisotope, ε = the detector efficiency for the decay gamma-ray, Γ = the branching ratio for the decay gamma-ray, T_i = the total irradiation time, and T_d = the length of decay.

From equations (7) and (8), one can see that the sensitivity of a particular reaction for DGNAA increases with increasing sample weight, detector efficiency, irradiation time, and neutron flux. A great advantage of delayed NAA is the ability to adjust the time of measurement in order to enhance the sensitivity for a particular radioisotope relative to other constituents in the sample.

Under certain irradiation conditions, simplifications to equation (8) are possible. If the irradiation time is long compared with the half-life of the radioisotope (i.e., $T_i \gg t_{1/2}$), the factor $[1\text{-exp}(-\lambda T_i)]$ reduces to an approximate value of 1. On the other hand, for extremely short irradiations (i.e., $T_i \ll t_{1/2}$) the factor $[1\text{-exp}(-\lambda T_i)]$ reduces to λT_i.

Although it is possible to perform an absolute activation analysis using the activation and decay equations to calculate the amount of target atoms in an unknown, the most common approach for DGNAA is the 'comparator' method. In this approach, a standard containing a known amount of the element of interest is irradiated with the unknown sample. After the irradiation, measurements on the standard and the unknown are performed under identical conditions. In addition to assuming that the neutron flux and detector efficiencies are the same for both, all other variables are usually the same (i.e., irradiation times, etc.). An equation used to compare count rates is given by:

$$\frac{R_{std}}{R_{unk}} = \frac{W_{std}(e^{-\lambda T_d})_{std}}{W_{unk}(e^{-\lambda T_d})_{unk}} \tag{9}$$

where the R values are the count rates for standard (std) and unknown (unk) samples, respectively; W = the weight of the element in the standard or the unknown; T_d = the individual decay time.

De Corte *et al.* [10, 11] and others developed a single element comparator method for multi-element NAA that has gained popularity in recent years. In this method, a so-called k_0 factor for each radionuclide is determined experimentally by comparison with a single-element monitor. The monitoring material employed most often is gold. The equation to calculate concentrations by the k_0 method is:

$$\rho = \frac{A_{sp}}{A^*_{sp}} \frac{1}{k_0} \left[\frac{f + Q^*_0(\alpha)}{f + Q_0(\alpha)} \right] \frac{\varepsilon^*}{\varepsilon} \tag{10}$$

where the asterisk refers to the coirradiated monitor; ρ = the concentration (ppm or μg/g); A_{sp} = the specific activity of the comparator element; k_0 = the k-zero factor for this isotope; f = the ratio of thermal to epithermal flux; Q_0 = the ratio of the resonance integral to the thermal neutron cross-section for this reaction; α = the deviation of the slope of the epithermal neutron flux from $1/E$; ε = the detector efficiency.

4.1.1 NAA with epithermal neutrons

For elements with large resonance integrals, epithermal neutrons provide an advantage by suppressing the activities of 'uninteresting' radionuclides with high yields due to high thermal neutron cross-sections or high concentrations in the sample. For example, in geological materials, elements such as Na, Al, Mn, Fe, and Sc at high concentrations are easily activated by thermal neutrons. As a result, the activities of trace or ultra-trace elements may not be detectable. For elements with large resonance integrals (i.e., I/σ is high), irradiation in an epithermal neutron spectrum often results in an enhancement of relative sensitivity.

Materials that are high thermal neutron absorbers, such as boron and cadmium, are used to shield the unknown samples for epithermal neutron activation analysis (ENAA). The shielding material selectively absorbs thermal neutrons and permits most epithermal neutrons to reach the sample.

The ENAA method is frequently useful for biological and geological specimens [12] to reduce high activities of Na and other matrix elements. Elements for which sensitivities may improve by ENAA include Ag, As, Au, Ba, Br, Cd, Cs, Ga, In, Mo, Rb, Sb, Se, Sm, Sr, Ta, Tb, Th, U, and W. Since the activity of the sample is much less than would occur with thermal neutrons, samples can often be handled sooner. In addition, they can be measured in closer sample-to-detector geometries which also enhance sensitivity.

4.1.2 Fast neutron activation analysis (FNAA)

High-energy neutrons from reactors and 14 MeV neutrons from neutron generators are sometimes employed for chemical analysis of elements not sensitive to lower energy neutrons. In general, the lower fluxes and smaller cross-sections for fast neutrons mean that FNAA has poorer relative sensitivity. However, a few of the light elements, such as oxygen and nitrogen, unavailable with thermal and epithermal neutrons are amenable to FNAA.

Oxygen measurements are possible using 14 MeV neutrons and the reaction $^{16}O(n, p)$ ^{16}N. The ^{16}N radionuclide has a half-life of 7.13 s and decays by emitting a pair of high-energy gamma-rays. Nitrogen can be determined using 14 MeV neutrons to induce the reaction ^{14}N (n, 2n) ^{13}N, which produces the positron emitting nuclide ^{13}N with a half-life of 9.97 min. Some elements frequently determined by FNAA are F, Mg, Al, Si, P, Cu, and Zr.

4.1.3 Sensitivities and applications for DGNAA

Tabulations of the nuclear parameters (cross-sections, half-lives, gamma-ray energies, abundances, etc.) employed by analysts to identify and quantify elements present in unknown samples are available from Erdtmann and Soyka [13], Tölgyessy and Bujdosó [14], and Glascock [15]. DGNAA has been applied to hundreds of different types of sample for single and multi-element determinations. Applications for DGNAA touch many disciplines, including archaeology, geology, agriculture, human nutrition, environmental studies, and industry. Sample matrices studied include pottery, rocks, water, biological speciments, air filters, and semiconductor materials. Recent reviews of applications for DGNAA are found in Alfassi [16], Ehmann and Vance [17], and Parry [18].

In general, the sensitivities available by DGNAA are superior to those for most other nuclear techniques. In addition to depending on the flux of neutrons, nuclear parameters, irradiation and decay conditions, and on detector efficiency, the sensitivities of elements may be affected by the sample matrix.

One ideal matrix for NAA is high-purity semiconductor silicon. Ultra-trace amounts (< 1 ppb) of certain impurity elements have adverse effects on the performance of semiconductor devices [19]. Thus, the identification of contaminating steps during the manufacture of semiconductor devices is important. Table 1 lists the nuclear parameters and detection limits by INAA for the analysis of semiconductor materials at the Missouri University Research Reactor [15].

4.2 Delayed gamma-rays following charged particle activation

Along their path through a target material, charged particles loss energy primarily from their interaction with atomic electrons. Thus, charged particles have a characteristic range, and particles of the same energy will stop after having traveled about the same (mean) path length. The range for any charged particle is

TABLE 1.
Nuclear parameters and detection limits for elements measured in semiconductor silicon by INAA at the Missouri University Research Reactor [15]

Target element	Nuclear reaction	Thermal or fast cross-section (barn)	Resonance integral (barn)	Half-life of product radionuclide	Gamma-ray energy (keV)	Brancing intensity (%)	Limit of detection[1] (ppb)
Ag	$^{109}Ag(n,\gamma)^{110m}Ag$	3.90	69	250 d	884.7	72.7	1.5×10^{-3}
As	$^{75}As(n,\gamma)^{76}As$	3.86	52.5	26.3 h	559.1	45.0	2.5×10^{-3}
Au	$^{197}Au(n,\gamma)^{198}Au$	98.6	1550	2.69 d	411.8	95.6	1.5×10^{-5}
Ba	$^{130}Ba(n,\gamma)^{131}Ba$	9.04	224	11.8 d	496.3	46.8	1.0×10^{-1}
Br	$^{81}Br(n,\gamma)^{82}Br$	2.58	49.8	35.3 h	554.3	70.8	3.0×10^{-3}
Ca	$^{46}Ca(n,\gamma)^{47}Ca$	0.62	0.81	4.54 d	1297.1	74.0	6.0×10^{1}
Cd	$^{114}Cd(n,\gamma)^{115}Cd(\beta^-)^{115}In$	0.22	8.7	53.5 h	336.2	46.1	2.5×10^{-2}
Ce	$^{140}Ce(n,\gamma)^{141}Ce$	0.575	0.48	32.5 d	145.4	48.2	2.0×10^{-2}
Co	$^{59}Co(n,\gamma)^{60}Co$	37.1	74	5.27 y	1173.2	99.9	4.0×10^{-4}
Cr	$^{50}Cr(n,\gamma)^{51}Cr$	15.2	8.1	27.7 d	320.1	9.8	2.0×10^{-2}
Cs	$^{133}Cs(n,\gamma)^{134}Cs$	30.7	390	2.06 y	795.8	85.4	1.0×10^{-4}
Cu	$^{63}Cu(n,\gamma)^{64}Cu$	4.28	4.88	12.7 h	511.0	35.7	3.0×10^{-2}
Eu	$^{151}Eu(n,\gamma)^{152}Eu$	5900	5564	13.3 y	1408.0	20.8	4.5×10^{-4}
Fe	$^{58}Fe(n,\gamma)^{59}Fe$	1.31	1.28	44.6 d	1099.2	56.1	8.0×10^{-1}
Ga	$^{71}Ga(n,\gamma)^{72}Ga$	4.61	30.6	14.1 h	834.0	95.6	4.0×10^{-3}
Hf	$^{180}Hf(n,\gamma)^{181}Hf$	13.5	34	42.4 d	482.2	80.6	1.5×10^{-3}
Hg	$^{202}Hg(n,\gamma)^{203}Hg$	4.35	3.8	46.6 d	279.2	82.8	4.0×10^{-3}
In	$^{113}In(n,\gamma)^{114m}In$	8.2	224	49.5 d	189.9	15.4	8.0×10^{-3}
Ir	$^{191}Ir(n,\gamma)^{192}Ir$	924	3750	73.8 d	316.5	82.8	1.5×10^{-5}
K	$^{41}K(n,\gamma)^{42}K$	1.45	1.41	12.4 h	1524.7	17.9	5.5×10^{-1}
La	$^{139}La(n,\gamma)^{140}La$	9.34	11.6	40.2 h	1596.2	95.5	1.0×10^{-3}
Mo	$^{98}Mo(n,\gamma)^{99}Mo$	0.131	6.96	66.0 h	140.5	87.2	1.5×10^{-1}
Na	$^{23}Na(n,\gamma)^{24}Na$	0.513	0.303	15.0 h	1368.6	100.0	3.5×10^{-2}
Ni	$^{58}Ni(n,p)^{58}Co$	0.113	–	70.9 d	810.8	99.4	4.0×10^{-1}
Pt	$^{190}Pt(n,\gamma)^{191}Pt$	150	67	2.96 d	538.9	13.4	7.0×10^{0}

(continued)

TABLE 1. *(continued)*

Target element	Nuclear reaction	Thermal or fast cross-section (barn)	Resonance integral (barn)	Half-life of product radionuclide	Gamma-ray energy (keV)	Brancing intensity (%)	Limit of detection[1] (ppb)
Rb	$^{85}Rb(n, \gamma)\,^{86}Rb$	0.494	7.31	18.7 d	1076.6	8.8	1.5×10^{-2}
Sb	$^{123}Sb(n, \gamma)\,^{124}Sb$	4.08	118	60.2 d	1691.0	47.6	2.0×10^{-3}
Sc	$^{45}Sc(n, \gamma)\,^{46}Sc$	26.3	11.3	83.8 d	889.3	100.0	6.5×10^{-5}
Se	$^{74}Se(n, \gamma)\,^{75}Se$	51.2	512	120 d	264.6	59.2	6.0×10^{-3}
Sn	$^{112}Sn(n, \gamma)\,^{113}Sn$	0.52	28.5	115 d	391.7	64.2	2.0×10^{-1}
Sr	$^{84}Sr(n, \gamma)\,^{85}Sr$	0.69	9.14	64.8 d	514.0	99.3	2.0×10^{-1}
Ta	$^{181}Ta(n, \gamma)\,^{182}Ta$	20.4	679	114 d	1221.3	35.3	2.5×10^{-4}
Tb	$^{159}Tb(n, \gamma)\,^{160}Tb$	23.8	426	72.1 d	879.4	30.4	2.0×10^{-4}
Th	$^{232}Th(n, \gamma)\,^{233}Th(\beta^-)\,^{233}Pa$	7.26	83.7	27.0 d	312.0	38.6	6.5×10^{-4}
Ti	$^{48}Ti(n, p)\,^{48}Sc$	0.0003	–	43.7 h	983.5	100.0	2.0×10^{2}
U	$^{238}U(n, \gamma)\,^{239}U(\beta^-)\,^{239}Np$	2.75	284	2.36 d	277.6	14.2	1.5×10^{-3}
W	$^{186}W(n, \gamma)\,^{187}W$	38.7	530	23.9 h	479.6	23.4	1.5×10^{-3}
Yb	$^{174}Yb(n, \gamma)\,^{175}Yb$	128	58.9	4.19 d	396.3	6.5	6.0×10^{-3}
Zn	$^{64}Zn(n, \gamma)\,^{65}Zn$	0.726	1.42	244 d	1115.5	50.7	4.5×10^{-2}
Zr	$^{94}Zr(n, \gamma)\,^{95}Zr$	0.053	0.268	64 d	756.7	54.5	3.0×10^{-1}

[1] Limits of detection in very pure silicon calculated according to Currie [20]. Experimental parameters were: undoped wafer, diameter 10 cm, weight 15 g; irradiation time 45 h in flux 2.5×10^{13} n cm^{-2} s^{-1}; measured twice on low background HPGe detector, 35% relative efficiency: (1) decay time = 48 h, measuring time = 1 h; (2) decay time = 15 days, measuring time = 6 h.

dependent on the Z of the target material and the energy, mass, and charge of the charged particle.

According to Friedlander *et al.* [4], the rate of energy loss or the stopping power $S(E)$ is given by:

$$S(E) = \frac{-\mathrm{d}E}{\mathrm{d}x} \tag{11}$$

where dx is the distance traveled for an energy loss dE. It can be shown that the stopping power for a given particle is directly proportional to the particle's charge and inversely proportional to the energy of the incident particle. The range of a particle with an initial energy E_{I} is defined by the equation:

$$R(E_{\mathrm{I}}) = \int_0^{E_{\mathrm{I}}} \frac{\mathrm{d}E}{S(E)} \tag{12}$$

Since the cross-section is zero below the threshold energy E_{T}, equation (12) can be written:

$$R(E_{\mathrm{I}}) = \int_{E_{\mathrm{T}}}^{E_{\mathrm{I}}} \frac{\sigma(E)\,\mathrm{d}E}{S(E)} \tag{13}$$

Particle ranges are usually expressed in units of $\mathrm{g\,cm^{-2}}$.

In CPAA, a sample (x) and a standard (s) are irradiated separately using charged particles of the same energy for the same length of time. Thus, the

TABLE 2.
Nuclear parameters for reactions analytically important by CPAA

Element	Nuclear reaction	Product half-life	Gamma-ray energy (keV)
B	$^{10}B(p,\alpha)\,^{7}Be$	53.3 d	477.6
	$^{10}B(d,n)\,^{11}C$	20.3 m	511.0
	$^{11}B(p,n)\,^{11}C$	20.3 m	511.0
C	$^{12}C(d,n)\,^{13}N$	9.97 m	511.0
	$^{12}C(^{3}He,\alpha)\,^{11}C$	20.3 m	511.0
N	$^{14}N(p,\alpha)\,^{11}C$	20.3 m	511.0
	$^{14}N(p,n)\,^{14}O$	70.6 s	2312.7
	$^{14}N(d,n)\,^{15}O$	122 s	511.0
O	$^{16}O(p,\alpha)\,^{13}N$	9.97 m	511.0
	$^{16}O(^{3}He,p)\,^{18}F$	1.83 h	511.0
P	$^{31}P(\alpha,n)\,^{34m}Cl$	32.2 m	2128.5
Ca	$^{40}Ca(\alpha,p)\,^{43}Sc$	3.89 h	372.8
Ti	$^{48}Ti(p,n)\,^{48}V$	16.0 d	983.5
Fe	$^{56}Fe(p,n)\,^{56}Co$	77.7 d	846.8
Cd	$^{111}Cd(p,n)\,^{111}In$	2.80 d	245.4
W	$^{182}W(p,n)\,^{182m}Re$	12.7 h	1121.5
Pb	$^{206}Pb(p,n)\,^{206}Bi$	6.24 d	803.1

concentration in the unknown, C_x, can be determined from the equation:

$$C_x = C_s \frac{I_s A_x R_s}{I_x A_s R_x} \tag{14}$$

where C_s = the concentration of the element in the standard; I_s/I_x = the ratio of beam intensities; A_x/A_s = the ratio of measured activities; R_x/R_s = the ratio of particle ranges.

The CPAA technique is frequently used to determine the lighter elements B, C, N, and O in metals and semiconductor materials. Table 2 lists several of the nuclear reactions and parameters that are important for CPAA determinations. Vandecasteele [21] describes a number of applications for CPAA.

4.3 Photon activation analysis

The photon activation analysis (PAA) technique plays a minor role in elemental analysis, in large part because of the small cross-sections for photonuclear reactions. However, it complements NAA by permitting the determination of several light and heavy elements insensitive to NAA. PAA has an advantage over NAA when the unknown sample contains elements such as C, N, O, F, and Pb with unsuitable nuclear parameters for analysis by NAA. The PAA technique is also useful for samples in which elements with large neutron absorption cross-sections (e.g., boron and cadmium) constitute the bulk of the matrix.

A target nucleus bombarded by photons may be transformed into a radionuclide by one of three possible reactions:

(1) a photoexcitation (γ, γ') into an isomeric state of the target nuclide;
(2) a photonuclear reaction which transforms the nuclide by particle emission. For example: (γ, n), (γ, p) and (γ, np);
(3) a photofission reaction (γ, f) with fissionable elements.

Photonuclear reactions have threshold energies below which the reactions will not occur. Reaction cross-sections are typically small when compared to thermal neutron reaction cross-sections. In general, photon beams must have high energies and intensities in order for PAA to be successful. Typically, a linac of about 30 MeV with a beam current of a few hundred μA is necessary to obtain satisfactory sensitivity.

The activation rate equation for PAA is similar to those for calculating other activities

$$R = N \int_{E_{min}}^{E_{max}} \sigma(E)\,\phi(E)\,dE$$

where $\phi(E)$ is the flux of photons per unit of energy interval.

Data on the cross-sections for photonuclear reactions are available in Forkman and Petersson [22]. A listing of common PAA reactions and element sensitivities is available in the work of Matsumoto, Kato and Suzuki [23]. Finally, Segebade, Weise and Lutz [24] and Kushelevsky [25] review various PAA applications.

4.4 Delayed neutrons following neutron activation

Although a majority of the radionuclides produced by neutron activation undergo beta decay followed by gamma-ray emission, there are a few analytically important reactions that involve the emission of delayed neutrons. Because the number of reactions that produce delayed neutron emitters is small, the technique is quite selective and potentially very sensitive. Some of the reactions that yield delayed neutron emitters are $^{9}Be(n,p)\ ^{9}Li$ and $^{17}O(n,p)\ ^{17}N$, providing a technique to determine beryllium and oxygen, respectively [26].

The thermal neutron-induced fission of ^{235}U and fast neutron-induced fission of ^{238}U and ^{232}Th are reactions that also produce delayed neutron emitters. Measurement of delayed neutrons permits determination of uranium and thorium in geological specimens. In some instances, it is possible to determine the ratio of $^{235}U/^{238}U$ using thermal and epithermal neutron irradiations. Automated irradiation and counting systems for measuring delayed neutrons have been developed by Minor *et al.* [27] and Schlechte [28].

5 PROMPT METHODS OF ANALYSIS

Prompt methods of analysis involve the measurement of particles or gamma-rays released by the compound nucleus immediately after interaction with the incoming particle. The prompt method depends on the reaction probability, the amount of target material, and the flux of incident particles, but is independent of the half-life of the product nucleus. Several prompt analysis methods are described below.

5.1 Prompt gamma-ray neutron activation analysis

The technique of prompt gamma-ray neutron activation analysis (PGNAA) utilizes prompt gamma-rays emitted immediately after the (n,γ) reaction with thermal neutrons. Since most nuclides have binding energies for capturing a neutron of ≈ 8 MeV, the prompt gamma-rays emitted from compound nuclei are usually higher in energy than gamma-rays from DGNAA. This higher initial excitation energy increases the number of nuclear levels available during de-excitation, so that prompt gamma-ray spectra are more complex than delayed gamma-ray spectra. The combination of a long range for neutrons in matter and minimal absorption of high energy gamma-rays makes PGNAA an ideal technique for bulk samples.

PGNAA overcomes the difficulties created by (n, γ) reactions that generate product nuclides with:

(1) very short or long half-lives: for example, $^{11}B\,(n,\gamma)\,^{12}B$ with $t_{1/2} = 0.02$ s or $^{9}Be\,(n,\gamma)\,^{10}Be$ with $t_{1/2} = 1.6 \times 10^6$ yr;
(2) stable ground-state configurations: for example, $^{1}H\,(n,\gamma)\,^{2}H$, $^{113}Cd\,(n,\gamma)\,^{114}Cd$ or $^{157}Gd\,(n,\gamma)\,^{158}Gd$;
(3) very few or no gamma-ray emissions: for example, $^{30}Si\,(n,\gamma)\,^{31}Si$, $^{31}P\,(n,\gamma)\,^{32}P$, $^{34}S\,(n,\gamma)\,^{35}S$ and $^{44}Ca\,(n,\gamma)\,^{45}Ca$.

A tabulation of elements suitable for analysis by PGNAA will be found in Table 3. The nuclear parameters and limits of detection by PGNAA for elements measured in geological samples are given [29]. Included are those elements with isotopes having large abundances, high cross-sections for the (n, γ) reaction, or

TABLE 3.
Nuclear reactions and detection limits for PGNAA in a geological reference material [29]

Element	Nuclear reaction	Prompt gamma-ray energy (keV)	Limit of detection[a] (ppm)
H	$^{1}H\,(n,\gamma)\,^{2}H^*$	2223.3	10
B	$^{10}B\,(n,\alpha)\,^{7}Li^*$	477.6	0.06
C	$^{12}C\,(n,\gamma)\,^{13}C^*$	1261.9	1.9%
N	$^{14}N\,(n,\gamma)\,^{15}N^*$	1884.8	0.2%
Na	$^{23}Na\,(n,\gamma)\,^{24}Na^*$	472.3	110
Mg	$^{24}Mg\,(n,\gamma)\,^{25}Mg^*$	585.1	1260
Si	$^{28}Si\,(n,\gamma)\,^{29}Si^*$	3539.2	800
P	$^{31}P\,(n,\gamma)\,^{32}P^*$	636.7	1600
S	$^{32}S\,(n,\gamma)\,^{33}S^*$	840.4	140
Cl	$^{35}Cl\,(n,\gamma)\,^{36}Cl^*$	517.0	6
K	$^{39}K\,(n,\gamma)\,^{40}K^*$	770.5	70
Ca	$^{40}Ca\,(n,\gamma)\,^{41}Ca^*$	1942.7	370
Ti	$^{48}Ti\,(n,\gamma)\,^{49}Ti^*$	1381.5	30
Fe	$^{56}Fe\,(n,\gamma)\,^{57}Fe^*$	352.3	490
Co	$^{59}Co\,(n,\gamma)\,^{60}Co^*$	277.1	16
Sr	$^{87}Sr\,(n,\gamma)\,^{88}Sr^*$	898.0	300
Cd	$^{113}Cd\,(n,\gamma)\,^{114}Cd^*$	558.3	0.07
Nd	$^{143}Nd\,(n,\gamma)\,^{144}Nd^*$	696.5	10
Sm	$^{149}Sm\,(n,\gamma)\,^{150}Sm^*$	333.9	0.03
Gd	$^{157}Gd\,(n,\gamma)\,^{158}Gd^*$	181.9	0.04

[a] Limits of detection in USGS reference material RGM-1 calculated according to Currie [20]. Experimental parameters were: 1 g of sample; neutron beam flux of 5×10^8 n cm^{-2} s^{-1}; measured with HPGe detector, 25% relative efficiency; measuring time 10 h.

stable products. Only a few elements—H, B, Cd, Sm, and Gd—have routine sensitivities at trace levels (< 100 ppm).

Three major inconveniences for PGNAA are background radiation, neutron scattering from the surrounding materials and neutron scattering within the sample. First, background radiation arises from interactions between the incident beam and materials other than the sample (i.e., beam tubes, shielding materials, sample holders). High backgrounds make determination of low intensity peaks difficult. Second, scattered neutrons may interact with the detector materials to complicate the spectrum, add to the background effects, distort peak shapes, and worsen detector resolution. According to Kraner, Chasman and Jones [30], degradation of detector resolution occurs after exposure to neutron fluences greater than 10^{10} neutrons cm^{-2}, resulting in damage to the detector. A third problem is the scattering of neutrons from hydrogenous samples. As reported by Copley and Stone [31] and Mackey *et al.* [32], difficulties with neutron scattering from samples containing hydrogen are least when using spherically shaped samples.

Higher fluxes using low energy 'cold' neutron beams are under development at some research reactor laboratories. Cold neutron beams for PGNAA reduce background effects and have the potential of enhancing sensitivities by one or two orders of magnitude.

Sensitivities for PGNAA are higher at reactor-based facilities than at those using portable neutron sources. As an analytical technique for routine sample analysis, reactor-based PGNAA facilities are complementary to the conventional delayed NAA facilities. In recent years, several PGNAA facilities have been installed at reactor-based laboratories in North America, including MIT [33], NIST [34], Missouri [35], Michigan [36], North Carolina State [37], and McMaster [38].

PGNAA has application to the same types of samples that are analyzed by DGNAA. Some of the elements reported in samples are present as major constituents, including Na, Al, Si, S, Ca, K, and Fe. An important industrial application of PGNAA is the analysis of low levels of B in semiconductor materials [39]. Boron concentrations are also of interest in biological samples [40], geochemistry [41], and cosmochemistry [42]. Application of PGNAA for authentication of archaeological specimens is described by Glascock *et al.* [43].

In addition to reactor-based PGNAA, several industrial and medical applications for PGNAA have been developed using portable neutron generators and radioisotopic sources coupled with detectors and shielding materials. Although these sources are less sensitive, the applications benefit from the capability of analyzing very large samples and transportability to hostile environments. Examples of these applications include: subsurface mineral exploration and well logging [44]; real-time monitoring and control of bulk process streams for electric power generation from the burning of coal [45]; analysis of total body composition [46]; and measurement of Cd burden levels in the liver and kidneys of humans [47]. Reviews by Anderson *et al.* [48], Glascock [49] and Lindstrom and Anderson [50] outline possible applications for PGNAA.

Measurement of prompt gamma-rays following the inelastic scatter of high energy neutrons is an analytical procedure that is similar to PGNAA. The energy of the emitted gamma-ray identifies the target nucleus and the intensity of the gamma-ray is related to the concentration. A few of the elements amenable to this technique are C, Mg, Al, Si, and Fe. Because of background interference problems, the technique is appropriate for samples weighing more than 10 kg and for the analysis of elements at concentrations greater than 1% [51]. An instrument combining a ^{238}Pu–Be source for neutrons with a ^{60}Co source for gamma-rays was developed by Sowerby [52]. It measures the neutron inelastic scatter prompt gamma-rays from carbon, the neutron-capture prompt gamma rays from hydrogen, and the amount of gamma-ray scattering. At coal-fired power plants, the instrument is capable of determining the specific energy, moisture and ash contents of coal. Other applications of neutron inelastic scattering include studies of Cr and Ni in metal alloys by Yates *et al.* [53] and the analysis of Al and Si in soils by Filo, Gilbert and Yates [54].

5.2 Particle-induced gamma-ray emission

Particle-induced gamma-ray emission (PIGE) is the charged particle analog of PGNAA, where charged particles induce nuclear reactions that produce prompt gamma-rays for chemical analysis. Protons, deuterons, tritons, alpha-particles and heavy ions can be employed for this purpose.

One of the earliest applications was the measurement of prompt gamma-rays following alpha-particle irradiation of beryllium in air-borne dust [55]. Another early application was the measurement of prompt gamma-rays from fluorine in glass samples bombarded by protons [56]. A more recent application has been the characterization of Li and P in thin films using a proton beam [57]. Most of these early applications involved studies with proton beams of less than 4 MeV for measuring the low-Z elements such as Li, Be, F, Na, Al, and Si. However, Demortier [58] reports on reactions using higher energy beams. A longer discussion of PIGE can be found in Chapter 8.

5.3 Particle-induced prompt particle emission

Particle-induced prompt particle emission (PIPPE) is a technique that becomes possible when the products of a direct reaction between an incident particle and a heavy target nucleus are a heavy product nucleus and a single low-mass particle. If the amount of energy available to the reaction is sufficient, the heavy product nucleus may be formed in different excitation states. Each of these states will have an associated low-mass product with a characteristic kinetic energy. The number of low-mass products at each energy is proportional to the cross-section and to the number of target nuclei present.

The PIPPE technique is a powerful tool for studying material surfaces by measuring element concentration variations as a function of depth. The application is also known as depth profiling. For incident charged particles, the depth to which the charged particle can penetrate the surface depends upon the range of charged particles within the material. Most applications using charged particles apply to the determination of low-Z elements, such as H, He, Li, Be, B, C, N, and O. A recent review of charged particle prompt spectrometry by Peisach [59] describes these applications.

5.4 Neutron depth profiling

Neutron depth profiling (NDP) is a prompt nuclear analysis technique used to determine element concentration depth profiles in semiconductor silicon. NDP employs a nuclear reaction that results in emission of charged particles with a specific kinetic energy. If the nuclear reaction that produced the charged particle occurs inside the sample, the charged particle loses some of its kinetic energy before arriving at the detector. The amount of energy loss is related to the distance it has traveled within the specimen. The NDP technique then provides a highly sensitive method for measuring the concentration depth profile of the element.

The NDP method was first developed by Ziegler, Cole and Baglin [60] to determine boron depth profiles using the reaction $^{10}B\,(n,\alpha)\,^{7}Li$, which has a very high thermal neutron capture cross-section of 3837 barn. In addition to boron, lithium, which has a thermal neutron capture cross-section of 940 barn for the $^{6}Li\,(n,\alpha)\,^{3}H$ reaction, can be studied by NDP.

The depth resolution and the main causes of measurement error in NDP are discussed by Biersack *et al.* [61]. Downing *et al.* [62] describe the NDP system at the NIST and summarize the characteristics of different NDP systems worldwide. A longer discussion of NDP is available in Chapter 9 of this volume.

6 ADVANTAGES AND DISADVANTAGES OF PROMPT VS. DELAYED ANALYSIS

Prompt and delayed analysis methods have advantages and disadvantages relative to one another and to other techniques. Some of the main advantages of prompt analysis are summarized below.

(1) Elements of low mass can be determined.
(2) Reactions producing stable product nuclides can be studied.

(3) Analysis depends only on cross-section, specimen mass, and incident flux, not on rate of decay.
(4) Higher energy gamma-rays are less likely to be absorbed.
(5) Single excited states can be enhanced by adjusting beam energies.
(6) Samples have lower residual activity after irradiation.
(7) Bulk samples can be analysed using neutron and photon beams.
(8) Neutron and charged particle beams can be used to investigate concentration depth profiles.

The main disadvantages of prompt analysis are:

(1) lower fluxes result in lower sensitivities;
(2) one cannot use half-lives to one's advantage;
(3) spectra are more complex;
(4) scattered neutrons cause detector degradation and require more shielding material;
(5) analysis is limited to single samples;
(6) particle spectroscopy requires thin targets.

The main advantages of delayed analysis are:

(1) half-lives of radioactive products can be exploited by adjusting irradiation and decay times;
(2) better sensitivity because delayed emissions are the sum of many individual excited states;
(3) when using neutrons, higher fluxes are possible to give greatly improved sensitivities;
(4) radiochemical separations can be used to enhance sensitivities for particular elements;
(5) it is possible to irradiate many samples at once and reduce cost per sample;
(6) samples being measured for long lived activities can be irradiated in one laboratory and sent to another for analysis.

The main disadvantages of delayed analysis are:

(1) some samples cannot be exposed to high fluxes;
(2) the analysis depends on radioactive decay and is not instantaneous;
(3) some reaction products are very short lived, long lived, stable, or emit inconvenient types of radiation;
(4) many low-mass elements are less sensitive.

7 CONCLUSIONS

Nuclear techniques for chemical analysis are among the most sensitive available for qualitative and quantitative analysis of materials. Applications for nuclear methods exist in a great number of disciplines, including archaeology, agriculture, biochemistry, geochemistry, environmental studies, forensics, and materials science. Prompt and delayed analysis techniques provide analysts with a wide range of methods and sensitivities in all areas.

REFERENCES

1. W. K. Chu, J. W. Mayer and M. A. Nicolet, *Backscattering Spectrometry*, Academic Press, New York (1978).
2. L. C. Feldman and J. W. Mayer, *Fundamentals of Surface and Thin Film Analysis*, Elsevier, New York (1986).
3. J. B. A. England, in *Elemental Analysis by Particle Accelerators*, edited by Z. B. Alfassi and M. Peisach, CRC Press, Boca Raton (1991), pp. 243–279.
4. G. Friedlander, J. W. Kennedy, E. S. Macias and J. M. Miller, *Nuclear and Radiochemistry*, 3rd Ed., Wiley, New York (1981).
5. R. Vandenbosch and J. R. Huizenga, *Nuclear Fission*, Academic Press, New York (1973).
6. P. Kruger, *Principles of Activation Analysis*, Wiley, New York (1971).
7. D. De Soete, R. Gijbels and J. Hoste, *Chemical Analysis, a Series of Monographs on Analytical Chemistry and Its Applications*, Vol. 34, *Neutron Activation Analysis*, edited by P. J. Elving and I. M. Kolthoff, Wiley, London (1972).
8. D. Brune and J. J. Schmidt (Eds.), *Handbook on Nuclear Activation Cross-Sections*, International Atomic Energy Agency, Vienna (1974).
9. S. F. Mughabghab, M. Divadeenam and N. E. Holden, *Neutron Cross Sections, Vol. I: Neutron Resonance Parameters and Thermal Cross Sections.* Part A: $Z = 1-60$, Academic Press, New York; Part B: $Z = 61-100$, Academic Press, Orlando (1984).
10. F. De Corte, S. Jovanovic, A. Simonits, L. Moens and J. Hoste, *Kerntechnik* **44**, 641 (1984).
11. F. De Corte, A. Simonits, A. De Wispelaere and J. Hoste, *J. Radioanal. Nucl. Chem.* **113**, 145 (1987).
12. E. Steinnes, in *Activation Analysis in Geochemistry and Cosmochemistry*, edited by A. O. Brunfelt and E. Steinnes, Universitetsforlaget, Oslo (1971), pp. 113–128.
13. G. Erdtmann and W. Soyka, *The Gamma Rays of The Radionuclides: Tables for Applied Gamma Ray Spectrometry,* Verlag Chemie, Weinheim (1979).
14. J. Tölgyessy and Ernö Bujdosó (Eds.), *CRC Handbook of Radioanalytical Chemistry*, Vol. I and II, CRC Press, Boca Raton (1991).
15. M. D. Glascock, *Tables for Neutron Activation Analysis*, Research Reactor, The University of Missouri (1991).
16. Z. B. Alfassi (Ed.), *Activation Analysis*, Vol. I and II, CRC Press, Boca Raton (1989).
17. W. D. Ehmann and D. E. Vance, *Radiochemistry and Nuclear Methods of Analysis*, edited by J. D. Winefordner, Wiley, New York (1991).
18. S. J. Parry, *Activation Spectrometry in Chemical Analysis*, edited by J. D. Winefordner, Wiley, New York (1991).
19. H. R. Huff, *Solid State Technol.* 89 (Feb 1983).
20. L. A. Currie, *Anal. Chem.* **40**, 586 (1968).
21. C. Vandecasteele, *Activation Analysis with Charged Particles*, edited by R. A. Chalmers, Ellis Horwood, Chichester (1988).
22. B. Forkman and R. Petersson in *Handbook on Nuclear Activation Data*, International Atomic Energy Agency, Vienna (1987).

23. K. Matsumoto, T. Kato and N. Suzuki, *Nucl. Instrum. Methods* **157**, 567 (1979).
24. C. Segebade, H.-P. Weise and G. J. Lutz, *Photon Activation Analysis*, Walter de Gruyter, Berlin (1987).
25. A. P. Kushelevsky, in *Activation Analysis*, Vol. II, edited by Z. B. Alfassi, CRC Press, Boca Raton (1990), pp. 219–237.
26. S. E. Binney and R. I. Scherpelz, *Nucl. Instrum. Methods* **154**, 413 (1978).
27. M. M. Minor, W. K. Hensley, M. M. Denton and S. R. Garcia, *J. Radioanal. Chem.* **70**, 459 (1982).
28. P. Schlechte, *J. Radioanal. Chem.* **61**, 55 (1981).
29. A. G. Hanna, Ph.D. dissertation (unpublished), University of Missouri (1980).
30. H. W. Kraner, C. Chasman and K. W. Jones, *Nucl. Instrum. Methods* **62**, 173 (1968).
31. J. R. D. Copley and C. A. Stone, *Nucl. Instrum. Methods Phys. Res.* **A281**, 593 (1989).
32. E. A. Mackey, G. E. Gordon, R. M. Lindstrom and D. L. Anderson, *Anal. Chem.* **63**, 288 (1991).
33. N. C. Rasmussen, Y. Hukai, T. Inouye and V. J. Orphan, *Report MITNE-85*, Massachusetts Institute of Technology—Department of Nuclear Engineering, Cambridge, MA (1969).
34. M. P. Failey, D. L. Anderson, W. H. Zoller, G. E. Gordon and R. M. Lindstrom, *Anal. Chem.* **51**, 2209 (1979).
35. A.G. Hanna, R. M. Brugger and M. D. Glascock, *Nucl. Instrum. Methods* **188**, 619 (1981).
36. J. D. Jones and M. A. Ludington, *Kerntechnik* **44**, 676 (1984).
37. G. D. Miller and B. W. Wehring, *Trans. Am. Nucl. Soc.* **53**, 161 (1986).
38. M. D. Higgins, M. G. Truscott, D. M. Shaw, M. Bergeron, G. H. Buffet, J. R. D. Copley and W. V. Prestwich, *Kerntechnik* **44**, 690 (1984).
39. J. E. Riley, Jr. and R. M. Lindstrom, *J. Radioanal. Nucl. Chem. Articles* **109**, 109 (1987).
40. T. Kobayashi and K. Kanda, *Nucl. Instrum. Methods* **204**, 525 (1983).
41. P. I. Nabelek, J. R. Denison and M. D. Glascock, *Am. Mineral.* **1**, 874 (1990).
42. D. B. Curtis and E. S. Gladney, *Earth Planet. Sci. Lett.* **75**, 311 (1985).
43. M. D. Glascock, T. G. Spalding, J. C. Biers and M. F. Cornman, *Archaeometry* **26**, 96 (1984).
44. F. E. Senftle, A. B. Tanner, P. W. Philbin, G. R. Boynton and C. W. Schram, *Min. Eng.* **30**, 666 (1978).
45. T. Gozani, D. R. Brown, G. Reynolds, H. Bozorgmanesh, E. Elias, T. Maung and V. Orphan, *Electric Power Research Institute Reports* CS-989, Vols. 1–8, Palo Alto, CA (1980).
46. M. A. Smith and P. Tothill, *Phys. Med. Biol.* **24**, 319 (1979).
47. I. K. Al-Haddad, D. R. Chettle, J. H. Fremlin and T. C. Harvey, *J. Radioanal. Chem.* **53**, 203 (1979).
48. D. L. Anderson, W. H. Zoller, G. E. Gordon, W. B. Walters and R. M. Lindstrom, in *Neutron-Capture Gamma-Ray Spectroscopy and Related Topics 1981*, edited by T. von Egidy, F. Gonnewein and B. Baier, Institute of Physics, London (1982), pp. 655–668.
49. M. D. Glascock, in *Neutron-Capture Gamma-Ray Spectroscopy and Related Topics 1981*, edited by T. von Egidy, F. Gonnewein and B. Baier, Institute of Physics, London (1982), pp. 641–654.
50. R. M. Lindstrom and D. L. Anderson, in *Neutron-Capture Gamma-Ray Spectroscopy and Related Topics 1984*, edited by S. Raman, American Institute of Physics, New York (1985), pp. 810–819.
51. B. D. Sowerby, *Nucl. Instrum. Methods* **166**, 571 (1979).
52. B. D. Sowerby, *Nucl. Instrum. Methods* **160**, 173 (1979).
53. S. W. Yates, A. J. Filo, C. Y. Cheng and D. F. Coope, *J. Radioanal. Chem.* **46**, 343 (1978).

54. A. J. Filo, J. W. Gilbert and S. W. Yates, *J. Radioanal. Chem.* **54**, 235 (1979).
55. R. Gold and C. A. Stone, *U.S.A.E.C. Rep.*, AECU-3505 (1957).
56. S. Rubin, T. O. Passel and L. E. Bailey, *Anal. Chem.* **29**, 736 (1957).
57. J. R. Bird and J. S. Williams (Eds.), *Ion Beams for Materials Analysis*, Academic Press, Sydney (1989).
58. G. Demortier, *J. Radioanal. Chem.* **47**, 459 (1978).
59. M. Peisach, in *Activation Analysis*, edited by Z. B. Alfassi, Vol. II, CRC Press, Boca Raton (1990), pp. 143–209.
60. J. F. Ziegler, G. W. Cole and J. E. E. Baglin, *J. Appl. Phys.* **43**, 3809 (1972).
61. J. P. Biersack, D. Fink, R. Henkelmann and K. Müller, *Nucl. Instrum. Methods* **149**, 93 (1978).
62. R. G. Downing, R. F. Fleming, J. K. Langland and D. H. Vincent, *Nucl. Instrum. Methods Phys. Res.* **218**, 47 (1983).

Chapter 5

RADIATION PROTECTION

Tuvia Schlesinger
Soreq Nuclear Research Center, Yavne, Israel 70600

1 INTRODUCTION

Chemical analysis by nuclear methods involves the use of a wide range of irradiation facilities and sources emitting ionizing radiation. Among these are neutron beams from nuclear reactors and other high intensity neutron sources, particle accelerators, X-ray machines and various radioactive materials. The utilization of this technology is therefore associated with the potential exposure of workers, and members of the public, to ionizing radiation. As explained below, such exposure may be detrimental in various ways to the health of the exposed person. To minimize this risk, strict safety measures have to be taken. The design of these measures is the task of the radiation safety officer (RSO) and/or the radiation safety adviser (RSA) of the radiation facility. He is authorized also to ensure that the safety regulations are followed and that the safety measures are applied on an everyday basis. However, the RSO and the RSA cannot carry out their work without full cooperation from the side of the 'user' (the person or persons who are engaged in applying the nuclear methods for chemical analysis described in this book).

The purpose of this chapter is to give to these 'users' an understanding of the basics of radiation protection, i.e. the types and routes of exposure, the special quantities and units used in radiation protection, the principles of radiation dosimetry, and practical radiation protection measures and procedures. The chapter includes also a short review of the biological effects of ionizing radiation and some details on the basic international standards for radiation protection.

Chemical Analysis by Nuclear Methods Edited by Z. B. Alfassi

2 TYPES AND ROUTES OF EXPOSURE

Persons engaged in performing chemical analysis by nuclear methods can be exposed to ionizing radiation of two major types:

(a) external exposure—due to radiation sources in their vicinity but outside their body;
(b) internal exposure—due to accidental intake of radioactive materials into the human body by ingestion and/or by inhalation. Such contamination may be caused by radioactive sources, or activation products associated with the nuclear techniques used.

There are radiation protection measures designed specifically for avoiding or reducing both types of exposure. These measures are outlined in Section 7 of this chapter.

3 PHYSICAL AND RADIOLOGICAL CONCEPTS, PARAMETERS, QUANTITIES AND UNITS USED IN RADIATION PROTECTION

3.1 Types of ionizing radiation

Ionizing radiation, in the context of radiation protection related to the subjects discussed in this book, includes the following: alpha- and beta-particles and gamma- and X-radiations emitted from radioisotopes, primary X-rays (characteristic and bremsstrahlung) from X-ray machines, secondary X-rays from materials exposed to electrons or to gamma- or X-radiations, neutrons (indirectly ionizing) from reactors, accelerators or other neutron sources, and various ion beams from accelerators.

3.2 Penetration range of ionizing radiations into matter

The ranges of the various types of ionizing radiation in biological tissue are of great importance in radiation protection, being related to the concepts of LET and QF explained in pa. Section 3.5.3.

3.2.1 Range of alpha-radiation

Alpha particles lose their energy via ionization and excitation in water or living tissue in a distance ranging up to tens of micrometers, depending in their energy.

3.2.2 Range of beta-radiation

Beta-particles lose their energy mainly via ionization in water or living tissue in a distance ranging up to tens of millimeters, depending on their energy.

3.2.3 Range of gamma-radiation and X-rays

Gamma-radiation and X-rays have no finite range in matter. The beam intensity of gamma-radiation and X-rays decreases exponentially with penetration into matter (see Section 3.8).

3.3 Special unit of energy

The unit normally used to measure the energy of nuclear particles or photons is the electron-volt (eV). This is the energy gained by an electron accelerated by an electric field through a potential drop of 1 V.

$$1\,\mathrm{eV} = 1.6 \times 10^{-12}\,\mathrm{erg} = 1.6 \times 10^{-19}\,\mathrm{joule}$$
$$1\,\mathrm{keV} = 10^{3}\,\mathrm{eV}$$
$$1\,\mathrm{MeV} = 10^{6}\,\mathrm{eV}$$

3.4 Units of activity and specific activity

The unit of activity is the becquerel, abbreviated as Bq. One Bq is the 'quantity' of radioactive material in which one disintegration (nuclear transformation) takes place per second (1 Bq = 1 d.p.s.). In the past, it was customary to measure activity in curie units (Ci). One Ci was defined as the quantity of radioactive material undergoing 3.7×10^{10} disintegrations per second. This is actually the number of disintegrations per second in 1 g of radium. Smaller units were the millicurie (mCi) and the microcurie (μCi).

$$1\,\mathrm{Ci} = 3.7 \times 10^{10}\,\mathrm{Bq}$$
$$1\,\mathrm{Bq} = 2.7 \times 10^{-11}\,\mathrm{Ci}$$
$$1\,\mathrm{MBq} = 27\,\mu\mathrm{Ci}$$
$$1\,\mu\mathrm{Ci} = 37\,\mathrm{kBq}$$

The specific activity is the activity per unit weight or per unit volume measured in units of, e.g., $\mathrm{Bq/cm^3}$, $\mathrm{Ci/cm^3}$, Bq/g, Ci/g etc.

3.5 Radiological quantities and units

3.5.1 Exposure

Exposure is a measure of the ability of gamma- and/or X-rays to cause ionization in air. The unit of exposure is the roentgen or R. One R is defined as the radiation intensity that will cause the production of electric charge, of either sign, equal to 1 e.s.u. per 1 cm^3 of dry air under standard pressure and temperature conditions. This is equivalent to the production of $\approx 2 \times 10^9$ ion pairs per 1 cm^3 of dry air under similar conditions.

$$1\ \text{R} = 2.58 \times 10^{-4}\ \text{C/kg air}$$
$$1\ \text{mR} = 0.258\ \mu\text{C/kg air}$$

The units of exposure rate are R/min, R/h, mR/h, etc.

3.5.2 Dose

The dose is defined as the energy absorbed in a unit mass due to its exposure to radiation (external or internal). The unit of dose is the gray (abbreviated as Gy). One Gy is equivalent to the deposition of 1 J per kilogram material. The former unit used for absorbed dose was the rad. This unit is equivalent to the deposition of 100 ergs of energy in 1 g of material, and 1 Gy is equivalent to 100 rad.

The exposure of a small mass of water (or soft tissue) to 1 R will cause a dose of approximately 1 rad.

$$1\ \text{Gy} = 100\ \text{rad}$$
$$1\ \text{rad} = 10^{-2}\ \text{Gy}$$
$$1\ \text{mGy} = 100\ \text{mrad}$$

The units of dose rate are Gy/min, Gy/h, mGy/h, rad/h, mrad/h, etc.

3.5.3 Dose equivalent

The dose equivalent (DE) is a quantity that measures the biological effect of ionizing radiation. The unit of the DE is the sievert (abbreviated as Sv). The dose equivalent, expressed in Sv units, is obtained by multiplying the absorbed dose in Gy units by a weighting factor specified for different types of ionizing radiation. This factor is called the quality factor (QF). Values of the QF for various types of radiation are given in Table 1.

TABLE 1.
The quality factor value for various types of radiation [1]; ICRP[a] 1977

Type of radiation	Quality factor (QF)
Gamma-rays and X-rays	1
Beta-particles	1
Thermal neutrons	3–6
Fast neutrons	20
Alpha-particles	20

[a] ICRP: International Commission on Radiation Protection.

The QF is related to the linear energy transfer (LET) of the radiation in the tissue, i.e. to the number of ion pairs produced by the particles or photons in unit length of the tissue.

The unit formerly used for the DE is the rem; 1 Sv is equivalent to 100 rem, and 1 millisievert (mSv) equals 100 millirem (mrem).

The units of DE rate are Sv/h, mSv/h, rem/h, mrem/h, etc.

3.5.4 Effective dose equivalent

The effective dose equivalent (EDE) is a quantity measuring the biological risk in exposing to ionizing radiation a single organ or tissue or several organs simultaneously. To calculate the EDE, a weighting factor W_i is defined for each organ. W_i is the relative risk of inducing late (delayed) effects (e.g. malignant tumors) in an organ or tissue T_i following its exposure to a certain DE, as compared to the risk of exposing the whole body to the same DE.

The effective dose equivalent is defined as:

$$EDE = \Sigma_i DE_i \times W_i$$

where DE_i is the dose equivalent to the i-th organ or tissue and W_i is the relative risk of that organ. The values of the weighting factor for some body organs recommended by the ICRP [1] are presented in Table 2.

TABLE 2.
Weighting factors for body organs (relative risk in exposing various body organs to ionizing radiation)

Organ or tissue (i)	Relative risk (W_i)
Gonads	0.25
Breast	0.15
Bone marrow (red)	0.12
Lung	0.12
Thyroid gland	0.03
Skeleton (without bone marrow)	0.03
Other organs[a] (five organs)	0.06/organ
Whole body	1.00

[a] Excluding skin, arms/legs, feet, hands, and ankles.

3.6 The K-factor

The K-factor relates the exposure rate in the vicinity of a radioactive source (emitting gamma-radiation) to the activity of the source. It is defined as the exposure rate (in units of Roentgen per hour) at a distance of 1 m from a source with an activity of 1 Ci. Examples are presented in Table 3.

TABLE 3.
Values of the K-factor[a] calculated for some radionuclides commonly used in chemical analysis by XRF and other nuclear methods

Radionuclide	K-factor ($R\text{-}m^2/Ci\text{-}h$)
^{24}Na	1.83
^{42}K	0.14
^{57}Co	0.10
^{60}Co	1.30
^{125}I	0.04
^{133}Ba	2.00
^{137}Cs	0.32
^{192}Ir	0.44
^{241}Am	0.016

[a] In some professional literature the K-factor is referred to as the specific gamma-ray dose rate constant.

3.7 Physical, biological and effective half-lives (or half-times)

The physical half-life $T_{1/2}(p)$ of a radionuclide is the time required for a given activity of this radionuclide to decrease by 50% due to physical decay.

The biological half life $T_{1/2}(b)$ of a chemical element or compound in the body (or in an organ or tissue) is defined as the time required for a given activity of this element or compound, in the respective organ or tissue, to decrease by 50% by biological excretion.

The effective half-life of a radioactive element or compound in the body (or in an organ or tissue) is defined as the time required for a given activity of this element or compound, in the respective organ or tissue to decrease by 50% due to both biological excretion and physical decay.

The effective half-life $T_{1/2}(\text{eff})$ is related to the physical and biological half-lives by the equation:

$$1/T_{1/2}(\text{eff}) = 1/T_{1/2}(p) + 1/T_{1/2}(b)$$

3.8 Shielding

3.8.1 Half value thickness

A useful parameter in designing shieldings for gamma-radiation and X-rays is the half value thickness (HVT). This is defined as the thickness of absorbing substance that reduces the intensity of a gamma-radiation or X-ray beam to half of its initial value. The HVT depends on the energy of the photons and the type of the shielding material. Some HVT values calculated for water and lead are presented in Table 4.

TABLE 4.
Half value thickness[a] for gamma-rays and X-rays in water and lead for photons of various energies

Radiation energy (MeV)	Shielding material	HVT (mm)
0.1	Water	40
0.5	Water	60
1.0	Water	100
2.0	Water	150
0.1	Lead	0.03
0.5	Lead	4
1.0	Lead	9
1.5	Lead	12
2.0	Lead	14

[a] Approximate values.

Radioactive elements often emit simultaneously several gamma-photons with energies and relative frequencies characteristic of the isotope; hence it is useful to determine HVT values not only for specific photon energies, but rather for specific radioisotopes. HVT values for several radioisotopes are presented in Table 5.

TABLE 5.
Half value thicknesses in lead and steel for some commonly used radioisotopes; NCRP 1976 [2]

Isotope	Energy	Half value thickness (mm)		
	(MeV)	Lead	Steel	Concrete
^{137}Cs	0.66	6.5	16	48
^{60}Co	1.17–1.33	12	21	62
^{192}Ir	0.3–0.6	6.0	13	43
^{226}Ra*	0.1–2.4	16.6	22	69

X-ray machines emit radiation with a continuous energy spectrum. The shape of this spectrum and the exposure rate in the direct beam depend on the maximum voltage (kV_p), the current and the filtration of the tube.

HVT values in lead and in concrete for radiation from X-ray equipment used for industrial and diagnostic purposes are presented in Table 6. Typical values of the radiation exposure rates at a distance of 1 m from the tube are also shown [2].

4 RELATION BETWEEN EXPOSURE, ACTIVITY, DISTANCE AND TIME

The exposure rate near a radioactive source is directly proportional to the activity of the source and inversely proportional to the square of the distance

TABLE 6.
Radiation output and HVT values for X-ray beams

Maximum voltage (kV_p)	Radiation intensity at a distance of 1 m (R/mA-min)	HVT (mm)[a]	
		Lead	Concrete
70	0.30	0.17	8.4
100	0.70	0.27	16
150	1.1	0.30	22
200	1.5	0.52	25
250	2.0	0.88	28
300	3.0	1.47	31

[a] For high attenuation.

from the source. The exposure is directly proportional to the exposure rate and to the time of exposure.

5 BIOLOGICAL EFFECTS OF IONIZING RADIATION

The biological effects of human exposure to ionizing radiation are usually of two main types:

(a) somatic effects (health detriment to the exposed person himself/herself);
(b) genetic effects (hereditary damage to the exposed person's offspring).

5.1 Somatic effects

5.1.1 Acute effects (also called deterministic or non-stochastic effects)

These are effects that appear as a consequence of the exposure of the whole body or single organs to high doses of radiation delivered within a short time (minutes to days). Acute effects have a threshold dose value that depends on the type of the effect. The severity of acute effects (above the threshold dose) increases with the increase in the dose.

Exposure of the whole body of very high doses of radiation is likely to cause death within minutes, hours or days following the exposure, the time delay depending on the dose.

Approximately 50% of those exposed to a whole body dose of ≈ 4 Gy (400 rad) will die from acute radiation illness within several days to a few weeks after the exposure (if left untreated).

Whole body doses exceeding 1 Gy (100 rad) are likely to cause acute radiation sickness (nausea, vomiting, diarrhea, loss of hair) commencing a number of minutes after the exposure (at doses of 10 Gy or more) up to a number of hours following exposure to lower doses.

Other acute effects, caused by exposure to high doses of ionizing radiation, include [3]:

- Temporary sterility in males, with a threshold acute dose of 0.15 Gy in the testes and a dose rate of 0.4 Gy y^{-1} for prolonged exposure; permanent sterility, with a threshold dose for acute exposure of the gonads of 3.5 Gy in males and 2.5 Gy in females.
- Opacities in the lenses of the eyes, sufficient to cause impairment of vision, with a threshold dose in the range 2–10 Gy for acute exposure to low LET radiation and a threshold dose 2 to 3 times lower for high LET radiations.
- Clinically significant depression of the blood forming process, with a threshold value for acute absorbed doses in the whole bone marrow of about 0.5 Gy. The dose rate threshold for protracted exposure over many years is 0.4 Gy y^{-1}.

5.1.2 Late (delayed) effects

Radiation doses below 0.1 Gy (10 rad) appear to have no immediate somatic effects, but there is a potential risk of late (delayed) effects.

Of all of the delayed somatic late effects, the most significant and prominent is the manifestation of malignant illnesses in the exposed person 5–40 years following a single exposure or following the onset of chronic exposure. Manifestation of the damage is stochastic (random), whereby the risk of actual manifestation of harm (and not the severity of harm) is proportional to the radiation dose.

There is a controversy among researchers regarding the dose–response relationship for these stochastic effects, but the vast majority agree [1, 4] that the relationship is linear–quadratic in form, with no threshold for low LET radiation (low LET radiation includes X-rays, gamma-rays and beta-rays) and for low doses, i.e., the damage is inherent even in the lowest doses but the probability of deleterious manifestation, per dose unit, increases with the increments of dose with a linear and quadratic component, i.e. $E = aD + bD^2$, where E = effect, D = dose, and a and b are constants depending on age at the time of exposure, genotype etc. The types of cancer attributed to ionizing radiation are primarily leukemia, breast cancer, lung cancer, cancer of the thyroid, and bone cancer. It is not possible to distinguish between cancers caused by radiation and those induced by chemicals and/or other agents. The chances that ionizing radiation will induce malignant disease depend on the age of the exposed person, his/her sex, and his/her affiliative genotype. Furthermore, there is an effect of the time span during which a particular cumulative dose has been delivered and the part of the body that absorbed the dose. The various organs differ in their susceptibility to radiation.

The quantitative approximation related to the cumulative excess number of cancers per unit radiation dose equivalent (i.e., the additional cancer morbidity and/or mortality in the exposed population per unit dose equivalent) ranges from a mortality of 70–230 and a morbidity up to three times these values in one million persons exposed to a dose equivalent of 10 mSv (one rem) (UN 1977; BEIR 1980). The borderlines for the estimate depend on the model employed to project the probability of cancer induction to the lifetime of the exposed population. Among many possibilities, the ICRP has suggested two principal models. One is the 'absolute risk' or additive projection model and the other is the 'relative risk' or multiplicative projection model. The former predicts a constant excess of induced cancer cases per year throughout the remaining life of the exposed population, whereas the latter predicts that the excess of induced cancers will increase with time as a constant multiple of the spontaneous (or natural) rate of cancer. This means, of course, that the excess rate will increase with the age of the exposed population.

According to the 1977 recommendations of the ICRP, a good basis for a quantitative risk assessment is to assume a cumulative increase of 125 cancer fatalities (during 5–50 years following the exposure) as a result of a short-term exposure of 1,000,000 individuals to a dose of 10 mSv (1 rem). The Commission did not refer to morbidity [1].

Partial exposure of the body and exposure of individual organs incurs a lower risk. For example, exposure of the lungs to a particular dose of radiation entails a risk eightfold smaller than the risk involved by exposing the entire body to the same dose (see Table 2 above). These estimates were based on the 'absolute risk' model.

NOTE. In the recent 1990 recommendations of the ICRP [3], the International Commission announced that the risk was apparently up to 3 times the risk anticipated in their 1977 recommendations [1]. They now assess that the short-term exposure of a worker population of 10^6 to an effective dose equivalent of 10 mSv (1 rem) will induce up to 400 additional fatal cancers in that population. This recent estimate is based on the 'relative risk' model and on new information related to the excess cancer mortality among the victims of Hiroshima and Nagasaki.

Based on this new estimate, we can conclude that the chronic exposure of a large portion of the population aged 18–50 years to an annual whole body dose of 1 rem will cause an increase of $\approx 20\%$ in the number of cancer fatalities, as compared with a population that is not exposed to ionizing radiation above the natural background. This represents an increase of about 40 cancer fatalities annually in a population of 100,000 workers exposed year after year to a dose of 1 rem.

5.2 Genetic damage

Among the genetic damage attributable to ionizing radiation, the two major types are gene mutations and chromosomal aberrations. These are phenomena that also occur spontaneously and as a result of environmental and other factors.

According to the assessments in the 1977 recommendations of the ICRP, the increase in genetic defects in the first generation of offspring of exposed persons is estimated at approximately 30% of the increase in the number of cancer fatalities attributable to radiation; in other words, in addition to an annual increase of 125 in cancer fatalities in a population of 1 million individuals exposed to a dose of 10 mSv (1 rem) each year, there will be an annual excess of approximately 40 cases of genetic mutations and severe birth defects in the same population.

6 RESTRICTIONS ON EXPOSURE TO IONIZING RADIATION—INTERNATIONAL STANDARDS

Based on the assumption that any radiation dose, no matter how tiny, is capable of causing somatic and genetic damage, as explained above, the international regulatory agencies have set an all-inclusive policy and quantifiable standards designed to reduce radiation damage as far as possible, while simultaneously enabling the effective use of nuclear and other technologies involving possible exposure of workers and the general population to ionizing radiation. These standards were jointly published in 1982 by the World Health Organization (WHO), the International Labour Organization (ILO), the International Atomic Energy Agency (IAEA), and the European Organization for Economic Cooperation (OECD) under the name 'Basic Safety Standards for Radiation Protection' (BSSRP). The standards [5] are based on the recommendations of the ICRP [1] and on the findings of the BEIR and UNSCEAR Committees [6, 7]. Besides the inclusion of quantified values for maximum annual radiation doses, the standards also prescribe three fundamental principles, as follows.

(a) Justification

Approval to use new technologies, instruments, devices and products, the implementation of which is likely to expose individuals, groups and/or a population to ionizing radiation, will be granted by the legal authorities only after it has been proven that the proposed use produces an actual (net) benefit to society.

(b) Optimization of radiation protection

After having proved clearly the justification for using a particular technology, instrument and/or device, all measures must be taken to keep human exposure to ionizing radiation as a result of the aforestated use as low as reasonably achievable (ALARA) and as low as reasonably practicable (ALARP). The minimum dose for targeting will be determined by performing cost–benefit analyses of the optimization of radiation protection, taking into consideration the economic and social cost of using this radiation protection on one hand and the quantitative risk averted on the other hand.

(c) Dose limitations

After having concluded the procedures of justifying and planning radiation protection, and having determined the optimally expected radiation dose (according to this planning), it must be verified that workers and individuals in the population will under no circumstances be exposed to radiation doses exceeding certain limits as specified below.

6.1 Annual dose limits for workers

These limits are presented in Table 7.

Exposure of the gonads, breast, bone marrow or lung to the DE shown in Table 7 is equivalent to the exposure of the whole body to a DE of 50 mSv (5 rem). If several organs are exposed simultaneously, the EDE must be calculated. The calculation is made by multiplying the DE in each of the exposed organs by the relative risk factor of that organ as given in Table 2 above, and summing the weighted dose equivalents of the organs exposed simultaneously (see Section 3.5.4 above).

Considering the fundamental principles (a) and (b), i.e. the requirement for justifying the practice and keeping doses as low as reasonably achievable, it is clear that the maximum doses specified in Table 7 are not 'permissible doses' and these values should not be used as a basis for planning routine work activity. The dose limits are primarily values representing the upper boundaries which are on no account to be exceeded. However, in advance planning, provision (b) is prescribed, namely, exposures must be kept as low as reasonably achievable. In practical terms, it is desirable to plan the work so that the whole body dose

TABLE 7.
Annual dose limits for persons engaged in radiation at their workplace according to the BSSRP [5]

	Annual dose limit	
	mSv	(rem)
Whole body	50	(5)
Body part:		
Eyes	150	(15)
Gonads	200	(20)
Breast	300	(30)
Bone marrow/lung	400	(40)
Thyroid gland/single organs	500	(50)

anticipated per worker during routine work activity will not exceed 5 mSv (500 mrem) per year for occupationally exposed workers or 0.5 mSv (50 mrem) per year for workers not involved with radiation at their workplace; hence, for work that is planned to be executed regularly, 40 hours per week, the dose rate should not exceed 2.5 μSv (0.25 mrem) per hour for occupationally exposed workers or 0.25 μSv (0.025 mrem) per hour for workers not involved with radiation at their workplace.

NOTE. In 1991, the ICRP published new recommendations [3] proposing the need to restrict the whole body radiation dose for occupationally exposed persons to an EDE of 20 mSv per year averaged over 5 years (100 mSv in 5 years), with the further provision that the EDE should not exceed 50 mSv in any single year.

6.2 Dose limits for members of the public

The dose limits for the general public (and for workers not involved with radiation at their workplace) are 50 times lower than the corresponding limits for occupationally exposed workers shown in Table 7.

6.3 Dose limits for pregnant women

According to the 1977 recommendations of the ICRP [1], the exposure of a pregnant woman must be limited to 10 mSv (1000 mrem) during the whole period of her pregnancy. The new ICRP recommendations [3] suggests lowering this limit to 2 mSv (200 mrem).

6.4 Medical exposures

The annual dose limits do not include medical exposure, e.g., diagnostic X-ray examinations. For example, a series of X-rays for gastrointestinal diagnostic purposes can cause a dose in the range of 0.5–1.5 rad to the organs involved. One single X-ray causes a partial body dose of a few tens to a few hundred mrad. In radiotherapy procedures, single organs may be exposed to doses of the order of a few tens of Gy (a few thousand rad).

6.5 Emergency exposures

In the event of an incident or accident such as the unintentional release of a sealed radioactive source from its shielding and/or spillage of radioactive material, the urgent need might arise to expose workers or rescue teams to radiation doses exceeding the annual dose limits so as to save lives, prevent physical injury, avert the spread of contamination, and/or salvage special and highly expensive property. In such cases it is permissible, according to the 1977 recommendations of the ICRP, to expose workers to a whole body DE of up to 0.1 Sv (10 rem). A worker should not accumulate from such exposures more than 0.25 Sv (25 rem) throughout his/her lifetime. Women of child-bearing age and adolescents under 18 should not participate in operations of this kind.

6.6 Internal contamination—limits of intake

When working with open (unsealed) radioactive materials, or in the case of incidents involving the dispersion of gaseous, liquid or particulate radioactive material, internal contamination is a potential hazard. The dose resulting from internal contamination depends on the metabolism of the substance, its biological half-life and the organs in which it concentrates. It is common practice to define a critical organ for each particular substance, the inference being to the organ most sensitive to radiation from a certain radioactive material (it is usually also the organ in which the material concentrates). The critical organ can be the entire body, e.g. for tritium (the whole body tissue excluding bones), the bones, e.g. for ^{241}Am, the thyroid gland, e.g. for iodine isotopes, etc.

Another significant factor in determining the radiation dose from internal contamination is the biological half-life, which is defined as the time it takes for the body or a particular organ to eliminate, by biological means, 50% of the quantity of the substance taken up by it.

TABLE 8.
ALI values for several radioisotopes according to ICRP Publication No. 30 (the most restrictive ALI is indicated for the radionuclides listed)

Radionuclide	ALI[a] for inhalation		ALI[a] for ingestion	
	Bq	μCi	Bq	μCi
^{3}H	3×10^{9}	8×10^{4}	3×10^{9}	8×10^{4}
^{14}C	9×10^{7}	2.4×10^{3}	9×10^{7}	2.4×10^{3}
^{24}Na	2×10^{8}	5.4×10^{3}	1×10^{8}	2.7×10^{3}
^{42}K	2×10^{8}	5.4×10^{3}	2×10^{8}	5.4×10^{3}
^{57}Co	2×10^{7}	5.4×10^{2}	2×10^{8}	5.4×10^{3}
^{60}Co	1×10^{6}	30	7×10^{6}	200
^{125}I	2×10^{6}	50	1×10^{6}	30
^{137}Cs	6×10^{6}	160	4×10^{6}	110
^{241}Am	2×10^{2}	0.005	5×10^{4}	1.4

[a] Rounded values.

In 1979–1980 the ICRP issued its Publication No. 30 [8] related to the quantities of radioactive material which, when taken up in the body, would cause an EDE (i.e., weighted radiation dose, according to the relative risk of the exposed organs) of 50 mSv (5 rem) or a radiation dose of 500 mSv (50 rem) to any single organ. Accordingly, limits were set on the annual intake of radioactive materials into the human body. The maximum quantities prescribed by means of this calculation are called annual limits of intake (ALI). Several ALI values are presented in Table 8.

7 METHODS OF RADIATION PROTECTION

7.1 External radiation—time, distance and shielding

(a) Time

Even when it is imperative that a person works in a radiation field, the external radiation can be minimized by keeping the exposure time as low as reasonably achievable. Staying for 2 h in a radiation field will cause an exposure that is twice that incurred by staying only 1 h in the same field.

(b) Distance

Keeping a distance from the radiation source will reduce exposure. At a distance of 0.5 m from a radiation source the exposure rate is fourfold greater than that incurred at a distance of 1 m from the same source. As a rule, the exposure rate due to a particular 'point' radiation source decreases as the square of the distance from that source. Hence it is always desirable not to handle the radiation source with one's hands. The use of tongs or forceps to handle the source will reduce exposure considerably. Tongs or forceps 10 times longer can reduce exposure 100-fold.

(c) Shielding

It is always possible to reduce exposure from an external source by means of shielding. A layer of material with a thickness equivalent to 10 HVTs will reduce the radiation intensity approximately 1000-fold ($2^{10} = 1024$). The shielding effect depends on the shielding material, as shown in Tables 4–6 above.

7.2 Internal contamination

Internal contamination must be avoided completely. Working with radio-isotopes in a laboratory that was not designed for operations with radioactive materials will almost invariably result in contamination. The use of protective clothing (disposable gloves, caps, overalls, special shoes etc.), elbow-operated taps, soap dispensers, foot-operated waste receptacles, disposable paper towels and easily washable smooth walls and working surfaces is essential for work with open (unsealed) radioactive materials. It is absolutely forbidden to eat, drink or smoke in laboratories of this kind.

8 MONITORING AND CONTROL

In order to ensure that a worker will not be exposed throughout the year to an accumulative radiation dose exceeding the annual limit, the employer must make provisions for means of monitoring and control. The monitoring provision will be personal if the worker is likely to be exposed throughout the year to a cumulative dose exceeding 10% of the annual dose limit. If work conditions assure that the worker will in no circumstances be exposed throughout the year to a cumulative radiation dose exceeding 10% of the annual dose limit, monitoring of the work area alone may suffice.

The radiation doses will be registered on the worker's personal file. The employer will notify the worker and the occupational physician of the level of exposure.

9 NATURAL BACKGROUND RADIATION

To put the risks of occupational exposure to ionizing radiation into perspective, we have to realize that all habitants of planet earth are exposed to ionizing radiations from natural sources. The following is a short review of the natural background radiation [7].

Natural radiation emanates from three major sources:

(1) cosmic radiation;
(2) radiation from terrestrial sources;
(3) internal radiation resulting from radioactive isotopes in the human body.

Cosmic radiation enters the atmosphere from interstellar space. It comprises mainly high energy protons. Upon reaching the atmosphere, these protons cause secondary radiation by ionization and nuclear interactions; high energy radiation composed of electrons, neutrons, mesons and electromagnetic radiation is thereby generated. The average effective dose equivalent (EDE) from such primary and secondary cosmic radiation ranges from 0.3–0.4 mSv (30–40 mrem) per year at sea level and increases approximately twofold for each 2000 m altitude, depending also on the terrestrial magnetic field.

The source of *terrestrial radiation* is natural radioactive isotopes, e.g. thorium, uranium and their decay products and potassium are found in rocks, and soil. Construction materials also contain various concentrations of natural radioactive ores. Some of these substances emit gamma-radiation. The average EDE for a human being from these sources is ≈ 0.5 mSv (50 mrem) per year, ranging from 0.3–1.5 mSv (30–150 mrem) per year according to the location and type of structure.

In addition to radiation from external sources, man is exposed to *internal radiation* resulting from natural radioactive isotopes inherent in his own body composition. The major contributor to this radiation is ^{40}K, a naturally occurring radioactive isotope of potassium. Also, tritium (^{3}H) and ^{14}C are commonly found in the human body. In addition, ^{226}Ra and its decay products are taken up by the body in meager quantities via food and drinking water.

The EDE due to internal exposure caused by potassium-40 (^{40}K) and the other radionuclides in the human body is approximately 0.2 mSv (20 mrem) per year. To this internal radiation dose we have to add an additional EDE of 0.8–1.2 mSv (80–120 mrem) per year as a result of inhaling radon gas and its decay products (caused by an average ^{222}Rn concentration of ≈ 40 Bq/m^3 in dwellings).

Therefore the natural radiation to which the popoulation of this planet is exposed reaches ≈ 2 mSv. (200 mrem) per year on the average. Approximately 0.8 mSv (80 mrem) is supplied by radiation sources outside the human body, whereas ≈ 1.2 mSv (120 mrem) more originates from internal radiation, from radon and its decay products, and from the disintegration of potassium-40. The range of exposure to natural radiation is normally between 1.5 and 3 mSv (150–300 mrem) per year, depending on the geomagnetic meridian, the altitude, deposits of uranium and thorium in the soil, and the construction materials in the immediate surroundings. In particular areas, background radiation may reach 5 mSv and even 8 mSv per year (500–800 mrem per year).

REFERENCES

1. International Commission on Radiological Protection, Recommendations, *Ann. ICRP* **1** (3), *ICRP Publ.* (26) (1977).
2. National Council on Radiation Protection and Measurements, Structural shielding design and evaluation for medical use of X-rays and gamma-rays of energies up to 10 MeV, *NCRP Rep.* (49) (1976).
3. International Commission on Radiological Protection, Recommendations, *Ann. ICRP* **21** (1–3), *ICRP Publ.* (60) (1991).
4. United Nations Scientific Committee on the Effects of Ionizing Radiation (UNSCEAR), Sources, effects and risks of ionizing radiation, *UNSCEAR Rep.*, United Nations, New York (1988).
5. International Atomic Energy Agency. Basic safety standards for radiation protection, *Saf. Ser.–I.A.E.A* (9) (1982).

6. BEIR, The Effects on Populations of Exposure to Low Levels of Ionizing Radiation, National Academy of Sciences, Washington, DC (1980).
7. United Nations Scientific Committee on the Effects of Ionizing Radiation (UNSCEAR), Sources and Effects of Ionizing Radiation, United Nations, New York (1977).
8. International Commission on Radiological Protection, Limits of intake of radionuclides by workers, *ICRP Publ.* (30); Part 1, *Ann. ICRP* **2** (3–4) (1979); Part 2, *Ann. ICRP* **4** (3–4) (1980).

Part 2

Elemental Analysis with Neutron Sources

Chapter 6

DELAYED INSTRUMENTAL NEUTRON ACTIVATION ANALYSIS

Sheldon Landsberger

Department of Nuclear Engineering, University of Illinois, 214 Nuclear Engineering Laboratory, 103 South Goodwin Ave., Urbana, IL61801, USA

1 INTRODUCTION

Neutron activation analysis (NAA) continues to be an essential tool for routine trace elemental determinations and in many areas of innovative research. During the 1960s, NAA was unsurpassed as a technique in achieving excellent sensitivity for a wide range of trace elements while still having the unique characteristic of being non-destructive. With the advent of newer multi-elemental chemical methods such as inductively coupled plasma–atomic emission spectroscopy (ICP–AES) and, more recently, ICP–mass spectrometry (ICP–MS), proton-induced X-ray emission (PIXE), and other advances in energy dispersive X-ray fluorescence (ED–XRF), there has been much competition with neutron activation analysis. Besides being capable of determining 40–50 elements, including many environmentally crucial ones (e.g. antimony, arsenic, cadmium, chromium, copper, mercury, nickel, selenium, vanadium, zinc, etc.), NAA is also capable of determining major elements such as sodium, chlorine, and potassium, as well as many rare earth elements (critically important in various geochemistry studies).

The determination of elemental concentrations by NAA is based on the measurement of induced radioactivity. The radioactive decay of each element emits a characteristic gamma-ray energy spectrum, and hence an individual nuclear 'fingerprint' can be measured and quantified. Activation analysis measures the total amount of an element in a material without regard to its chemical or physical form and has the following advantages: (1) samples for NAA

Chemical Analysis by Nuclear Methods Edited by Z. B. Alfassi

can be liquids, solids or powders; (2) NAA is non-destructive and, since no pre-chemistry is required, reagent-introduced contaminants are completely avoided; (3) NAA is a multi-elemental analytical technique in that many elements can readily be determined simultaneously; (4) the final significant advantage of NAA is its sensitivity to trace elements. These factors have pushed the detection limits of many elements of interest to very low levels not readily achievable by other analytical techniques. NAA is also totally unaffected by the presence of organic material in the sample. Organic material is a significant matrix problem in many types of conventional chemical methods.

A detailed description of the theory and application of NAA is beyond the scope of this chapter. However, a concise description is given in the first part of this book. Since the middle 1980's several well written and informative books have been published on this technique and other related topics. These include Kolthoff and Elving [1], Heydorn [2], Csikai [3], Tolgyessy and Klehr [4], Das, Faanhoff and Van der Sloot [5], Alfassi [6] and Tolgyessy and Kyrs [7]. More recently, two books, one on radiochemistry and nuclear methods by Ehmann and Vance [8] and *Activation Spectrometry in Chemical Analysis* by Parry [9], have appeared. Special mention should be given to the in-depth reviews on nuclear and chemical analysis compiled by Ehmann and co-workers [10–13]. The International Atomic Energy Agency (IAEA) [14] released a document on the practical aspects of operating a NAA laboratory, as well as other documents on activation analysis techniques [15, 16]. Two books pertaining to recent developments and overviews on radiation detection and measurements [17] and on gamma-ray and X-ray spectrometry [18] have also been published.

The rest of this chapter is to introduce the basic theory of delayed neutron activation analysis while focussing on several specific aspects, including nuclear and spectral interferences, accuracy and precision, gamma-ray self-absorption, sensitivity and detection limits, quality control, and the use of reference materials. A few examples showing the various gamma-ray spectra produced by neutron-induced reactions are also presented, with a list and discussion of the most commonly used radionuclides and their respective half-lives and gamma-ray energies. No attempt is made to deal with any aspects of fast neutron, prompt gamma or charged particle NAA, or radiochemical NAA with either pre- or post-separation methods.

2 THEORY OF NEUTRON ACTIVATION ANALYSIS

NAA can be divided into non-destructive and destructive analysis, i.e. instrumental (INAA) and with radiochemical separation, respectively. Although the detection limits in the latter are usually at least one order of a magnitude lower than in the former, instrumental NAA still plays an important role in the field because no time-consuming separation work is needed. In particular, NAA is ideally suited for the analysis of small sample sizes, such as air-borne particulate matter, which can commonly have weights less than a milligram.

Neutrons used in NAA can be produced in several ways: by a nuclear reactor, or a particle accelerator or from artificial isotopes, such as plutonium-beryllium. The most common source is a fission reactor, owing to its high neutron flux. Usually the reactor neutrons can be used at thermal and epithermal energies. Epithermal neutrons can be obtained by using a filter which is made of material with a high neutron absorption for low energy neutrons, such as cadmium and boron, so that virtually all low energy neutrons are absorbed. Therefore only higher energy neutrons, known as epithermal neutrons, activate the sample. Fast neutrons from particle accelerators can also induce reactions; this technique has been adequately covered by Csikai [3] and will not be considered further in this chapter.

2.1 Neutron-induced reactions

When a neutron collides with a nucleus, the following reactions may occur: elastic scattering (n, n), inelastic scattering (n, n′), radiative capture (n, γ), charge particle reaction (n, α) or (n, p), (n, 2n), and fission (n, f). The most useful reaction in NAA is radiative capture, which can be denoted as:

$$n + {}^{A}Z \rightarrow {}^{A+1}Z^{*} \rightarrow {}^{A+1}Z + \gamma \tag{1}$$

where ${}^{A}Z$ is the target nucleus and ${}^{A+1}Z^{*}$ is called the compound nucleus and is usually in an excited state. It can de-excite with the emission of a gamma-ray called a prompt gamma. In normal NAA this gamma-ray is not used in the analysis; instead, delayed gamma-radiation from the product ${}^{A+1}Z$ is employed. This means that the product ${}^{A+1}Z$ is required to be radioactive, emitting at least one gamma-ray photon. It can be detected by a high resolution gamma-ray detector for both energy and intensity. If the product ${}^{A+1}Z$ is a stable isotope, it cannot be detected. For example, although magnesium has in nature three stable isotopes, ${}^{24}Mg$, ${}^{25}Mg$, and ${}^{26}Mg$, only the last one, which has an abundance of only 11%, can be used in neutron activation analysis employing the ${}^{26}Mg$ (n, γ) ${}^{27}Mg$ reaction. Neutron-induced reactions of ${}^{24}Mg$ and ${}^{25}Mg$ give the stable isotopes ${}^{25}Mg$ and ${}^{26}Mg$, respectively.

Since stable isotopes of an element can undergo different reactions depending on the neutron energy, it is possible that the same product ${}^{A+1}Z$ can be produced via different nuclear reactions. For example, ${}^{28}Al$ is produced by the following three reactions:

$${}^{27}Al + n \rightarrow {}^{28}Al + \gamma \tag{2}$$

$${}^{28}Si + n \rightarrow {}^{28}Al + p \tag{3}$$

$${}^{31}P + n \rightarrow {}^{28}Al + \alpha \tag{4}$$

The first reaction (2) is the expected reaction in thermal NAA; (3) and (4) are referred to as primary nuclear interference reactions.

2.2 Reaction rate

Neutron capture cross-section is defined as the probability of a radiative capture reaction occurring in a neutron collision with a nucleus given in terms of an area, which depends on the incident neutron energy. The reaction rate R of a particular element in the sample can be evaluated from:

$$R = N \int_{\substack{\text{whole}\\ \text{energy}\\ \text{range}}} \sigma(E)\Phi(E)\,\mathrm{d}E \tag{5}$$

where $\Phi(E)\,\mathrm{d}E$ is the neutron flux of neutrons with kinetic energy between E and $E + \mathrm{d}E$, in n $\mathrm{cm}^{-2}\,\mathrm{s}^{-1}$, $\sigma(E)$ is the neutron capture cross-section, in cm^2, and N is the number of atoms of the element in the sample.

Equation (5) can be separated into three terms: thermal, epithermal, and fast neutron ranges:

$$R = R_{\text{th}} + R_{\text{epi}} + R_{\text{fast}} \tag{6}$$

For many nuclides, particularly those of low atomic number, the neutron capture cross-section decreases linearly with increasing velocity of the neutron (known as $1/v$ absorbers) for low energies (<1 eV). Thermal neutrons have a velocity of 2200 m/s, corresponding to an energy of 0.025 eV. The thermal reaction rate is:

$$R_{\text{th}} = N \int_{\text{thermal}} \sigma(E)\Phi(E)\,\mathrm{d}(E)$$
$$= N\sigma_{\text{th}}\Phi_{\text{th}} \tag{7}$$

where Φ_{th} is the thermal neutron flux in n $\mathrm{cm}^{-2}\,\mathrm{s}^{-1}$. At medium energies (0.5 eV $< E <$ 0.5 MeV), the neutron capture cross-section varies rapidly, with resonance peaks. It is assumed that the neutron flux follows a $1/E$ distribution in the epithermal range (see Chapter 1). In practice, the neutron radiative capture resonance integral (I_0) is used to refer to the neutron capture cross-section in the medium energy range. The epithermal reaction rate can be evaluated by:

$$R_{\text{epi}} = N \int_{\text{epithermal}} \sigma(E)\Phi(E)\,\mathrm{d}E$$
$$= N \int_{\text{epithermal}} \sigma(E)\Phi_{\text{epi}}\,\mathrm{d}E/E$$

$$= N\Phi_{epi}\left[\int_{epithermal} \sigma(E)\,dE/E\right]$$

$$= N\Phi_{epi}\left[\int_{epithermal} \sigma(E)\,d(\ln E)\right]$$

$$= N\Phi_{epi} I_0 \qquad (8)$$

where Φ_{epi} is the epithermal flux per unit (ln E). At high energy, the neutron capture cross-section is usually very small, whereas the cross-sections for other reactions such as (n,p) are dominant and the fast radiative capture reaction rate R_{fast} is negligible. Values for the thermal neutron capture cross-section and the resonance integral are given by Mughabghab, Divadeenam and Holden [19] and IAEA [20]. The reaction rate can thus be expressed as:

$$R = N\sigma_{th}\Phi_{th} + NI_0\Phi_{epi} \qquad (9)$$

In neutron irradiations, the reaction rate is governed by the two terms shown in equation (9). The first term is usually dominant in thermal irradiations, but in epithermal irradiations this term becomes less important. Some elements which have high ratios of I_0/σ_{th} include antimony, arsenic, barium, cadmium, indium, gallium, molybdenum, selenium, tungsten and uranium. The epithermal technique is often useful for samples containing sodium, since its activity is greatly reduced in epithermal NAA, resulting in lower background and better sensitivities. A detailed evaluation of the benefits of epithermal NAA has been made by Rowe and Steinnes [21], and several other authors have successfully used this method in a wide variety of investigations [22–27].

3 NEUTRON ACTIVATION ANALYSIS CALCULATIONS

As we mentioned in Section 2.1, neutron activation analysis is based on the measurement of characteristic gamma-rays from a radionuclide formed by a specific neutron reaction. The corresponding radioactivity is governed by the usual radioactive decay law. If other factors are considered in the measurement, the final equation would be [7]:

$$A = \sigma\Phi(m/M)N_A S D C \theta P_\gamma \varepsilon \qquad (10)$$

where A is the measured activity (Bq) from the product of an expected reaction, σ is the activation cross-section of the reaction (cm^2), Φ is the activating flux (n cm^{-2} s^{-1}), m is the amount of the element determined (g), M is the atomic weight of the element to be determined (g/mol), N_A is the Avogadro constant (6.022×10^{23} molecules/mol), $S = [1 - \exp(-\lambda t_1)]$ is the saturation factor (λ is the decay constant of the radioactive product, t_i is the duration of irradiation),

$D = \exp(-\lambda t_d)$ is the decay factor (t_d is the duration of the decay), $C = [1 - \exp(-\lambda t_c)]$ is the correction factor for nuclide decay during the counting time (t_c is the duration of counting), θ is the relative natural isotopic abundance of the activated isotope, P_γ is the probability of emission of a photon with energy E, and ε is the detector efficiency for the measured radiation energy.

The above equation can be greatly simplified by using the comparator method. If equal weights of both sample and standard (with a known concentration of the element of interest) have the same irradiation, decay and counting times, then equation (10) becomes:

$$C_{sam} = C_{std}\,(A_{sam}/A_{std}) \tag{11}$$

where C_{sam} is the unknown concentration of the element in the sample, C_{std} is the known concentration of the element in the standard, A_{sam} is the activity of the sample, and A_{std} is the activity of the standard.

In this case, the unknown concentration of the element in the sample can be obtained by comparing the activities of the gamma-ray peaks. However, usually the terms D and C in equation (10) have to be used as well, since they often vary. Normalization of the weights between standards and unknowns must also often be done. The overall equation is then as follows:

$$C_{sam} = C_{std}(A_{sam}/A_{std})(D_{std}/D_{sam})(C_{std}/C_{sam})(W_{std}/W_{sam}) \tag{12}$$

where W_{sam} and W_{std} are the weights of the sample and the standard, respectively.

There are essentially two different ways to calibrate the NAA system for elemental analysis. The first involves the use of liquid standards which are prepared from either commercially available solutions or the basic chemicals. If the basic chemicals are chosen, extreme care should be given to the stoichiometry of the chemical, including any hydrated forms. Inorganic chemicals recommended for use in calibration have been reviewed [28]. All chemicals used should be over-dried in order to remove moisture. Both polyethylene and quartz vials can be utilized for the encapsulation of liquid samples. Liquid standards are pipetted into acid-washed vials, which are heat-sealed. If polyethylene vials are chosen, liquid samples should be doubly encapsulated for irradiation owing to inherent leakage. The heat-sealed vial needs to be placed in a larger vial which is then also heat-sealed. If a larger vial cannot be used, the heat-sealed smaller vial can be wrapped and taped in polyethylene sheeting or heat-sealed in a small polyethylene bag. After irradiation, all liquid standards are transferred to labelled inert vials for counting. In some reactors there are no facilities available for liquid irradiations, or the neutron flux is so high as in turn to cause evaporation and extensive leaking due to thermal heating.

In the comparator method, calibration can also be done by pipetting a known amount of one or more elements onto a filter paper. This method is commonly used by several factors must be taken into consideration. The filter paper used should contain only minimal concentrations of contaminants when compared with the amount pipetted (hence the need for trace element characterization of the filter paper before use). When combining two or more standard solutions of two elements, it has to be ascertained that the solution of one element does not contain trace or minor quantities of the second element. For instance, it is quite conceivable that commercially bought or laboratory prepared compounds may contain unspecified quantities of other elements. Filters need to be placed in acid-washed vials for irradiation, then transferred to labeled inert vials prior to counting. However, some types of filter paper may not be transferable because of their condition following irradiation, particularly for high neutron flux conditions. Filter papers can also be prepared in the form of a pellet, but care must be taken to prevent contamination from the apparatus used for the procedure. The advantage of pelletization is that an exact geometry can be reproduced for irradiation and counting procedures.

Another means of calibration in the comparator method involves the use of reference materials. In general, reference materials are not used for the calibration procedure but rather for testing the analytical accuracy of the methods and techniques. Many geological or biological reference materials have trace and minor elements certified with errors of up to $\pm 20\%$ (and in some cases even more), or may not be certified with known elemental concentrations. This means that the final derived elemental concentrations in the actual samples will have accuracies no better than the reported errors of the individual elemental concentrations. The existence of spectral interferences may also limit the application of certified reference material made from geological material. For instance, the 136.0 keV and 264.7 keV gamma-rays coming from the $^{74}Se(n,\gamma)^{75}Se$ activation product are susceptible to substantial interference from the same photopeaks arising from the $^{180}Hf(n,\gamma)^{181}Hf$ and $^{181}Ta(n,\gamma)^{182}Ta$ reactions, respectively. Other elements, such as nickel, exhibit poor sensitivities in thermal NAA and therefore should not be used in any calibration procedure.

The second method of calibration is the k_0 method. Essentially, this is an absolute technique in which the amount of material is calculated from the measured counts by means of equation (10). The nuclear constants of each nuclide are combined together to give the k_0 constant. As there are some uncertainties in the nuclear constants, the k_0 values are determined experimentally [29]. Several activation groups world-wide are pursuing this method; however, there are important considerations to be taken into account. These include abnormal epithermal neutron flux density distributions, rigidly standardized counting procedures, and coincidence counting effects.

4 IRRADIATION AND COUNTING PROCEDURES

Depending on the neutron flux density, the mass of the sample and the efficiency of the germanium detector, irradiation, decay and counting times may vary but generalized schemes can be adhered to. Typically two irradiations are performed:

one to determine short-lived radionuclides (several minutes) and one for medium/long lived radionuclides (several hours to several weeks). If any epithermal NAA is required to augment the sensitivities of some radionuclides, then an additional irradiation is necessary. For activation analysis to produce short lived radionuclides, irradiation times varying from 5s to 10 min, depending on the type of sample, are employed. The radionuclides produced are shown in Table 1. For many samples a decay period of at least five minutes is needed to let the usually high ^{28}Al activity decrease by two half-lives. High backgrounds due to the Compton effect are caused by this high activity. Counting periods of 5–15 min are normally used to determine the rest of the elements. To improve detection limits for barium, iodine, indium, potassium and strontium, longer decay (30–60 min) and counting (30–60 min) periods are utilized. Detection limits for these elements and silicon are significantly improved by using epithermal neutrons.

For the determination of medium and long lived activities, irradiations of one to several hours are needed to activate the samples sufficiently to achieve good counting statistics. The other main limiting factor for good detection limits is the high background resulting from ^{24}Na and ^{82}Br photopeaks. Improved detection limits for arsenic, antimony, gallium, molybdenum and tungsten can be achieved

TABLE 1.
Properties of reactions producing short lived radionuclides

Element	Isotope	Half-life	Gamma-ray energies (keV)
Ag	^{110}Ag	24.6 s	657.8
Al	^{28}Al	2.24 min	1778.9
Ba	^{139}Ba	83.2 min	165.9
Br	^{80}Br	17.7 min	616.2
Br	^{80m}Br	4.42 h	37.1
Ca	^{49}Ca	8.7 min	3084.4
Cl	^{38}Cl	37.3 min	1642.4, 2167.5
Co	^{60m}Co	10.48 min	58.6
Cu	^{66}Cu	5.1 min	1039.4
Dy	^{165}Dy	2.33 h	94.7
F	^{20}F	11.0 s	1633.8
I	^{128}I	25.0 min	442.3
In	^{116m}In	54.2 min	416.9, 1097.3
K	^{42}K	12.36 h	1524.7
Mg	^{27}Mg	9.45 min	843.8, 1014.4
Mn	^{56}Mn	2.58 h	846.7, 1810.7
Na	^{24}Na	15.0 h	1368.6, 2754.1
Se	^{77m}Se	17.4 s	161.7
Sb	^{122m}Sb	4.15 min	61.5
Si	^{29}Al[a]	6.6 min	1273.0
Sr	^{87m}Sr	2.81 h	388.4
Ti	^{51}Ti	5.8 min	320.1
U	^{239}U	23.5 min	74.6
V	^{52}V	3.76 min	1434.1

[a] Si is detected by the ^{29}Si(n,p)^{29} Al reaction.

TABLE 2.
Properties of reactions producing medium lived radionuclides

Element	Isotope	Half-life	Gamma-ray energies (keV)
As	^{76}As	26.3 h	559.1
Au	^{198}Au	2.7 d	411.8
Br	^{82}Br	35.3 h	554.3, 776.5
Cd	^{115m}In[a]	53.5 h	336.3
Ga	^{72}Ga	14.1 h	834.0, 629.9
Ge	^{77}Ge	11.3 h	264.4
Hg	^{197}Hg	64.1 h	77.4
Ho	^{166}Ho	26.8 h	80.6
K	^{42}K	12.36 h	1524.7
La	^{140}La	40.23 h	1596.2, 328.8, 487.0
Mo	^{99}Mo	66.02 h	140.5
Na	^{24}Na	15.02 h	1368.6, 2754.1
Pd	^{109}Pd	13.7 h	88.0
Sb	^{122}Sb	2.72 d	564.0
Sm	^{153}Sm	46.7 h	103.2
U	^{239}Np[b]	2.35 d	277.7
W	^{187}W	23.9 h	85.8
Zn	^{69m}Zn	13.8 h	438.6

[a] Cd is determined from the ^{114}Cd(n, $\gamma\beta$) ^{115}Cd → ^{115}In reaction.
[b] U is determined from the ^{238}U (n, $\gamma\beta$)^{239}U → ^{239} Np reaction.

by epithermal NAA, as the Compton background due to ^{24}Na is significantly decreased. This, however, is at the expense of not having enough activity for the long-lived isotopes (since the epithermal flux density can be one order of magnitude less than the thermal flux density). If the samples can be split, with one portion being used for the medium-lived nuclides and one for the long-lived activation products, then epithermal NAA may be advantageous. Elements that can be determined by detection of medium-lived radionuclides are shown in Table 2.

The list of long-lived radionuclides is presented in Table 3. In general, one-half to several hours of counting time are needed to achieve the required statistics. For the measurement of long-lived radionuclides, shielding of the detector with cadmium- and copper-lined lead bricks is a desirable means of reducing the effect of background radiation coming from the counting room, which can otherwise result in the impairment of the detection limits. Furthermore, 'cross-talking' with other samples counted on other detectors and radiation from stored sources is thereby significantly reduced or eliminated.

A list of the salient features of the elements that can be determined in many samples, their spectral and nuclear interferences has been recently presented [30]. This is by no means an exhaustive list but just an indication of the neutron-induced radionuclides used in NAA.

Gamma-ray spectra for short-lived thermal and epithermal NAA are shown in Figs. 1 and 2, respectively. As can be seen, the use of epithermal neutrons

TABLE 3.
Properties of reactions producing long lived radionuclides

Element	Isotope	Half-life	Gamma-ray energies (keV)
Ag	^{110m}Ag	249.8 d	657.8
Ce	^{141}Ce	32.5 d	145.4
Cr	^{51}Cr	27.72 d	320.0
Cs	^{134}Cs	2.06 y	795.8
Co	^{60}Co	5.27 y	1173.2, 1332.4
Eu	^{152}Eu	13.4 y	1408.0
Fe	^{59}Fe	44.5 d	1099.2, 1291.6
Hf	^{181}Hf	42.4 d	482.2
Hg	^{203}Hg	46.6 d	279.2
Lu	^{177m}Lu	160 d	378.5
Nd	^{147}Nd	10.99 d	91.1
Ni	^{58}Co [a]	70.9 d	810.8
Rb	^{86}Rb	18.7 d	1076.6
Sb	^{124}Sb	60.2 d	1691.0
Sc	^{46}Sc	83.8 d	889.3, 1120.5
Se	^{75}Se	119.8 d	136.0, 264.7, 400.7
Sn	^{113}Sn	114.4 d	391.7
Sr	^{85}Sr	64.84 d	514.0
Ta	^{182}Ta	115.0 d	1221.4
Tb	^{160}Tb	72.1 d	879.4,
Th	^{233}Pa [b]	27.0 d	311.9
Tm	^{170}Tm	129 d	84.3
Yb	^{175}Yb	4.19 d	396.3
Zn	^{65}Zn	243.8 d	1115.5
Zr	^{95}Zr	64.0 d	756.7

[a] Ni is determined using the $^{58}Ni(n, p)\,^{58}Co$ reaction.
[b] Th is determined using the $^{232}Th(n, \gamma\beta)\,^{233}Th \rightarrow {}^{233}Pa$ reaction.

increases the signal-to-noise ratio for uranium, iodine, indium and silicon, resulting in better detection limits. A gamma-ray spectrum for medium-lived NAA is shown in Fig. 3. Usually only As, Br, K, La, Lu, Mo, Na, Sb, U and W photopeaks are determined. A typical gamma-ray spectrum for long-lived NAA is shown in Fig. 4. A multitude of photopeaks with good ratios of signal to background is usually seen. Depending on irradiation, decay and counting times and the neutron flux, several other elements can also be detected in the short, medium and long-lived NAA. These include the platinum group elements; however, these usually require chemical pre- or post-separation.

5 INTERFERENCES IN NAA

When properly performed, neutron activation analysis can be among the most precise and accurate of all analytical methods for the determination of trace ele-

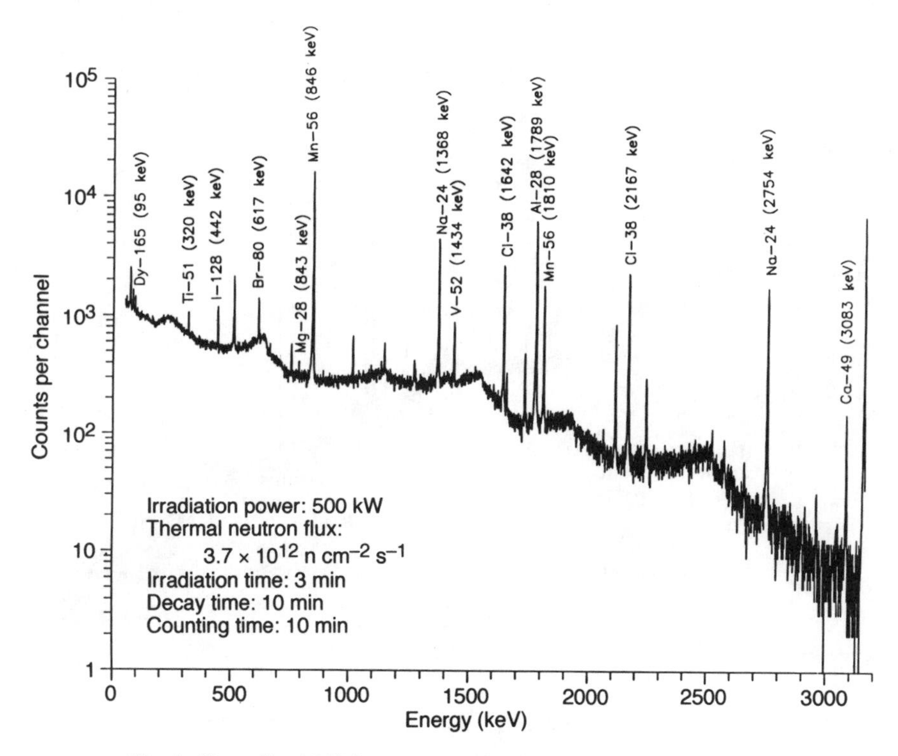

Fig. 1. Short lived NAA spectrum of airborne particulate matter.

ments. Several matrix interferences in NAA result in increased analytical errors. The primary nuclear interference reaction discussed in Section 2.1 is potentially a major problem, particularly if the target element has a high cross-section. Therefore, the method of choosing a suitable reaction, reducing the interference reaction, or making the proper correction must be seriously considered.

Another common nuclear interference reaction arises from uranium fission. This is particularly true for many geological samples containing tens of parts per million of uranium. In uranium fission, some of the fission products are exactly the same radionuclides as those produced by the desired (n,γ) reaction. This is particularly important for elements with atomic masses around 95 and 140. For example, analysis of Mo and La involves ^{99}Mo and ^{140}La, which are two common radionuclides produced from uranium fission. Thus, a correction is necessary if the concentration of uranium is high in the sample. An overview of such interferences and updates of published data have been compiled by Landberger [31, 32].

A gamma-ray spectral interference occurs when two radionuclides emit gamma-rays of the same, or nearly the same, energy. For example, ^{56}Mn has an

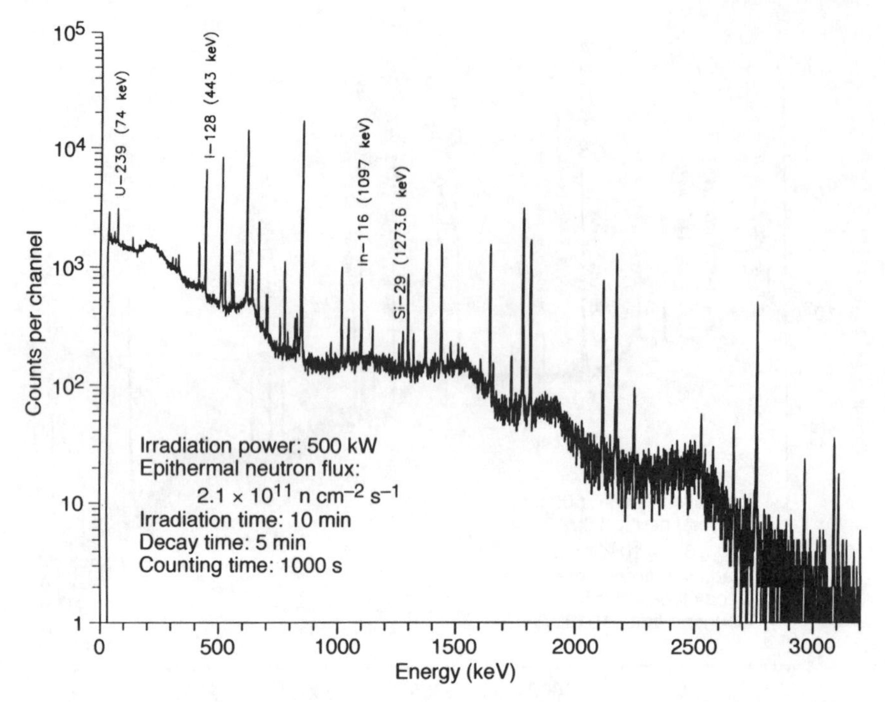

Fig. 2. Short lived epithermal NAA spectrum of airborne particulate matter.

846.8 keV gamma-ray and ^{27}Mg has an 843.8 keV gamma-ray. This problem can be solved in several ways: using a high resolution germanium detector, choosing suitable decay and counting times, or finding another photopeak. In the case of ^{203}Hg, which emits a single gamma-ray at 279 keV, its separation from the 279 keV gamma-ray belonging to ^{75}Se is impossible, so a correction based on other selenium photopeaks must be made. Uranium and thorium interferences in the detection of samarium have also been experimentally examined by Landsberger and Simsons [33].

It is often wrongly assumed that NAA methods are completely independent of matrix. The presence of high-Z elements, such as various combinations of zinc, iron, cobalt, lead, nickel, tin, copper, silver, antimony, etc., in ore concentrates, meteorites, and archaeological artifacts can severely limit the detection of lower energy gamma-rays owing to the self-absorption by the matrix. Bode, De Bruin and Korthoven [34] and Zikovsky [35] have described correction factors for these effects, and Jaegers and Landsberger [36] have published a PC-based program for the determination of the self-absorption fractions of gamma-rays for three types of geometrical configuration. Other interferences, such as neutron self-shielding owing to the high neutron capture cross-sections may occur if there

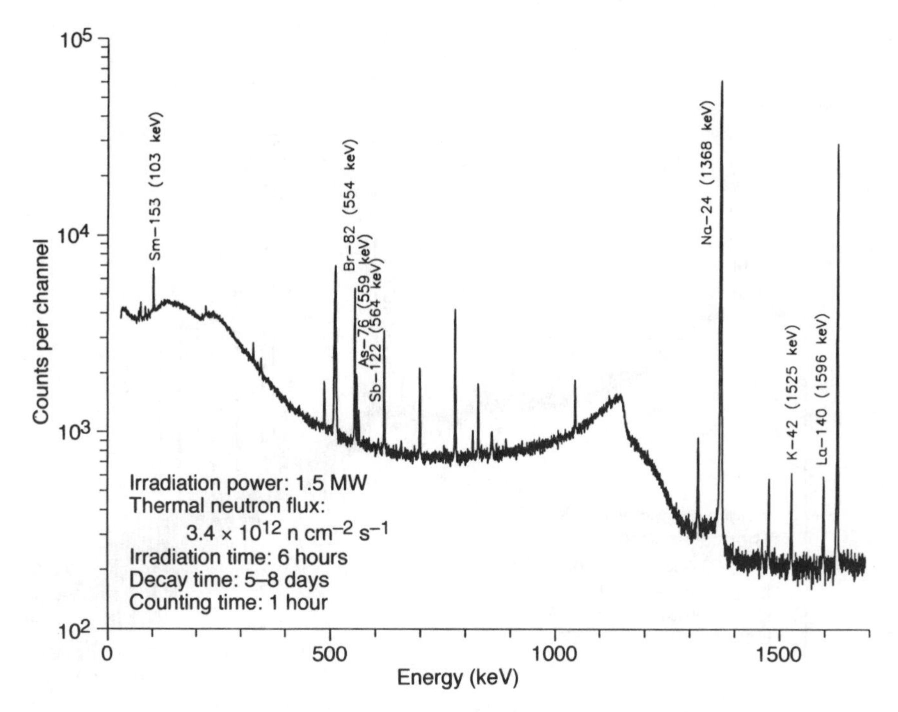

Fig. 3. Medium lived NAA spectrum of airborne particulate matter.

are elevated concentrations of boron, cadmium, gold, silver or certain other elements in the sample.

6 ACCURACY AND PRECISION

The definitions of accuracy and precision are often misused or interchanged. A simple but well depicted explanation of accuracy and precision is shown in Fig. 5. In NAA, counting statistics is the dominant factor for the precision of the analytical results. Therefore, to make an estimate of the analytical precision, several measurements of the sample must be made. Associated with the results of replicate determinations to estimate precision is the average standard deviation σ. Precision can thus be reported as 1σ, 2σ or 3σ, representing 68%, 95% and 99% confidence limit. This is illustrated in Fig. 6. As in other chemical techniques, there are numerous pitfalls which can hamper reliable measurements. An estimate of the accuracy of the measurement is achieved by analyzing certified reference materials. This indicates how good one's techniques and methods are

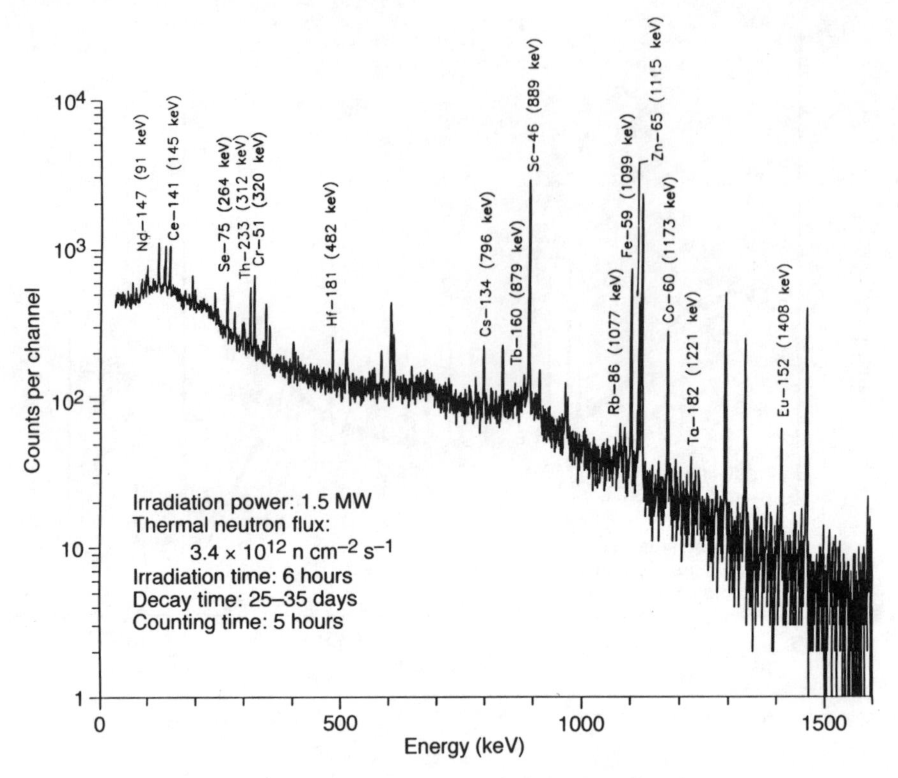

Fig. 4. Long lived NAA spectrum of air-borne particulate matter.

when the results are compared with known concentrations of major, minor, and trace elements in similar or other matrices. An extremely well documented overview of NAA precision and accuracy has been published by Heydorn [37].

There are several factors which can severely limit the accuracy of NAA, including:

(1) changes in neutron flux profile which are not carefully monitored.

(2) inaccurate placement of samples in front of detectors, including volume variations between standards and samples (see Fig. 7.) Radiation from the sample to the detector decreases a $1/r^2$. This is particularly significant when counting at short distances from the detector. Sample positioning is crucial, and every effort should be made to retain a nearly identical geometry between the samples and standards. Vertical germanium detectors are usually more susceptible to effects of small changes in the counting position than are the horizontal ones. Automatic turning of the sample in front of the detector can also be used to ensure better reproducibility of the counting. A quality control check can be performed by

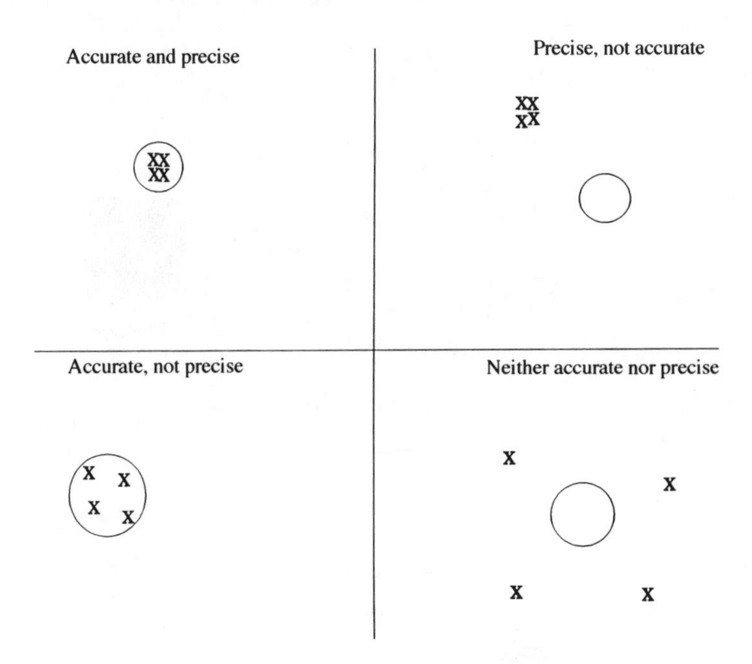

Fig. 5. Illustration of the differences between accuracy and precision.

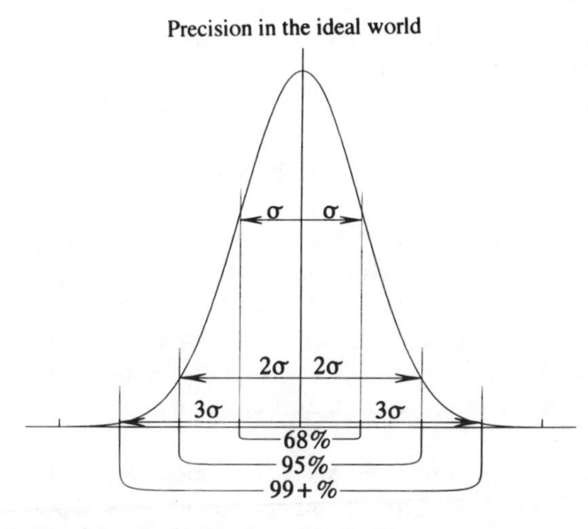

Fig. 6. Precision in the ideal world, showing standard deviations.

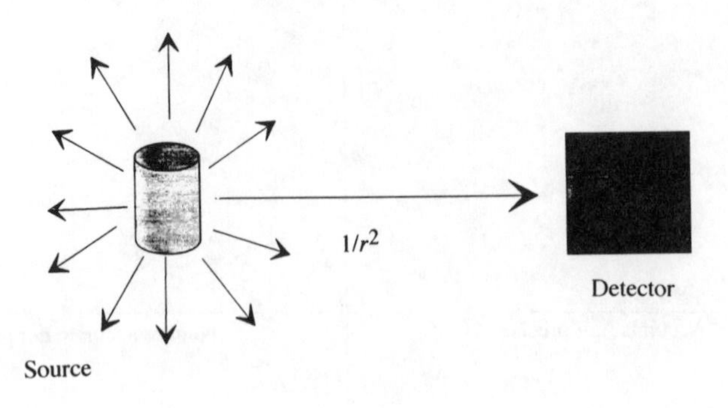

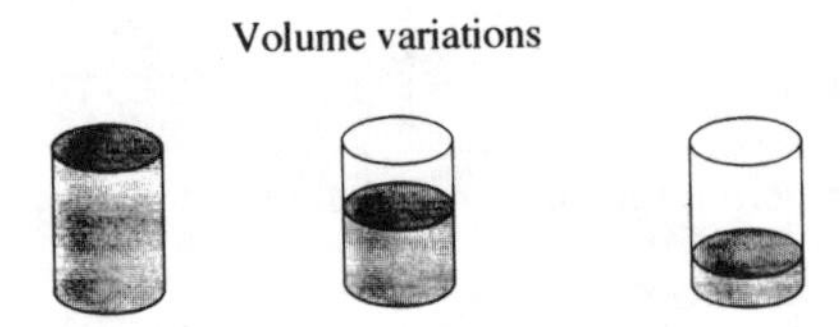

Fig. 7. Geometrical placement and volume variations affecting accuracy.

turning the sample 180° and seeing if the new calculated concentrations are significantly different. With the use of pelletized samples either vertical or horizontal oriented detectors can be used;

(3) unknown changes in relative natural abundances of certain elements such as boron, sulphur, and lead;

(4) inaccurate correction for deadtime losses and pulse pile-up by either software or hardware methods;

(5) poor deconvolution of overlapping peaks by computer programs;

(6) unforeseen nuclear or spectral interferences;

(7) human factor errors, including incorrect readings of irradiation, decay or counting times and sample weights. Exact times for short irradiations are crucial;

(8) trace impurities in counting vials which have been activated along with the samples (this is particularly crucial for biological and some environmental samples);

(9) volatilization of certain elements such as mercury during irradiation;

(10) incorrect drying procedures for bioenvironmental samples which may have a high moisture content.

For investigations of airborne particulate matter, accuracy and sensitivity can be severely compromised by an inherent trace element contamination from filters and counting vials. Before any sampling is undertaken, a detailed analysis of the chosen filter type must be made. At least five to ten filters should be analyzed in order to determine trace concentrations and the variability of the results. A large standard deviation may be unacceptable if the element of environmental interest has concentrations in the same range as the 'blank' element. Another consideration in choosing filters is the presence of elements in high amounts and those which have large activation cross-sections. For instance, high Compton backgrounds in short lived activation analysis can be attributed to reactions of aluminum, $^{27}Al(n,\gamma)^{28}$ Al; chlorine, $^{37}Cl(n,\gamma)^{38}Cl$; manganese, $^{55}Mn(n,\gamma)^{56}Mn$; and sodium, $^{23}Na(n,\gamma)^{24}Na$, resulting in poor analytical detection limits for such elements as vanadium and copper. Reactions of bromine, $^{81}Br(n,\gamma)^{82}Br$, and sodium, $^{23}Na(n,\gamma)^{24}Na$, give photopeaks that have the same effects in long irradiations on the determination of some of the radionuclides of medium half-life (e.g. ^{76}As, ^{122}Sb, ^{72}Ga and ^{42}K). In fact, the many ^{82}Br peaks usually seen in such spectra contribute to the high background resulting from the Compton effect.

7 SENSITIVITY

There have been numerous definitions for sensitivity and detection limits. Generally speaking, as the sensitivity increases, the detection limit decreases, and conversely. Ultimately it is the background radiation in a gamma-ray spectrum that determines how little material is needed to be adequately detected. The theory of detection limits as it applies to gamma-ray spectroscopy has been well covered in the classic paper by Currie [38]. In many NAA applications, detection of trace constituents can be limited by the complexity of the gamma spectrum or by the masking effect of bremsstrahlung radiation from beta emitting radionuclides present at high levels (e.g. phosphorus in biological samples) or Compton scattering. Lower energy peaks can often exhibit a poor signal-to-noise ratio, making the analyses of these peaks difficult. In particular, this is a result of Compton scattering due to the presence of elements at the percent level (e.g. aluminium, chlorine, iron and sodium) as well as at minor levels (e.g. bromine, manganese, scandium, zinc and cobalt). Most methods to enhance elemental sensitivities are straightforward and usually revolve around optimizing irradiation, decay and counting times (including cyclic NAA), employing loss-free counters, larger detectors or X-ray detectors, using alternate photopeaks, or ultimately using radiochemical means to isolate the radionuclide of interest. Increases in analytical sensitivity for certain elements can also be augmented by using epithermal NAA, employing cadmium or boron or a combination of the two, as filters.

Ideally one would like to diminish the effect of Compton scattering so that photopeaks which cannot be seen, or have poor statistics, can be analyzed with relatively good precision. Two non-destructive techniques which are able to accomplish this are Compton suppression (also referred to as anti-Compton) and gamma–gamma coincidence. While Compton suppression methods are not new (extensively used in on-line nuclear physics experiments and prompt gamma NAA), very few groups anywhere in the world have judiciously employed this technique in either research or routine delayed NAA environmental work [39–41]. While reducing the Compton effect, Compton suppression also provides an excellent shield for external background, owing to the thick sodium iodide detectors needed. A typical set-up can be seen in Fig. 8. Recently this technique has been successfully used to determine very low levels (1–2 ng/g) of arsenic and cadmium in biological reference materials [42, 43] and cadmium in environmental tobacco smoke [44]. A spectrum illustrating the very low background attainable using Compton suppression is seen in Fig. 9.

Gamma–gamma coincidence methods have also been utilized in numerous nuclear physics experiments, but only in a very few cases for delayed NAA. Previous work has included the detection of selenium in biological samples [45], copper in geological samples [46] and, more recently, various trace elements in rice flour [39]. There has also been some interest in gamma–gamma directional correlation measurements to obtain information on the chemical environment of

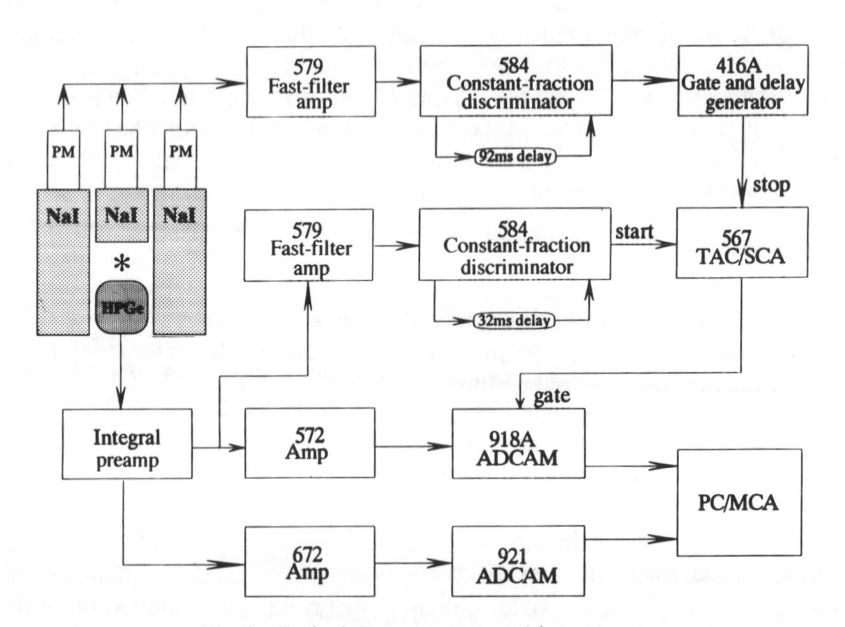

Fig. 8. Typical Compton suppression set-up.

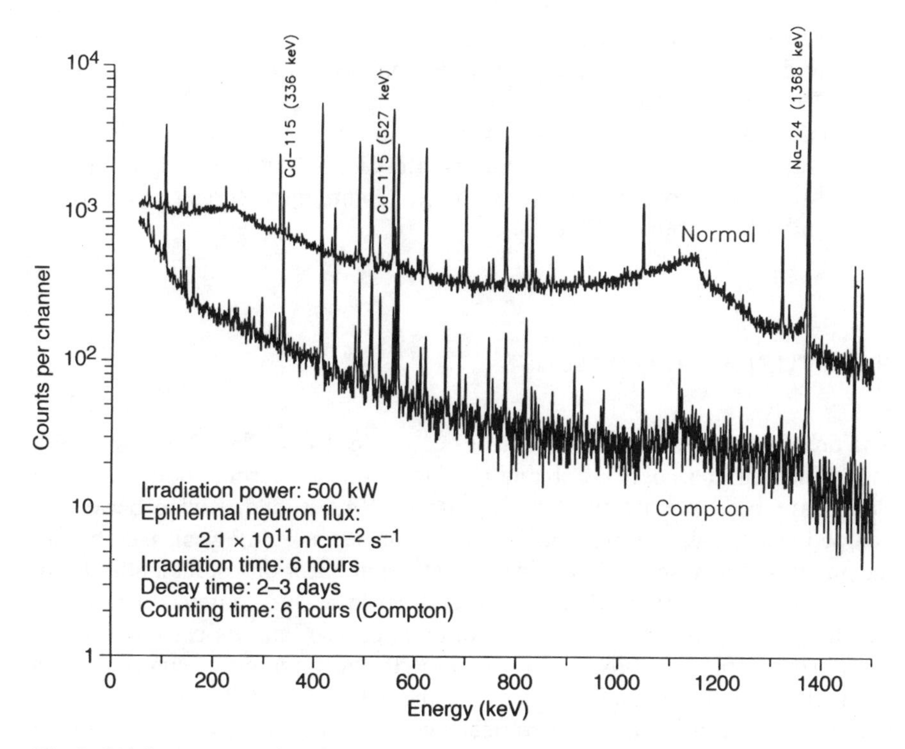

Fig. 9. NAA spectrum of environmental tobbaco smoke using Compton suppression.

trace elements in solutions [47]. This technique has many more possibilities which remain to be explored.

8 QUALITY CONTROL AND REFERENCE MATERIALS

Reference materials are needed to test the accuracy of analytical methods and not as calibration standards. An on-going protocol to ensure the reliability of analytical data must be established in each NAA laboratory. This includes:

(1) The analysis of certified reference materials.
(2) Re-irradiation of selected samples to determine short-lived radionuclides so as to ascertain the analytical reproducibility of the methods used.
(3) A careful evaluation of each individual spectrum to see if any potentia interferences exist (e.g. uranium fission).

(4) The analysis of at least two gamma-rays from the same nuclide (if possible) to determine if any spectral or nuclear interferences exist.
(5) The participation in inter-laboratory comparison studies. Numerous reference materials are available from international organizations including the National Institute for Standards and Technology, the International Atomic Energy Agency, the European Community Bureau of Reference, and other national laboratories.

9 FUTURE DEVELOPMENTS

The field of neutron activation analysis continues to be of vital importance in many different areas of research and routine chemical determinations. Although there have been predictions that the use of NAA would decline because of competing methods, new innovative research in medical, biological, and environmental areas has resulted in a wider variety of applications of its methods and techniques. Furthermore, there are several world-wide commercial companies which analyze tens of thousands of samples per year, and its use continues to grow in developing nations. It still remains as the only non-destructive technique available to determine up to 50 elements and, because of this, NAA plays a key role in the certification of standard reference materials. Along with cold neutron sources, more sophisticated counting equipment, the re-emergence of coincidence and Compton suppression methods, and imaging techniques, NAA will continue to have an important place in chemical analysis.

REFERENCES

1. I. M. Kolthoff and P. J. Elving, Nuclear activation and radiosotopic methods of analysis, in *Treatise on Analytical Chemistry*, Wiley, New York (1986).
2. K. Heydorn, *Neutron Activation Analysis for Clinical Trace Element Research*, Vol. 1 and 2, CRC Press, Boca Raton (1984).
3. J. Csikai, *Handbook of Fast Neutron Generators*, Vol. 1 and 2, CRC Press, Boca Raton (1987).
4. J. Tolgyessy and E. H. Klehr, *Nuclear Environmental Chemical Analysis*, Series in Analytical Chemistry, Ellis Horwood, Chichester (1987).
5. H. A. Das and H. A. van der Sloot, Radioanalysis in geochemistry, in *Developments in Geochemistry*, Vol. 5, Elsevier, Amsterdam (1989).
6. Z. B. Alfassi, *Activation Analysis*, Vol. 1 and 2, CRC Press, Boca Raton (1989).
7. J. Tolgyessy and M. Kyrs, *Radioanlytical Chemistry*, Vol. 1 and 2, Ellis Horwood, Chichester (1989).

8. W. D. Ehmann and D. E. Vance, *Radiochemistry and Nuclear Methods of Analysis, Chemical Analysis: A Series of Monographs on Analytical Chemistry and its Applications*, Vol. 116, edited by J. D. Winefordner and I. M. Kolthoff, Wiley, New York (1990).
9. S. J. Parry, *Activation Spectrometry in Chemical Analysis, Chemical Analysis: A Series of Monographs on Analytical Chemistry and its Applications*, Vol. 119, edited by J. D. Winefordner and I. M. Kolthoff, Wiley, New York, (1990).
10. W. D. Ehmann and S. W. Yates, *Anal. Chem.* **58**, 49R (1986).
11. W. D. Ehmann and S. W. Yates, *Anal. Chem.* **60**, 42R (1988).
12. W. D. Ehmann, J. D. Robertson and S. W. Yates, *Anal. Chem.* **62**, 50R (1990).
13. W. D. Ehmann, J. D. Robertson and S. W. Yates, *Anal. Chem.* **64**, 1R (1992).
14. International Atomic Energy Agency, *Practical Aspects of Operating a Neutron Activation Analysis Laboratory*, IAEA-TECDOC-564, IAEA, Vienna (1990).
15. International Atomic Energy Agency, *Comparison of Nuclear Analytical Methods with Competitive Methods*, IAEA-TECDOC-435, IAEA, Vienna (1987).
16. International Atomic Energy Agency, *Applications of Short-Lived Activation Products in Neutron Activation Analysis of Bio-Environmental Samples*, IAEA/RL/141, IAEA, Vienna (1987).
17. G. F. Knoll, *Radiation Detection and Measurement*, 2nd Ed., Wiley, Chichester (1989).
18. K. Debertin and R. G. Helmer, *Gamma- and X-ray Spectrometry with Semiconductor Detectors*, North-Holland, Amsterdam (1988).
19. S. F. Mughabghab, M. Divadeenam and N. E. Holden, *Neutron Cross Section*, Vol. 1: *Neutron Resonance Parameters and Thermal Cross Section, Part A*: $Z = 1-60$, Academic Press, New York; *Part B*: $Z = 60-100$, Academic Press, Orlando (1984).
20. International Atomic Energy Agency, Handbook on nuclear activation data, *Tech. Rep. Ser.–I.A.E.A.* (273) (1987).
21. J. J. Rowe and E. Steinnes, *Talanta* **24**, 433 (1977).
22. S. Landsberger, *Anal. Chem.* **60**, 1842 (1988).
23. S. Parry, *J. Radioanal. Chem.* **72**, 195 (1982).
24. L. Tobler, V. Furrer and A. Wyttenbach, *Biol. Trace Elem. Res.* **26–27**, 623 (1990).
25. R. Dowalti and R. E. Jervis, *J. Radioanal. Nucl. Chem.* **150**, 455 (1991).
26. N. Lavi, F. Lux and Z. B. Alfassi, *J. Radioanal. Nucl. Chem.*, **129**, 93 (1989).
27. R. Zaghloul, W. H. El-Abbady and N. S. Ghoma, *J. Radioanal. Nucl. Chem.* **116**, 235 (1987).
28. J. R. Moody, R. R. Greenberg, K. W. Pratt and T. C. Rains, *Anal. Chem.* **11**, 1230A (1988).
29. F. De Corte and A. J. Simonits, *J. Radioanal. Nucl. Chem.* **133**, 43 (1989).
30. S. Landsberger, *J. Trace Microprobe Tech.* **10**, 1 (1992).
31. S. Landsberger, *Chem. Geol.* **57**, 415 (1986).
32. S. Landsberger, *Chem. Geol.* **77**, 65 (1989).
33. S. Landsberger and A. Simsons, *Chem. Geol.* **62**, 223 (1987).
34. P. Bode, M. de Bruin and P. J. M. Korthoven, *J. Radioanal. Chem.* **64**, 559 (1981).
35. L. Zikovsky, *Nucl. Instrum. and Methods Phys. Res., Sect. B* **B4**, 421 (1984).
36. P. Jaegers and S. Landsberger, *Nucl. Instrum. Methods Phys. Res., Sect. B* **B44**, 479 (1990).
37. K. Heydorn, Aspects of precision and accuracy in neutron activation analysis, *Risoe Rep.* R-419 (1980).
38. L. A. Currie, *Anal. Chem.* **40**, 586 (1968).
39. S. Suzuki and S. Hirai, *Bunseki Kagaku* **39**, 255 (1990).
40. J. B. Cumming, P. P. Parekh, A. V. Murali, *Nucl. Instrum. Methods Phys. Res., Sect. A* **A265**, 468 (1988).

41. H. T. Millard, *Trans. Am. Nucl. Soc.* **56**, 195 (1988).
42. M. Petra, G. Swift and S. Landsberger, *Nucl. Instrum. Methods Phys. Res., Sect. A* **A299**, 85 (1990).
43. S. Landsberger, *J. Radioanal. Nucl. Chem.* **161**, 5 (1992).
44. S. Landsberger, S. Larson and D. Wu, *Trans. Am. Nucl Soc.*, **65**, 176 (1992).
45. L. E. Wangen, E. S. Gladney and W. K. Hensley, *Anal. Chem.* **52**, 765 (1980).
46. O. B. Michelsen and E. Steinnes, *Talanta* **15**, 574 (1968).
47. M. de Bruin and P. Bode, *Analusis.* **11**, 49 (1983).

Chapter 7

RADIOCHEMICAL NEUTRON ACTIVATION ANALYSIS

Zeev B. Alfassi
Department of Nuclear Engineering, Ben-Gurion University, Beer-Sheva, Israel

and

Isotope Division, Risϕ Research Center, DK-4000 Roskilde, Denmark

1 INTRODUCTION

The previous chapter deals with instrumental neutron activation analysis (usual acronym INAA). In this method of analysis no chemical separation or treatment is needed. Radiochemical neutron activation analysis (acronym RNAA) shares with INAA the activation of the test sample by irradiation with neutrons and the measurement of the induced activities. However, it differs from INAA in that RNAA also includes the dissolution of the sample (if it is not already in aqueous solution) and chemical separation of the analyte from solution. Two questions immediately arise: (1) if the INAA method can determine many elements without separation, why use RNAA at all? This question becomes more significant when recalling that one of the main advantages of INAA is its non-destructive character, which allows the determination of trace elements in objects which must not be destroyed (precious materials or materials intended for use) or in materials that are difficult to dissolve or decompose; (2) why do we have to deal with chemical separation specially in neutron activation analysis, and why is it is different from chemical separation in any other analytical method?

The answer to the first question lies in the fact that, although INAA is a multi-element method, it cannot determine all the elements. Some elements, such as C, H, and O, cannot be determined at all by NAA, since their activation products are not radioactive. However, even elements that can be determined in

Chemical Analysis by Nuclear Methods Edited by Z. B. Alfassi

principle by neutron activation analysis cannot be determined in matrices in which their concentration is too low or interfering elements are abundant. In some cases this problem can be solved if the interfering elements are shorter lived than the required element. In this instance, sufficiently long 'cooling' (decay time) will remove the interference. However, in many cases the interfering elements have half-lives longer than or similar to that of the studied element. In other cases the wanted elements can be measured after sufficient 'cooling', but this decay time is too long for many practical purposes.

The use of radiochemical separations in neutron activation analysis complements the analysis by INAA, enabling the measurement of additional elements at low concentration. Thus, for example, Chatt *et al.* [1], in their measurement of trace elements in food samples, measured by INAA the elements As, Br, Ca, Cl, Co, Cr, Fe, I, K, Mg, Mn, Na, Rb, Sc, Se, Sn and Zn, and they used RNAA to measure the concentration of As, Au, Co, Cu, Fe, Hg, Mo, Sb, Se and Zn. Some of the elements were determined also by RNAA although they had already been determined by INAA, since higher precision is obtained for these elements by the RNAA method. Similarly, Zeisler, Greenberg and Stone [2], in their measurement of trace elements in human liver, used INAA for many elements but for some elements they had to use RNAA. They measured several elements by RNAA even though only one of them, Sn, could not be measured at all by INAA in this system. For the other elements application of RNAA improved the precision of the measurement. Yinsong *et al.* [3] determined trace elements in human hair and some internal organs; Br, Ca, Cl, Co, Cr, Fe, Mg, Mn, Na, Rb, S and Se were determined by INAA, whereas RNAA was used for the determination of As, Cd, Cu, Hg and Sn. Tian and Ehmann [4] used RNAA to measure the concentration of As, Cd, Cu and Mo after at least 17 other elements had been determined by INAA of the same sample. To summarize, in order to achieve the ultimate sensitivity of the NAA technique, it is frequently necessary to separate the element(s) of interest from the other radionuclides in the sample.

It should be remembered that the radiochemical separation depends strongly on the matrix, since the major elements of the matrix are the main interferences in the determination of the activity of each of the nuclides in the activated sample. A radionuclide in the activated sample can interfere, not only if it has γ-rays with similar or close energy to a radionuclide of the studied elements, but also if its activity is too high. High activity leads to high background at energies lower than its γ-ray photopeak, owing to Compton scattering and to high deadtime, which force the counting to be carried out at a longer distance from the detector, so decreasing the counts due to the trace elements. A good example of the effect of the matrix can be seen in the determination of trace elements in two important semiconductor materials—silicon and gallium arsenide [5]. Activation of Si leads to the formation of only short lived radionuclides: Mg (half-life = 9.5 min), ^{28}Al (2.31 min) and ^{31}Si (3.05 h). Thus, after cooling for, at most, 2–3 days, the activity due to the trace elements in the sample can be measured. In contrast to the silicon matrix, in GaAs the long-lived ^{76}As (17.8 days) is formed, by the

$^{75}As(n, 2n)^{76}As$ reaction, and hence the activity of the trace elements can be determined only after chemical removal of the As [6]. Similarly, many trace elements can be determined instrumentally (i.e. by INAA) in biological and geological matrices, since their major elements are either not activated (C, H, O) or activated to short lived nuclides (^{28}Al, ^{27}Mg, ^{31}Si). However, the determination of trace elements in pure metals cannot be carried out instrumentally for many metals [7]. Krivan and co-workers, in recent years, developed methods for RNAA of many metals, for example tungsten and molybdenum [8]. Although for some metals, such as pure Al metal, cooling for over 1 week will allow instrumental measurement of many elements which produce long lived nuclides (the radionuclide used in activation analysis for measurement of an element is called the indicator radionuclide—acronym IRN), the most effective means of improving the limits of detection is radiochemical separation [9].

The answer to the second question, on the difference between chemical separation in NAA and other chemical methods, has three main components.

(1) The interferences from the major elements are different for various measuring techniques. Thus, whereas silicon interferes with ICP–AES measurements [10], it does not interfere with NAA measurements. Another example is the GaAs matrix mentioned before. The only long-lived radionuclide formed from the matrix in GaAs activation is ^{74}As [6]. Thus, for measuring the activities of the trace element IRNs (indicator radionuclides), it is sufficient to remove only the As. This was done [6] by evaporation of $AsCl_3$. The gallium remaining in the solution did not inhibit the measurement of the activities of the IRNs, whereas in many other methods of trace analysis Ga also has to be removed.

(2) The multi-element character of NAA means that the trace elements need not be separated one by one; it is sufficient to separate them into several groups. The groups are designed so that in each there is no interference between the various radionuclides. In many cases, even separation into groups is not needed, and it is sufficient to remove the major sources of interference. One such case, which has already been mentioned, is trace element determination in GaAs. Similarly, Egger and Krivan [9] in their determination of trace elements in aluminum removed only the interference of ^{24}Na (by sorption on a hydrated antimony pentoxide—HAP—column) and were then able to measure 38 trace elements, each in a concentration of less than 1 ppb.

(3) This difference is indicated in the name *radiochemical* separation. We are separating radionuclides, with the attendant disadvantage of radiation hazards and sometimes the need to work partially in hot cells. However, from the analytical point of view, it has two enormous advantages not shared by any other analytical method. In conventional (non-nuclear) analytical chemistry of trace elements, there are two main sources of error: (1) contamination by the reagents, which leads to high results; and (2) adsorption of the trace elements on the vessel walls, a phenomenon characteristic of solutes at very low concentration, which results in too low a value of measured concentration. Although these two error sources have opposing effects, very rarely do they compensate one another, and in

many cases they lead to erroneous results. The first source of error, i.e. contamination, can be diminished and in many cases almost eliminated by working in 'clean' rooms with purified reagents; however it requires lengthy work, time and money. Contamination and sorption errors are relatively easily overcome in RNAA. In RNAA, the sample is irradiated without the addition of any reagent. The dissolving and separating reagents are added only after the end of the irradiation. Any contamination in the vessels, reagents or the surrounding atmosphere is not radioactive, and consequently has no influence on the measurement results. However, it should be remembered that the statement that RNAA is free from 'blank' problems refers primarily to blanks due to reagents. An almost blank-free system occurs only when the sample can be irradiated as a single integral piece, with subsequent etching of the surface to eliminate surface contamination (which is not characteristic of the total material). It should be recalled that the surface etching is done after the irradiation. This procedure is usually applied in the RNAA of ultra-pure metals and some composite materials. Sometimes, it is also done for geological materials [11], but in that case there are problems of homogeneity. It is recommended that the etching should be deep and to remove not only the surface layers but also materials beneath the surface, in order to remove surface contaminants which recoil, because of the nuclear reaction, to subsurface layers.

When the sample is dissolved (after irradiation), together with the sample, there are added to the dissolution vessel carriers for all the elements to be measured. This increases considerably the concentration of these elements, and although it does not reduce the absolute amount adsorbed on the surface of vessels and tubes, it renders their fraction negligible. The absolute amount adsorbed is fairly constant, owing to saturation of the adsorption sites, so increasing the total amount makes the fraction adsorbed unimportant. The addition of carriers not only reduces considerably the fraction lost by adsorption (or by volatilization during dissolution), it also causes the trace elements to behave as expected at larger concentrations. Many elements behave differently on the micro scale (concentration of mg/ml or less) and at the macro level (higher concentration). Elements that normally are not adsorbed at all on an ion-exchange column may at very low concentrations be leached only slowly from the resin, and hence require a larger elution volume than is necessary for higher concentration. Elements that on the macro scale are concentrated in one phase in liquid–liquid extraction may at the micro level be split between the two fractions. Addition of carrier materials overcomes these problems, which exist in other methods of trace analysis. The carriers added should not be those for only the analyte elements, but also those of the interfering elements (referred to as holdback carriers), to maximize the separation of the measured elements from interfering ones.

Although it was stated that chemical treatment of the sample takes place after the irradiation, there are some cases of pre-irradiation treatment. Ehmann, Robertson and Yates [12], in their comprehensive bibliographic review on nuclear and radiochemical analysis, called NAA with pre-irradiation chemical separation by the name CNAA (chemical neutron activation analysis), whereas the name RNAA is specific to post-irradiation chemical separation, since only in this case does the work involves radiochemicals. There are several reasons for using pre-irradiation chemical separation, despite the described advantages of post-irradiation chemical separation. The main reason is the short half-life of the IRN, which prevents its dissolution and chemical separation before its decay. The main elements studied in this way are aluminum [12–14], selenium [15–17], vanadium [18–21] and sometimes also titanium [19]. Aluminum and vanadium are activated only to short-lived IRNs (^{28}Al—2.24 min, and ^{52}V—3.76 min). Se has also a long lived IRN, ^{75}Se (120 days); however, if a rapid result is needed, the use of ^{77m}Se (half-life 17.4 s) is preferable, even though the results may be less accurate. A chemical separation can sometimes be done even in this short time; however, the dissolution of most studied materials takes too long [22]. Another reason for pre-irradiation chemical treatment is to preconcentrate elements from very dilute solution, mainly natural waters [23–25], but also geological samples [26–29] and metals [30]. An additional reason for pre-irradiation chemical separation is to separate the same element into different chemical species [31–33]. For yet another reason, King, Kerrich and Daddar [34] separated pre-irradiation boron from boron-rich minerals. They wanted to avoid neutron flux suppression due to the exceptionally high cross-section of boron for absorption of thermal neutrons. Another reason caused Chatt *et al.* [1] to undertake pre-irradiation separation of U and Th.

However, it should be remembered that the pre-irradiation chemical separation NAA, suffers from the problems of contamination and loss by adsorption to walls, similarly to all other analytical methods. In many cases, it is preferable to use other analytical methods, non-nuclear ones to measure the concentration in the separated sample. In this chapter we will deal only with RNAA, i.e. with post-irradiation chemical treatment. More material on pre-irradiation chemical separation can be found in the book on environmental analysis by Das, Faanhof and Van der Sloot [35].

1.1 The different steps in RNAA

The sequence of processes in RNAA is given in the following flow chart:

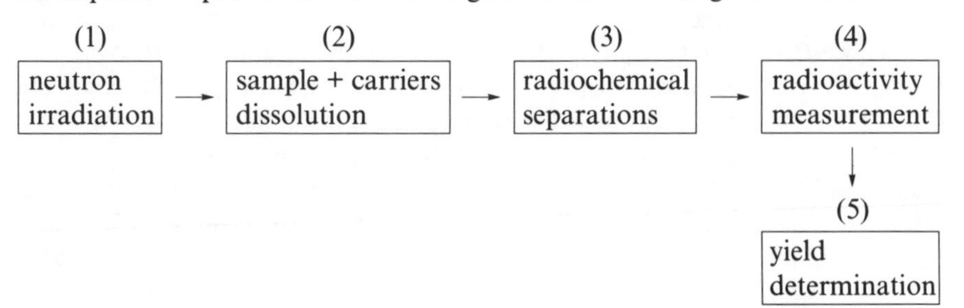

Steps 1 and 4 are done also in INAA and were discussed in the preceding chapter. We will treat only steps 2, 3 and 5, which are characteristic of RNAA.

2 SAMPLE DISSOLUTION

Samples that have been irradiated for a long time may suffer considerable radiolysis, leading to the formation of gaseous products. It is advisable for this reason to cool the irradiation container before opening it. Most long irradiations are done in quartz ampules (quartz is chosen because it contains very few activatable elements, whereas glass contains sodium, and hence the glass and metal ampules will be too radioactive and will be radiation hazards, polyethylene becomes brittle after long irradiation). It is usual to cool the ampule in liquid nitrogen before opening. The irradiation sample, together with carriers, is transferred to the decomposition vessel.

It should be remembered that we are dissolving, and later separating, radioactive materials. In many cases a faster solubilization or separation method may be preferred in order to minimize the decay of the radionuclides. In many other cases this factor is not important, since the measurements involve mainly medium to long lived radionuclides, and a few hours or even a day will not be significant. Because of the high radiation hazard of some samples, work should be done behind a lead shield, with as little bodily exposure to radiation as possible.

The sample decomposition method depends on the type of sample analyzed. The various methods for sample dissolution were reviewed recently by Bajo [22], and only a relatively short description will be given here.

2.1 Geological samples

Geological samples are decomposed either by fusion with $Na_2O_2 + NaOH$ in Zr or Ni crucibles or by dissolution with strong acids. It is usually preferred to decompose silicate samples by fusion, which takes a relatively short time. However, dissolution with mineral acid can be speeded up considerably by the use of a microwave oven [36]. The procedure for fusion is as follows: carrier solutions are introduced into a Zr or Ni crucible and dried. Then the irradiated sample, ≈ 10 g of Na_2O_2 and 2 g of NaOH are added. The crucible is covered by a lid and heated over a gas burner for about 10–20 min, until fusion is complete. After cooling, the crucible and lid are placed in a 250 ml beaker (covered with a watch glass). Dilute HCl solution (50–100 ml) is used to dissolve the fusion cake.

Acid mixtures (HNO_3, $HClO_4$, H_2SO_4, HF), sometimes with additional small amounts of H_2O_2, are also used to mineralize geological samples. The dissolution is usually done either in a platinum crucible or in pressurized decomposition vessel. The use of a microwave oven [36] shortens the dissolution time to 5–10 min.

Rocks and soils have also been decomposed by burning in oxygen in a dynamic system [37, 38] for the measurement of very low trace element concentrations. This method, which uses very small amounts of reagents (for dissolution after combustion), and an apparatus with a small surface, is very important in other methods of analysis to minimize blanks. In RNAA, where blank errors are not important, this method has not yet been used. This method is also used to separate selenium in the digestion step. Other elements which form sparingly volatile oxides remain in the ash holder, whereas selenium dioxide volatilizes and condenses on a cold finger.

2.2 Metal samples

Metals are usually dissolved in a mixture of acids. Aqua regia can dissolve almost all metal. However, in many studies they can be dissolved also in $HF + HNO_3$ mixture [8, 9]. A $HF + HNO_3$ mixture with a high proportion of HF is used also for surface etching.

2.3 Biological samples

Biological (organic) samples are digested by several methods, which can be divided into two groups—dry and wet ashing. Dry ashing in furnaces at temperatures between 400 and 700 °C is used mainly because of its simplicity and suitability for large samples and large numbers of samples. They samples are placed in porcelain or quartz crucibles, the crucibles are inserted into a furnace and the temperature is raised gradually to 400–700 °C. The ash is then dissolved with a few ml of dilute acid, chiefly HCl. However, it should be remembered that considerable losses of several elements may occur during this ashing procedure. The losses may be due either to volatilization or to retention on the crucible surface, even after dissolving the ash with more concentrated acid [39]. In many cases it was found that addition of certain salts considerably reduced these losses [40]. For example, the addition of magnesium nitrate to biological materials was found to reduce the losses of Se to negligible levels [41]. Losses due to volatilization can be avoided by dry ashing in a closed system and collecting the volatile compounds on a cold finger [42]. In a typical experiment of dry ashing, 1–2 g of sample was placed in a quartz vial and heated gradually to 400 °C in a quartz oven [14].

Another way to dry digest biological materials in a closed system is to burn them in oxygen by the Schöniger technique (oxygen flask method) [43–45] (Fig. 1). This method is especially suitable for small samples. However, the

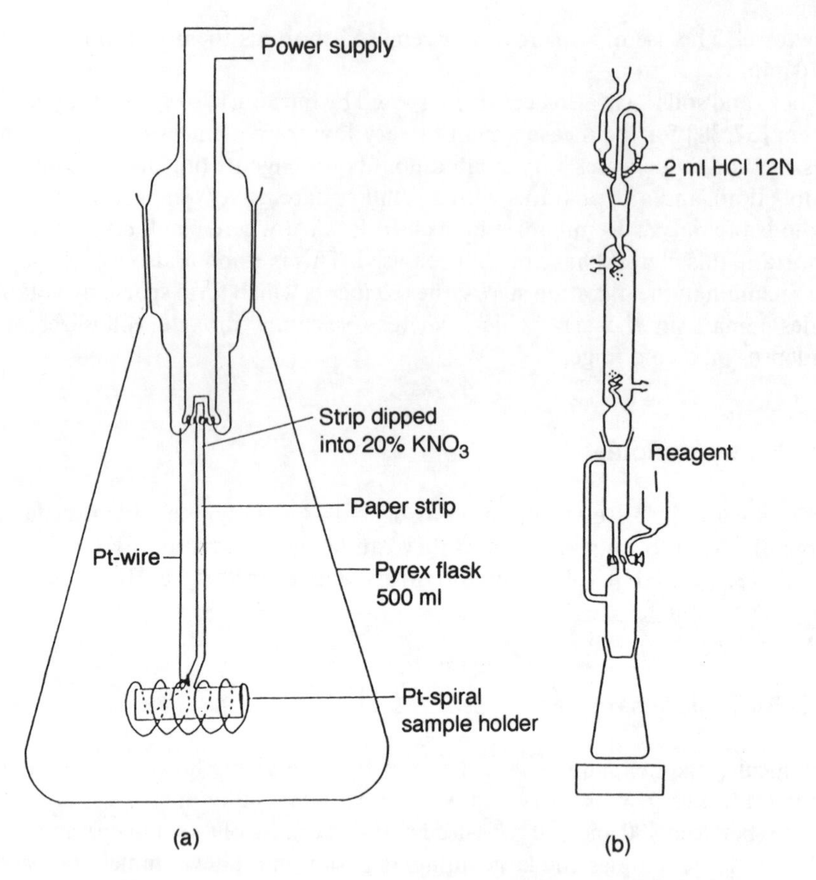

Fig. 1. Closed system digestion for biological material: (a) dry ashing by burning with oxygen (Schöniger technique); (b) wet digestion with acids in Bethge apparatus.

method is not suitable for As, Pb and Bi, which react with the platinum holder [46] or even with a quartz holder [47]. Desai *et al.* [47] found also that Mn, Cu, Zn, Sb, Na and Hg were deposited on the holder and were not recovered completely.

Burning in oxygen in a static system suffers from the large surface of the apparatus, and hence from the risk of losing trace elements by adsorption. The disadvantage is overcome by a dynamic system of combustion in oxygen. The static system is restricted to small amounts of sample, owing to the small amount of oxygen in the vessel, whereas the dynamic method allows combustion of large quantities of sample, and consequently a larger ratio of the trace elements to be determined relative to the surface of the apparatus. The combustion can be done in a commercially available apparatus, the Trace-O-Mat (Anton Paar, Graz,

Austria; H. Kurner, Neuberg, Germany). The description of the apparatus is given in Fig. 2(a). The combustion chamber is made wholly of quartz, and it permits complete mineralization of 1 g of organic or biological solid sample. The combustion takes place in pure oxygen in a very small combustion chamber ($\approx 75\,cm^3$). The controlled incineration is started with an IR lamp. All volatile trace elements are condensed, together with the products of the combustion process, in a cold-finger cooling system filled with liquid nitrogen and mounted on top of the combustion chamber. Subsequent refluxing with a suitable acid in a quartz test tube mounted below the combustion chamber collects both the volatilized elements from the cooled areas and non-volatile elements in the ashing residue. The volume of acid is ≈ 2 ml. The decomposition process takes ≈ 50–60 min. This sytem was found to give recoveries $\geq 95\%$ for all elements studied in biological standard reference materials [48] and also in oil and fats [49]. This method is used mainly in other methods of analysis where the problems of reagent blanks are more acute. However, it is the preferred method also for RNAA if the system is available in the laboratory. This system was used for the determination of trace elements in human serum by RNAA with combustion of the irradiated sample in the Trace-O-Mat apparatus [50]. The method suffers from the disadvantage that only one sample can be digested each time whereas with dry ashing in an oven, several samples can be processed simultaneously. In the second generation of the apparatus the Trace-O-Mat II, higher sample throughput is available and it is possible to process 10 samples per hour [51]. The various steps of operation are seen in Fig. 2(b).

Another possibility for ashing organic materials with oxygen is to use an oxygen plasma which is excited by high frequency or microwaves [52–54]. The sample temperature is barely above 100 °C and hence the method is referred to as 'cold ashing'. The decomposition is performed in a closed quartz vessel equipped with a cooling finger. In spite of these low temperatures As, Sb, Se and Hg will be partially volatilized, and the use of cold finger prevents their loss. In order to shorten the time required for decomposition of organic samples of high mineral content, a magnetic stirrer is used to provide steady sample circulation during the decomposition process [54]. All ashing residues and elements condensed on the surface of the decomposition vessel or cooling finger are collected in 1–2 ml of high purity acid by refluxing. It was found [54] that the system provides a universal apparatus for quantitative recovery of most elements and is characterized by low blanks. Even difficult to decompose organic materials like charcoal, graphite, all plastics (including PTFE), sugars etc. can be ashed in this system without problems. As of now, no RNAA study has been reported involving this system. Its main advantage of low blanks is of no importance in RNAA.

More popular than dry ashing in RNAA is wet ashing. It is common to dissolve dry biological material rapidly by boiling nitric acid with $HClO_4$ or H_2SO_4, or both. Sulfuric acid is added sometimes to reduce the danger of explosion due to $HClO_4$. Care should be taken not to allow the digestion mixture to evaporate to dryness, otherwise it may explode, although in several papers it is reported that the solution was heated to dryness [55]. However, most laboratories avoid the use of $HClO_4$, and instead use HNO_3 or HNO_3/H_2SO_4 mixtures. If the solution obtained is not clear, H_2O_2 is added dropwise to the hot solution until the solution is clear [56, 57]. The use of microwave for heating the acid mixture with the sample leads to rapid dissolution in 5–10 min [58]. Bajo, Suter and Aeschliman [59] described a simple apparatus for wet ashing of any organic material by H_2SO_4/HNO_3 mixture. The digested material is treated with H_2SO_4 at a temperature of 300 °C (such a high temperature cannot be used with HNO_3 owing to its lower boiling temperature), and regular small amounts of HNO_3 are fed automatically into the reaction mixture. Nearly all the species vaporized at a higher temperature than nitric acid are trapped in nitric acid solution. The dissolution is done overnight automatically and no supervision is required. We found also that, for microwave oven digestion of biological material with acids, it is preferable to use $H_2SO_4 + H_2O_2$ rather than HNO_3 or $HNO_3 + H_2O_2$. Because of the higher temperature possible with H_2SO_4, more prolonged dissolution is possible without complete loss of the solvent.

Hot concentrated sulfuric acid decomposes most organic compounds primarily by charring (dehydration); however, in the presence of H_2O_2 a clear colorless solution is obtained. In a typical experiment, $\approx$ 1 g of biological material is treated with $\approx$ 10 ml of conc. H_2SO_4 + 5 ml of 30% H_2O_2, and the microwave oven is operated for $\approx$ 5–10 min. By avoiding the use of HNO_3 the emission of NO_2/N_2O_4 is prevented. However, nitrates are more soluble than sulfates, and the use of H_2SO_4 has the disadvantage of possible formation of insoluble sulphates. This is specially true for Ca-rich samples. For elements which form insoluble sulfates it is preferable to use $HNO_3 + H_2O_2$ mixtures.

Heating in open systems can lead to vaporization of the more volatile elements, such as Hg and Os. The use of refluxing in a Bethge apparatus is recommended [60] (see Fig. 1). Other methods of wet ashing in closed systems involve the use of teflon-covered stainless steel pressure bombs [61–63] or a closed quartz tube [63]. In some cases, the vaporization of some elements are even used, to obtain partial separation already in the digestion step. Guzzi, Pietra and Sabbioni [64] suggested the following procedure for digestion and distillation of biological materials.

Fig. 2. (a) Quartz combustion device of the Trace-O-Mat. (1) Combustion chamber (*ca.* 75-ml capacity); (II) cooling system (liquid N_2); (III) quartz test tube; (1) sample holder; (2) oxygen inlet; (3) cooling mantle; (4) cold finger; (5) condenser; (6) i.r. lamps (from Ref. 3,). (b) Trace-O-Mat II®. Flow diagram of the process. (I) Sampling, (II) cooling, (III) burning, (IV) refluxing. (1) Sample transfer into the burning chamber, (2) liquid nitrogen, (3) burning apparatus, (4) oxygen, (5) refluxing heater, (6) cooling water (from Ref. 51,).

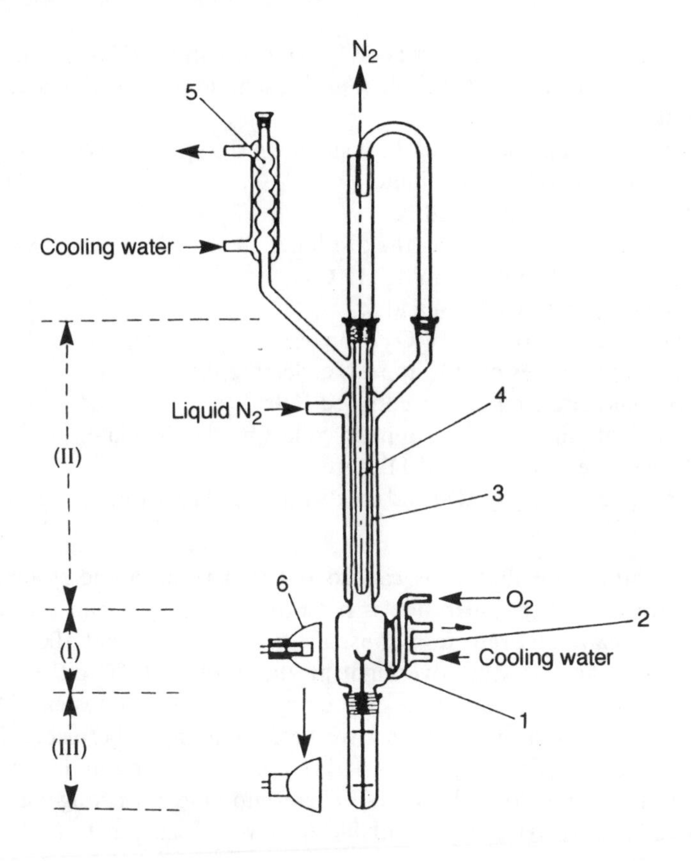
N2
5
Cooling water
Liquid N2
4
(II)
3
6
O2
2
(I)
Cooling water
1
(III)

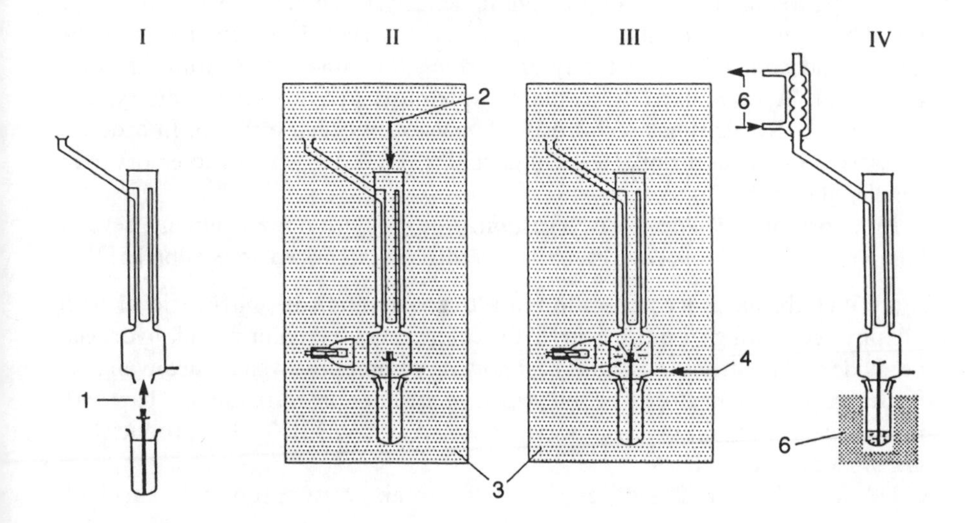
I
II
III
IV
2
6
1
4
6
3

(1) Add 2 ml of concentrated H_2SO_4, 5 ml of fuming HNO_3 and 10 mg of carrier Au to 100–300 mg of irradiated sample contained in a dissolution flask.
(2) Boil for 10 min and collect the distillate in the upper reservoir of the flask.
(3) Cool and recycle the distillate.
(4) Boil for 10 min under reflux.
(5) Boil for 10 min, collecting the distillate.
(6) Cool and add 5 ml of fuming HNO_3.
(7) Repeat operations 2, 3, 4 and 5.
(8) Cool and add 10 ml of H_2O_2 dropwise.
(9) Distil until fumes of SO_3 appear, collecting the distillate.
(10) Cool and add 5 ml of concentrated HBr.
(11) Distil until fumes of SO_3 appear, collecting the distillate.
(12) Repeat operations 10 and 11 twice.
(13) Wash the cooling system and the distillation flask three or four times with 6 M HCl.

The dissolution–distillation is carried out in an automatic system for six samples. However, if the operation is not done routinely it can be performed manually. In many laboratories the usual technique is to heat 100–500 mg of sample with an acid mixture overnight at about 140–170 °C either in a high pressure Teflon decomposition vessel (bomb) or in a reflux system. Most wet dissolutions of biological materials have been done in acidic media. Shimizu, Cho and Murakami [65] suggested the use of tetramethylammonium hydroxide for solubilization of biological material. This compound is a strong base equivalent to NaOH or KOH and is available with very high purity (ppb levels of contaminants).

In order to digest many samples simultaneously, it is preferable to have a large metal block in which all the sample vessels are inserted. The metal block can be heated and controlled constantly and ashing time and temperature are thus controlled. A commercially available apparatus (Tecator digestion system, Tecator AB, Sweden) has 40 borosilicate glass tubes each of 80 ml. In order to digest small samples, Frank [66] equipped this apparatus with an accessory to fit smaller glass tubes.

In laboratories where many digestion processes are done routinely, several commercially available devices may be used, as summarized by Knapp [51].

(1) Wet chemical decomposition in a closed system with nitric acid or with mixtures consisting of nitric acid, hydrochloric acid, perchloric acid, hydrogen peroxide and hydrofluoric acid. The following commerical systems are available.

Decomposition in closed Teflon vessels at medium pressure (up to 8 bar) with microwave heating in the microwave digestion system MDS-81D, produced by GEM Corporation, USA.

Decomposition in closed Teflon vessels at high pressure (up to 85 bar) with

TABLE 1.
Pressurized decomposition systems

Property	System		
	Microwave digestion system MDS-81D	Microwave acid digestion bomb	High-Pressure-Asher HPA
Inorganic sample dry weight	1–2 g	1–2 g	0.2–0.5 g
Organic sample dry weight	0.1–0.5 g	0.2 g	1.0–1.5 g
Maximum temperature	200 °C	250 °C	320 °C
Maximum pressure	8 bar	85 bar	120 bar
Ashing time	2–60 min	2–60 min	45–150 min
Bomb volume	120 ml	45 ml	85 ml
Vessel material	Teflon	Teflon	quartz

microwave heating in the microwave acid digestion bomb, produced by the Parr Instrument Company, USA.

Decomposition in closed quartz or glassy carbon vessels at high pressure (up to 120 bar) with conventional heating in the High-Pressure-Asher HPA, produced by the Anton Paar Company, Austria.

(2) Low pressure ashing with a radio frequency-induced oxygen plasma in the Cool-Plasma-Asher CPA-4, produced by the Anton Paar Company, Austria.

(3) Combustion in a dynamic system with complete recovery of all volatile elements. The system is the Trace-O-Mat VAE-II, produced by the Anton Paar Company, Austria.

A comparison of the three pressurized decomposition systems is given in Table 1.

Teflon is more subject to adsorption than quartz. However, this is not so important in RNAA, owing to the presence of carriers.

3 DETERMINATION OF CHEMICAL YIELD

A common pitfall in radiochemical separation is a yield of less than 100%. The best radiochemical separation is a method with a reproducible yield close to 100%. Usually the recovery yield measurements are done with radiotracers. However, even if radiotracer experiments show that the recovery is near 100%, it is recommended that the yield be measured also in the unknown sample. If the recovery yield in this case is close to 100%, it need not be measured in every experiment. However, if the yield is lower it will never be precisely reproducible, and the recovery yield should be measured in each analysis. One of three methods is usually used for this purpose. The first two are the same in principle. We have mentioned that, before starting the digestion procedure, carriers are added to the irradiated material. These are macroamounts and can be determined more easily and more accurately than trace amounts. As the carriers behave in the same way as the test element in the sample, the recovery yields should be the same. Thus, the recovery yield of the carrier element is taken as the recovery yield of the element being determined. In the first method, the carrier, which is not in a trace amount, can be separated and determined by classical analytical methods, such as gravimetry, colorimetry or some titration methods. In the second method, advantage is taken of laboratory activation analysis techniques. After measuring the radioactivity induced by the original activation of the sample, the separated fraction (carriers together with the trace elements from the sample) is reactivated by irradiation with neutrons, and the amount of the carriers is determined by INAA. The amount originating from the sample is negligible compared with the amount of added carrier, and INAA of the separated fraction yields the amount of the separated carrier. As the amount of added carrier is known, the ratio gives the recovery yield.

A more direct methods is the use of radiotracers, added with the carriers to the sample before digestion. This requires the use of radioisotopes which are not the same as those produced by the neutron activation, and which can be measured simultaneously by γ-ray spectrometry owing to their different γ-ray energies. Thus, for example, Co is determined in NAA via ^{60}Co as indicator radionuclide (IRN), which has γ-rays of 1173 and 1332 keV. Its recovery can be measured with the use of ^{57}Co ($E_\gamma = 122$ or 136 keV) or ^{56}Co ($E_\gamma = 847$, 1238 or 2598 keV). Zirconium is determined either from ^{95}Zr or ^{97}Zr, and ^{88}Zr can be used as a radiotracer for determination of the recovery yield. As the radiotracers are not the radionuclides produced by neutron activation, they have to be produced in a cyclotron or, sometimes, produced by fission (as for, e.g. ^{131}I for radiotracing of I) or by other methods (e.g. ^{131}I produced from neutron irradiation of Te). It is preferable to use cyclotron-produced ^{123}I, and not ^{131}I, which can be formed in the neutron activated sample either by fission or from Te. Thus, ^{56}Co is formed by cyclotron via the ^{56}Fe (p, n) ^{56}Co reaction or the ^{55}Mn (^{3}He, 2n) ^{56}Co reaction, and ^{88}Zr is produced via the ^{89}Y(p, 2n) ^{88}Zr reaction. However, in some cases it is possible to use neutron-produced radionuclides as radiotracers in RNAA. This is so when neutron activation of the naturally occurring isotopes of one element leads to formation of at least two radioisotopes with considerably different half-lives. For example, the irradiation of Zr leads to formation of ^{95}Zr ($t_{1/2} = 64$ d, $E_\gamma = 757$ keV) and ^{97}Zr (16.8 h, 508 or 1148 keV). Formation of ^{93}Zr also occurs, but since its half-life is so long (10^6 years) it can be treated for our purpose as a stable isotope. Irradiation of our studied sample for several days and letting the sample decay ('cool') for two weeks leads to complete decay of ^{97}Zr and allows measurement of Zr content with ^{95}Zr as IRN (only 10^{-6} of the ^{97}Zr activity remains). After these two weeks of 'cooling' ^{97}Zr can be added to the sample together with the carriers and the digestion–separation process is started. Here, ^{97}Zr will be the radiotracer used to measure the recovery yield of Zr in the digestion–separation process. There are several elements for which solely neutron-produced radionuclides can be used. Other examples inclued Mo, Ru, Cd, In, Sn and Sb.

4 RADIOCHEMICAL SEPARATIONS

This section will deal only with general definitions and short descriptions of the various methods used for separation. These are the same methods used for separation in inorganic chemistry. More extensive treatment of each of these methods can be found in many modern textbook on analytical chemistry.

A measure of the efficiency of a radiochemical separation scheme is usually the decontamination factor, defined as the ratio of the activities of the studied nuclide (A_s) and the interfering nuclide (A_i) after and before separation. The decontamination factor (DF) is given by the equation:

$$DF = (A_s/A_i)_a/(A_s/A_i)_b$$

where the subscripts a and b stand for after and before the separation process. Some authors [67] used instead of the decontamination factor its reciprocal value, which is called the separation factor. Some authors consider that the separation removes only the interfering element, and define the decontamination factor as the ratio of the activities of the interfering element before and after the separation [68]. However, it is not always true that $(A_s)_a$ is equal to $(A_s)_b$. The ratio of these values is defined as the recovery R of the studied element.

The main methods used for separation are (1) separation on columns, (2) solvent extraction, (3) precipitation and (4) distillation.

The most popular method is separation on columns. The use of columns, which is liquid–solid phase extraction, is preferable to liquid–liquid extraction because of its rapidity, higher efficiency, ease of work and less danger of radiation hazard and radioactive contamination. As an example of rapidity, we found [13] that $>90\%$ of interfering ^{24}Na can be removed from a solution of digested biological sample in less than 5 s by passage through a 1 cm HAP (hydrated antimony pentoxide) column (flow rate 1 ml/s). Slower passage results in removal of 99.9% of the ^{24}Na but is not always necessary, and for some cases rapidity is more important. Most of the columns used are ion exchange columns; however, other columns contain ion retention media, such as active carbon [69, 70], or C_{18}-bonded silica gel [71]. Hydrated antimony pentoxide is the most popular packing material for removal of ^{24}Na (which is a very important process in many samples); it is also called polyantimonic acid (V). However, the mechanism of its operation is not clear despite several studies [72–75]. It is clear that it is not ion exchange, since the process is irreversible. One study suggested that sodium ions are precipitated as the insoluble salt sodium oxyantimonate [75].

Separation columns are prepared by packing the appropriate material—resin, oxide, active carbon etc. —in a glass or plastic tube 0.5–1.0 cm in diameter to a height of few cm. The lower end of the tube is narrower than the rest and is filled with quartz, glass or Teflon sieves to retain the packing material (Fig 3). Broken pipettes are very suitable for this use. Some columns have a large upper region as a reservoir for the eluting solvent (eluant). Several columns and mini columns are also available commercially from various companies. A solution of the mixed radionuclides is poured into (loaded onto) the column. As the solution passes through the column bed (material), some of the radionuclides are retained on the column material, by various mechanisms, while other radionuclides remain in the solution (eluate). If only the interfering elements are retained on the column, the subsequent process will deal only with the effluent liquid (eluate). If the desired elements are retained on the column, they will either be counted together with the column or eluted from the column with suitable solutions and either counted or processed for the next step.

The packing materials are mainly oxides, such as acidic aluminum oxide (AAO), hydrated manganese dioxide (HMD), hydrated antimony pentoxide (HAP), tin dioxide (TDO), or synthetic ion exchange resins, which are made of polymeric organic materials with a large number of acidic or basic functional groups. Cation exchange resins usually contain the sulfonic acid group ($R-SO_3H$) or carboxylic group (R–COOH), where R is the polymeric hydrocarbon. The resin can be in the H^+ form, or the H^+ can be exchanged with some other cation, e.g. Na^+, to give the Na^+ form. Anion exchange resins contain secondary, tertiary or quaternary amines. They are supplied usually in the Cl^- form, but the anion can be changed by washing with the appropriate acid. Anion exchangers are often preferred over cation exchangers, owing to their higher selectivity for many cations as a results of complex formation. For example, many metal cations form negative ionic complexes with Cl^-. The use of a high concentration of Cl^- in the original solution will transform many of the metal cations to negative complex ions, which will be retained on the anion exchange column. Gradual decrease of the Cl^- concentration in the eluting solution will transorm different metals back to the cation form, and consequently they will be eluted from the column. Elution with solutions of different Cl^- concentration will elute different metal ions. However, cations which do not form strong negative complexes are separated on cation exchange columns. Girardi, Petra and Sabbioni [76] summarized the data for retention of many metal ions on 11 different column materials with various eluting solutions. The results are presented schematically in Periodic table form.

Solvent extraction is very selective and is used frequently for separation of single elements although procedures for multi-element separation based on solvent extraction have also been devised [77]. Some of the procedures involved both column chromatography and solvent extraction [78]. Girardi and Pietra [79] reviewed many of the schemes for multi-element separations based on solid–liquid (column) and liquid–liquid extractions. Solvent extraction is based

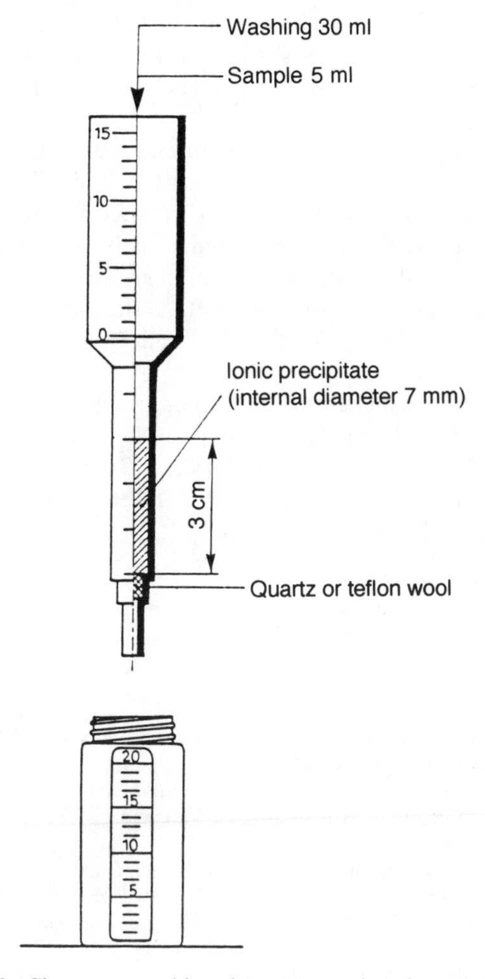

Fig. 3. Chromatographic column separation (from Ref. 76).

on transforming some of the metal ions into compounds that are more soluble in water-immiscible organic solvents than in water. In most cases this is done by addition of organic compounds which form metal chelates. Examples of commonly used chelating agents are dithizone, 8-hydroxyquinoline and various carbamates. Most of the metal chelates are soluble in chloroform or cyclohexane and insoluble in water. The addition of a carbamate with chloroform to an aqueous solution will transfer all the chelate-forming metals to the chloroform phase, whereas Na^+, K^+, Mg^{2+} and similar ions will remain in the aqueous phase. This method can be very selective if, instead of using chloroform with sodium carbamate, a chloroform solution of a mercury carbamate without free carbamate is used for extraction. Only elements that form stronger chelates than mercury (larger formation constants) will be transferred into the organic phase. By using chloroform solutions of various chelates, gradual separation of the radionuclide mixture can be achieved [80].

The precipitation agents are mainly the same ones used in the basic qualitative inorganic analysis course in first year chemistry: Cl^- is used for Ag and Hg; CO_3^{2-} and SO_4^{2-} are used for Ba, Sr and Ca; OH^- is used for Fe, Sc and rare earth elements (REE). The REE are precipitated also by F^-, and S^{2-} is used to precipitate Cu, Ni, Zn, Cd, Au, Ag and platinum group elements. However, some organic chelates are also used to precipitate certain elements, as, e.g., in the precipitation of Cd, Co, Cr, Fe, Ni, Se, Ti, V and Zn from solutions of digested biological material [81].

Several radionuclides which form volatile compounds can be separated from a mixture of IRNs by distillation of these compounds. Mercury, Cl, Br and I can be separated by volatilization of the element itself; Cl, Br and I can also be distilled as HCl, HBr, and HI. Arsenic, Se and Sb can be distilled as halides from $HCl-H_2SO_4$ or $HBr-H_2SO_4$ solutions [6].

REFERENCES

1. A. Chatt, H. S. Dang, B. B. Fong, C. K. Jayawiekreme, L. S. McDowell and D. L. Pegg, *J. Radioanal. Nucl. Chem.* **124**, 65 (1988).
2. R. Zeisler, R. R. Greenberg and S. F. Stone, *J. Radioanal. Nucl. Chem.* **124**, 47 (1988).
3. W. Yinsong, Z. Guisun, T. Mingguang, Z. Min and C. Yuandi, *J. Radioanal. Nucl. Chem.* **151**, 301 (1991).
4. W. Z. Tian and W. D. Ehmann, *J. Radioanal. Nucl. Chem.* **89**, 104 (1985).
5. Z. B. Alfassi (Ed.), *Activation Analysis*, Vol. II, CRC Press, Boca Raton, (1990), p. 579.
6. R. S. Liu, P. Y. Chen, Z. B. Alfassi and M. H. Yang, *J. Radioanal. Nucl. Chem.* **141**, 317 (1990).
7. V. Krivan, *Pure Appl. Chem.* **54**, 787 (1982).
8. R. Caletka, R. Hausbeck and V. Krivan, *J. Radioanal. Nucl. Chem.* **120**, 305 (1988); K. H. Theimer and V. Krivan, *Anal. Chem.* **62**, 2722 (1990).
9. K. P. Egger and V. Krivan, *Fresenius' Z. Anal. Chem.* **323**, 827 (1986).
10. C. C. Chu, P. Y. Chen, M. H. Yang and Z. B. Alfassi, *Analyst* **115**, 29 (1990).
11. D. A. Becker, H. L. Rook and P. D. LaFleur, in *Nondestructive Activation Analysis*, edited by S. Amiel, Elsevier, Amsterdam (1981), p. 237.
12. W. D. Ehmann, J. D. Robertson and S. W. Yates, *Anal. Chem.* **62**, 50R (1990).
13. Z. B. Alfassi, and B. Rietz, *Analyst*, in press.
14. M. Speziali, M. de Casa and E. Orvini, *Biol. Trace Elem. Res.* **17**, 271 (1988).
15. N. Lavi, M. Mantel, and Z. B. Alfassi, *Analyst* **113**, 1855 (1988).
16. J. Kalouskova, K. Drabek, L. Pavlik and F. Hodik, *J. Radioanal. Nucl. Chem.* **129**, 59 (1989).
17. G. D. McOrist, J. J. Fardy and T. M. Florence, *J. Radioanal. Nucl. Chem.* **119**, 449 (1987).
18. K. Heydorn and H. R. Lukens, *Risøe Rep.* (138) (1966).

19. N. Lavi and Z. B. Alfassi, *J. Radioanal. Nucl. Chem.* **126**, 361 (1988).
20. A. J. Blotcky, W. C. Duckworth, A. Ebrahim, F. K. Hamel, E. P. Rack and R. B. Sharma, *J. Radioanal. Nucl. Chem.* **134**, 151 (1989).
21. J. Kucera, M. Simkova, J. Lener, A. Mravcova, L. Kinova and I. Penev, *J. Radioanal. Nucl. Chem.* **141**, 49 (1990).
22. S. Bajo, in *Preconcentration Techniques of Trace Elements*, edited by Z. B. Alfassi, CRC Press, Boca Raton (1992), p. 3.
23. J. M. Lo, J. C. Wei and S. J. Yeh, *Anal. Chem.* **49**, 1146 (1977); S. J. Yeh, J. M. Lo and C. L. Tseng, *J. Radioanal. Nucl. Chem.* **124**, 157 (1988).
24. N. Lavi and Z. B. Alfassi, *J. Radioanal. Nucl. Chem.* **130**, 71 (1989); C. R. Lan, Y. C. Sun, J. H. Chao, C. Chung, M. H. Yang, N. Lavi and Z. B. Alfassi, *Radiochim. Acta* **50**, 225 (1990).
25. J. C. Laul, E. A. Lepel and M. R. Smith, *J. Radioanal. Nucl. Chem.* **123**, 349 (1988).
26. N. R. Das, *J. Radioanal. Nucl. Chem.* **122**, 13 (1988).
27. S. J. Parry, M. Asif and I. W. Sinclair, *J. Radioanal. Nucl. Chem.* **123**, 593 (1988).
28. Y. Terakado, T. Fujitani and J. Takada, *J. Radioanal. Nucl. Chem.* **129**, 23 (1989).
29. E. L. Hoffman, *J. Geochem. Explor.* **32**, 301 (1988).
30. R. Caletka, R. Hausbeck and V. Krivan, *J. Radioanal. Nucl. Chem.* **120**, 319 (1988).
31. A. J. Blotcky, A. Ebrahim and E. P. Rack, *Anal. Chem.* **60**, 2734 (1988).
32. J. T. Van-Eltern, J. Hoegge, E. E. van der Hoek and H. A. Das, *J. Radioanal. Nucl. Chem.* **154**, 343 (1991).
33. C. R. Lan, C. Tseng, M. H. Yang and Z. B. Alfassi, *Analyst* **116**, 35 (1991).
34. R. W. King, R. W. Kerrich and R. Daddar, *Am. Mineral.* **73**, 424 (1988).
35. H. A. Das, A. Faanhof and H. A. van der Sloot (Eds.), *Environmental Analysis*, Elsevier, Amsterdam (1983), p. 83.
36. P. J. Lamothe, T. L. Fries and J. J. Consul, *Anal. Chem.* **58**, 181 (1986).
37. H. B. Han, G. Kaiser and G. Tölg, *Anal. Chim. Acta* **128**, 9 (1981).
38. H. B. Han, G. Kaiser and G. Tölg, *Anal. Chim. Acta* **134**, 3 (1982).
39. E. I. Hamilton, M. J. Minski and J. J. Cleary, *Analyst* **92**, 257 (1967).
40. T. T. Gorsuch, *Analyst* **84**, 135 (1959).
41. S. S. Krishnan and D. R. Crapper, *Radiochem. Radioanal. Lett.* **20**, 287 (1975).
42. G. Knapp, S. E. Raptis, G. Kaiser G. Tölg, P. Schramel and B. Schreiber, *Fresenius' Z. Anal. Chem.* **308**, 97 (1981).
43. W. Schoniger, *Fresenius' Z. Anal. Chem.* **181**, 28 (1961).
44. A. M. G. Macdonald, in *Advances in Analytical Chemistry and Instrumentation*, edited by C. N. Reily, Vol. 4, Wiley, New York (1965), p. 75.
45. J. Basset, R. C. Denny, G. H. Jeffery and J. Mendham, *Vogel's Textbook of Quantitative Inorganic Analysis*, 4th Ed., Longman, London (1978), p. 115.
46. T. T. Gorsuch, *International Series of Monographs in Analytical Chemistry*, Vol. 39, Pergamon Press, Oxford (1970).
47. H. B. Desai, R. Kayasth, R. Parthasarathy and M. S. Das, *J. Radioanal. Nucl. Chem.* **84**, 123 (1984).
48. G. Knapp, S. E. Raptis, G. Kaiser, G. Tölg, P. Schramel and B. Schreiber, *Fresenius' Z. Anal. Chem.* **308**, 97 (1981).
49. S. E. Raptis, G. Kaiser and G. Tölg, *Anal. Chim. Acta* **138**, 93 (1982).
50. D. van Renterghem and R. Cornelis, in *Trace Elements Analytical Chemistry in Biology and Medicine*, edited by P. Brätter and P. Schramel, Walter de Gruyter, Berlin (1988), p. 55.
51. G. Knapp, in *Trace Elements Analytical Chemistry in Medicine and Biology*, edited by P. Brätter and P. Schramel, Vol. 5, Walter de Gruyter, Berlin (1988), p. 63.
52. C. E. Gleit and W. D. Holland, *Anal. Chem.* **34**, 1454 (1965).
53. G. Kaiser, P. Tschöpel and G. Tölg, *Fresenius' Z. Anal. Chem.* **253**, 177 (1971).

54. S. E. Raptis, G. Knapp and A. P. Schalk, *Fresenius' Z. Anal. Chem.* **316**, 482 (1983).
55. H. R. Ralston and E. S. Sato, *Anal. Chem.* **43**, 129 (1971).
56. A. J. Krynitsky, *Anal. Chem.* **59**, 1884 (1987).
57. N. Lavi, F. Lux and Z. B. Alfassi, *J. Radioanal. Nucl. Chem.* **129**, 93 (1989).
58. H. M. Kingston and L. B. Jassie, *Anal. Chem.* **58**, 2534 (1986).
59. S. Bajo, U. Suter and B. Aeschliman, *Anal. Chim. Acta* **149**, 321 (1983).
60. R. Bock and I. L. Marr, A *Handbook of Decomposition Methods in Analytical Chemistry*, Halstead Press (1979), p. 115.
61. R. Uhrberg, *Anal. Chem.* **54**, 1906 (1982).
62. K. Okamoto and K. Fuwa, *Anal. Chem.* **56**, 1758 (1984).
63. A. Faanhof and M. A. Das, *Radiochem. Radioanal. Lett.* **30**, 405 (1977).
64. G. Guzzi, R. Pietra and E. Sabbioni, *J. Radioanal. Chem.* **34**, 35 (1976).
65. S. Shimizu, T. Cho and Y. Murakami, in *Trace Elements Analytical Chemistry in Medicine and Biology*, edited by P. Brätter and P. Schramel, Vol. 5, Walter de Gruyter, Berlin (1988), p.72.
66. A. Frank, in *Trace Elements Analytical Chemistry in Medicine and Biology*, edited by P. Brätter and P. Schramel, Vol. 5, Walter de Gruyter, Berlin (1988), p. 78.
67. A. Mizuike, *Preconcentration Techniques of Trace Elements*, Springer, Berlin (1988), p. 51.
68. K. Heydorn, *Neutron Activation Analysis for Clinical Trace Element Research*, Vol. 1, CRC Press, Boca Raton (1984), p. 113.
69. H. A. van der Sloot, G. D. Wals, C. A. Weers and H. A. Das, *Anal. Chem.* **52**, 112 (1980).
70. C. I. Siripone, G. D. Wals and H. A. Das, *J. Radioanal. Chem.* **79**, 35 (1983).
71. J. J. Fardy and T. Mingguang, *J. Radioanal. Nucl. Chem.* **123**, 573 (1988).
72. F. Girardi and E. Sabbioni, *J. Radioanal. Nucl. Chem.* **1**, 169 (1968).
73. M. Abe and T. Ito, *Bull. Chem. Soc. Jpn.* **41**, 333 (1968).
74. C. Konecny and I. Hartl, *Z. Phys. Chem. Leipzig* **256**, 17 (1975).
75. S. K. Nyarku, *Anal. Lett.* **17**, 2213 (1984).
76. F. Girardi, R. Pietra and E. Sabbioni, *J. Radioanal. Chem.* **5**, 141 (1970).
77. R. F. Coleman and G. C. Goode, *J. Radioanal. Chem.* **15**, 367 (1973).
78. S. May and G. Pinte, J. Radioanal. Chem. **3**, 329 (1969).
79. F. Girardi and R. Pietra, *At. Energy Rev.* **14**, 3 (1976).
80. A. Wyttenbach and S. Bajo, *Anal. Chem.* **11**, 1813 (1976).
81. N. Lavi and Z. B. Alfassi, *Analyst* **115**, 817 (1990).

Chapter 8

PROMPT ACTIVATION ANALYSIS

Chien Chung

School of Nuclear Science, National Tsing Hua University, Hsinchu 30043, Taiwan, ROC

1 INTRODUCTION

With the substantial increase in scientific and technological activities, considerable interest exists in the development and application of new analytical techniques. Many severe limitations of conventional delayed neutron activation analysis (NAA) exist with respect to its capability (on which research communities rely heavily) for on-line characterization of a process sample to give an instant result. The need for more and better analytical techniques has resulted in a quite spectacular development and diversification of new methods in chemical analysis. Particularly noteworthy has been the development of nuclear techniques, which have an unusually high degree of rapidity, elemental selectivity and sensitivity. The penetrating radiations usually involved in nuclear methods enable measurements to be made that are characteristic of the bulk sample instead of just its surface condition. It is often possible to avoid chemical separation; more importantly, nuclear methods have also shown adaptability to extremely harsh field conditions which might preclude the use of other chemical analytical methods. This has led to the development of nuclear analytical techniques for in situ, in vivo, and on-line analyses of particular samples, such as hidden explosives in airport security inspection of luggage. It has stimulated further search for and application of nuclear phenomena as analytical probes.

Among the nuclear methods which have proved useful in chemical analysis is prompt gamma-ray activation analysis (PGAA), in addition to conventional NAA. Although the PGAA technique has been developed using various neutron sources, and the system is usually installed around a nuclear facility such as a research reactor, many demands create field applications such as in vivo

Chemical Analysis by Nuclear Methods Edited by Z. B. Alfassi

(IVPGAA) medical diagnosis, in situ (ISPGAA) geological surveys, and on-line (OLPGAA) detection of sensitive objects. These real-time, on-line techniques do not rely on induced radioactivity and thus avoid the lengthy period of counting during sample decay long after the irradiation.

Despite the extensive utilization of chemical analysis, no single method, nuclear or chemical, has been found to satisfy all of the sensitivity requirements for the elements of interest to all scientific and engineering disciplines. Each method is complementary to others.

Chemical analysis by nuclear methods is certainly not limited to taking the sample to a fixed laboratory for measurement. In the last decade, many field applications using the PGAA technique have been developed. Among the latest developments is the installation of a PGAA facility in an airport for explosives detection [1]. The idea is to detect the unique, very high energy prompt gamma-rays emitted from nitrogen, of which the explosive has a high content. This nuclear method to detect explosives hidden in luggage out-performs chemical techniques to analyze explosive vapor or traditional methods such as hand searches and animal olfaction [2].

A knowledge of the neutron reaction and the prompt gamma-rays emitted from it is prerequisite for the PGAA method, and the neutron source and gamma-ray spectrometer are indispensable integral parts of a PGAA facility. With the high resolution gamma-ray spectrometer commercialized in the 1960s and prompt gamma-rays well documented in the 1970s, PGAA was rapidly adopted by the analytical community as a reliable nuclear method in the 1980s. Future expansion of PGAA, in particular for field applications, is probable owing to its mobility, convenience, and versatility.

2 PRINCIPLES

The principles of PGAA have been introduced in previous chapters in detail and are therefore described here only briefly. PGAA contains three stages involving nuclear phenomena: (1) the sample to be investigated is bombarded with neutrons, a neutral ionizing radiation; (2) a nuclear reaction caused by neutron interaction with the sample may eject characteristic prompt gamma-rays; and (3) quantitative measurement of prompt gamma-rays using a spectrometer.

2.1 Nuclear reactions

Neutrons, which possess no electric charge, have proved to be very effective in penetrating positively charged nuclei, thereby producing nuclear reactions. Not only are high energy neutrons capable of penetrating the nucleus, but comparatively low energy (thermal) neutrons have also been found to be extremely effective. A great deal of work has been done with thermal neutrons, and the information so obtained on the thermal neutron-induced reaction is the basis of the PGAA technique. When a nucleus in the sample is bombarded with neutrons, it may capture the incident neutron and subsequently form a compound nucleus at one of the excited states. The excitation energy (E^*) is provided by the kinetic energy of the bombarding neutron (KE_n) and the binding energy of the neutron in the compound nucleus (B_n):

$$E^* = KE_n + B_n, \quad \text{MeV} \tag{1}$$

A typical example is the nitrogen nuclide bombarded with thermal neutrons ($KE_n < 4 \times 10^{-7}$ MeV) by which a ^{15}N compound nuclide is formed:

$$^{14}N + {}^{1}n_{th} \rightarrow {}^{15}N^*, \quad E^* \cong B_n \tag{2}$$

One possible way to complete the reaction process is for ^{15}N to return to its normal ground state, emitting a series of prompt gamma-rays with total energy equivalent to E^*:

$$^{14}N + {}^{1}n_{th} \rightarrow {}^{15}N^* \rightarrow {}^{15}N + E_r \tag{3}$$

This process, also known as the (n, γ) reaction, has been observed with most elements, particularly the heavy ones. In most cases, the emission of these gamma-rays is extremely rapid, depending upon the lifetime of the excited state, and typically occurs $\approx 10^{-12}$ s after the formation of the compound. Thus these photons are given the name of 'prompt' gamma-rays.

Sometimes the capture of a neutron is accompanied by the emission of a charged particle, in particular for energetic neutron bombardment. For example, the nitrogen nuclide may also have the following reactions induced:

$$^{14}N + {}^{1}n \rightarrow {}^{15}N^* \rightarrow \begin{array}{l} ^{1}H + {}^{14}C \\ ^{3}H + {}^{12}C \\ ^{4}He + {}^{11}B \\ ^{4}He + {}^{4}He + {}^{7}Li \end{array} \tag{4}$$

The above reactions, with the exception of the (n, p) process, require non-thermal neutrons with $KE_n > 0.16$ MeV to make them happen. Competition certainly exists among various nuclear reactions, and sometimes the reaction product is radioactive, like the 5730-year ^{14}C.

The ability to produce prompt gamma-rays from almost any element has made PGAA techniques available to scientists, engineers, and technicians for the detailed study of various samples. In order experimentally to determine the elemental concentration, the production rate of compound nuclei and the emission rate of prompt gamma-rays have first to be evaluted.

The probability of occurrence of an (n, γ) neutron capture reaction is conveniently expressed in terms of the concept of cross-section. Since interactions in an

(n, γ) reaction take place with individual target nuclei independently of each other, it is useful to refer the probability of a reaction to one target nucleus. Assume that a sample as a target material is bombarded by a flux of neutrons, consisting of ϕ neutrons per unit time per unit area. One can suppose that with such a target nucleus there is associated an area σ, such that if the neutron strikes inside σ, there is an (n, γ) reaction happening. The quantity σ is defined as the reaction cross-section, representing the (n, γ) reaction probability per target nucleus.

Unfortunately, both σ and ϕ are neutron energy dependent, with drastic variation across many orders of magnitude of neutron energy. For instance, the $Cd(n, \gamma)$ reaction cross-section, compiled from numerous investigators [3], in the neutron energy range 10^{-10} to 10^2 MeV, varies from 10^{-19} down to $10^{-24}\,cm^2$. On the other hand, the bombarding neutrons originating from a research reactor, a neutron generator or an isotopic source, may contain neutrons from thermal to high energy ($KE_n > 10$ MeV). Thus, the reaction rate (R) producing the (n, γ) compound nucleus becomes:

$$R = \int m/m_0 \, N_a \, d\phi(E)/dE \cdot \sigma(E)/dE = m/m_0 \, N_a \, \bar{\phi} \cdot \bar{\sigma}, \quad \text{reactions/s} \qquad (5)$$

where m_0 is the atomic weight of the element of interest and N_a is the Avogadro constant; m is the elemental weight to be analyzed and $m/m_0 N_a$ is the number of target nuclei. For simplicity, one can use the weighted average of neutron flux $\bar{\phi}$ and reaction cross-section $\bar{\sigma}$ to replace the complex energy integration.

2.2 Prompt gamma-rays

Once the compound nucleus is formed after the bombarding neutron strikes the target nucleus, the excitation energy can be evaluated according to the mass difference. For example, in the thermal neutron capture reaction of (3), the total energy of prompt gamma-rays becomes:

$$E^* = B_n = \{[m(^{14}N) + m_n] - m(^{15}N)\} c^2 = \Delta mc^2, \quad \text{MeV} \qquad (6)$$

where $m(^{14}N)$ = the mass of the ^{14}N nuclide, $m(^{15}N)$ = the mass of the ^{15}N nuclide, m_n = rest mass of the neutron, and c = speed of light.

Using the experimental data of atomic mass unit (amu) compiled in a reference book [4]: $m(^{15}N) = 15.000109$ amu, $m(^{14}N) = 14.003074$ amu, $m_n = 1.008665$ amu, and 1 amu $= 931.5\,MeV/c^2$, one can obtain $E^* = 0.01163\,\text{amu} \times c^2 = 10.83$ MeV.

The total excitation energy E^*, experimentally determined as 10.8292 ± 0.0001 MeV, is the fuel for subsequent de-excitations to the ground state; the nuclide may de-excite in one shot, utilizing all E^*, or de-excite through numerous excited states and eject many low energy prompt gamma-rays. As many as 50 prompt

TABLE 1.
Prompt gamma-rays from thermal neutron capture reactions [6]

Element	σ_{th} (10^{-24} cm^2)	E_r (MeV)	I_r [a]	Element	σ_{th} (10^{-24} cm^2)	E_r (MeV)	I_r [a]
H	0.332	2.2232	100	Rh	150	0.2172	8.71
Li	0.036	2.0325	89.33	Ag	63.6	0.1984	23.88
B	755 [b]	0.4776	93	Cd	2450	0.5586	72.73
N	0.0747	10.8293	14.12	In	194	1.2934	17.59
S	0.520	5.4205	59.08	Te	4.70	0.6031	9.34
Cl	33.2	6.1109	20.00	I	6.20	0.4429	4.38
Ar	0.678	4.7450	55.00	Xe	24.50	2.2252	12.56
K	2.10	0.7703	51.48	Cs	29.0	1.6189	1.50
Ca	0.43	1.9420	72.55	La	9.14	1.5962	15.36
Sc	2.65	8.1747	11.83	Nd	50.5	0.6973	73.29
Ti	6.10	1.3815	69.08	Sm	5800	0.7376	11.04
V	5.04	6.5172	17.83	Eu	4600	1.5640	0.40
Cr	3.10	8.8841	26.97	Gd	49000	1.1865	10.83
Mn	13.3	7.2438	12.13	Dy	930	2.7034	2.48
Fe	2.55	7.6311	28.51	Ho	66.5	0.5435	2.97
Co	37.2	6.8768	8.21	Er	162	0.8160	40.96
Ni	4.43	8.5334	16.98	Tm	103	0.6874	3.77
Cu	3.79	7.6366	15.71	Yb	36.6	5.2657	5.78
Zn	1.10	1.0774	18.93	Lu	77.3	0.4577	6.95
Ge	2.30	0.5964	33.10	Hf	102	1.2064	4.60
As	4.30	1.5343	7.18	Ta	21.1	0.4029	27.18
Br	6.80	0.7769	4.13	Os	15.3	2.2233	25.02
Kr	25.0	0.8818	84.00	Ir	426	5.6670	1.15
Sr	1.21	1.8361	57.79	Au	98.8	1.2018	12.17
Y	1.28	6.0798	77.49	Hg	376	0.3681	81.35
Mo	2.65	0.8488	14.52				

[a] Number of prompt photons per 100 neutrons captured.
[b] $B(n_{th}, \alpha)$ reaction instead of capture reaction.

gamma-rays associated with the $^{14}N(n, \gamma)\,^{15}N$ reaction have been reported, each one with different energy and probability [5]. These prompt gamma-rays may serve for quantitative identification of the element to be analyzed in the sample.

There are many data sets of prompt gamma-rays compiled from numerous experimental investigations conducted in the past three decades; the one widely referred to is that compiled by Lone, Leavitt and Harrison [6], in which 1915 prompt gamma-rays emitted from 84 elements in thermal neutron capture reactions are tabulated for quantitative identification purposes; the prompt gamma-ray information for those elements measurable by PGAA is shown in Table 1, in which the thermal neutron capture cross-section as well as the gamma-ray energy and the intensity recommended for identification are listed. Thus, the emission rate of the specific prompt gamma-ray from the neutron capture reaction becomes:

$$S_r(E_r) = R\, I_r(E_r)/100 = mN_a\bar{\phi}\cdot\bar{\sigma}\, I_r(E_r)/(100\, m_0), \quad \text{photons/s} \tag{7}$$

where $I_r(E_r)$ is the prompt gamma-ray intensity at E_r in terms of photons per 100 neutrons captured.

2.3 Prompt gamma detection

Both the energy and the intensity of the gamma-ray should be rapidly detected in order to allow quantitative measurement of the elemental concentration in the target material. A gamma-ray spectrometer, equipped with either a scintillation or a semiconducting detector, may be placed next to the target sample to catch the emitted prompt gamma-rays. Since the prompt gamma-rays are emitted isotropically to all directions, the detector can be positioned anywhere except in the neutron beam path. A gamma-ray detector has an absolute efficiency ε, defined as the ratio of the number of counts in the full-energy photopeak area in the spectrum, $C(E_r)$, to the total number of emitted gamma-rays; ε is measured as counts per photon and is a function of both detector's intrinsic character and the geometric arrangement of the detector relative to the sample. Thus, the count rate for the specific full-energy prompt gamma-ray becomes:

$$\mathrm{d}C(E_r)/\mathrm{d}t = \varepsilon(E_r)\, S_r(E_r), \quad \text{counts/s} \tag{8}$$

If a counting period T_c is preset for PGAA measurement in which the neutron flux does not change during the irradiation, the elemental weight (m) can be evaluated by combining equations (7) and (8):

$$m = 100\, C\, m_0/[N_a\, \varepsilon(E_r)\, T_c\, \bar{\phi}\cdot\bar{\sigma}\, \mathrm{I}_r(E_r)], \quad \mathrm{g} \tag{9}$$

Let us again take the nitrogen nucleus as an example. Assuming that a single bismuth germanate scintillation detector, placed 25 cm away from an item of luggage, has an efficiency of $\varepsilon = 10^{-3}$ counts/photon for the 10.829 MeV prompt gamma-ray emitted from the N(n, γ) reaction. In a 100-s irradiation and counting period, the luggage is bombarded with thermalized neutrons with a flux of 10 000 n cm^{-2} s^{-1}; the total counts collected in the 10.829 MeV photopeak area in the gamma-ray spectrum is 300 counts. The weight of nitrogen in the luggage becomes:

$$m = \frac{100 \times (300\ \text{counts}) \times (14.01\ \text{g/mol})}{(6.022 \times 10^{23}/\text{mol})(10^{-3}\ \text{count/photon})(100\ \text{s})(10^{4}/\text{cm}^{-2}\,\text{s}^{-1})(0.0747 \times 10^{-24}\ \text{cm}^{2})(14.1\ \text{photon})}$$
$$= 663\ \text{g of nitrogen}$$

The nitrogen content in the widely used plastic explosive HMX ($C_4H_8O_8N_8$) is 37.8% by weight, implying that 0.663 kg/37.8% = 1.8 kg of HMX explosive could be hidden in the inspected luggage if other personal belongings in the luggage contain very little nitrogen.

Some of the target nuclei may become radioactive after PGAA. For instance, a food of high salt content placed in the luggage subject to airport security inspection may become radioactive after PGAA interrogation. We further assume that 10 kg of food carried in the luggage contains 230 g of sodium. The induced activity, described in the previous chapter, of the radioactive ^{24}Na with 15 h half-life becomes:

$$A_0 = N\phi_{th}\sigma_{th}[1 - \exp(-\lambda T_c)], \quad \text{Bq} \tag{10}$$

where A_0 = activity of ^{24}Na at the end of the PGAA scan, Bq; N = total number of sodium nuclides $= m/m_0 N_a = 6.022 \times 10^{24}$; σ_{th} = thermal neutron capture cross-section $= 0.53 \times 10^{-24}\ \text{cm}^2$; λ = decay constant of $^{24}Na = 0.693/15$ h.

Therefore, with $\phi_{th} = 10^4/\text{cm}^2.\text{s}$, and $T_c = 100$ s, the activity becomes:

$$\begin{aligned} A &= (6.022 \times 10^{24}) \times (10^4\ \text{cm}^{-2}\,\text{s}^{-1}) \times (0.53 \times 10^{-24}\ \text{cm}^2) \\ &\quad \times (0.693 \times 100)/(15 \times 3600) \\ &= 40\ \text{Bq} = 1.1\ \text{nCi}, \quad \text{or } 0.11\ \text{nCi/kg of food.} \end{aligned}$$

This radioactive concentration, 0.11 nCi/kg, is far below the natural radioactivity in foodstuffs, around 2.1 nCi/kg, caused primarily by natural ^{40}K decay [7]. The minute amount of induced radioactivity in PGAA causes no health concerns of radiation safety, allowing PGAA field applications such as in vivo studies on the living body and on-line scans for foodstuffs acceptable by the authorities.

3 QUANTITATIVE DETERMINATION

The absolute measurement of elemental weight in equation (9) requires the absolute determination of $\varepsilon(E_r)$, $\bar{\phi}$ and $\bar{\sigma}$, whereas $I_r(E_r)$ can be obtained from references. The spectrometric detector efficiency $\varepsilon(E_r)$ can be measured for a point sample up to a certain gamma-ray energy; unfortunately, most prompt gamma-rays used for identification purposes have energy greater than 2 MeV, making the high energy calibration extremely stringent. In field applications, such as explosives detection in luggage, the irregular shape of the sample matrix, with various densities and heterogeneous distribution of elements of interest, further hampers the absolute determination of efficiency $\varepsilon(E_r)$. Furthermore, considering the energy spread of bombarding neutrons, very few experimenters even try to make absolute determination of $\bar{\phi}$ and $\bar{\sigma}$ in the sample.

Much as in the conventional NAA technique, a 'comparator' with known concentrations of elements of interest in a similar (preferably identical) sample matrix is used in PGAA for calibration purposes, leading to the quantitative determination of the elements of interest in the sample. Assuming that a comparator, containing the element of interest with certified or known weight $m(R)$, is analyzed in the PGAA facility, the total number of counts $C(R)$ collected during the irradiation period T_c becomes:

$$C(R) = m(R)N_a\varepsilon(E_r)T_c(\bar{\phi}\cdot\bar{\sigma})_r\cdot I_r(E_r)/100\,m_0 \tag{11}$$

Thus, the unknown elemental weight in the sample can be evaluated by combining equations (9) and (11):

$$\begin{aligned} m &= m(R) \times \frac{100\,C\varepsilon(E_r)T_c(\bar{\phi}\cdot\bar{\sigma})_r\cdot I_r(E_r)m_0N_a}{100C(R)\varepsilon(E_r)T_c(\bar{\phi}\cdot\bar{\sigma})\,I_r(E_r)m_0N_a} \\ &= m(R)C(\bar{\phi}\cdot\bar{\sigma})_r/[C(R)(\bar{\phi}\cdot\bar{\sigma})] \end{aligned} \tag{12}$$

If the irradiation condition of the neutrons is unchanged, that is, neutron flux intensity and neutron energy spectrum have no variation during the two experimental irradiations, the quantitative determination can be further simplified as:

$$m = m(R)C/C(R) \tag{13}$$

A flux monitor such as a neutron counter is usually positioned behind the PGAA set-up to give on-line, real-time readings of the flux intensity, providing the necessary correction if indeed a variation occurs in the irradiation conditions.

3.1 Spectroscopic analysis

In the measurement of prompt gamma-ray energies above 100 keV, there are only two detector categories of major importance: the inorganic scintillation detector and the semiconducting detector. The choice in a given PGAA application resolves into a trade-off between detector efficiency and spectral resolution. Both scintillation and semiconducting detectors have been extensively utilized in the PGAA set-up to acquire gamma-ray spectra for quantitative analysis.

The semiconducting detector has much superior energy resolution; that used in most PGAA facilities is the high purity germanium (HPGe) detector. A typical prompt gamma-ray spectrum measured by a HPGe detector is shown in Fig. 1. A well resolved spectrum not only helps separate neighboring photopeaks but aids in the detection of weak photons on the spectral continuum. This is particularly important for multi-elemental analysis when many elements in the sample matrix are irradiated by neutrons and emit several hundred prompt gamma-rays simultaneously. As the HPGe detector is placed as close as possible to the irradiated sample to acquire the maximum counts, the detector is inevitably subject to neutron bombardment; thus the spectrum in Fig. 1 can also be used as a reference for the spectral background during prompt gamma counting.

Although the HPGe has the advantage of high spectroscopic resolution, the detector efficiency is rather poor in comparison with the scintillation detector. The high efficiency sodium iodide [NaI(Tl)] and bismuth germanate (BGO) scintillators have the advantage of high atomic number and high density,

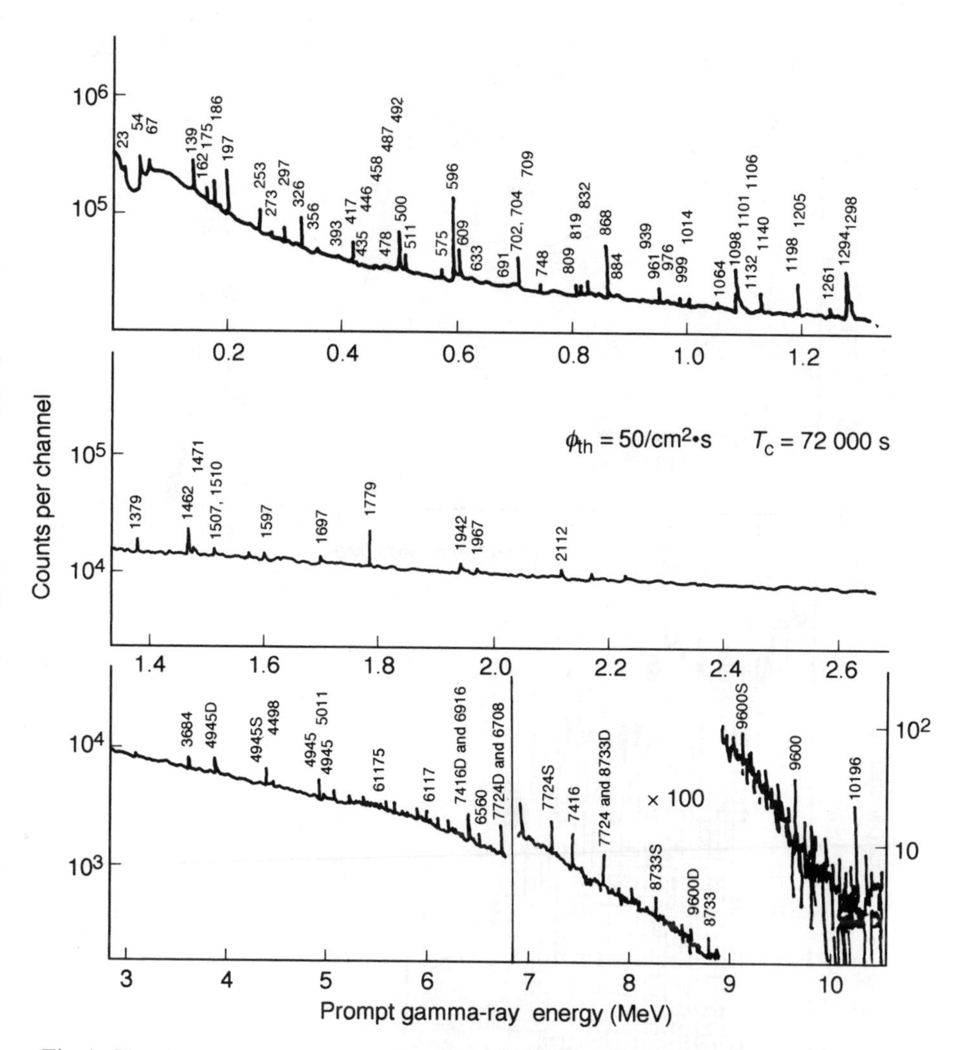

Fig. 1. Prompt gamma-ray spectrum of a 30% relative efficiency HPGe detector itself under irradiation by a low flux of thermal neutrons.

resulting in very high interaction probability for gamma-rays. For instance, the detecting efficiency of a NaI(Tl) scintillator outperforms the HPGe detector by at least a factor of 10 for $E_r > 8$ MeV. Therefore a rapid PGAA scan is better conducted by use of a scintillation detector. Although NaI(Tl) detectors are available in large size (as large as 24 in. diameter in gamma-ray scanning), the detector itself may become radioactive in PGAA irradiation, as the sodium and iodine in the scintillator may be activated to form 15 h ^{24}Na and 25 min ^{128}I, respectively, through (n, γ) reactions. The high density BGO scintillator

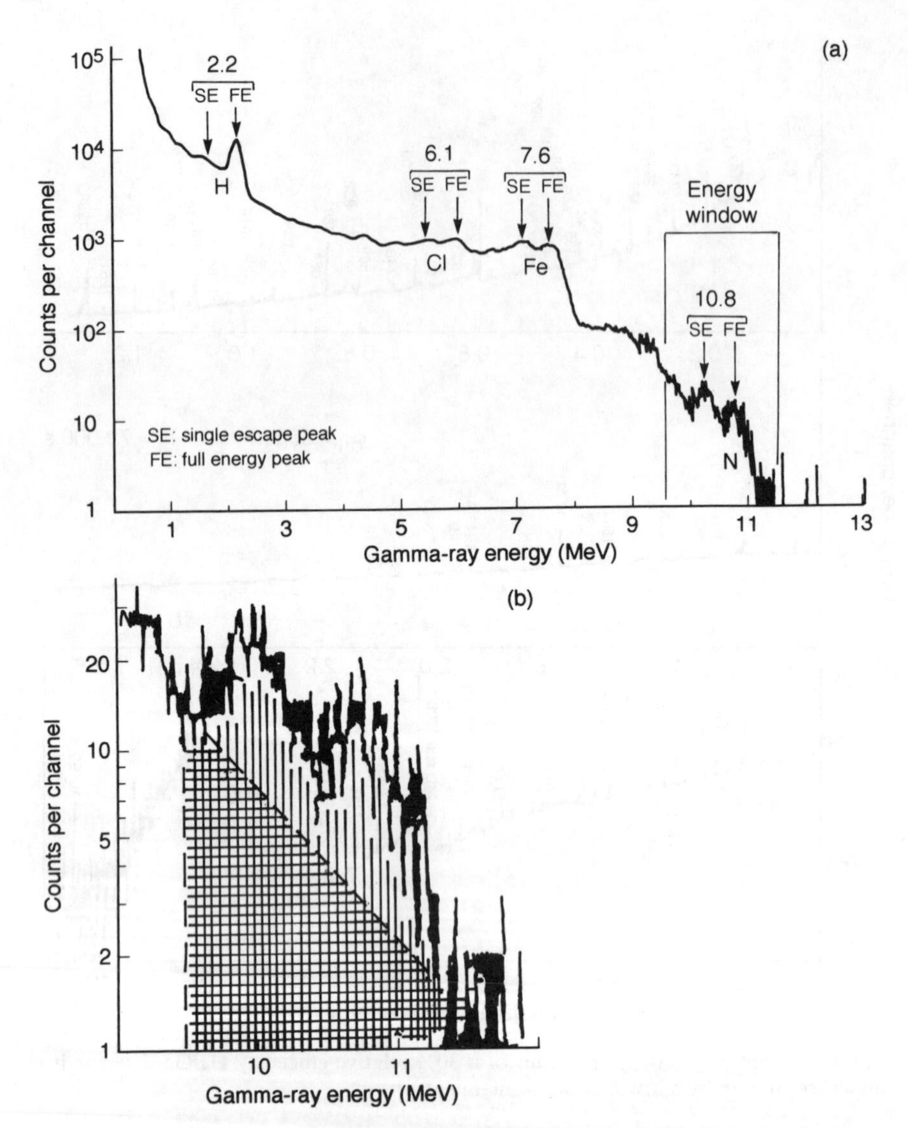

Fig. 2. (a) Prompt gamma spectrum measured by 2 in. × 2 in. BGO detector with explosive-like material hidden in the irradiated luggage, and (b) photopeak area deduction on the 10.829 MeV full-energy and single-escape peaks in the N(n, γ) reaction.

[7.13 g/cm^3, versus 3.67 g/cm^3 for NaI(Tl)], however, acquires no such internal radioactive contamination, and its detecting efficiency further exceeds that of the NaI(Tl) scintillator by another factor of 10 at $E_r > 8$ MeV.

An example of a prompt gamma-ray spectrum measured by a BGO scintillator is illustrated in Fig. 2(a), which represents 2 kg of explosive-like material hidden in luggage and subject to a PGAA scan. Although the detecting efficiency is high for the BGO detector, its energy resolution at 10.829 MeV is poor, $\approx 5\%$, compared with 0.05% for the HPGe detector. The poor resolution of the BGO scintillator is offset by its rugged, shock resistant and temperature insensitive character in addition to high efficiency, making the scintillator an ideal choice for field applications such as on-line PGAA.

3.2 Counting statistics

The weight of an element in a sample can be determined using equation (13), by which the total counts in the prompt gamma-ray photopeak area can be determined in the multichannel spectrum. Let us take again the nitrogen investigation as an example. In Fig. 2(a) the full-energy and single-escape peaks of the 10.829 MeV prompt gamma-ray appear explicitly in the high energy region. The single escape peak results from the escape of one of the 0.511 MeV annihilation photons after pair production in the BGO detector. The area can be determined by subtracting the spectral background (B) from the gross counts (G) between the indicated limits:

$$C = G - B \tag{14}$$

In Fig. 2(b), the measured G and B of the twin photopeak area are indicated by vertical and horizontal bars, respectively. The spectral background is caused by complicating effects such as Compton scattering and accidental summing among gamma-rays with various energies and intensities. The standard deviation of the total counts C can be calculated as:

$$\Delta C = [\Delta G^2 + \Delta B^2]^{1/2} = [G + B]^{1/2} = [2B + C]^{1/2} \tag{15}$$

The background under the photopeak area in this example is measured as $B = 214$ with gross counts measured as $G = 514$, resulting in a total count of $C = 300$. Therefore the associated standard deviation is:

$$\Delta C = ([514^{1/2}]^2 + [214^{1/2}]^2)^{1/2} = [2 \times 214 + 300]^{1/2} = 27$$

and

$$\Delta C/C = 27/300 = 0.09 = 9\%$$

That is to say, the relative error associated with the photopeak area of interest is 9%. Following the rule of error propagation, the fractional error of the elemental weight, $\Delta m/m$, can be evaluated as the square root of the sum of the square of the fractional error of each term presented in equation (13):

$$\Delta m/m = \{ [\Delta m(R)/m(R)]^2 + (\Delta \dot{C}/C)^2 + [\Delta C(R)/C(R)]^2 \}^{1/2} \tag{16}$$

Here $\Delta m(R)$ denotes the standard deviation of the calculated value $m(R)$. We further assume that a certified 2 kg of HMX explosive has a nitrogen content

$m(R) \pm \Delta m(R) = 0.757 \pm 0.008$ kg with $C(R) \pm \Delta C(R) = 340 \pm 28$ counts under the identical irradiation condition in PGAA; the fractional error then becomes:

$$\Delta m/m = [(8/757)^2 + (27/300)^2 + (28/340)^2]^{1/2} = 0.122 = 12.2\%$$

and

$$\Delta m = m \times \Delta m/m = 663 \text{ g} \times 12.2\% = 81 \text{ g}$$

This is an acceptable standard deviation for PGAA; the reported amount of nitrogen in the inspected luggage is thus 663 ± 81 g, equivalent to 1.75 ± 0.21 kg of HMX explosive, sufficient enough to blow up a commercial airliner in mid-air [8].

3.3 Detection limit

For analytical purposes, the minimum detectable amount for elements of interest, whether it is bulk analysis for a large sample or trace analysis of a small sample, has to be evaluated under specific analytical conditions. All the factors to the right of the equals sign in equation (9) affect the detection limit. For instance, by careful choice of the appropriate indicator prompt-gamma ray of high intensity, a high efficiency detector, a longer counting period, and a high flux of thermalized neutrons, the detectable amount of an unknown element can be minimized. On the other hand, a low background in the prompt gamma-ray multichannel spectrum allows easy identification of weak photopeaks. The detection limit for the element of interest, DL, can be evaluated [9] as:

$$\text{DL} = 3.29(\text{d}B/\text{d}t/T_c)^{1/2}/S, \quad \text{g} \tag{17}$$

where $\text{d}B/\text{d}t$ = background count rate, cps, and S = elemental counting sensitivity, cps/g.

The elemental counting sensitivity S is directly related to the term $S = C/T_c m$, or combining equations (9) and (17):

$$\text{DL} = \frac{329 m_0 (\text{d}B/\text{d}t/T_c)^{1/2}}{N_a \varepsilon(E_r) \bar{\phi} \cdot \bar{\sigma} I_r(E_r)}, \quad \text{g} \tag{18}$$

Thus, the detection limit for nitrogen in luggage is 106 g, equivalent to 0.28 kg of HMX explosive.

One can estimate the detection limit if the irradiation conditions are preselected. Let us consider a typical PGAA facility attached to the external neutron beam of a research reactor, and further assume the following irradiation conditions:

The sample receives neutrons with a flux of $\phi = 10^7 \text{ n}_{th} \text{ cm}^{-2} \text{ s}^{-1}$.
A HPGe detector is used with efficiency indicated in Fig. 1.
The spectral background count rate is 7.2 times those displayed in Fig. 1.

TABLE 2.
Detection limit for elements analyzed by PGAA technique with preselected irradiation conditions

Detection limit	Elements
0.1–1.0 μg	B, Gd, Cd
1.0–10 μg	Sm, Hg
10–100 μg	Er, Ag, Nd, Rh, In, Kr, H
100 μg–1 mg	Eu, Ti, Dy, Au, Ta, Cl, Lu, Tm, Hf
1–10 mg	Ho, Co, Xe, Os, Mn, La, Ir, Cr, Ge, Ca, Sr, V, Mo, Ni, Te, Br, I, S, Fe, K, Y, Zn, As, Cs, Cu, Li, Yb, Ar, Sc, N
10–100 mg	Se, Ga, Si, W, Ru, Na, Pr, Zr, Al, Re, Pt, Sn, Mg, P, Tb, Ba, Pb, Ce
100 mg–1 g	Nb, Be, Tl, Sb, Ne, C, F, Bi, Rb
1–10 g	O

The indicator prompt gamma-rays are those listed in Table 1.
The counting period is set at 10 000 s.

The detection limits for the 79 elements measurable by PGAA can be estimated using equation (18), and are listed in Table 2. From this table it can be seen that a number of important elements appear among the first 18 elements. Note that H, Li, B, C, N, F, Ne, Si, P, S and Ar, can only be determined by PGAA rather than NAA. Boron is particularly difficult to determine by any other analytical method. Furthermore, PGAA is a powerful method for measuring most rare earth elements down to trace level, although a few (notably Ce, Pr and Tb) have poor detection limits owing to weak prompt gamma-ray intensities.

By changing the preselected irradiation conditions, the detection limits shown in the table can be varied accordingly. For instance, by replacement of the HPGe detector by a larger one with three times the detecting efficiency and irradiation for a longer period of 100 000 s, the detection limit for the elements listed in Table 2 can be lowered by a factor of 10. The performance can be further improved by use of an anti-Compton device to suppress the spectral background; a suppression factor of up to 11 has been reported for a PGAA multichannel spectrum [10]. Thus, all elements listed in Table 2, apart from those in the last two lines, can be measured down to 1 mg or below; for Ag, B, Cd, Er, H, Gd, In, Kr, Hg, Nd, Rh and Sm the detection limit can be lowered to 1 ng to 1 μg by using an optimized PGAA set-up. In practice, the higher spectral background at $E_r < 2$ MeV (due mainly to the Compton scattering mechanism) makes the detection limit worse for some elements, so that the higher energy part of the gamma-ray spectrum may be the more useful region for the identification and measurement of certain elements.

4 PGAA FACILITIES

As mentioned earlier in this chapter, the PGAA method requires neutrons to bombard the target and a spectrometric detector to count the emitted prompt gamma-rays for quantitative analysis; therefore a neutron source and a gamma-ray spectrometer, together with the sample handling device, are the major parts of a PGAA facility. Since the PGAA facility is usually attached to a radiation control zone such as a nuclear research reactor, researchers are inevitably required to perform on-site sample changing and trouble-shooting in an environment of an intense neutron beam as well as scattered neutrons and prompt gamma-rays. Before the details of the PGAA facility are discussed, personal radiation safety for PGAA operation should be recalled.

Both neutrons and gamma-rays are ionizing radiations, causing human tissue damage and organ dysfunction; prolonged exposure to radiation, in particular to high energy neutrons and gamma-rays, may lead to instant injury or even fatality. An upper limit of an accumulated radiation dose of 50 milli-Sievert (mSv, a radiation dose equivalent unit) each year is recommended for radiation workers [11]. A 1 mSv/h dose rate approximates to an exposure to a thermal neutron flux of 26 000 n $cm^{-2} s^{-1}$, or to a fast neutron flux of 670 n $cm^{-2} s^{-1}$, or to an 8 MeV high energy gamma flux of 12 000 photons/cm^2.s. In the previous section, a thermal neutron flux of 10^7 n $cm^{-2} s^{-1}$ was assumed to be delivered to the sample; accidental exposure to these neutrons for 8 min can cause researchers to exceed the annual dose allowance, facing the risk of acute symptoms of radiation injury. Therefore, in a PGAA facility, maximum efforts should be made to shield off the neutrons and gamma-rays and confine them within the facility itself to ensure biological protection.

4.1 General layout

A basic PGAA facility contains (1) a neutron source, (2) a neutron collimator and shield, (3) a sample handling device, and (4) a gamma-ray spectrometer; each is vitally important for the success and safe operation of PGAA.

4.1.1 Neutron sources

There are three kind of neutron source available for the PGAA set-up: a nuclear reactor, an isotopic neutron source, and a neutron generator. Most nuclear reactors with attached PGAA facilities utilize neutrons from the sustained fission chain reactions among fuel elements in the reactor core. Neutron flux in the core center ranges from 10^6 to 10^{15} n $cm^{-2} s^{-1}$, depending upon the thermal output level of the reactor. Neutrons can also be moderated (slowed down) to thermal energy for research purpose; this can be accomplished by placing moderating material such as heavy water or graphite around the reactor core. The sample can be placed either inside the reactor core or at the external station to which the neutrons are extracted from the core; in the latter case, the neutron flux in the external geometry is reduced by a factor of up to 10^6.

As a local laboratory, for instance a diagnostic division in a hospital, may have great difficulty in acquiring an expensive nuclear reactor, portable neutron sources such as isotopic sources and neutron generators are popular for PGAA field applications. The most frequently used is the ^{252}Cf isotopic neutron source, which emits neutrons by its spontaneous fission decay process with strength of 2.34×10^6 n s^{-1} μg^{-1}. Other isotopic neutron sources commercially available all utilize the (α, n) reaction mechanism to eject neutrons from α-particle bombardment of a blanket sheathing material. A typical source of this kind is the ^{241}Am/Be assembly which can emit 2×10^6 neutrons/s/Ci; the radioactive ^{241}Am provides α-particles by α decay, which strike the sheathing Be, causing neutrons to be ejected via the ^{9}Be(α, n)^{12}C reaction. Popular α-parents include ^{210}Po, ^{226}Ra, ^{227}Ac, ^{238,239}Pu and ^{242,244}Cm in addition to ^{241}Am, and light elements (Li, Be, B, C and F) in the form of a thin foil are suitable sheathing materials.

Unlike the nuclear reactor operation, neutrons can not be started up or shut down from isotopic neutron sources. Although these sources are light in weight, easy to handle, and highly flexible for field applications, they should be carefully shielded when not in use. Considering the long life of these isotopes (2.64 years for ^{252}Cf and 433 years for ^{241}Am), they should be permanently enclosed in a neutron/gamma shield to ensure radiation safety. A 'switchable' neutron generator is an alternative for PGAA field applications. The neutron generator has the advantage of producing a high neutron yield, being operated in pulse mode, and can be switched off when not in use. The most popular type of neutron generator is based on the reaction ^{3}H(d, n)^4 He, producing monoenergetic 14.7 MeV neutrons, by which inelastic scattering reactions with the surrounding sample are readily initiated and the subsequently emitted prompt gamma-rays can be utilized to detect elements which are hardly detectable in thermal neutron reactions. However, the lifetime of a field neutron generator is limited to several hundred hours, making frequent replacement of both the generator and spectroscopic gamma-ray detectors damaged by fast neutrons. These disadvantages mean that neutron generators find only limited use in PGAA field applications.

4.1.2 Neutron collimator and shield

For radiation safety and health concerns, as well as the reduction of neutron damage to sophisticated components such as the spectrometric detector, neutrons should be collimated and shielded properly so that only the sample is exposed to the intense bombarding neutrons. There are two necessary confinements for neutrons in the PGAA facility: a collimation of neutrons prior to bombardment on the sample and the shielding of scattered neutrons afterwards. A neutron collimator is a device by which neutrons from one end are allowed to pass through a channel and come out at the other end in a well defined direction toward the irradiation position. Neutrons striking material other than the hollow channel will eventually be absorbed. A neutron shield absorbs neutrons that impinge on it and generates minimal prompt gamma-rays and induced radioactivity.

The design of neutron collimator and shield is performed by first setting up the maximum permissible radiation dose rate at various locations of the PGAA facility. This implies a knowledge of the physical layout of the facility and its associated equipment. The neutron flux intensities at various locations must be established in order to determine the source distribution. With knowledge of the PGAA facility available, a preliminary choice can be made of collimator and shield layout, so that neutron attenuation and transport calculation can be carried out.

A maximum permissible radiation dose rate of 0.025 mSv/h, for both neutrons and gamma-radiation together, is adopted outside the PGAA facility. This is calculated from the recommended accumulated annual dose of 50 mSv for a radiation worker involved in on-site PGAA work 40 hours a week and 50 weeks a year. For both neutron collimation and absorption design, a material that can slow down neutrons with a large scattering effect is required first, followed by thermal neutron absorption with minimal induced radioactivity. Ideal slowing down materials should contain light elements with large scattering and small absorption cross-sections for fast neutrons; the best choices, taking account of economic considerations, are deionized water, heavy water, beryllium, graphite, and zirconium hydride of nuclear grade. Thermal neutron absorption materials are those containing elements with high capture cross-section, low emission rate of prompt gamma-rays, and negligible amount of induced radioactivity. Among the best choices are materials containing ^{6}Li, ^{10}B, ^{113}Cd and some rare earth elements, such as Sm, Eu, Gd, or Dy. The neutron collimator and neutron shield can be made up from combinations of both the slowing down and the absorption materials mentioned above.

4.1.3 Sample handling device

Since the sample to be investigated is exposed to a high neutron flux and emits intense prompt gamma-rays, it should be handled by an automatic control device. Alternatively, if the sample is handled by manual procedures, the neutron source should be shut off by a mechanical device before manual sample handling. Radiation safety for the researcher, convenience, and economic considerations are prime requirements for sample handling design.

The sample to be analyzed can be fixed on a stand; the criteria of choosing the stand material are the same as those for the neutron collimator and shield—minimal induced prompt gamma-rays and residual radioactivity. For long irradiations, a fixed sample holder made from polyethylene or Lucite together with an automatic changing device is adopted by many investigators. For short irradiation with large sample turnover, a conveyer belt driven by a motor is preferred.

To add to the safety of sample handling, most PGAA facilities have an interlock of sample handler to neutron beam shutter in order to avoid accidental exposure of the researcher to the intense neutron beam. The shutter device contains a driven unit to eject neutron absorbing material into the neutron beam or into the collimator. For instance, saturated boric acid (H_3BO_3) solution can be flooded into the neutron beam line to stop the neutrons completely when the irradiation station is opened for sample handling [12].

4.1.4 Gamma-ray spectrometer

A prompt gamma-ray spectrometric system, containing either a scintillation or a semiconducting detector, is the heart of the PGAA facility. Both types of radiation detector are subject to neutron irradiation during prompt gamma-ray analysis and various degrees of damage can result from prolonged exposure to neutrons. The NaI(Tl) detector generates intense internal radioactivity of ^{24}Na and ^{128}I when it is exposed to even a low flux of neutrons, severely hampering the collection of the prompt gamma-ray spectrum. The BGO detector is not subject to induced radioactivity, but irradiation with thermal neutrons beyond 10^{15} n/cm^2 may disable it as a spectrometer owing to intense radioactivity in the detector case [13]. On the other hand, exposure of a HPGe detector to a fast neutron fluence of 10^9 n/cm^2 is sufficient to risk the high resolution performance and the detector may become unusable with exposure beyond 10^{10} n/cm^2 owing to destruction of the crystal lattice [14]. Therefore a neutron shield surrounding the spectrometer is mandatory to prevent scattering of neutrons into the detector.

A spectrometric detection system also contains complex electronics in addition to the gamma-ray detector. These electronic signal processing units, such as the amplifier (AMP), analog-to-digital converter (ADC), multichannel analyzer (MCA), power supply (HVPS) and associated computer with analyzing software, should be placed in a counting room away from the PGAA facility and preferably with temperature and humidity control. Since hundreds of prompt gamma-rays appear in the multichannel spectrum and many of them overlap one another, software with automatic data reduction and sophisticated photopeak analyzing ability is demanded. Many software packages with a prompt gamma-ray library are commercially available for use with a PGAA facility.

4.2 Typical PGAA facility

In the reactor-based PGAA facility, irradiation can be conducted internally in the core or externally outside the reactor by extraction of a neutron beam through a collimator. The trade-off between high flux with low detecting efficiency for internal irradiation and low flux with high detecting efficiency in external irradiation approximately cancel in the two configurations; however, external geometry is superior for the following reasons:

- lower interferences from fission gamma-rays;
- no restriction on the material (such as an explosive) being investigated;
- larger space for special samples, such as an inert atmospheric chamber;
- larger space to accommodate bulk samples;
- minimal sample deterioration due to heating and radiation damage;
- flexible design for multi-sample station;
- flexible design for multi-detector arrangenment;
- little induced radioactivity in the irradiated sample;
- flexible neutron beam profiles using neutron/gamma filters.

Hence most reactor-based PGAA facilities are in external geometry, and the one attached to the Tsing Hua Open-pool Reactor (THOR) in Taiwan, a typical PGAA set-up using an external reactor beam, is introduced briefly as an example.

Neutrons from the 1 MW THOR facility are extracted horizontally by means of the through port ≈ 30 cm from the nearest fuel elements, as shown in Fig. 3(a).

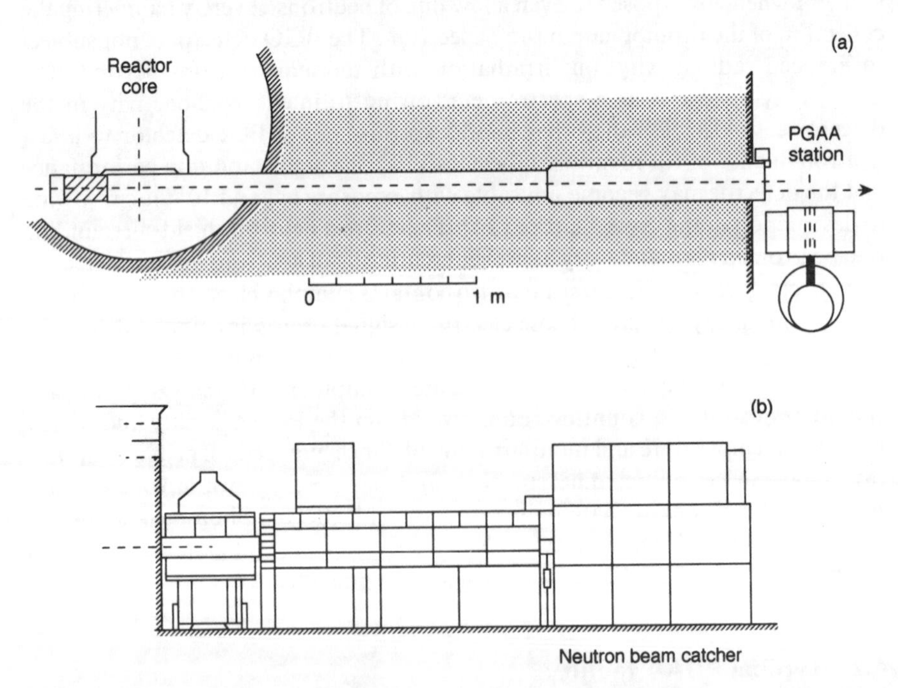

Fig. 3. Neutron beam tubes used for PGAA facility at THOR: (a) internal through port in reactor shielding; (b) external beam port on the experimental floor.

A two-section beam tube is inserted in the through port to collimate and deliver the neutron beam at targets with a cross-section of 5 cm. The thermal neutron flux measured in the beam using gold foil is 1.3×10^6 n cm^{-2} s^{-1} at the target with a cadmium ratio of 26.4:1. The beam passing through the target is captured with a beam catcher consisting of 40% weight of BO_2 in polyethylene (PE) matrix blocks surrounded by heavy concrete blocks, as illustrated in Fig. 3(b). The concrete blocks also serve as a biological shield for those who have to work around the facility.

The main gamma-ray detectors are HPGe detector and an annular NaI(Tl) detector shield to perform Compton suppression and to detect pairs of 511 keV

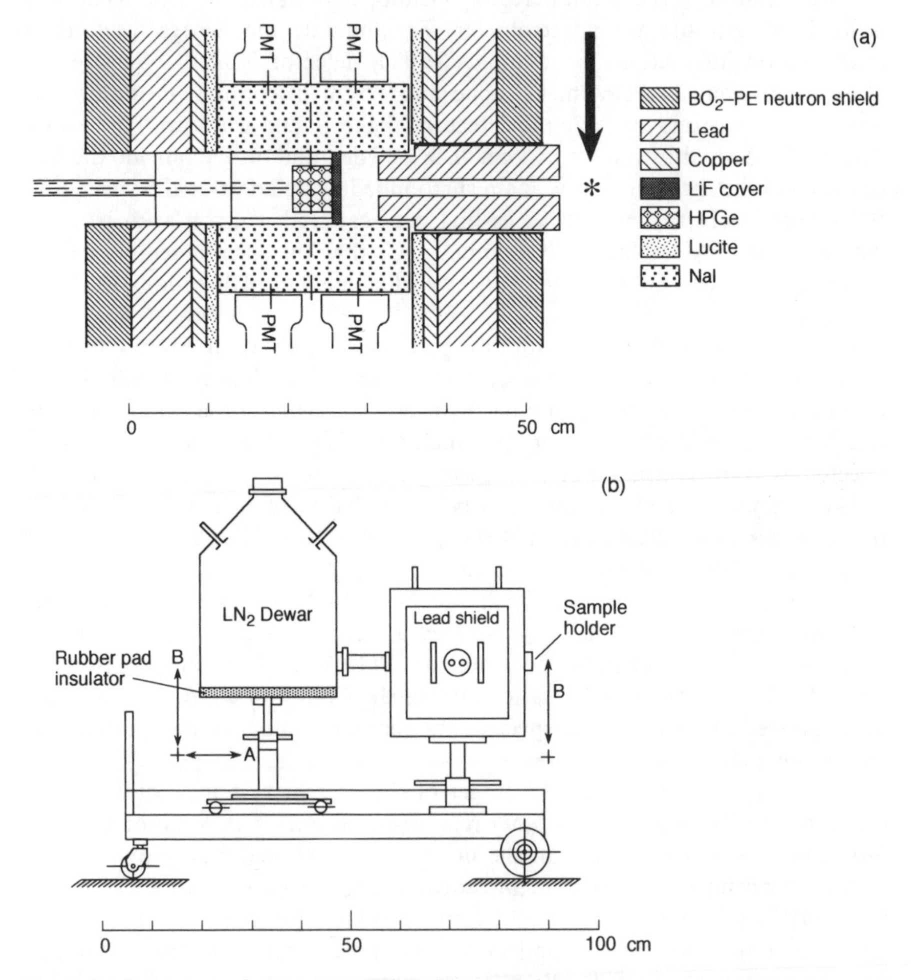

Fig. 4. (a) Top view of the cross-sectional layout of the gamma-ray spectrometer and shielding; (b) side view of the one-ton detector assembly carried by trolley.

gammas. The 145 cm^3 n-type HPGe detector has 30% relative efficiency and a resolution of 2 keV FWHM (full width at half maximum) at 1332 KeV, and a peak-to-Compton ratio of 46:1 at the same photopeak energy. The detector shield surrounding the primary HPGe detector is an annual NaI(Tl) detector having a diameter of 9 in. and a length of 10 in. A 3 in. diameter lateral hole can accommodate the 38 cm long HPGe detector can. Eight 3 in. photomultiplier tubes (PMT), four on each optically isolated half, view the NaI(Tl) detector. The sample for irradiation can be placed as close as 25 cm from the HPGe detector head. The geometric arrangement of the two detectors, together with the shielding materials associated with them, is shown in Fig. 4(a).

Lead is chosen as the major detector shielding material, and a layer 8 cm thick is used to surround the spectrometer. To eliminate the X-rays induced by lead–photon interaction, the annular NaI(Tl) detector is further covered by absorption layers of 1 cm thick Lucite and 1.5 cm of copper. In addition, to prevent the neutrons reaching the detectors, the front end-window of the HPGe detector is covered by a layer of 1 cm ^{6}LiF powder, and on the outside the lead shielding of the spectrometer is again surrounded by an additional layer of 5 cm BO_2-loaded polyethylene blocks as first-line scattered neutron absorber. The detector assembly is placed on a trolley such that the 1 ton assembly can be moved around the PGAA facility with easy adjustment in all directions, as shown in Fig. 4(b).

The electronic block diagram for this spectrometer is shown in Fig. 5. The basic electronic set-up is for the two optional data accumulation modes. In the Compton suppressed mode, gamma-ray events collected in the HPGe detector which have coincident events in the annular NaI(Tl) detector within 0.1 μs are rejected at ADC 1 input. For pair spectrometric mode, gamma-ray events of 511 keV signal occurring simultaneously in both halves of the NaI(Tl) crystal are treated as an allowable gate for a HPGe pulse at ADC 1. The effect of Compton suppression, judged at 662 keV for the ^{137}Cs photopeak's peak-to-Compton ratio, is improved by a factor of 5.2; at 2754 keV (^{24}Na photopeak) the peak-to-Compton ratio is further improved by a factor of 7 under anti-Compton operation. In a complex radioenvironmental PGAA test, Compton suppression not only yields a much better ratio (11:1) of the Compton continuum between unsuppressed and suppressed spectra, but eliminates most contaminating and interfering gamma-rays.

The neutron beam flux at the target of the present system is quite low in comparison with that in many other reactor-based PGAA installations, or two orders of magnitude lower than the most intense external flux. However, improvements can be made by reducing the sample-to-detector distance and using a larger HPGe main detector. The detection sensitivities for some elements are thus sensitive enough for many applications. However, extending the irradiation period is expected to provide better detection limits, especially for low interaction rate elements, but not all elements give good results. Further developments are

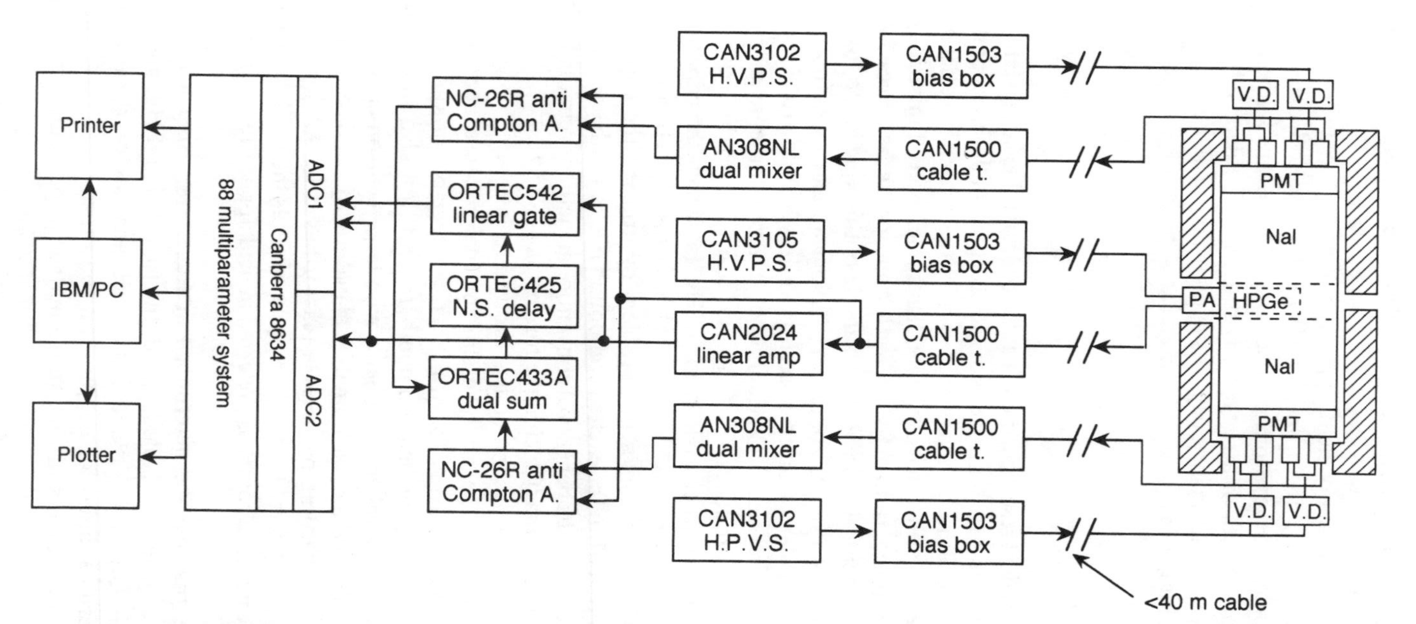

Fig. 5. Block diagram of the electronic and logic modules used with Compton suppressed or pair spectrometer for PGAA facility at THOR.

expected to improve the reactor-based PGAA facility, making it even more useful.

On the other hand, many field applications of PGAA utilize the isotopic neutron sources. Among them, in vivo (IVPGAA) medical diagnosis, in situ (ISPGAA) geological surveys and on-line (OLPGAA) scans frequently employ portable neutron sources in a rugged environment. The detailed set-up, from mobile unit to permanent facility, depends upon the particular usage, as reported by Alfassi and Chung [15].

4.3 Applications

In the field of analytical research, there is still a considerable interest in improving the techniques leading to sensitive, non-destructive, multi-elemental analysis. Conventional NAA, although a powerful tool for the quantitative determination of numerous elements in trace amounts, has some severe limitations, since it depends upon induced radioactivities with appropriate half-lives for post-irradiation measurement. An interesting example of this kind is the toxic element cadmium at trace level in environmental, biological, and living-body samples. The enormous thermal neutron cross-section (nearly $20\,000 \times 10^{-24}$ cm^2) for the 12.2% abundance isotope ^{113}Cd cannot be utilized for conventional NAA because the reaction product of ^{113}Cd(n, γ) ^{114}Cd is non-radioactive. A similar problem arises with another toxic element, Hg, for which the capture reaction product of high cross-section (2000×10^{-24} cm^2), the 17% abundance ^{199}Hg, is again stable. PGAA overcomes this problem by utilizing the capture prompt gamma-rays emitted from the excited compound nucleus with energy above that of the stable isotope. As the benefits of conventional methods remain, PGAA is an indispensable supplementary technique for chemical analysis.

In the last decade, the technique has evolved to become a powerful nuclear analytical method. It is generally useful for major constituent analysis, in particular for those elements for which no appreciable amount of induced radioactivity is produced by neutron bombardment. It is also suitable for trace analysis for a limited number of elements with large neutron cross-sections. As the elemental sensitivity in nuclear analytical methods depends directly upon the intensity of the bombarding particles, a reactor-based PGAA facility having a high neutron flux is preferred in trace analysis, whereas the PGAA set-up using a low flux isotopic neutron source is preferred in field applications for determining major elements.

The elements that can be determined at trace level using a reactor-based PGAA facility, as listed in Table 2, are B, the toxic Cd and Hg, and some rare earth elements—Gd, Sm, Er and Nd. Trace analysis of these elements is sometimes the critical index for the problem concerned.

Some major constituents in the sample matrix, in particular those elements with Z up to 18, cannot be activated by nuclear methods and therefore cannot be determined thus. For instance, the hydrogen content in most environmental and biological samples is high; however, the reaction $^1H(n, \gamma)\ ^2D$ produces a stable isotope, making conventional NAA impossible for determining hydrogen. On the other hand, PGAA allows convenient determination via the 2.223 MeV prompt gamma-ray emitted from the $^1H(n, \gamma)\ ^2D$ reaction, with a detection limit down to the sub-milligram level. Thus, in many trace analyses by reactor-based PGAA, major element analysis or bulk analysis is an added benefit, supplementing analyses that cannot be performed readily by other analytical methods. A typical example is the rapid determination of the major elements H and C as well as the trace elements B and Cd using a reactor-based PGAA facility, whereas conventional NAA cannot achieve this because the necessary radioactivity is not induced.

The frontier of major element analysis lies in field applications. Some critical determinations of major element content have to be performed on a very large sample matrix, for instance a rock formation around a borehole, and such a complex facility as a nuclear reactor is difficult to adapt to field applications. Convenient PGAA arrangements using portable detection systems and isotopic neutron sources have been developed in the past decade for field applications such as in vivo medical diagnosis, in situ geological surveys and, more recently, on-line scanning for sensitive objects, such as suspected explosives and unidentified chemical warfare agents.

4.4 Comparison with other nuclear methods

Except for the nuclear methods used only for surface analysis, several nuclear analytical techniques mentioned elsehwere in this book can compete with PGAA. Among the nuclear techniques which have proved very useful are conventional NAA, which may be effected using radiochemical separations (RNAA) or completely by instrumental means (INAA), as well as the instrumental methods of photon activation analysis (IPAA) and charged particle activation analysis (ICPAA). All these 'off-line' techniques rely on the production of radioactive nuclides and the subsequent detection and measurement of the decay of the induced radioactivity. These methods demand a waiting of up to a month, or more, to permit short lived nuclei to decay and allow measurement of the longer lived species with lower specific activities. With PGAA, in contrast, the data are immediately available during irradiation.

A problem often encountered with nuclear methods is the escape of volatile components from the sample after irradiation. This is especially true for Hg, which is quite volatile in the elemental state. The 'hot atom' Hg is released from its compounds owing to the recoil from the emission of its prompt gamma-rays following the (n, γ) reaction. The recoil is sufficient to rupture the chemical bonds of the Hg compound, and Hg then diffuses out of the container after irradiation, thus precluding the measurement of its radioactive decay. With PGAA, however, this effect is of little concern, since by the time recoil occurs the prompt gamma measurement has already been completed.

There is a synergism between PGAA and other nuclear methods such as INAA in that these techniques employ the same equipment and technology to determine complementary sets of elements. INAA is best known as a method of trace analysis, insensitive in general to the major and minor elements that make up many materials. On the other hand, some of the strongest lines in PGAA Compton suppressed and pair spectra often come from the major constituents, especially C, H, N, O, P, S and Si. If the PGAA is applied to an unknown material before non-destructive INAA, the difficult question of neutron self-shielding by strongly absorbing elements in the sample is directly answerable, using the same sample repeatedly if necessary. The two methods together are capable of measuring non-destructively, in a wide variety of materials of scientific, economic or regulatory interest, all the major elements except perhaps oxygen, all minor elements, and many trace elements,—up to 78 in all. In oxide matrices such as rocks, a partial check is possible on the accuracy and completeness of the analysis by adding the concentrations of the elements as oxides and comparing the sum with 100%. This synergism of prompt and delayed radioactive analysis is a major justification of PGAA with the neutron intensity attainable from presently available neutron sources.

In field applications, IVPGAA, ISPGAA and OLPGAA meet with little competition from other analytical techniques. Numerous prototype PGAA set-up have been installed by many users and clients from various scientific and engineering disciplines. Once the radiation safety and reliability for quantitative determination have been satisfactorily resolved, the rapid expansion of field applications of PGAA can be anticipated in the near future.

REFERENCES

1. P. Shea, T. Gorden and H. Bororgmanesh, *Nucl. Instrum. Methods Phys. Res., Sect. A* A299, 444 (1990).
2. J. Yinon and S. Zitrin, *The Analysis of Explosives*, Pergamon Press, Oxford (1981).
3. D. I. Garber and R. K. Kinsey, Neutron Cross Sections. Vol. II. Curves, *Brookhaven Natl. Lab.*, [*Rep.*] BNL-325 (1976).
4. C. M. Lederer and V. S. Shirley, *Table of Isotopes*, 7th Ed., Wiley, New York (1978).
5. R. C. Greenwood and R. E. Chrien, *Neutron Capture Gamma Ray Spectroscopy*, Plenum Press, New York (1979), p. 621.
6. M. A. Lone, R. A. Leavitt and D. A. Harrison, *At. Data Nucl. Data Tables* **26**, 511 (1981).
7. S. M. Liu, C. Chung and C. C. Chan, *J. Radioanal. Nucl. Chem., Articles* **162**(2), 363, (1992).

8. L. Grodzins, *Nucl. Instrum. Methods Phys. Res., Sect. B* **B56/57**, 829 (1990).
9. T. G. Dzubay, *X-ray Fluorescence Analysis of Environmental Samples*, Ann Arbor, Michigan (1978), Chapter 5.
10. C. Chung, L. J. Yuan and K. B. Chen, *Nucl. Instrum. Methods Phys. Res., Sect. A* **243**, 102 (1986).
11. International Commission for Radiological Protection, Recommendation, *ICRP Rep.* No. 26 (1977).
12. C. Chung and C. Y. Chen, *Nucl. Technol.* **92**, 159 (1990).
13. C. J. Lee and C. Chung, *Appl. Radiat. Isot.* **42**(8), 729 (1991).
14. P. H. Stelson, J. K. Dickens, S. Raman and R. C. Trammell, *Nucl. Instrum. Methods* **98**, 481 (1972).
15. Z. B. Alfassi and C. Chung, *Prompt Gamma Neutron Activation Analysis*, CRC Press, Boca Raton, in press.

Chapter 9

CHEMICAL ANALYSIS BY THE THERMALIZATION, SCATTERING AND ABSORPTION OF NEUTRONS

C. M. Bartle
Institute of Geological and Nuclear Sciences, Lower Hutt, New Zealand

1 INTRODUCTION

Applications of the thermalization, scattering and absorption of neutrons are widespread in the field, in laboratories and in industry. This chapter is a review which indicates where the 'cutting edge' of this research topic lies and enables the reader to share in the adventure taking place. Recent novel applications include the measurement of snow thickness in remote terrains and prospecting for oil using portable neutron generators down boreholes. During the 1980s, fast neutron transmission gauges have been increasingly adopted by industry. The simultaneous transmission of fast neutrons and gamma-rays (Neugat) allows measurements on chemical mixtures taking account of any change in the mass per unit area in the beam. Applications are diverse, including moisture measurements in coke and fat measurement in meat. The process of neutron absorption, based on elemental cross-sections, can be used to deduce elemental concentrations from changes in neutron transmission. Neutron gauges based on transmission are becoming important for detection of explosives in packages and luggage.

Hubbell [1] has presented a survey of industrial, agricultural and medical applications of radiometric gauging and process control. Oelgaard [2] and Michalik [3] have discussed neutron and gamma-ray gauges. A previous account of chemical analysis by thermalization, scattering and absorption of neutrons

Chemical Analysis by Nuclear Methods Edited by Z. B. Alfassi

was given by Tölgyessy and Kyrš [4]. Radiation detection and measurement are discussed in Chapter 2 and, more extensively, by Knoll [5].

'Thermalization' [6] involves the slowing down of neutrons until they reach thermal equilibrium with the material. Thermalization of a neutron occurs more rapidly (taking $\approx 10\,\mu s$ after neutron production) than neutron absorption (which occurs with emission of gamma-rays, protons or alpha-particles ≈ 1 ms after thermalization). The equilibrium concentration of thermal neutrons is sensitive to the scattering and absorbing properties of the material, and therefore to its chemistry.

In a reactor core, an equilibrium is reached between neutron production, scattering, thermalization and absorption. The concentration of neutrons per unit energy range is dependent on the reactor materials. Chemical analysis can be performed by introducing new chemicals. Other sources of neutrons include radioactive sources such as ^{241}Am–Be (Am–Be) [2, 5] and ^{252}Cf [2, 5, 7], and accelerator sources, which use a range of neutron producing reactions such as $^{2}H(d, n)^{3}He$ and $^{3}H(d, n)^{4}He$ [2, 8]. Radiation sources are discussed in Chapter 3.

Neutrons are named to indicate their energy range: *fast* neutrons (> 500 keV); *intermediate* neutrons (1–500 keV); *slow* neutrons (0–1000 eV); *thermal* neutrons (in thermal equilibrium—0.025 eV at 300 K); *epithermal* neutrons (energies above thermal to ≈ 10 keV). The intensity of a neutron beam (neutron flux) and reaction probabilities (cross-sections) are defined in Refs. 2 and 9.

This chapter incorporates separate discussions on the topics of thermalization, scattering and absorption of neutrons. Innovation in scattering methods is particularly emphasized.

1.1 Analysis by the thermalization of neutrons

The operation of a gauge involves three steps. Fast neutrons penetrate into the material, the neutrons are slowed to thermal energies and thermal neutron concentration is inferred from a representative detector count rate. Thermalization of a fast neutron is illustrated in Fig. 1. Most of the distance traversed from the source (A) to the thermalization region occurs in the fast region. At thermal energies (B), the neutron's translational progress is relatively small. For light nuclei, thermalization of the neutrons occurs more rapidly and higher concentra-

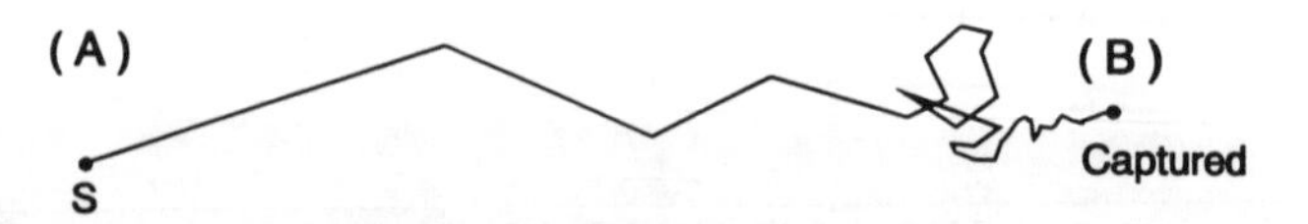

Fig. 1. Trajectory of a fast neutron during thermalization.

tions are achieved close to the detector. These gauges are used mostly to detect water. The range of moisture measuring methods available is discussed by Pyper [10].

The mean 'crow fly' distance travelled by a fast neutron from the source to the point of thermalization is the 'slowing down length'. In practical materials, and with Am–Be or ^{252}Cf sources, the slowing down length is 0.1–0.6 m. Instruments usually have portable dimensions.

There are three main sources of error with gauges based on the thermalization. First, there is a variation in the count rate due to variation in the fast neutron emission from the source. In practice, this error can be reduced to insignificance by counting for a sufficiently long time. Second, there are errors associated with the detection and counting components of the gauge. Finally, there are errors associated with establishment of the calibration curve. The calibration curve is sensitive to the presence in the material of elements such as ^{10}B and ^{6}Li which have large absorption cross-sections. Detector designs [5] often incorporate these elements or another strong thermal neutron absorber, ^{3}He, as a facet of their design, and have excellent stability [11].

1.1.1 Measurement of the moisture content of soil and materials in field locations

Gauges consist of a fast neutron source and a thermal neutron detector in a shielded housing (the 'probe'), with a small electronic package which can be carried by one person. The fast neutron source is usually Am–Be. Surface gauges are placed on the soil surface, and subsurface or insertion gauges are lowered into a hole in the soil.

Irrigation management is a routine use of such a gauge. In the 1985–86 cotton season, 45% of the Australian cotton crop, or 72 000 ha, was irrigation-scheduled using measurements with a neutron probe. In one irrigation area, after introduction of neutron gauges, the production increased from 3.3 bales/ha to 5.9 bales/ha [12]. Tölgyessy and Kyrš [4] and Oelgaard [2] discuss gauge designs in detail.

Most commercial gauges have a linear calibration:

$$n = b\theta + a \tag{1}$$

where n is the ratio of the count rate in the soil to the count rate in a standard material (usually water). The quantity θ is the volume concentration of free water, usually defined as the water released by drying at 105 °C, and b and a are constants.

Figure 2 shows a typical set of calibration curves. A detailed description is given by Greacen and coworkers [13] Much has been written about the theory and operation of these gauges [14–33].

Thermal neutron detection techniques have an established role in exploration and production associated with the drilling of oil wells [34]. Porosity measure-

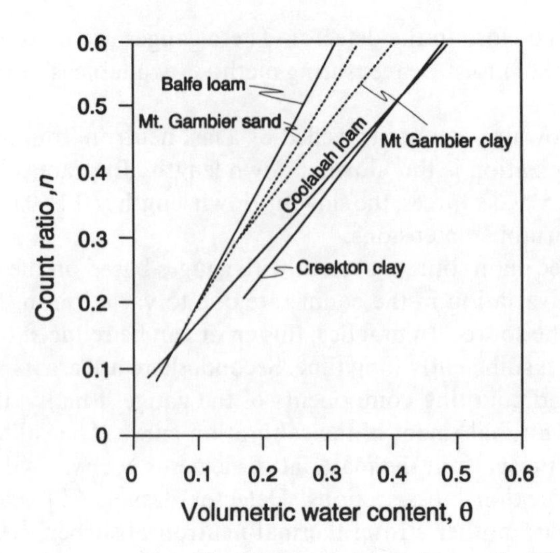

Fig. 2. Soil moisture gauge calibration curve for various soils. (From Ref. 13.)

ments, called 'porosity logs', are made down boreholes with the equivalent of a subsurface gauge. The greater the porosity, the greater the water content, and therefore the higher the thermal neutron flux near the detector. The use of a ^{252}Cf source reduces environmental concerns for a source should it become stuck in a bore hole [35]. Proportional counters (see Chapter 2) containing ^{3}He, and 'ruggedized' for logging applications, are the principal neutron detectors used [36]. Two neutron detectors sited at a 'far' and a 'near' distance from the source compensate for environmental perturbations, such as borehole roughness and mud cake [37]. Recently other detector combinations have been used [38]. Several authors [39–41] have reported correction algorithms for absorbers and salinity when calibrating neutron logs. Recently, miniature accelerator sources which generate 14 MeV neutrons have been used for porosity log measurements [42–45]. In prompt fission neutron (PFN) uranium logging, fission neutrons are detected following irradiation with a pulsed neutron generator [46].

Cosmic ray protons produce neutrons in the atmosphere. The neutron flux is about 0.33 n/m^2/s over land and 0.11 n/m^2/s over the sea [47]. These neutrons can be used to estimate moisture in snow cover, soil and air. The new field of study called 'applied cosmic ray physics' has been studied by Kodama [48] and by Avdyushin *et al.* [49]. For snow cover measurement, thermal neutron detectors are positioned above and below the accumulating snow cover. A ratio a which is sensitive to the snow thickness is measured:

$$a = \left(\frac{I_{\mathrm{w}}}{I_{\mathrm{w0}}}\right)\left(\frac{I_{\mathrm{r0}}}{I_{\mathrm{r}}}\right) \tag{2}$$

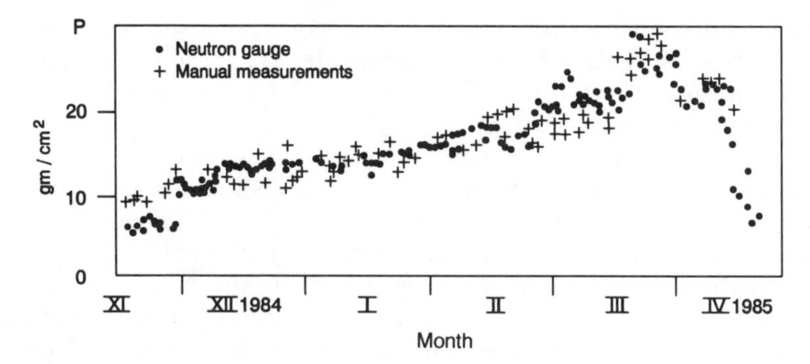

Fig. 3. Comparison of the results from manual and cosmic ray-based snow gauge. (From Ref. 49.)

where I_w is the neutron count above the snow layer, I_r is the neutron count below the snow layer, I_{w0} is the neutron count in the above-snow position when no snow is present, and I_{r0} is the neutron count in the below-snow position when no snow is present.

Radiotransmissions from remote sites have been collected at a data reception centre [49]. Results from 1984 to 1985 are shown in Fig. 3.

The treatment of waste water by the percolating filter process depends on the presence of an active microbial slime or film. In winter, the film thickens and may block the flow. An insertion gauge can be used to measure the moisture content in the filter medium and hence the film accumulation. Such a gauge is rapidly becoming indispensable in filter management [50].

1.1.2 Water and hydrogen measurement in industrial materials

Non-invasive analysis includes mainly capacitance [51, 52], microwave [53, 54] and nuclear-based methods. Traditionally, in industry, production economics are based on small samples analysed by conventional chemistry or gravimetric methods.

Chemical analysis of samples is too slow to allow adjustments to be made during production, or to detect rapid changes in composition. Also, the results for a small sample may not be representative of the product as a whole. The test procedure is destructive and large amounts of the product cannot be analysed. The procedures are labour intensive and may require the use of dangerous or polluting chemicals.

Where this can be achieved, chemical analysis using non-invasive methods overcomes these difficulties. The method can be applied automatically to all the product on a conveyer at production speed. Non-invasive methods are being progressively adapted to the needs of industry.

Wood chips contain up to 50 wt % of water, and have a commercial value based on their dry wood content. Korell, Kurth and Mette [55] and Koditz *et al.* [56] describe an on-site neutron method for measuring woodchip moisture in piles or hoppers based on the use of insertion gauges.

For coal and coke, moisture measurement based on capacitance or on microwave transmission is difficult owing to the high electrical conductivity. Williams [57] discusses a surface mounted neutron gauge and an insertion gauge for the coal industry. Similar gauges are described by other authors [11, 58–60].

Dinuzzo *et al.* [61] describe a system for the in-stream determination of solids-weight fraction and ash content of coal in flotation circuit slurries of a coal preparation plant. Thermal neutron moisture gauges have been developed for concrete and building materials [62–65]. Measurement can be applied to the various constituents during the mixing process and to the final mixture. Moisture determinations have been made on live trees, lactose used in pharmaceutical products, agricultural products, fireclay and plaster. Asphalt with a high hydrocarbon content in road surfaces has also been measured [4].

Kühn [59] and Czembor and Hartmann [66] describe a system for measurement of the concentration of carnallite ($KCl \cdot MgCl_2 \cdot 6H_2O$) in rock salt (NaCl, KCl) based on the high water of crystallization in carnallite. Mirowicz and Lis [67] have devised a gauge which measures the concentration of sulphuric acid flowing in a pipeline.

Mathew *et al.* [68] describe a method to determine the levels of solids and liquids in containers and also to measure moisture levels of iron ore samples to 0.5 wt%. The backscattering of thermal neutrons has been used by Pekarskii [69, 70] to measure liquid levels at a gas/liquid boundary through a wall thickness. Blockages in pipelines have also been detected through metals (steel, cast iron, copper and aluminium) or non-metals (ceramics, glass, and glass-fibre materials). These gauges have been used in the manufacture of inorganic fertilizers and in the cellulose and paper industry. A similar gauge can measure coating thicknesses on metals and moisture in layers of thermal insulation.

1.2 Analysis by the scattering of neutrons

For fast neutrons, the principal reaction processes are elastic and inelastic scattering. In this section, chemical analysis is based mostly on neutron scattering above thermal energies, as illustrated in Fig. 1.

1.2.1 The kinematics of scattering

A neutron scattered through an angle ϕ in the centre-of-mass system by a nucleus of atomic weight A has a corresponding laboratory scattering angle θ (Fig. 4)

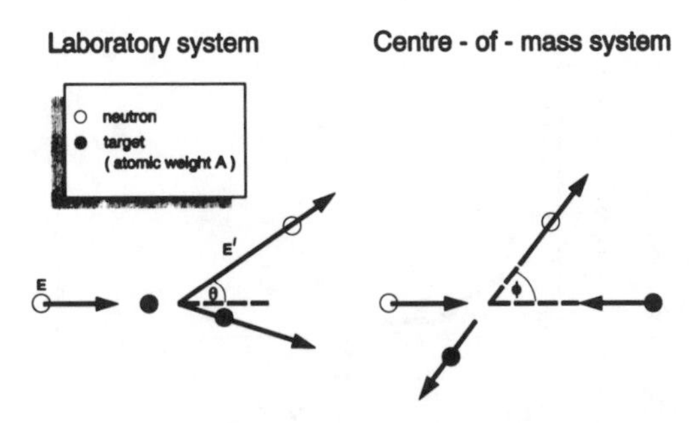

Fig. 4. Neutron scattering kinematics.

given by:

$$\cos\theta = (1 + A\cos\phi)/(A^2 + 1 + 2A\cos\phi)^{1/2} \tag{3}$$

The relationship relating the incoming energy E and the scattered neutron energy E'[6] is:

$$E' = E(A^2 + 1 + 2A\cos\phi)/(A + 1)^2 \tag{4}$$

The range of E' is from $E'_{max} = E$ at $\theta = 0°$ to $E'_{min} = [(A - 1)/(A + 1)]^2 E$ at $\theta = 180°$. For single scatterings, measurement of E' and E yields A, and thus chemical information.

Neutron scatterers incorporated in a detector as a design facet show their chemical nature through the observed recoil energy spectrum $(E - E')$, which mirrors the scattering cross-section in the centre-of-mass system [71]. Organic scintillators such as NE213 [72,73], based on proton recoil detection, are used as neutron detectors mounted on fast photomultipliers (see Chapter 2) such as the Philips XP 4512B [74] or the Burle 8854 [75]. Inelastic scattering identifies the level structure(s) and chemistry of the scatterer. Neutron groups detected from inelastic scattering on an external scatterer allow quantitative chemical measurements to be made. An organic scintillator response to several incident neutron energies [76] is illustrated in Fig. 5(a). Each component recoil spectrum extends from the maximum neutron energy to zero. The numbers of events in the individual neutron groups, and hence the scatterer cross-sections, are determined by an unfolding procedure (Fig. 5(b)). Apart from the measurement of chemical concentrations, neutron interactions with nuclei allow the measurement of detailed nuclear structure.

1.2.2 Analysis based on neutron transmission

In a transmission (attenuation) gauge, the detector is aligned with a collimated beam of neutrons. The sample is positioned between the source and the detector,

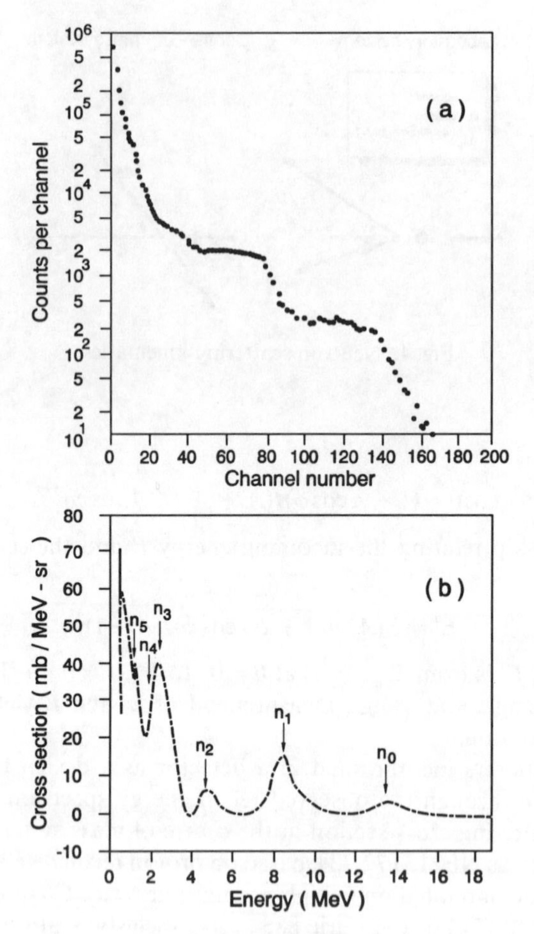

Fig. 5. (a) Response of an organic scintillator to several neutron groups; (b) identification of the groups by unfolding. (From Ref. 76.)

usually with its face normal to the beam direction. Insertion of the sample attenuates the detector signal from I_0 to I.

For monoenergetic neutrons the relationship between I and I_0 is:

$$I = I_0 \exp\left[-\sum(N_j \sigma_j t)\right] \tag{5}$$

where the sample thickness is t, N_j is the concentration of scattering or absorbing nuclei of chemical species j in the sample, and σ_j is the cross-section for the jth chemical species.

In practice, the mass attenuation coefficient (MAC) μ is defined according to:

$$\mu_j = N_0 \sigma_j / A_j \tag{6}$$

and

$$I = I_0 \exp[-\sum(\mu_j m_j)] \tag{7}$$

where N_0 is Avogadro's number, m_j is the mass per unit area of the jth chemical species and A_j is the atomic weight of the jth chemical species. The detector count rate I depends on the properties of the neutron source, the properties of the sample (e.g. density, composition and geometry) and the structural details of the experimental apparatus itself. It is common to define an apparent mass attenuation (or absorption) coefficient (AMAC) μ_a, applicable to a particular gauge for a chosen material with the value:

$$\mu_a = -(1/m)\ln(I/I_0) \tag{8}$$

where m is the sample mass per unit area in the radiation beam.

Because of multiple scattering [77], the AMAC values decrease with increasing thickness [78]. The AMAC values are significantly higher for water than for most other materials [79].

Williams [57] describes a transmission moisture gauge for sinter mix maintained at a constant thickness (0.12 m) on a conveyer belt. The success of this gauge depends on maintaining the sinter mix thickness constant, since variations in the detector count rate with thickness would mask the dependence on water content. In general, if changes in areal density are variable, the mass per unit area must be independently assessed. In the next section, independent assessment based on simultaneous measurements of neutron and gamma-ray transmission is discussed.

1.2.3 Analysis based on the simultaneous transmission of neutrons and gamma-rays (Neugat)

The combined use of neutrons and gamma-rays (Neugat) allows corrections for variations in mass per unit area. Several authors [80–82] describe the use of gamma-rays alone for weighing materials such as ores, coal, coke and fertilizers on conveyer belts. The Neugat method overcomes the difficulties of thermal neutron gauges discussed in Section 1.1.1. These difficulties include restricted sample penetration, density corrections, temperature sensitivity and the presence of high thermal neutron absorbers. Several authors discuss the simultaneous use of thermal neutrons and gamma rays [83–86]. Others discuss applications for fast neutrons [78, 79, 87, 88].

(a) The fundamental equations

The mathematical description is similar to that for dual energy gamma-ray (Gamgat) gauges described by several authors [89–92]. Beams of neutrons and gamma-rays are transmitted through a sample composed of a mixture of two components, x and y. From equations (7) and (8) it follows that:

$$I_n = I_{n0} \exp[-(\mu_{nx} m_x + \mu_{ny} m_y)] \tag{9}$$

$$I_g = I_{g0} \exp[-(\mu_{gx} m_x + \mu_{gy} m_y)] \tag{10}$$

where I_n and I_g are the transmitted neutron and gamma-ray intensities, respectively, and I_{n0} and I_{g0} are the corresponding intensities with the sample removed. By solving the equations (9) and (10) for m_x and m_y, one can write the weight fraction of m_x, w and the density, ρ, as:

$$w = m_x/(m_x + m_y) \tag{11}$$

or

$$w = (\mu_{gy} R - \mu_{ny})/[(\mu_{nx} - \mu_{ny}) - R(\mu_{gx} - \mu_{gy})] \tag{12}$$

and

$$\rho = (m_x + m_y)/t \tag{13}$$

where t is the sample thickness, μ_{nx} and μ_{ny} are neutron AMAC values, μ_{gx} and μ_{gy} are gamma-ray AMAC values, and R is the logarithmic ratio defined as:

$$R = \ln\left(\frac{I_n}{I_{n0}}\right) \bigg/ \ln\left(\frac{I_g}{I_{g0}}\right) \tag{14}$$

The wt% measurement $100w$ is insensitive to thickness [see equation (11)] for constant AMACs. The method is insensitive to the way the material x is distributed in material y. For example, equations (9) and (10) can both be equally derived for a sample consisting of thin alternate layers of materials x and y, or for two quite separate layers of materials x and y stacked one on top of the other.

The errors in the measured values of w and ρ can be written:

$$\Delta w = \frac{1}{\rho t F}\left\{\left(\frac{\Delta I_n}{I_n}\right)^2 [\mu_{gy} + w(\mu_{gx} - \mu_{gy})]^2 + \left(\frac{\Delta I_g}{I_g}\right)^2 [\mu_{ny} + w(\mu_{nx} - \mu_{ny})]^2\right\}^{1/2} \tag{15}$$

$$\Delta \rho = \frac{1}{tF}\left[\left(\frac{\Delta I_n}{I_n}\right)^2 (\mu_{gx} - \mu_{gy})^2 + \left(\frac{\Delta I_g}{I_g}\right)^2 (\mu_{nx} - \mu_{ny})^2\right]^{1/2} \tag{16}$$

where ΔI_n is the error in I_n, and ΔI_g is the error in I_g. Errors in I_{n0} and I_{g0} are assumed to be small compared with errors in I_n and I_g. ΔI_n and ΔI_g are statistical errors, $\sqrt{I_n}$ and $\sqrt{I_g}$, in the source output, given stable electronic performance. The quantity $F = \mu_{nx}\,\mu_{gy} - \mu_{ny}\,\mu_{gx}$.

For most materials, the values of μ_{gx} and μ_{gy} are approximately the same, and F is given approximately by $(\mu_{nx} - \mu_{ny})\,\mu_{gx}$. Large factors F, product densities ρ and count rates are desirable. There are large differences in the values of μ_n for water and common mineral materials [78, 79].

(b) The optimum sample thickness

By differentiating equation (15), t can be freely chosen to minimize the error in w. For a parallel collimated beam, where the count rate is independent of the source/detector separation, the optimal thickness is given by:

$$t_{\text{optimal}} = \frac{2\left(\frac{A_n}{I_n} + \frac{A_g}{I_g}\right)}{\rho\left[\frac{A_n\mu_n}{I_n} + \frac{A_g\mu_g}{I_g}\right]} \tag{17}$$

where $A_n = [\mu_{gy} + w(\mu_{gx} - \mu_{gy})]^2$, $A_g = [\mu_{ny} + w(\mu_{nx} - \mu_{ny})]^2$, $\mu_n\rho t = \mu_{nx}m_x + \mu_{ny}m_y$, and $\mu_g\rho t = \mu_{gx}m_x + \mu_{gy}m_y$.

Net optimal thicknesses for measurements of fat in meat (Section 1.2.4(b)) are illustrated by the calculated curves in Fig. 6. Curves are shown for high fat and low fat samples. Optimal thicknesses are commonly in the range 0.1–0.3 m. In practice, the optimal thickness cannot be too large, since the differences in the neutron AMAC reduce with thickness [78].

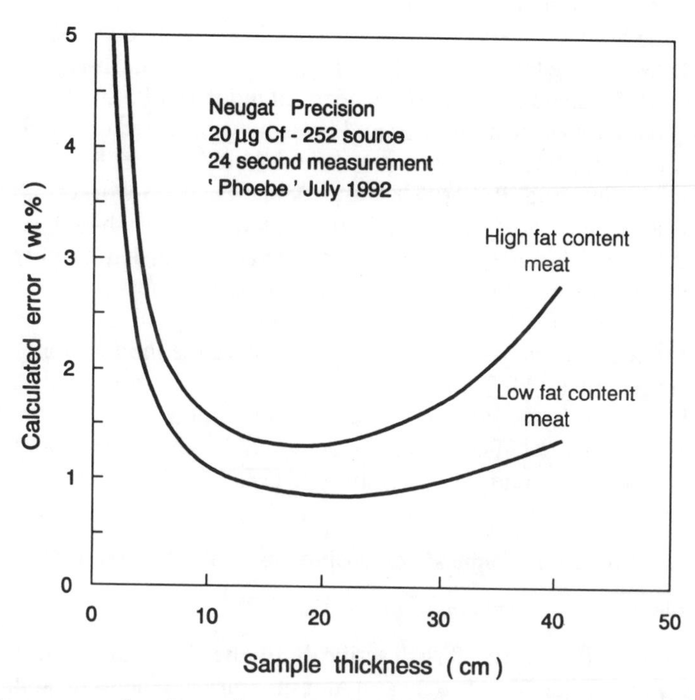

Fig. 6. Error curves showing optimum material thicknesses for fat measurement in meat.

(c) Radiation sources and detectors

This topic is discussed in Chapter 3. Sources such as Am–Be and ^{252}Cf emit both neutrons and gamma-rays. Tominaga *et al.* [79] prefer to use ^{252}Cf sources for industrial applications. The lifetime of practical sources is discussed by Lorch [93]. A Neugat pulsed accelerator source based on a standard neutron time-of-flight (TOF) system with a modified target has been developed [94, 95]. In recent applications, organic scintillators [96] detect both fast neutrons and gamma-rays efficiently. Liquid organic scintillators such as NE213 [73] exhibit pulse shape discrimination (PSD), a mechanism for separating neutron and gamma events in the same detector. PSD systems are of two types: the zero crossing method [97–99] and the charge comparison method [100–102].

In the newer systems [103], organic scintillators operate at high count rates. Detector units are installed in a temperature controlled tank. The PSD unit used (Link Systems 5020 [104]) is maintained at a constant temperature in a closed box using an electronically controlled [105] Peltier unit [106]. Using a 20 μg ^{252}Cf source, total no-sample count rates (I_0) of 400 kHz (100 kHz neutron rate and 300 kHz gamma rate) are achieved. Such rates allow chemical measurements on chemical mixtures where F and ρ are small (e.g. $F < 10^{-5}\ m^4/kg^2$).

Tominaga and coworkers [79, 107] describe a system based on organic scintillators and PSD. Kumahara and Tominaga [108] describe an industrial PSD analyser using Sony CX20052 flash-type ADCs. The system operates at rates up to 100 kHz with long-term stability of 0.5%. The use of flash-type or charge integrating ADCs offers an alternative means of industrial PSD [109–113]. The ADC conversion times are about 5 μs and the achievable detector count rates are lower than those for the charge comparison method.

Sowerby, Millen and Rafter [78] and Millen and co-workers [114, 115] employ several detector systems including a lithium-glass-based scintillator enriched in ^{6}Li. The fast neutrons are thermalized in a high density polyethylene shroud and detected in the glass through the $^6Li(n,\alpha)^3H$ reaction (see Chapter 7). Gamma-ray events have lower pulse heights than the neutron-induced events, and are separated on the basis of pulse height. Since the moderation process is temperature-sensitive [116], the detector temperature is controlled. Although lithium glass has a lower efficiency than organic liquid scintillators, the electronics are simpler [78]. The stabilities achieved for the neutron and gamma count rates were 0.16% and 0.12% relative, respectively.

1.2.4 Measurements of chemical composition using the Neugat method

(a) Measurement of water in materials

Gardner and Calissendorff [83] describe both the dual gamma-ray method (Gamgat) and Neugat for water and density measurement in soil. Corey, Boulogne and Horton [88] also describe moisture measurement in soil, and

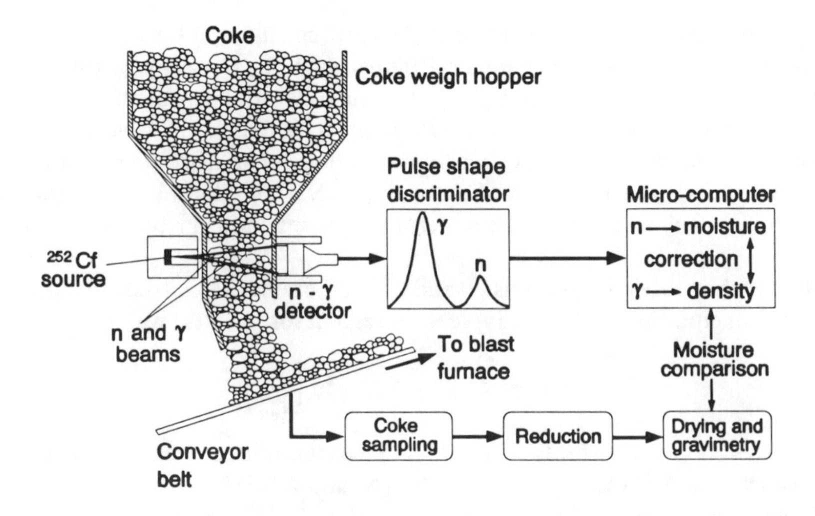

Fig. 7. A Neugat coke hopper gauge used at the Nippon Steel Corporation, Kimitsu Works. (From Ref. 79.)

Corey and Hayes [87] describe moisture and density measurement in marine sediment.

Patents have been awarded to the Japanese Atomic Energy Research Institute and to Hitachi Ltd. [117,118]. Tominaga *et al.* [79] describe an on-line coke moisture hopper gauge for the iron-making process in a blast furnace at the Kimitsu Works of the Nippon Steel Corporation (Fig. 7). Equations (9) and (10) are solved moment by moment under computer control. Over two months of operation, the precision achieved for moisture, with a 99% confidence level, was 0.5 wt%. As discussed in Section 1.1.2, Williams [57] achieved a precision of 3 wt% using fast neutron transmission only. In the case of coal, moisture measurement using Neugat is unsatisfactory because of variations in the non-water hydrogen in coal [119].

To achieve simultaneous chemical determination of C, H, O and Si in simulated coal samples, Tominaga [120] has investigated energy selection of the detected neutrons and gamma-rays. This approach is similar to that of Overley [121], who used a 'white' accelerator-based neutron source to analyse for C, H, O and N in samples (see section 1.3.1).

Sowerby and co-workers [78,114,115] use Neugat (which they call the FNGT technique) to determine moisture in lump coke, both on the conveyer belt and passing through a hopper at the No. 3 blast furnace at the BHP Newcastle Steel-works (Australia). These gauges have a narrow beam, typically 5° deg wide, and coke moisture profiles involve integrated measurements over several minutes. As a result of their work in the mineral industry, Watt, Sowerby, Cutmore and Howarth [122] were awarded the 1992 Australia Prize consisting of a US $200,000 international award for outstanding achievement in a selected area of science and technology promoting human welfare. The instruments produced from their work generate sales of more than US $7.5 million a year under the name 'Coalscan', and benefit the mineral processing industry by about US $65 million a year.

Rather than solving equations (9) and (10) as Tominaga *et al.* did, an empirical calibration equation is used. A typical expression for the wt% water, 100*w*, is:

$$100w = k_1 + k_2 R + k_3 \ln\left(\frac{I_0}{I_{n0}}\right) \tag{18}$$

where k_1, k_2, k_3 are constants, I_n and I_{n0} are the transmitted and incident neutron intensities, and R is the logarithmic ratio [equation (14)].

Neugat, using both thermal [123] and fast [124] neutrons, shows good sensitivity to the moisture content of wood chips. The merits of three different Neugat radiation sources have been compared [95] for moisture measurement in wheat. Neugat can be used to measure water in wool. Measurements have also been made of the composition for mixtures of water with latex rubber, acrylic paint or alcohol [124].

(b) Measurement of non-water materials

A Neugat gauge for fat measurement in boneless meat packed in 27 kg boxes was successfully demonstrated in a meat processing plant in 1990 (New Zealand Patent 213,777/214,666). Called 'Phoebe', its features include a 40° radiation beam, large NE213 scintillation detectors, and PSD separation of neutrons and gamma-rays [103, 125]. For these measurements, the higher neutron AMAC is for fat and not for water (as in previous gauges) and, as both component materials have high hydrogen concentrations, F is small (about 10% of that for water in coke). High count rate operation is being developed, but the dose to the meat is small, typically 0.01 mGy (similar to the natural dose every two days). Neugat provides the only non-invasive fat measuring system for frozen boneless meat. TOBEC (total body electrical conductivity) machines have applications [126, 127] for unfrozen boneless meat only. In Fig. 8 is shown a comparison between the distribution of fat measurements with Phoebe for a 10 wt% sample and the expected theoretical distribution based on equation (15). Good agreement is obtained.

Cream/skim milk mixtures have also been measured using Neugat methods with a ^{252}Cf source and an accelerator source. Greasy (raw) wool has a diverse chemical nature depending on the environment. Neutron activation (see Chap-

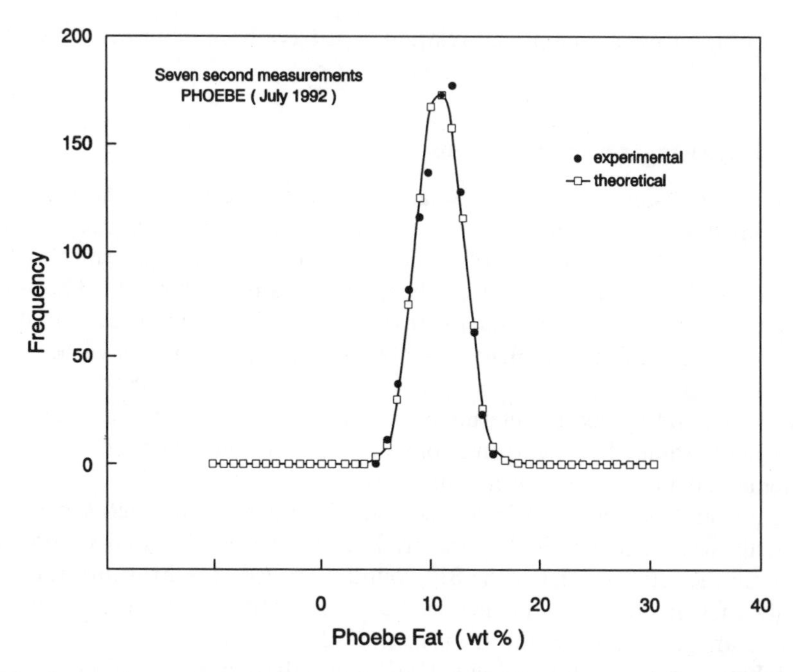

Fig. 8. Comparison of experimental and theoretical precision of measurement on a standard 10 wt% fat meat box. The theoretical curve is based on equation (15) in the text.

ters 6 and 7) has been used to measure trace elements in wool [128, 129]. Neugat has been used to measure wool fat in wool [124].

1.3 Analysis by the absorption of neutrons

The classical procedure of thermal neutron absorption analysis involves elements such as boron and lithium which have a large absorption cross-section (Section 1.1). Recently, the method has been extended to determine elements with small absorption cross-sections, using a method of chemically combining atoms of high and low cross-sections. Some elements are strong neutron absorbers at energies higher than thermal energies. For example, cadmium has a large cross-section (7200 barn) at 0.176 eV which allows its concentration to be determined in alloys, zinc solutions and zinc ores [130].

The precise determination of elemental concentrations using neutron absorption is complicated by neutron scattering [131, 132]. To simplify the relationship between the transmission of neutrons in the sample and the concentration of the absorbing element, empirical relationships have been described by several authors [133–135].

1.3.1 Polyenergetic neutron transmission

Relative and absolute methods of analysis are used. Relative methods involve measuring count rates, first with a sample of known absorbing element concentration, then with the sample of unknown but comparable concentration. In general, the best conditions for analysis are obtained when the detector is surrounded by the sample and the neutron source and the sample are placed in a moderator. Several arrangements for absorption experiments have been discussed [4]. For a given source and experimental geometry, a calibration curve is constructed, and is used to determine the composition of the unknown sample based on the value of the neutron attenuation. The absolute method is based on mathematical analysis using elemental cross-sections [4,133–135].

Frazzoli and Mancini [136] describe a simple apparatus for relative measurements, illustrated in Fig. 9. The material to be measured, liquid or powder, surrounds the active volume of the BF_3 counter. The use of neutron absorption to identify ancient silver-plated coins has been described [137, 138]. Seshadri and Venkatesan [139] discuss the transmission characteristics of ^{252}Cf neutrons through rare earth and boron-loaded concrete slab shields.

Systems in the UK, which continuously monitor boron content in nuclear fuel element cooling ponds at concentrations of 750–2000 ppm, have been discussed [140].

The state of reactivity of a nuclear reactor, called the power reactivity method [141], can be used as an indication of the presence and concentration of a neutron absorber. An associated method is the pile oscillator (Pomerance [142]). For uranium analysis, the sample is activated with reactor neutrons and the delayed neutrons emitted are detected once the sample has been removed [143, 144]. Krinninger, Wiesner and Faber [145] describe a method for the simultaneous determination of ^{235}U and ^{239}Pu contents of irradiated and non-irradiated reactor fuel elements, based on employing two slow neutron irradiation energies

Neutron resonant absorption measurements have been made with a 'white' neutron source. Schrack *et al.* [146] describe a source based on bombarding tungsten with electrons from the NBS Linac. They measured the isotopic abundance of reactor fuel elements. The energies of the transmitted neutrons were separated by TOF. Figure 10 shows a time scan of the energies of transmitted neutrons for a fresh nuclear fuel element. The absorption of neutrons is analysed collectively in several ^{235}U and ^{238}U resonances. It was deduced that the relative abundance of ^{235}U was $4.01 \pm 0.08\%$. This compared with $3.96 \pm 0.005\%$ from scintillation spectroscopy. Anderson [147] has measured the concentrations of In, Hf, Au, Gd, Cd, Eu, Dy, Co, Ir, Mn, Rh, Sm, Ag, U and Ta to 5% precision.

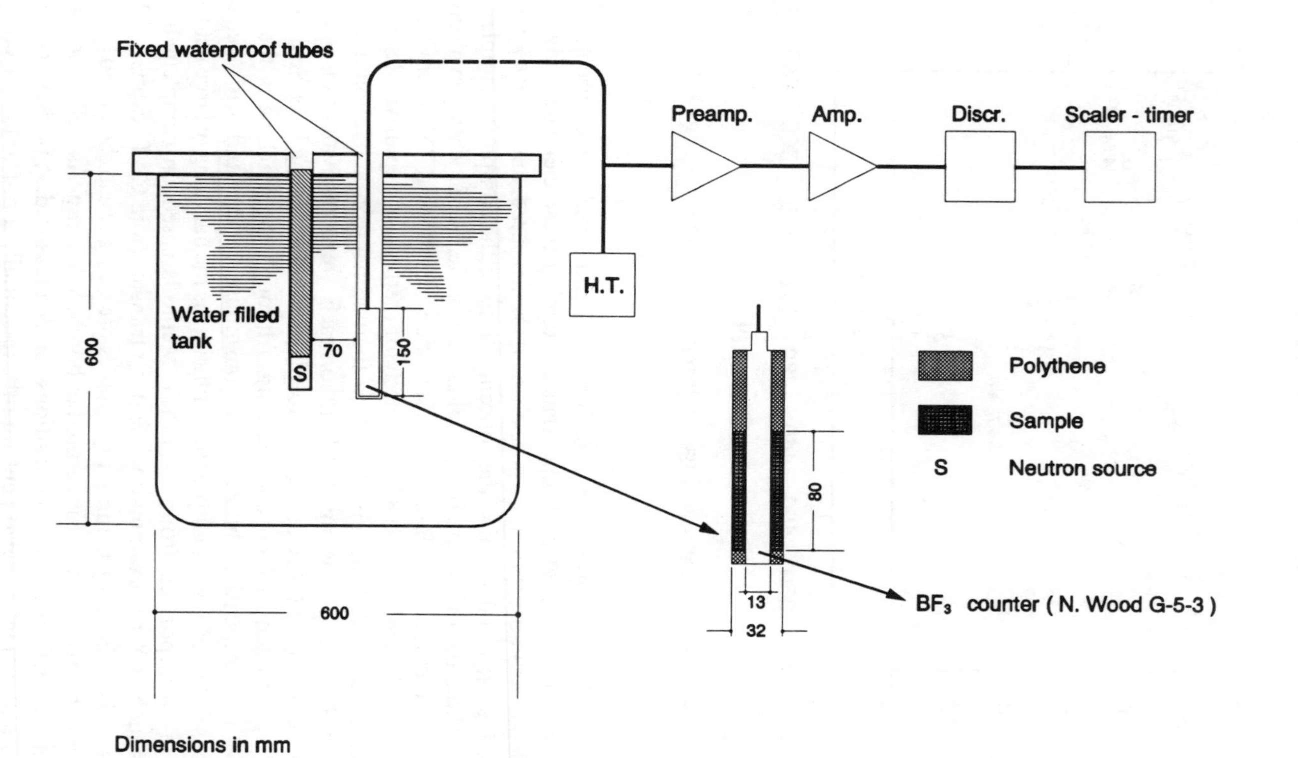

Fig. 9. Typical apparatus for neutron absorption experiments. (From Ref. 136.)

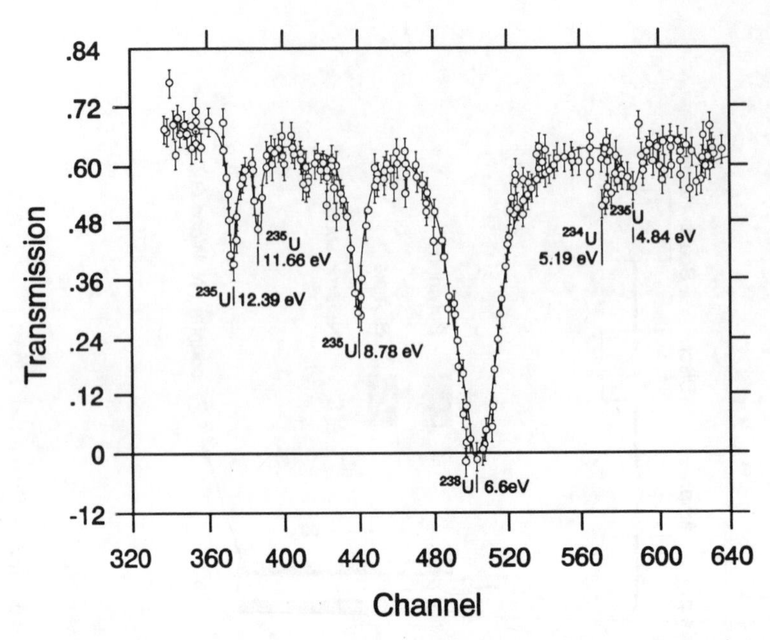

Fig. 10. Radiograph time scan of a reactor fuel pellet transmission. (From Ref. 146.)

Overley [121] describes a method for the simultaneous measurement of H, C, N and O content of bulk materials containing these elements only. The energy dependent absorption for pure elements is measured, and a least squares fit method is used to determine the concentrations in the composite sample.

This method and others are being evaluated as a means of locating explosives in packages and suitcases. Since the destruction of a Boeing 747 aircraft over Lockerbie on December 21, 1988, the search for methods of explosives detection have been intensified. Machines based on thermal neutron activation analysis (TNAA, see Chapter 7), costing US $1 million each, are under evaluation at a few US airports [148]. TNAA sensitivity to plastic explosives such as Semtex is masked by nitrogen in food and clothing. Although the machine is the best available, it is expensive, bulky, time consuming to use and with definite limitations [149, 150]. A major practical difficulty is the handling of large neutron sources in an airport environment [151]. Methods based on complementary methods such as X-ray examination, neutron transmission, CAT scanners, PET scanners and dual energy gamma-ray sources are being researched [150].

Neutron absorption can be measured in borehole logging. As well as porosity logs (Section 1.1.1), chemical compositions can be measured. A package consisting of a small diameter sealed neutron tube (essentially a portable accelerator), typically $\approx 1.2\,\mathrm{m}$ long and $0.04\,\mathrm{m}$ in diameter, with a pulsed output, and a thermal neutron detector are lowered down a borehole. In a neutron 'die-away log', the time rate of decay of thermal neutrons in a formation can be used to distinguish the absorption of salt water from that of oil (via the large thermal absorption cross-section of chlorine in the salt water). This is a means of discovering oil in prospecting [107].

1.3.2 Monoenergetic neutron transmission

Gauges are available in which the strong resonant absorption of nuclei to monoenergetic neutrons is utilized. The measurements are similar to those in the previous section but at a single neutron energy. Accelerators are used to generate the monoenergetic neutrons. Boldeman and Walsh [152] describe an experimental method of measurement of protein (wt%) in meat by measuring the transmission of neutrons at the 432 keV resonance of nitrogen. The amount of protein in the meat is deduced by fitting the measured resonance with calculated curves for nitrogen and a nearby oxygen resonance.

REFERENCES

1. J. H. Hubbell, *J. Res. Natl. Inst. Stand. Technol.* **95**(6), 689 (1990).
2. P. L. Oelgaard, Lecture notes on neutron and gamma gauges, Technical University of Denmark, Lyngby, Afdelingen for Electrofysik (1981); INIS-MF-8592.
3. J. Michalik, Industrial Applications of Radioisotope Techniques in Poland, Institute of Atomic Energy, Otwock-Swierk (1985); NTIS DE90610933/HDM.
4. J. Tölgyessy and M. Kyrš, *Radioanalytical Chemistry 2*, Wiley, New York (1989), Chapter 9.6, p. 227.
5. G. F. Knoll, *Radiation Detection and Measurement*, Wiley, New York (1989).
6. P. J. Grant, *Elementary Reactor Physics*, Pergamon Press, Oxford (1966), p. 52.
7. E. J. Hall and H. H. Rossi, *Tech. Rep. Ser. – I.A.E.A.* (159) (1974).
8. M. Drosg, *Nucl. Sci. Eng.* **106**, 279 (1990).
9. A. U. Austin, *NBS Handb. (U.S.)* (72) (July 15, 1960).
10. J. W. Pyper, *Anal. Chim. Acta* **170**, 159 (1985).
11. K. Papez, J. F. Cameron and B. Machaj, Proceedings of a Symposium on Nuclear Techniques in the Basic Metal Industries, Helsinki, IAEA (1972), p. 183.
12. P. Cull, *NZ J. Agric.* **152** (10), 12 (1987).
13. E. L. Greacen, R. L. Correll, R. B. Cunningham, G. G. Johns and K. D. Nicolls, in *Soil Moisture Measurement by the Neutron Method*, edited by E. L. Greacen, CSIRO, East Melbourne, Australia (1981), p. 50.
14. J. F. Cameron (Ed.), Tech. Rep. Ser. – *I.A.E.A.* (112) (1970).
15. D. J. Wilson, *Nucl. Technol.* **60**, 155 (1983).
16. D. J. Wilson and A. I. M. Richie, *Aust. J. Soil Res.* **24** (1), 11 (1985).
17. M. Vauclin, R. Haverkamp and G. Vachaud, Proceedings of the International Symposium on Isotope and Radiation Techniques in Soil Physics and Irrigation Studies, IAEA (1983); INIS 15: 027624.
18. A. V. Lakshmipathy and P. Gangadharan, *Nucl. Instrum. Methods* **142** (3), 577 (1977).

19. D. B. McCulloch and T. Wall, *Nucl. Instrum. Methods* **137**, 577 (1976).
20. J. K. Ruprecht and N. J. Schofield, *Aust. J. Soil Res.* **28** (2), 153 (1990).
21. S. Kasi and H. Koskinen, *Nucl. Eng. Des.* **3**, 74 (1966).
22. S. S. H. Kasi, *Int. J. Appl. Radiat. Isot.* **33**, 667 (1982).
23. S. S. H. Kasi, J. Immonen and K. Saikku, *Res. Rep. – Helsinki Univ. Technol. Lab. Phys.* (120) (1983).
24. P. Votava, D. Castek and Z. Smejkal, Czech Patent Document 253040/B1/; INIS 20:076782.
25. D. J. Wilson, Australian Nuclear Science and Technology Organisation, ANSTO/E-669 (1988); ISBN 0 64259879 7, INIS 20:068091.
26. D. J. Wilson, *Aust. J. Soil Res.* **26** (1), 87 (1987).
27. D. J. Wilson, Australian Nuclear Science and Technology Organisation, ANSTO/E-674 (1988); ISBN 0 642598843, INIS 20:044544.
28. D. J. Wilson, *Aust. J. Soil Res.* **26** (1), 97 (1987).
29. G. LeR. Stewart, 30729; Ph.D Thesis, Washington State University, Pullman (1962); *Nucl. Sci. Abstr.* **17** (18), 30729.
30. G. A. Buzlov, V. P. Ivanov, V. Yu. Kitov and A. B. Podrugin, *Meas. Tech. (Engl. Transl.)* (3), 229 (1986).
31. Proceedings of a Symposium on Isotope and Radiation Techniques in Soil Physics and Irrigation Studies, Istanbul, IAEA (1967).
32. Proceedings of the French-Polish Colloquium on Soil Physics, Lublin, Poland (1983); NTIS DE84750711/HDN.
33. G. Csom and S. Benedik, *Isotopenpraxis* **9**, 284 (1973).
34. R. L. Caldwell, K. P. Desai and W. R. Mills, *Jr., Nucl. Tech. Geochem. Geophys. Proc. Panel*, Vienna (1974), p 3.
35. W. R. Mills, L. S. Allen and D. C. Stromswold, Proceedings of a Symposium on Nuclear Techniques in the Exploration and Exploitation of Energy and Mineral Resources, IAEA, Vienna (1990), p. 247.
36. F. L. Glesius and T. A. Kniss, *IEEE Trans. Nucl. Sci.* **35** (1), 867 (1988).
37. J. E. Galford, C. Flaum, W. A. Gilchrist, Jr., P. D. Soran and J. S. Gardner, *Soc. Pet. Eng. Form. Eval.* **3**, 371 (1988).
38. M. Oraby, K. Verghese and R. P. Gardner, *Nucl. Instrum. Methods, Sect. A*, **A299**, 674 (1990).
39. D. V. Ellis, C. Flaum, J. E. Galford and H. D. Scott, Proceedings of the 62nd Annual Technical Conference, Society of Petroleum Engineers, Dallas (1987), Paper 16814.
40. M. P. Smith, Proceedings of the 27th Annual Logging Symposium, Society of Professional Well Log Analysts, Houston (1986), Paper YY.
41. M.P. Smith, Proceedings of the 28th Annual Logging Symposium, Society of Professional Well Log Analysts, Houston (1987), Paper R.
42. M. Gartner, C. Schnoor and P. Sinlair, Proceedings of the 27th Annual Logging Symposium, Society of Professional Well Log Analysts, Houston (1986), Paper UU.
43. P. Albats and S. Antkiv, U.S. Patent 4,760,252 (1988).
44. W. R. Mills, L. S. Allen and D. C. Stromswold, *Nucl. Geophys.* **2** (2), 81 (1988).
45. W. R. Mills, D. C. Stromswold and L. S. Allen, *Log. Anal.* **30**, 119 (1989).
46. R. W. Barnard, W. A. Stephenson and J. H. Weinlein, *IEEE Trans. Nucl. Sci.* **30** (2), 1664 (1983).
47. H. J. M. Bowen, *Environmental Chemistry of the Elements*, Academic Press, London (1979), p. 198.
48. M. Kodama, *Jpn. J. Appl. Phys.* **23** (6), 726 (1984).
49. S. I. Avdyushin, E. V. Kolomeets, I. M. Nazarov, V. V. Oskomov, A. N. Pegoev and Sh. D. Fridman, *Meteorol. Gidrol.* **3**, 51 (1986).

50. N. F. Gray, *Effluent Water Treat. J.* **24** (5), 203 (1984).
51. N. G. Cutmore, R. J. Holmes, B. D. Sowerby and J. S. Watt, Proceedings of the 13th Congress of the Council of Mining and Metallurgy, Singapore (May 1986).
52. UB100 Series Moisture Monitors, Mineral Control Instrumentation Ltd, PO Box 64, Unley, South Australia.
53. N. G. Cutmore, D. Abernethy and T. G. Evans, *J. Microwave Power Electromagn. Energy* **24** (2), 79 (1989).
54. N. G. Cutmore, S. I. Doumit and T. G. Evans, Proc. Eur. Microwave Conf. 20th, Budapest (1990).
55. U. Korell, H. Kurth and H. J. Mette, *Isotopenpraxis* **21**, 220 (1985).
56. K. Koditz, G. Scheidewig, J. Mey, M. Rasch, U. Korell and A. Zenner, DD Patent 228901 A1 (1985); INIS 17:047342.
57. R. B. Williams, *J. Radioanal. Chem.* **48**, 49 (1979).
58. J. M. Eskes, *Can. Control Instrum.* **11**, 30 (1972).
59. W. Kühn, *Atompraxis*, **5**, 133 (1959).
60. A. K. Stroikovskii, A. A. Pershin and A. N. Sheikin, *Meas. Tech. (Engl. Transl.)* **23**, 351 (1980).
61. P. Dinuzzo, V. L. Gravitis, A. Toop, J. S. Watt and H. W. Zastawny, *Nucl. Geophys.* **1** (4), 329 (1987).
62. V. Vorišek, *Jad. Energ.* **3**, 258 (1957).
63. H. Cermak, *Isotopentechnik* **1**, 140 (1960/61).
64. H. Cermak, *Kernenergie* **7**, 557 (1964).
65. L. Nedeljković, *Technika (Belgrade)* **18**, 621 (1963); *Chem. Abstr.*, 19905 (June 30, 1964).
66. R. Czembor and W. Hartmann, *Isotopenpraxis* **13**, 252 (1977).
67. J. Mirowicz and L. Lis, *Nukleonika* **34**, 19 (1989).
68. P. J. Mathew, C. Ceravolo, P. Huppert and J. G. Miles, *Int. J. Appl. Radiat. Isot.* **34**, 1377 (1983).
69. G. Sh. Pekarskii, *Sov. J. Nondestr. Test. (Engl. Transl.)* **23** (5), 307 (1988).
70. G. Sh. Pekarskii, *Sov. J. Nondestr. Test (Engl. Transl.)* **23** (7), 512 (1988).
71. H. H. Barschall and T. L. Powell, *Phys. Rev.* **96**, 713 (1954).
72. A. Bertin and A. Vitale, *Nucl. Instrum. Methods* **91**, 649 (1971).
73. NE Technology Ltd, Bath Rd, Beenham, Reading, UK.
74. Philips Photonics photomultiplier tube preliminary specification, XP4512B (June 1992).
75. Type 8854 PMT Data Bulletin, Burle Industries Inc., 1000 New Holland Ave, Lancaster, PA.
76. J. A. Harvey and N. W. Hill, *Nucl. Instrum. Methods* **162**, 507 (1979).
77. J. B. Marion and J. L. Fowler, *Fast Neutron Physics*, Interscience, New York, (1963).
78. B. D.Sowerby, M. J. Millen and P. T. Rafter, *Nucl. Geophys.* **2** (1), 55 (1988).
79. H. Tominaga, N. Wada, N. Tachikawa, Y. Kuramochi, S. Horiuchi, Y. Sase, H. Amano, N. Okubo and H. Nishikawa.; *Int. J. Appl. Radiat. Isot.* **34** (1), 429 (1983).
80. J. F. Cameron, *Phys. Bull.* **24**, 605 (1973).
81. S. Teller, Proceedings of a Symposium on Nuclear Techniques and Mineral Resources, IAEA, Vienna (1977), p. 621.
82. N. Kuper, Proceedings of a Conference on Applications of Radioisotopes and Radiation Technology, Grenoble (1982), p. 379.
83. W. H. Gardner and C. Calissendorff, Proceedings of the Symposium on Isotope and Radiation Techniques in Soil Physics and Irrigation Studies, Istanbul, IAEA and FAO (1967), p. 101.
84. B. Machaj, *Nukleonika* **19** (2), 24 (1974).

85. T. E. Reim, *Instrum, Soc. Am. Conf. Prepr.* **14**, 47 (1967).
86. E. Frevert and G. Stehno, *IEE Conf. Publ.* **84**, 215 (1972).
87. J. C. Corey and D. W. Hayes, *Deep-Sea Res.* **17**, 917 (1970).
88. J. C. Corey, A. R. Boulogne and J. H. Horton, *Water Resour. Res.* **6** (1), 223 (1970).
89. M. Glaser, H. W. Thummel and G. Korner, *Isotopenpraxis* **23**, 58 (1987).
90. R. A. Fookes, V. L. Gravitis, J. S. Watt, P. E. Hartley, C. E. Campbell, E. Howells, T. McLennan and M. J Millen. *In. J. Appl. Radiat. Isot.* **34** (1), 63 (1983).
91. M. D. Rebgetz, J. S. Watt and H. W. Zastawny, *Nucl. Geophys.* **5** (4), 479 (1991).
92. J. S. Watt. H. W. Zastawny, M. D. Rebgetz, P. E. Hartley and W. K. Ellis, *Nucl. Geophys.* **5** (4), 469 (1991).
93. E. A. Lorch, *Radiol. Prot. Bull.* (34) (May 1980).
94. C. M. Bartle and C. R. Purcell, *Nucl. Instrum. Methods Phys. Res., Sect. A* **A254**, 219 (1987).
95. C. M. Bartle, C. R. Purcell and A. Wilson, *Nucl. Instrum. Methods. Phys. Res., Sect. A* **A291**, 655 (1990).
96. F. D. Brooks, *Nucl. Instrum. Methods* **162**, 477 (1979).
97. J. Biakowski, M. Moszyński and D. Wolski, *Nucl. Instrum. Methods Phys. Res., Sect. A* **A275**, 322 (1989).
98. R. A. Winyard, J. E. Lutkin and G. W. McBeth, *Nucl. Instrum. Methods* **95**, 141 (1971).
99. D. B. C. B. Syme and G. I. Crawford, *Nucl. Instrum. Methods* **104**, 245 (1972).
100. F. D. Brooks, *Nucl. Instrum. Methods* **4**, 151 (1959).
101. T. Elevant, *Nucl. Instrum. Methods Phys. Res., Sect. A* **A278**, 774 (1989).
102. J. M. Adams and G. White, *Nucl. Instrum. Methods* **156**, 459 (1978).
103. C. M. Bartle, *Appl. Radiat. Isot.* **42** (11), 1115 (1991).
104. Pulse shape discriminator, model 5020, Link Systems, Halifax Rd, High Wycombe, UK.
105. Shinko Microcomputer Based Temperature Indicating Controller, MCS-100 series, Osaka, Japan.
106. Thermoelectric Refrigeration Ltd. model TC2000, PO Box 1827, Christchurch, New Zealand.
107. H. Tominaga, *Radioisotopes* **33**, 52 (1984).
108. T. Kumahara and H. Tominaga, *IEEE Trans. Nucl. Sci.* **NS-31** (1), 451 (1984).
109. Z. W. Bell, *Nucl. Instrum. Methods* **188**, 105 (1981).
110. J. Kasagi, T. Murakami and T. Inamura, *Nucl. Instrum. Methods Phys. Res., Sect. A* **A236**, 426 (1985).
111. J. H. Heltsley, L. Brandon, A. Galonsky, L. Heilbronn, B. A. Remington, S. Langer, A. V. Molen and J. Yurkon, *Nucl. Instrum. Methods Phys. Res., Sect. A* **A263**, 441 (1988).
112. S. Pai, W. F. Piel, Jr., D. B. Fossan and M. R. Maier, *Nucl. Instrum. Methods Phys. Res., Sect. A* **A278**, 749 (1989).
113. R. Aleksan, J. Bouchez, M. Boussicut, T. Desanlis, D. Jourde, J. Mullié, F. Pierre, L. Poinsignon, R. Praca, G. Roussel, J. F. Thomas, J. Collot, A. Stutz and E. Kajfasz *Nucl. Instrum. Methods Phys. Res., Sect. A* **A273**, 303 (1988).
114. M. J. Millen, P. T. Rafter and B. D. Sowerby, *Nucl. Geophys.* **2** (4), 207 (1988).
115. M. J. Millen, P. T. Rafter, B. D. Sowerby, M. T. Rainbow and L. Jelenich, *Nucl. Geophys.* **4** (2), 215 (1990).
116. J. Keinert, *Temperature Dependence of Thermal Neutron Scattering Cross Sections for Hydrogen Bound in Moderators*, Faehinformationszentrum Energie, Physik, Mathematik GmbH, Karlsruhe (1982); NTIS DE85780157/HDM.
117. Moisture determination using radiation transmission, Japanese Patent 82, 63, 440 (April 16, 1982); *Chem. Abstr.* **97**, 103626r (1982).

118. Hitachi Ltd., Moisture measurement (with the elimination of the bulk density effect), Japanese Patent 81, 22, 941, (March 14, 1981); *Chem. Abstr.* **95**, 117642p (1981).
119. B. D. Sowerby, *Nucl. Geophys.* **5**(4), 491 (1991).
120. H. Tominaga, in *Artificial Radioactivity*, edited by K. N. Rao and H. J. Arnikar, McGraw-Hill, New Delhi (1985), p. 480.
121. J. C. Overley, *Int. J. Appl. Radiat. Isot.* **36**, 185 (1985).
122. Science and Education in Australia and New Zealand, 2, in *New Scientist* **133**, 3 (25 Jan 1992).
123. E. Frevert, *Isotopenpraxis* **14**, 308 (1978).
124. C. M. Bartle, in preparation.
125. C. M. Bartle, Proceedings of the 26th Meat Industry Research Conference, MIRINZ, Hamilton, New Zealand (1990), p. 277.
126. N. J. Newby, N. L. Kein and D. L. Brown, *Am. J. Clin. Nutr.* **52** (2), 209 (1990).
127. I. J. Eustace, W. J. Butt, R. A. Gibbons, M. J. Rowland and F. F. Thornton, *Meat Research Report* 1/91, CSIRO Meat Research Laboratory, Cannon Hill, Australia (1991).
128. W. B. Healy, L. C. Bate and T. G. Ludwig, *N. Z. J. Agric. Res.* **7**, 603 (1964).
129. H. Özyol, *J. Radioanal. Nucl. Chem.* **139**, 339 (1990).
130. L. De Norre, J. Op de Beeck and J. Hoste, *J. Radioanal. Chem.* **78**, 137 (1983).
131. J. R. D. Copley and C. A. Stone, *Nucl. Instrum. Methods Phys. Res., Sect. A* **A281**, 593 (1989).
132. D. Copley, *Trans. Am. Nucl. Soc.* **61**, 105 (1990).
133. D. D. Deford and R. S. Braman, *Anal. Chem.* **30** (11), 1765 (1958).
134. J. E. Strain and W. S. Lyon, Proceedings of a Symposium on Radiochemical Methods of Analysis, Salzburg (1964), Vol. 1, IAEA, Vienna (1965), p. 245.
135. J. Tölgyessy, Š. Varga and P. Dillinger, *Sb. Pr. Chemickotechnol. FaK. SVST* 97 (1967).
136. F. V. Frazzoli and C. Mancini, *Kerntechnik* **15**, 136 (1973).
137. C. Mancini and P. Petrillo Serafin, *Archaeometry* **18** (2), 214 (1976).
138. C. Mancini, *Isotopenpraxis* **13** (7), 247 (1977).
139. B. S. Seshadri and R. Venkatesan, *Nucl. Eng. Des.* **117**, 325 (1989).
140. J. F. Cameron, *Nucl. Tech. Geochem. Geophys., Proc. Panel* Vienna (1974), p. 175.
141. T. I. Taylor, R. H. Anderson and W. W. Havens, Jr., *Science* **114** (7), 341 (1951).
142. H. Pomerance, *Phys. Rev.* **88** (2), 412 (1952).
143. G. Carrard, *AAEC Nucl. News* **14** (1982).
144. R. Ashrafi, Proceedings of a Symposium on Nuclear Techniques in the Exploration and Exploitation of Energy and Mineral Resources, IAEA, Vienna (1991), p. 437.
145. H. Krinninger, S. Wiesner and C. Faber, *Nucl. Instrum. Methods* **73**, 13 (1969).
146. R. A. Schrack, J. W. Behrens, R. Johnson, and C. D. Bowman, *IEEE Trans. Nucl. Sci.* **NS-28**, 1640 (1981).
147. R. H. Anderson, *Diss. Abstr.* **16**, 1336 (1956).
148. G. Stix, *Sci. Am.* **266** (1), 116 (Jan. 1992).
149. B. Cook, *Technol. Update* **47** (20), 107 (May 13, 1991).
150. C. Joyce, *New Sci.*, **129**, 4 (26 Jan. 1991).
151. L. Grodzins, *Nucl. Instrum. Methods Phys. Res., Sect. B* **B56/57**, 829 (1991).
152. J. W. Boldeman and R. L. Walsh, Proceedings of the 5th AINSE Nuclear Techniques of Analysis Conference, Sydney (1987).

Part 3

Elemental Analysis with Particle Accelerators

Chapter 10

CHARGED PARTICLE ACTIVATION ANALYSIS

Karel Strijckmans

Laboratory of Analytical Chemistry, Institute for Nuclear Sciences, Universiteit Gent, Proeftuinstraat 86, B-9000 Gent, Belgium

1 INTRODUCTION

Charged particle activation analysis (CPAA) is an analytical method for determining the elemental concentration (or more exactly the stable isotope concentration) of trace elements in the bulk of a sample. Speciation (i.e. discrimination between different oxidation states of the analyte element or between different compounds) is not possible. CPAA is based on charged particle-induced nuclear reactions leading to radionuclides. Identification of the radiation emitted (i.e. measurement of the energy and/or half-life) leads to qualitative analysis, measurement of the amount of particles or photons emitted (i.e. the activity) yields quantitative analysis. CPAA provides excellent detection limits, accuracy and precision. Its major advantages are: solid samples are not dissolved before irradiation, surface contamination can be removed after irradiation so that only the bulk concentration is determined, and the method is not subject to reagent blank errors. CPAA has proved its excellent capabilities for the determination of light elements (boron, carbon, nitrogen and oxygen) in metals and semiconductors. At low concentration levels (μg/g), it is now considered as the only analytical method to determine the bulk concentration. Its disadvantages are: inherent complexity and cost, less suitable for liquid samples, and heating of the sample during irradiation. Special precautions have to be taken for samples that are not massive and are poor thermal conductors.

Chemical Analysis by Nuclear Methods Edited by Z. B. Alfassi

2 THEORY

2.1 Interaction of charged particles with matter

When a sample is irradiated with charged particles, most of the particles do not induce nuclear reactions. Charged particles passing through matter lose energy mainly by interaction with electrons, leading to excitation or ionization of the target atoms or molecules (at least in the energy interval of interest for CPAA).

2.1.1 Stopping power of elements

The stopping power of a target for a charged particle is defined as the energy loss per unit thickness, and is commonly expressed as [in MeV/(g/cm^2)]

$$S = -\mathrm{d}E/(\rho\,\mathrm{d}l) \tag{1}$$

where E = energy (MeV), l = thickness (cm), and ρ = mass density (g/cm^3).

For a sufficiently high energy (>1 MeV) the stopping power can be calculated by the Bethe formula [1]. The stopping power depends on the charged particle and its energy and the target material. The stopping power is roughly proportional to $Z_a^2\, m_a/E_a$, with Z_a the atomic number, m_a the mass and E_a the energy of the charged particle. The stopping power increases for decreasing atomic number (Z_A) of the target material. Figure 1 gives the stopping power of titanium for protons, deuterons, helium-3 and helium-4 particles, i.e the charged particles commonly used in CPAA. The stopping power is lower for protons and deuterons and much higher for helium-3 and helium-4 particles. Figure 1 shows also the stopping power of hydrogen, helium, titanium and thorium for protons. At very low Z_A there is a very steep increase of the stopping power: the stopping power of hydrogen for protons is about 5 times the stopping power of targets with an atomic number higher than titanium but the stopping power of helium is only twice as high.

2.1.2 Stopping power of mixtures and compounds

For mixtures, the stopping power can be calculated from stopping power data for the elemental components by the additivity rule of Bragg and Kleeman [2]. If stopping powers are expressed in MeV/(g/cm^2) this rule can be written as

$$S = \sum w_i S_i \tag{2}$$

with w_i = mass fraction of component i, and S_i = stopping power of component i.

For compounds, Bragg's rule can be applied. However, deviations from this simple additivity rule may occur owing to chemical binding effects, i.e. differences in the outer-shell electronic structure of a free atom and the same atom bound in a compound. Deviations from Bragg's rule should be expected mainly at low

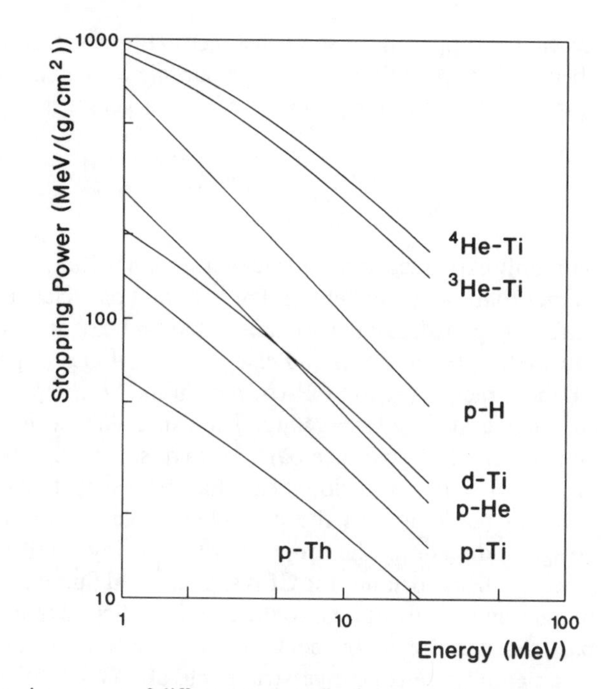

Fig. 1. Stopping power of different targets (hydrogen, helium, titanium and thorium) for different charged particles (protons, deuterons, helium-3 and helium-4 particles) as a function of energy.

energy (where the contribution of the valence electrons to the total stopping power becomes large) and, especially, for light target elements (where the outer-shell electrons constitute a major fraction of the total number of electrons). Therefore, stopping powers for protons and helium-4 particles at energies between 0.1 and 1 MeV have been measured very accurately (statistical and systematic error $\approx 1\%$) in a large number of saturated and unsaturated hydrocarbons [3]. The results obtained were compared with stopping power data calculated according to Anderson and Ziegler (Section 2.1.4). It has been found that these deviations are only of the order of a few percent for energies (0.1–1 MeV) far below the energy of interest in CPAA, i.e. above the Coulomb barrier ($\gg 1$ MeV, Section 2.2.2). For CPAA this deviation from Bragg's rule is negligible, even for compounds with components of low atomic number.

2.1.3 Range

A charged particle loses only a very small fraction of its energy in a single interaction with an electron. As it is virtually undeflected from its original direction, charged particle paths are straight lines. Charged particles with the same initial energy will be stopped at the same distance in the target, i.e. the range. The relationship between the range R of a charged particle with an initial energy E_i in a particular target and the stopping power S of that target for that particular charged particle is given by

$$R = \int_{E_i}^{0} \frac{dE}{-dE/(\rho dl)} = \int_{0}^{E_i} \frac{dE}{S} \quad (3)$$

The range is commonly expressed as mass thickness (g/cm^2). Ranges of protons or deuterons are larger than ranges of helium-3 or helium-4 particles, and ranges in the very light targets (e.g. hydrogen) are smaller than in heavier targets.

This simple picture (straight line paths and *all* charged particles stopped at the same distance R in the target) is used in CPAA, but should be completed by the ideas of projected range and range straggling. The range defined is the linear range, i.e. the total path length the charged particle traverses through the target. Small deviations from its original direction mean that the penetration depth of a charged particle should be defined as the projected range, i.e. the trajectory projected on the original direction. The projected range, being smaller than the linear range, is of negligible importance for CPAA. Statistical fluctuations in the energy loss per interaction and the number of interactions dictate that the range for all charged particles is not exactly the same but shows a spread of a few percent. This phenomenon is called range straggling, and is also of negligible importance for CPAA. Stopping power and range should thus be defined as average values.

2.1.4 Calculation of stopping power and range data for elemental matter, mixtures and compounds

The stopping power of all elemental matter for protons, deuterons, helium-3 and helium-4 particles can be calculated using the formulae and data of Andersen and Ziegler [4–5]. In the energy interval of interest for CPAA, i.e. between the Coulomb barrier and up to 5 to 20 MeV higher, these compilations make use of the Bethe equation and experimentally determined parameters. They claim an accuracy of a few percent. The energy is expressed in keV/amu (keV/u) for protons and deuterons and in keV or MeV for helium-3 and helium-4 particles. The stopping power is expressed in $eV/(10^{15}\ atoms/cm^2)$. Here the number of atoms per cm^3 (the number density of entities or number concentration) is used in the definition of stopping power instead of the mass density as equation (1).

The compilations of Andersen and Ziegler provide data on the linear (i.e. total path length) range for protons, deuterons and helium-4 particles (but not for helium-3 particles) only in graphical form. The formulae given for the projected range of helium-4 particles are of no importance for CPAA. Therefore equation (3) is applied to calculate the (linear) range (in g/cm^2) from the stopping power data [in $MeV/(g/cm^2)$] by numerical integration. It should be mentioned that, if no stopping power data are calculated for lower energies, a small, positive and systematic error will be made in the calculation of the range. As will become clear later (Section 2.4.2), quantitative calculations performed in CPAA do not require ranges but merely differences of ranges, so that this systematic error is cancelled. (The only exception is a standardization method called as 'Ricci II', which will prove to be the worst choice). There is thus no need to calculate stopping powers or ranges in the energy region below 1 MeV.

For mixtures and compounds for which the main matrix composition is known, the stopping power data can be calculated using equation (2) and from the stopping power data for the elemental compounds of the mixture or compound. Similarly, from the stopping power data, the range can be calculated by equation (3).

2.2 Nuclear reaction

CPAA is based on a nuclear reaction A(a, b)B, where a stable target nucleus A (at rest) is irradiated by accelerated charged particles a to form a radionuclide B with emission of one (or several or a cluster of) particle(s) b. Important characteristics are the minimum particle energy required to induce a nuclear reaction and the probability that this reaction will proceed. As a charged particle is slowed down when traversing matter (Section 2.1), the latter characteristic should be known as a function of the energy. Then the question arises concerning a decrease of the beam intensity due to nuclear reactions. Finally, some practical considerations are given as to which activity is induced in a particular matrix by irradiation with a particular charged particle.

2.2.1 *Q*-value and threshold energy

A nuclear reaction A(a, b)B is characterized by a Q-value, i.e. the energy released by one reaction. The Q-value is related to the difference between the rest masses of reactants and products.

$$Q = (m_A + m_a - m_b - m_B)c^2 \tag{4}$$

If masses are expressed in unified atomic mass units (u), the Q-value in MeV is given by

$$Q = 931.5(m_A + m_a - m_b - m_B) \tag{5}$$

Although masses of the nuclei should be used in equations (4) and (5), they can be replaced by atomic masses, as electron masses are cancelled, at least if this is done also for the charged particle a and the emitted particle(s) b if b is not a neutron (the probability of photon emission is very low for charged particle-induced reactions). A nuclear reaction can be exoergic ($Q > 0$) or endoergic ($Q < 0$), similar to exo- and endothermic chemical reactions. Endoergic reactions are only energetically possible for charged particles with a minimum energy, i.e. the threshold energy E_t. The threshold energy is slightly higher than $-Q$. The compound nucleus, formed by collision of the charged particle a and the target nucleus A, retains a fraction of the kinetic energy of the charged particle, $m_a/(m_A + m_a)$, as kinetic energy. This fraction is thus 'lost' to compensate for the shortfall of mass/energy for an endoergic reaction. Thus, for endoergic reactions ($Q < 0$) the threshold energy is given by

$$E_t = -\frac{m_A + m_a}{m_A} \cdot Q \approx -\frac{A_A + A_a}{A_A} \cdot Q \tag{6}$$

where the mass m can be replaced by the mass number A. For exoergic reactions ($Q > 0$) the threshold energy is zero by definition.

Q-values were compiled by Keller, Lange and Münzel [6] for all reactions of interest in CPAA: such as (x, n), (x, 2n), (x, α), (x, d), $\cdots$ with x = p, d, ^{3}He or α (^{4}He). The accuracy of the Q-values calculated from experimental data is claimed to be within 0.1 MeV, while for the others (marked with *) the accuracy is 1 Mev for target nuclides with $Z_A > 30$ and 5 MeV for target nuclides with $Z_A < 30$. Emission of clusters of particles (e.g. deuterons) is energetically more favorable than emission of several particles (p and n), the difference being the binding energy of the cluster. Therefore the threshold energy for (x, d), (x, t) and (x, ^{3}He) reactions has to be calculated from the Q-value for the (x, pn), (x, p2n) and (x, 2pn) reactions increased by the binding energy for d (2.2 MeV), t (8.5 MeV) and ^{3}He (7.7 MeV), thus leading to a lower threshold energy for endoergic reactions.

2.2.2 Coulomb barrier

The Coulomb barrier also determines the minimum charged particle energy needed to induce a nuclear reaction. The coulombic repulsive force between the target nucleus and the charged particle dominates at 'large' distances and increases with decreasing distance of separation between the charged particle and the target nucleus until the charged particle comes within the range of the attractive nuclear forces of the target nucleus. At some particular distance (i.e. the sum of the radii of the charged particle and the target nucleus) the forces balance each other, and at shorter distances the attractive nuclear force dominates. The decrease in kinetic energy is given by the Coulomb barrier

$$E_C(\text{MeV}) \approx 1.02 \cdot \frac{A_A + A_a}{A_A} \cdot \frac{Z_A Z_a}{A_A^{1/3} + A_a^{1/3}} \tag{7}$$

where the nuclear radius is proportional to $A^{1/3}$. The kinetic energy, the charged particle lost by the Coulomb barrier, is released again when nuclear reaction occurs. Consequently the Coulomb barrier influences the energetics of a nuclear reaction only in that a charged particle must have a kinetic energy higher than the

Coulomb barrier before the reaction can occur. Figure 2 approximates the Coulomb barrier for protons, deuterons, helium-3 and helium-4 particles as a function of the atomic number of the target. Quantum mechanical treatment of the problem explains that the reaction probability for particles with kinetic energies lower than the Coulomb barrier is not exactly zero but very low, and rapidly drops as the energy of the charged particle decreases.

2.2.3 Nuclear reaction cross-section

The probability of a nuclear reaction is expressed as the (nuclear reaction) cross-section, which has the dimensions of area. This originates from the simple picture that the probability for a reaction between the target nucleus and the incident charged particles is proportional to the geometric cross-section that the target nucleus presents to a beam of charged particles. The average geometric cross-section is somewhere near 10^{-28} m^2. The barn (1 b = 10^{-28} m^2) is used as a unit for nuclear reaction cross-section.

The cross-section for a particular reaction (also called partial reaction cross-section) depends on the energy of the charged particle. It is zero for charged particles below the threshold energy of that reaction. Below the Coulomb barrier

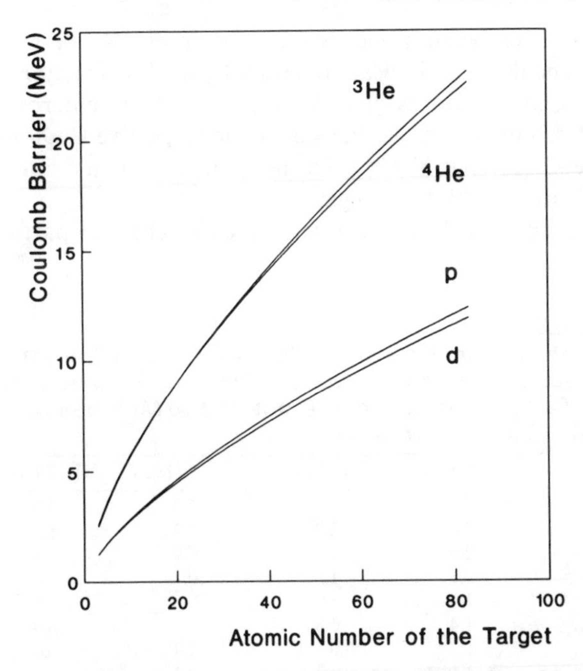

Fig. 2. Coulomb barrier (best fit) for protons, deuterons, helium-3 and helium-4 particles as a function of the atomic number of the target.

the probability for tunneling (and also the cross-section) is very low. At higher energies the cross-section increases up to a maximum (typically 1 b) and decreases as more complex reactions becomes competitive. The cross-section as a function of the charged particle energy is called the excitation function or curve. As charged particles are slowed down in matter, this function may be of interest in CPAA. It will be shown later that knowledge of the absolute cross-section is not required. Therefore, to distinguish it from the excitation function, the relative cross-section as a function of the charged particle energy will be indicated as the activation curve.

Several compilations with experimental data for excitation functions exist: Keller *et al.* [7] (graphical form), McGowan and Milner [8] (references only), the IAEA handbook of nuclear activation data [9] (graphical form) and, especially, for reactions of interest for medical radioisotope production [10–12]. There is a considerable shortage of data, and the data available are often in very poor agreement. For reactions where no experimental data exist, Keller, Lange and Münzel provide a semi-empirical procedure to estimate the excitation functions for a large number of reactions of interest in CPAA [13]. Computer codes have also been developed to calculate the functions on a theoretical basis [14–16].

2.2.4 Beam intensity reduction

A charged particle beam traversing matter loses its energy by collisions (Section 2.1). The decrease in beam intensity by nuclear reaction, calculated for $\sigma = 10^{-28}\,m^2$ and the ranges of 15 MeV protons and deuterons and 30 MeV helium-3 and helium-4 particles (i.e. above the respective Coulomb barriers) in different targets, is given in Table 1. The beam intensity reduction is mostly much lower than 1% and can thus be neglected.

In CPAA, a simplified picture can be used of charged particles traversing

TABLE 1.
Relative beam intensity reduction (%) in different targets, calculated for $\sigma = 10^{-28}\,m^2$ and the range of 15 MeV protons and deuterons and 30 MeV helium-3 and helium-4 particles

Target	p, 15 MeV	d, 15 MeV	^{3}He, 30 MeV	^{4}He, 30 MeV
H	6.7	3.8	2.3	1.8
He	3.8	2.2	1.2	1.0
Li	2.6	1.5	0.9	0.7
Be	2.1	1.2	0.7	0.6
B	1.7	1.0	0.6	0.5
C	1.4	0.8	0.5	0.4
N	1.3	0.7	0.4	0.4
O	1.1	0.7	0.4	0.3
F	1.0	0.6	0.4	0.3

matter along straight lines, losing their energy by collisions and all stopped at exactly the same distance, called the range. Phenomena such as projected range, range straggling and beam intensity reduction should not be further considered.

2.2.5 Radionuclides formed by nuclear reaction

The most probable radionuclides formed in a particular matrix by irradiation with charged particles of a certain energy can be studied using a chart of nuclides [17–18] and knowledge of the threshold energy and Coulomb barrier. The cross-section can, in a first approximation, be considered as almost equal for identical reaction types. For proton activation, for example, the reaction types given in Fig. 3 should be applied to all stable nuclides of the elemental components of the matrix. The higher the proton energy, the more complex the reactions that have to be considered. For protons below 10 MeV, in general, only (p, n) and (p, α) reactions have to be considered; for 20 MeV protons, (p, 2n), (p, d), (p, t), (p, ^{3}He) and (p, αn) reactions have also to be considered. Reactions leading to stable nuclides are, of course, of no importance. In the inventory of possible reactions, those reactions with a threshold energy higher than the charged particle energy can be cancelled. Reactions on target nuclides which provide a Coulomb barrier higher than the charged particle energy induce only a low activity (but cannot be excluded). For a 'thick' target, i.e. a target thicker than the range, the induced activity is also roughly proportional to the thickness activated, i.e. the range for the incident energy minus the range for the threshold energy or Coulomb barrier (whichever is the higher). The induced activity is proportional

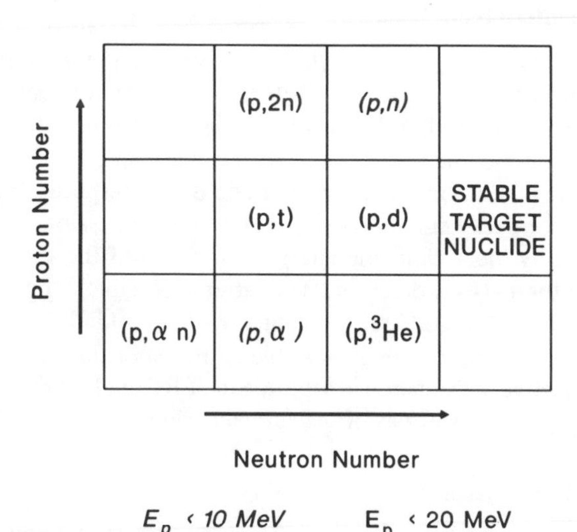

Fig. 3. Possible proton-induced reactions.

to the number of target nuclides (the concentration of the element in the target and the isotopic abundance of the target nuclide) and the number of charged particles. The induced activity increases (but not linearly) with irradiation time and decreases with higher half-life of the radionuclide.

2.3 Interferences

For the determination of an analyte element A based on the reaction A(a,b)B, where B is a radionuclide, three different types of interference can be distinguished: nuclear, spectral and matrix interference.

2.3.1 Nuclear interference

If the radionuclide B is formed from an element C other than the analyte element A, the determination of A by reaction A(a, b)B can be subject to nuclear interference from C by reaction C(c, d)B. Mostly the particles a and c are identical. However, nuclear interference by secondary particles [e.g. fast or thermalized neutrons produced by (x, n) reactions, where x is the charged particle] should also be considered. If the threshold energy of the interfering reaction is higher than the threshold energy of the analyte reaction, nuclear interference can be avoided by a proper choice of the charged particle energy. To obtain high sensitivity, the charged particle energy is chosen just below the threshold energy of the interfering reaction.

2.3.2 Spectral interference

If a radionuclide D is formed from an element C other than the analyte element A, the determination of A by reaction A(a, b)B can be subject to spectral interference from C by the reaction C(c, d)D. To avoid spectral interference, three methods can be applied: (1) proper choice of the charged particle energy (see nuclear interference); (2) selective measurement of radionuclide B with respect to radionuclide D by spectrometry or decay curve analysis. The former is possible if the energies of the γ-lines differ by more than the energy resolution of the spectrometer used (typically 2 keV for a HPGe detector). The latter is possible if the half-lives of the radionuclides C and D are different. For largely different half-lives, a sufficiently short or long decay time makes it possible to measure the short or long lived radioisotope selectively; (3) chemical separation of B from D. This is only possible if B and D are not radioisotopes of the same element.

2.3.3 Matrix interference

Matrix interference is a special case of spectral interference. As CPAA is primarily a method for trace element determinations, the concentration of the analyte element is much lower (typically by a factor of 10^6) than the concentration of the major matrix element(s). If a major matrix element (C) is activated, it often happens that the activity of D is much higher (also 10^6-fold) than that of B. Even for quite different γ-energies or half-lives, selective measurement becomes quite difficult. For γ-ray spectrometry, it is no longer sufficient that the γ-lines should differ by more than the energy resolution of the detector. The Compton plateau, ADC dead-time and/or pulse pile-up mostly prevent γ-ray spectrometry of such a low activity in the presence of a 10^6 times higher activity. The same holds for decay curve analysis. Establishment of decay conditions such that almost equal activities of B and D can be measured should be possible. In practice, matrix interference can be avoided only by a proper choice of the incident energy or by chemical separation of B from D. In the former case an instrumental analysis (i.e. without chemical separation) is possible.

2.4 Standardization

In CPAA a relative method is mostly used, whereby a sample and a standard are irradiated and the activity formed from the analyte element is measured in both the sample and the standard. The concentration in the sample can then in principle be calculated from the concentration in the standard and the ratio of the activities measured in the sample and the standard, the latter being corrected for different irradiation and measuring conditions. Especially when pure compounds are used as standards, the stopping power of the sample versus the standard can be different, so that an additional correction factor F is necessary. This section deals with the different standardization methods that have been proposed to calculate F (or to avoid it), a critical evaluation of their accuracy and an intercomparison to reveal their resemblances or differences. For that purpose, the different methods will be explained using the concept of the ratio of stopping powers [19], [20]; rather than the ways in which the respective authors have devised their methods.

2.4.1 Introduction

The number of radionuclides produced per unit time by irradiation of an infinitesimally thin target is determined by the increase due to nuclear reactions (proportional to the beam intensity, the cross-section and the number of target nuclides per cm^2) and by the decrease due to radioactive decay ($-\lambda N$)

$$\mathrm{d}N/\mathrm{d}t = I \cdot \sigma \cdot n \cdot \mathrm{d}l - \lambda \cdot N \tag{8}$$

with N = number of radionuclides*; t = time (s); I = number of charged particles per unit time or beam intensity (s^{-1}); σ = partial nuclear reaction cross-section* (cm^2); n = number of target nuclides* per unit volume (cm^{-3}); l = thickness of the target (cm); λ = decay constant of the radionuclide* formed (s^{-1}); the asterisk means 'for the nuclear reaction considered'.

The number of radionuclides formed after an irradiation time t_i is found by integration of equation (8) on condition that the beam intensity I is constant during irradiation. The activity is then

$$A = \lambda \cdot N = I \cdot \sigma \cdot n \cdot dl(1 - e^{-\lambda t_i}) \tag{9}$$

For a 'thick' target, i.e. thicker than the range R of the charged particles used, the activity is obtained by integration of equation (9) between zero and R

$$A = I \cdot n \cdot \rho^{-1} \cdot (1 - e^{-\lambda t_i}) \int_0^R \sigma \rho \, dl \tag{10}$$

as the beam intensity I is nearly constant as a function of the depth l (Section 2.2.4) and on condition that the analyte element is homogenously distributed in the sample. Using the stopping power S [equation (1)], the integral in equation (10) can be expressed as a function of the energy

$$\int_0^R \sigma \rho \, dl = \int_{E_i}^0 \frac{\sigma dE}{dE/(\rho dl)} = \int_{E_t}^{E_i} \frac{\sigma dE}{S} \tag{11}$$

with E_i = the incident energy, where the integration limit $E = 0$ (corresponding to $l = R$) may be replaced by the threshold energy E_t, as the cross-section is zero for an energy below E_t.

Although, in principle, the concentration of the analyte element can be calculated from equation (10), in CPAA a relative method is usually applied, whereby a standard (s) and a sample (x) are irradiated separately but with the same incident energy E_i. From equations (10) and (11), the concentration of the analyte element in the sample is then

$$c_x = c_s \cdot \frac{A_x I_s (1 - e^{-\lambda t_{i,s}})}{A_s I_x (1 - e^{-\lambda t_{i,x}})} \cdot F \tag{12}$$

$$F = \int_{E_t}^{E_i} \frac{\sigma dE}{S_s} \Bigg/ \int_{E_t}^{E_i} \frac{\sigma dE}{S_x} \tag{13}$$

The number of target nuclides per unit volume (n) can be replaced by the concentration of the analyte element (c, in e.g. μg/g), at least if the isotopic abundance of the target nuclide is equal for sample and standard (which is mostly the case).

The concentration of the analyte element in the sample (c_x) can thus be calculated by equation (12) from:

- the concentration of the analyte element in the standard c_s;
- the ratio of the activity in sample and standard at the end of irradiation A_x/A_s;
- the ratio of the beam intensity for standard and sample I_s/I_x;
- the irradiation time t_i;

- the decay constant $\lambda = \ln 2/t_{1/2}$, with $t_{1/2}$ = half-life;
- the F-factor of equation (13), representing the correction for different stopping powers of sample versus standard.

The F-factor can be calculated according to equation (13) if (1) the stopping power of sample (S_x) and standard (S_s) and (2) the cross-section (σ) are known, both as a function of the energy in the interval between the threshold energy (E_t) and the incident energy (E_i) and at least in a relative way. The first condition can be fulfilled in a large number of cases. If the main matrix composition of the sample is known, and if the matrix does not change during irradiation, the stopping power of the sample can be calculated accurately as described in Section 2.1. It is obvious that the stopping power of the standard can be calculated. If the second condition is also fulfilled, the most obvious method to calculate the F-factor is numerical integration of equation (13). This method is further called the 'exact' method. As the second condition is not always fulfilled, several approximative standardization methods have been presented [21–23].

2.4.2. Approximative standardization methods

First, the concept of the ratio of stopping powers will be introduced [19, 20]. The ratio of the stopping power of one element to that of a second element as a function of energy is observed to be nearly constant for elements of similar atomic number and/or for rather high energies. This is shown in Fig. 4 for protons, where this ratio is normalized to 1 for 25 MeV protons. If

$$K(E) = S_x(E)/S_s(E) \tag{14}$$

is nearly independent of energy, equation (13) can be approximated by

$$F = \frac{\displaystyle\int_{E_t}^{E_i} \frac{\sigma \, dE}{S_s}}{\displaystyle\int_{E_t}^{E_i} \frac{\sigma \, dE}{K S_s}} \approx K \frac{\displaystyle\int_{E_t}^{E_i} \frac{\sigma \, dE}{S_s}}{\displaystyle\int_{E_t}^{E_i} \frac{\sigma \, dE}{S_s}} = K \tag{15}$$

The most obvious choice to assess K is to take equation (14) for an energy

$$E_M = (E_i + E_t)/2 \tag{16}$$

which is the method proposed by Chaudhri *et al.* [23].

Using equations (15) and (3), the first method of Ricci and Hahn [21] (subsequently called Ricci I) is obtained.

$$F \approx K = \frac{\displaystyle\int_{E_t}^{E_i} \frac{dE}{S_s}}{\displaystyle\int_{E_t}^{E_i} \frac{dE}{K S_s}} = \frac{\displaystyle\int_{E_t}^{E_i} \frac{dE}{S_s}}{\displaystyle\int_{E_t}^{E_i} \frac{dE}{S_x}} = \frac{\displaystyle\int_{0}^{E_i} \frac{dE}{S_s} - \int_{0}^{E_t} \frac{dE}{S_s}}{\displaystyle\int_{0}^{E_i} \frac{dE}{S_x} - \int_{0}^{E_t} \frac{dE}{S_x}} = \frac{R_s(E_i) - R_s(E_t)}{R_x(E_i) - R_x(E_t)} \tag{17}$$

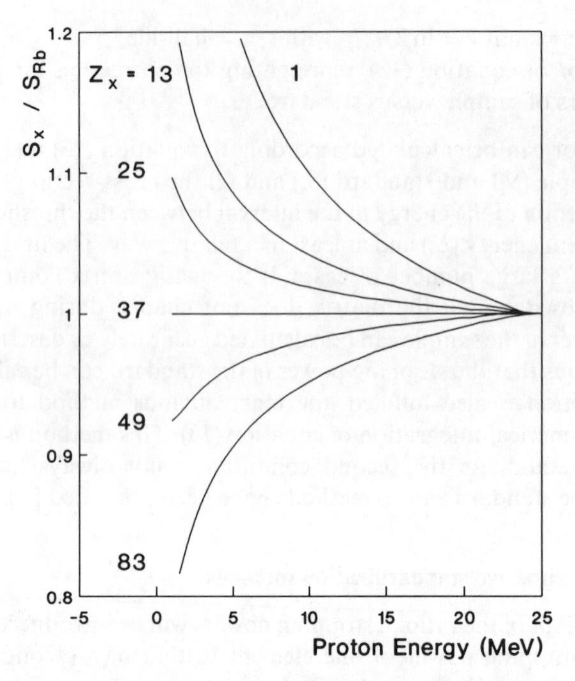

Fig. 4. The ratio of the stopping power of several elements (hydrogen, aluminum, manganese, indium, bismuth) to that of rubidium as a function of the proton energy. For 25 MeV the ratio is normalized to 1.

The same assumptions have to be made as for the method of Chaudhri, i.e. equation (14) should be nearly constant in the energy interval between the threshold and the incident energy. The two approximative methods are thus equivalent.

The second method of Ricci and Hahn [22] (subsequently called Ricci II) can be obtained in the same way if E_t is replaced by zero.

$$F \approx R_s(E_i)/R_x(E_i) \tag{18}$$

The assumption to be made, however, is now that equation (14) should be constant in the energy interval between zero and the incident energy. It is clear from Fig. 4 that this assumption holds only for a sample and a standard with nearly equal atomic number.

The maximum error that will be made by the approximative methods of Chaudhri (or Ricci I) can be assessed by [19]

$$[K(E_M) - K(E_t)]/K(E_M) \tag{19}$$

Although the error is largely overestimated, equation (19) provides a simple tool to identify those cases where the approximative methods are essentially sufficient to provide accurate results.

The approximative methods have been compared to the 'exact' method for a number of nuclear reactions of which the excitation function was known and for a number of sample–standard pairs with similar or with quite different atomic numbers[19].

In conclusion, the approximative methods of Ricci I and Chaudhri are equivalent and yield accurate results if one or more of the following conditions are fulfilled:

- the atomic number of sample and standard are similar;
- the threshold energy of the nuclear reaction used is high;
- the incident energy is high.

The method of Ricci II should not be used. The 'atomic number' of a mixture or compound should be understood as the atomic number of that element with the same stopping power as the mixture or compound. It is also good practice to use the Coulomb barrier instead of the threshold energy if the former is higher than the latter (this is always the case for exoergic reactions), as the cross-section can be considered as nearly zero below the Coulomb barrier. The 'exact' method (or the 'average stopping power' method; see below) has to be used if one or more of the conditions given above are not fulfilled, and in any case if the cross-section is known in the energy interval between the threshold energy and the incident energy, at least in a relative way. None of the methods requires stopping power or range data below 1 MeV (Section 2.1.4) except the method of Ricci II.

2.4.3 The average stopping power method

The 'average stopping power' method, proposed by Ishii, Valladon and Debrun, requires knowledge of the cross-section and is in principle also an approximative method, but with a negligible systematic error, which can be calculated [24]. To understand this method one can consider equation (15). As, in general, the cross-section increases with increasing energy, assessment of K by simply taking the average of the threshold and the incident energy [equation (16)] yields a K value for too low an energy. If one uses the weighted mean, with the cross-section as a weighting factor

$$E_m = \int_0^{E_i} E\sigma \mathrm{d}E \bigg/ \int_0^{E_i} \sigma \mathrm{d}E \tag{20}$$

the 'average energy' of the method of Ishii is obtained. It is clear that this method should provide better results than that of Chaudhri. The Ishii method requires the same information as the 'exact' method but calculations are much simpler. The 'exact' method requires a numerical integration calculation for each nuclear reaction used and for each sample–standard pair, and requires stopping power data in the energy interval between the threshold and the incident energy. The method of Ishii requires only one numerical integration calculation for each particular reaction. The F-factor is then calculated from only two stopping power values.

Ishii *et al.* also proposed a method to calculate the 'average energy' E_m from the thick target yield, which is easier to obtain experimentally [25].

2.4.4 The internal standardization method

The different methods for calculating the F-factor, reviewed in Sections 2.4.2 and 2.4.3, all require knowledge of the stopping power of the sample. The stopping power can be accurately calculated for all elemental matter and also for any mixture or compound with known matrix composition (Section 2.1). Many samples, however, have a very complex matrix composition which is not always accurately known. Irradiation with charged particles can cause substantial heating of materials with low thermal conductivity, e.g. silicates, such as environmental and geological materials. This can give rise to evaporation of, e.g., the organic fraction present in the sample. There is a risk of introduction of positive errors into the analysis owing to enrichment of the analyte element during irradiation and because the stopping power calculated for the sample is too high. Indeed, the stopping power of a material, taking into account also the organic fraction, is higher than the stopping power of the inorganic fraction of that material (the stopping power of hydrogen for protons is about three times the stopping power of, e.g., silicon dioxide). If the organic fraction (or part of it) evaporates during irradiation, too high a stopping power is used in the calculations and consequently too high a concentration is obtained. Therefore an internal standardization method has been proposed by Casteleyn *et al.* [20] for materials with unknown or variable matrix composition. Approximative methods not requiring knowledge of the cross-section have also been proposed.

The idea of internal standardization is that, by using the known concentration of one element (i.e. the internal standard) and by measuring the induced activity of the internal standard, this correction can be calculated for all elements to be determined (i.e. the analyte elements) without exact knowledge of the matrix composition of the sample. The internal standard is an element with an accurately known concentration, either because the concentration is certified or determined by another analytical method or because an exactly known amount of it is added to the sample. The activity produced from this element, should be measured accurately, precisely and also preferably simultaneously with the activity produced from the analyte elements.

In the following treatment, the index 1 is used for the internal standard or a reaction on the internal standard, and index 2 is used for the analyte element or a reaction on that element. Since the concentration of the internal standard in the sample (c_{x_1}) is known by definition, the F_1-factor can be determined experimentally using equation (12) by irradiation and measurement of a sample and a standard (s_1) for the internal standard element. It is obvious from equation (13) that the F_2-factor for the analyte element is not equal to F_1. To determine the relationship between F_1 and F_2, a theoretical sample (x′) is introduced with a matrix composition for which the stopping power can be calculated. A function

$$K_x(E) = S_x(E)/S_{x'}(E) \tag{21}$$

gives the relationship between the (unknown) stopping power of the sample (x) and that of the theoretical sample (x′). It has been demonstrated (Section 2.4.2, Fig. 4) that this function is nearly constant in the energy interval between the threshold and the incident energy if one or more of the following conditions is fulfilled:

- the atomic number of the sample x is similar to that of the theoretical sample x′ (e.g. x′ = SiO_2 for x = a geological sample);
- the threshold energy for both the reaction on the internal standard (E_1) and that on the analyte element (E_2) is high;
- the incident energy is high.

Applying equation (13) for both the internal standard and the analyte element

$$F_2 = \frac{\displaystyle\int_{E_2}^{E_i} \frac{\sigma_2 \, dE}{S_{s2}} \cdot \int_{E_1}^{E_i} \frac{\sigma_1 \, dE}{S_x}}{\displaystyle\int_{E_2}^{E_i} \frac{\sigma_2 \, dE}{S_x} \int_{E_1}^{E_i} \frac{\sigma_1 \, dE}{S_{s_1}}} \cdot F_1 \tag{22}$$

and if $K_x(E)$ [equation (21)] is nearly constant in the interval between threshold energies and incident energy, F_2 can be written as

$$F_2 = \frac{\displaystyle\int_{E_2}^{E_i} \frac{\sigma_2 \, dE}{S_{s2}} \cdot \int_{E_1}^{E_i} \frac{\sigma_1 \, dE}{S_{x'}}}{\displaystyle\int_{E_2}^{E_i} \frac{\sigma_2 \, dE}{S_{x'}} \int_{E_1}^{E_i} \frac{\sigma_1 \, dE}{S_{s_1}}} \cdot F_1 \tag{23}$$

Since F_1 can be determined experimentally, the factor F_2 for the analyte element can be calculated now by numerical integration on condition that the cross-sections are known.

If that is not the case, equation (23) can be approximated [in the same way as equation (13) was approximated by equations (14) and (16)] by

$$F_2 \approx \frac{S_{x'}(E_{M_2})}{S_{s_2}(E_{M_2})} \cdot \frac{S_{s_1}(E_{M_1})}{S_{x'}(E_{M_1})} \cdot F_1 \tag{24}$$

with $E_{M_1} = (E_i + E_1)/2$ and $E_{M_2} = (E_i + E_2)/2$ or [in the same way as equation (13) was approximated by equation (17)]

$$F_2 \approx \frac{R_{s_2}(E_i) - R_{s_2}(E_2)}{R_{x'}(E_i) - R_{x'}(E_2)} \cdot \frac{R_{x'}(E_i) - R_{x'}(E_1)}{R_{s_1}(E_i) - R_{s_1}(E_1)} \cdot F_1 \tag{25}$$

if one or more of the following conditions is fulfilled:

- the atomic number of the theoretical sample x′ is similar to that of both the standard sample of the internal standard (s_1) and the analyte element (s_2);
- the threshold energy for the reaction on both the internal standard (E_1) and the analyte element (E_2) is high;
- the incident energy (E_i) is high.

2.4.5 The two reactions method

The 'two reactions' method, proposed by Ishii *et al.*, is derived from the 'average stopping power' method (Section 2.4.3) and also makes use of an internal standard of known concentration [26]. In contrast to the internal standardization method (Section 2.4.4), activation of the analyte element and the internal standard element can be carried out at two different energies or even with two different charged particles. The main advantage is that no stopping power data at all are necessary, so that the accuracy of the analytical method is not affected by the accuracy of the stopping power data. The main disadvantage is that the method requires, for each analyte element, a rigorous definition of the charged particle, its energy and the standards used by calculation of the systematic error involved. The method can be reduced to the experimental simplicity of the internal standardization method, i.e. irradiation of the sample, the standard (s) for the analyte element(s) and a standard for the internal standard element, all with the same charged particle of the same energy. If both

$$K_1(E) = S_x(E)/S_{s_1}(E) \tag{26}$$

$$K_2(E) = S_x(E)/S_{s_2}(E)$$

are nearly constant and if $K_1 = K_2$, then equation (22) yields $F_2 = F_1$ and, from equation (12), the 'two reactions' method is obtained for the case of 'one particle–one energy'

$$c_{x2} = c_{s2} \frac{A_{x2}}{A_{s2}} \frac{c_{x1}}{c_{s_1}} \frac{A_{s_1}}{A_{x_1}} \tag{27}$$

If the internal standardization method is compared with the 'two reactions' method in the case of 'one particle–one energy', the latter is an approximation of the former.

2.4.6 Standard addition method

Up to now, a relative method has been applied, whereby a standard and a sample are irradiated separately but with the same incident energy. In the standard addition method, however, a sample is irradiated and so too, separately, is the same sample to which an accurately known amount of the analyte element has been added, i.e. the comparative standard. As the two matrices are almost identical (only trace amounts of analyte are added), the F-factor is unity. Then it is possible to use an internal standard as a monitor for the beam intensity. This method has been proposed and applied to CPAA by Masumoto and Yagi [27, 28]. The authors call this method the 'internal standard method'. In this work, 'standard addition' is preferred, as it better denominates the standardization method applied. Replacing c_s in equation (12) by $c_x + c_a$ (i.e. the concentration in the comparative standard) and F by unit, one obtains for equal irradiation times for the sample and the comparative standard

$$c_x = c_a \Big/ \left(\frac{A_s}{I_s} \frac{I_x}{A_x} - 1 \right) \tag{28}$$

with c_a = the concentration added in the comparative standard; s = the comparative standard; I = the activity of the internal standard, which is measure of the beam intensity.

3 EXPERIMENTAL

3.1 Irradiation

3.1.1 Accelerator

For CPAA, charged particles have to be accelerated to an energy higher than the Coulomb barrier (to obtain a high cross-section and thus high sensitivity) and lower than the threshold energy of more complex reactions than (x, n) or (x, α) reactions, with x = the charged particle (to avoid nuclear interferences). Energies between a few MeV and 20–30 MeV should be attainable for protons and deuterons and twice as high for helium-3 or helium-4 particles. An isochronous cyclotron, designed as a variable energy accelerator and envisaged with a helium-3 recovery system, covers this energy range of interest for CPAA. For the CPA analyst the cyclotron is a 'black box' that provides a charged particle beam of variable energy. The energy should be well defined (an accuracy of 0.1 MeV is largely sufficient, as this is the uncertainty in the threshold energy) and reproducible (to compare irradiations on different days). The beam intensity should be tunable from 0.05 to 5 μA and kept constant during irradiation [equation (9) was calculated from equation (8) assuming that the beam intensity is constant]. The unfavorable beam characteristics of a cyclotron, such as pulsed beam and energy resolution, are thus of no importance for CPAA. The reader is referred to Chapter 3 for the principles of operation of the isochronous cyclotron.

3.1.2 Target

Irradiation with a charged particle beam can cause substantial heating of the sample. If a 'thick' sample (Section 2.2.5) is irradiated, the heat release is given by

$$Q(\mathrm{W}) = E(\mathrm{J}) \cdot I(\mathrm{A})/e(\mathrm{C}) = k \cdot E(\mathrm{MeV}) \cdot I(\mu\mathrm{A}) \qquad (29)$$

where Q = heat release (W), E = charged particle energy ($1\ \mathrm{eV} \approx 1.6 \times 10^{-19}$ J), I = beam intensity (A or μA), e = elementary charge ($\approx 1.6 \times 10^{-19}$ C), k = constant ($= 1\ \mathrm{JeV^{-1}C^{-1}}$).

The energy is released mainly at a depth slightly less than the range because the stopping power increases with decreasing charged particle energy (i.e. the Bragg peak). Therefore efficient cooling of the target is the major concern in target design.

The simplest target design is possible for solid samples with good thermal conductivity, not powdered and available as a foil or sheet with a thickness slightly larger than the range. Irradiation under vacuum is then possible with the sample mounted on a water-cooled sample holder. For powdered samples or samples with poor thermal conductivity, irradiation in a helium atmosphere has been developed [29]. Application of CPAA for the analysis of liquids is less common, as many optical (AAS, ICP–OES) or mass spectrometric (ICP–MS) methods of analysis exist for diluted aqueous solutions.

Figure 5 shows an experimental set-up for the irradiation of powdered or disk samples in a helium atmosphere [30]. A titanium foil separates the vacuum of the cyclotron from the helium. A powdered or disk sample is loaded in a dedicated sample holder (B or C) together with a beam intensity monitor foil and an aluminum foil (Section 3.2). Powders can be loaded from the right side and quantitatively unloaded from the left side. Disk samples that are good thermal conductors (e.g. metals) can be water-cooled from the back. The sample holder (B or C) is brought into the water-cooled target (A), and the target is evacuated and filled with helium.

3.1.3 Direct beam intensity monitoring

Although quantitative beam intensity monitoring is effected with monitor foils (Section 3.2), direct beam intensity measurement is necessary as a feedback to

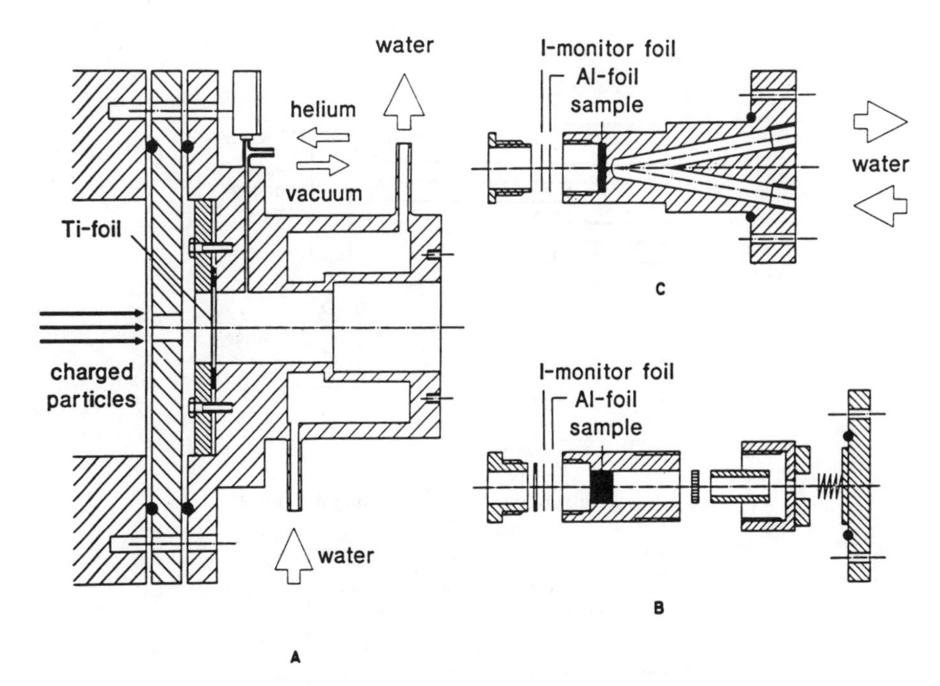

Fig. 5. Target system for irradiation in a helium atmosphere: A: target, B: sample holder for powders, C: sample holder for disk samples. (From Ref. 30.)

keep the intensity constant during irradiation. Therefore the target holder is electrically insulated from the cyclotron beam line and connected to the ground by a microammeter, as shown in Fig. 6. A suitable diaphragm (grounded) stops the beam hitting the target holder, which would cause not only avoidable activation of the target holder, but also a systematic positive error in the beam intensity measurement. During charged particle irradiation, secondary electrons are released at the surface of the target. If they reach any grounded part of the beam line (e.g. the diaphragm) a positive systematic error occurs in the beam intensity measurement. To avoid this, two experimental set-ups are used. (1) A ring at negative potential (typically 100 V) is inserted between the target holder and the diaphragm. The electric field prevents escape of the secondary electrons from the positively charged target to ground. (2) The target holder is equipped with a tube preventing escape of secondary electrons over a large spatial angle. Both devices are protected from the beam by the diaphragm. It is clear that electrical insulation between diaphragm, negative potential ring and target holder should be perfect, and that water cooling should not cause any electrical leakage to the ground. Deionized cooling water in plastic tubing is thus necessary. The whole assembly for irradiation under helium should be considered as

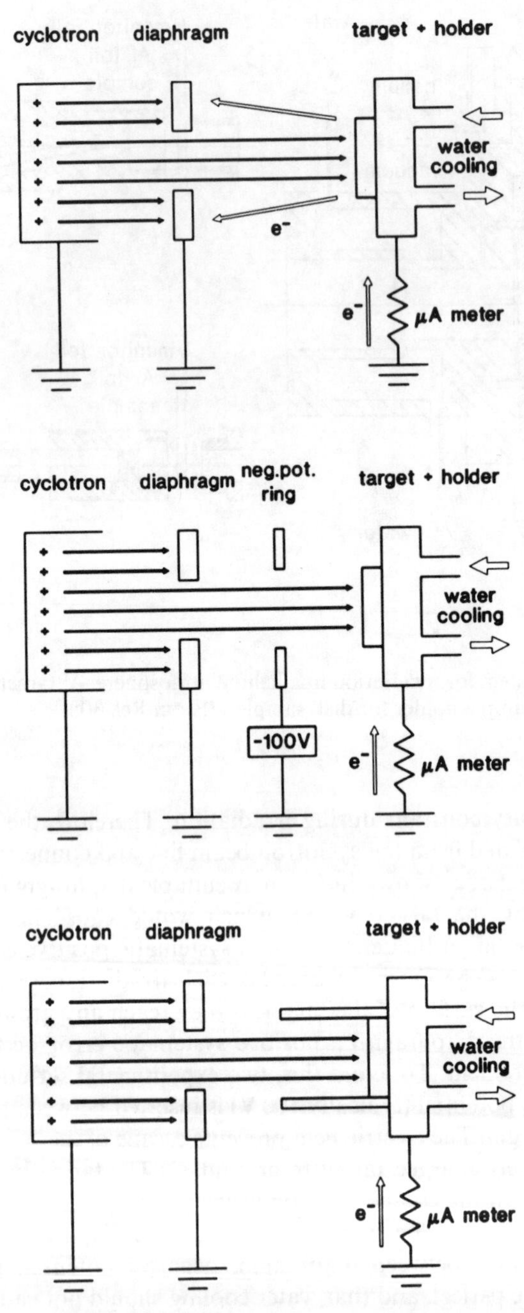

Fig. 6. Direct beam intensity monitoring.

the target in this context. The microammeter can be a current meter as well as a current integrator (μC meter).

3.2 Quantitative beam intensity monitoring

When a relative method is used in CPAA [Section 2.4.1, equation (12)], the ratio of the beam intensity for standard and sample has to be determined experimentally. As a knowledge of the absolute beam intensity is not required, it is common practice to cover the sample and the standard with a foil thinner than the range, subsequently called the I-monitor, of exactly the same thickness for both sample and standard. If equation (9) is integrated from zero to the thickness of the I-monitor, the induced activity in the I-monitor is a measure of the beam intensity

$$\frac{I_s}{I_x} = \frac{A_s(1 - e^{-\lambda t_{i_x}})}{A_x(1 - e^{-\lambda t_{i_s}})} \tag{30}$$

with A = activity induced in the I-monitor foil at the end of irradiation; λ = decay constant of the radionuclide induced in the monitor foil; t_i = irradiation time.

Pure metal foils are the obvious choice as I-monitors: they are good thermal conductors, monoelemental and available in different thicknesses. Table 2 summarizes a few possibilities and related nuclear data [31].

The use of I-monitor foils provides a simple tool to quantify the ratio of beam intensities in an accurate and precise way. Compared with direct measurement of current, this approach is not subject to possible systematic errors such as production of secondary electrons and the beam hitting the target holder. Possible disadvantages are: an additional activity measurement is necessary; there is an uncertainty in the incident energy on sample and standard; recoil radionuclides may enter the sample or standard. Because irradiation of monoelemental I-monitors mostly lead to very simple γ-ray spectra, the additional measurement does not introduce any problem except that the induced activity for standard and sample can be quite different. The thickness of the I-monitor should be exactly the same for sample and standard, so that the initial assumption of

TABLE 2.
Beam intensity monitors

Reaction	Threshold energy (MeV)	Half-life	γ-Ray energy (keV)
^{63}Cu (p, n) ^{63}Zn	4.2	38.1 min	669.6, 962.1
^{65}Cu (p, n) ^{65}Zn	2.1	243.8 d	1115.5
^{60}Ni (d, n) ^{61}Cu	$Q > 0$	3.408 h	283.0, 656.0
^{65}Cu (^{3}He, 2n) ^{66}Ga	5.0	9.4 h	1039.4
^{63}Cu (α, n) ^{66}Ga	8.0		

'a standard and a sample irradiated separately but with the same incident energy' holds (Section 2.4.1). The absolute thickness of the *I*-monitor should also be known in order to assess the effective incident energy on sample or standard (Section 3.4). To avoid recoil nuclides from the *I*-monitor entering the sample (or the standard), a foil (called the recoil foil) is inserted between the *I*-monitor and the sample (or the standard). Aluminum is used as a recoil foil, because it is essentially not activated by charged particle-induced reactions. The range of the recoil nuclides in the recoil foil is much lower than the range of the charged particles commonly used in CPAA (protons, deuterons, helium-3 and helium-4 particles). An aluminum foil a few μm thick thus stops the recoil nuclides completely, but the energy of the charged particles is reduced by a minor (but not negligible) fraction (Sections 3.4 and 3.5).

3.3 Chemical etch

Although charged particles have a limited range in matter, CPAA is considered as an analytical method to determine the concentration in the bulk of a sample rather than at the surface. Therefore interference from the analyte element at the surface has to be avoided. This is possible for solid samples that are not powders, simply by etching the sample before irradiation, at least if further contamination of the surface can be avoided until the end of irradiation (contamination after irradiation does not interfere). However, for the determination of light elements (boron, carbon, nitrogen and oxygen) at low concentration the latter condition can not always be fulfilled. This is the case for, e.g., the determination of the bulk oxygen concentration in aluminum. A freshly etched aluminum sample acquires in air an aluminum oxide surface layer which will be activated also. Chemical etch *after* irradiation is then the method of choice. The activated surface layer will be removed and replaced by an inactive surface layer that will no longer interfere. However, the thickness to be removed is not determined by the thickness of the oxide layer alone. By recoil, radionuclides formed from surface oxygen can penetrate into the sample. The penetration depth of recoil nuclides will be estimated in Section 3.5.

Chemical etching after irradiation reduces the effective incident energy on the sample, i.e. the sample as it will be measured. If the standards are not etched in the same way, the initial assumption of 'a standard and a sample irradiated separately but with the same incident energy' (Section 2.4.1) is no longer fulfilled. If a 'monoelemental' standard (a pure element or a single compound) is used (e.g. silicon dioxide, and not an aluminum reference material with certified oxygen content) etching is not necessary. Moreover, etching of the standard may be inconvenient or even impossible (e.g. for powdered standards or polymers). The activity of a standard, irradiated at exactly the same energy as the sample after etching, is obtained as shown schematically in Fig. 7. Different series of stan-

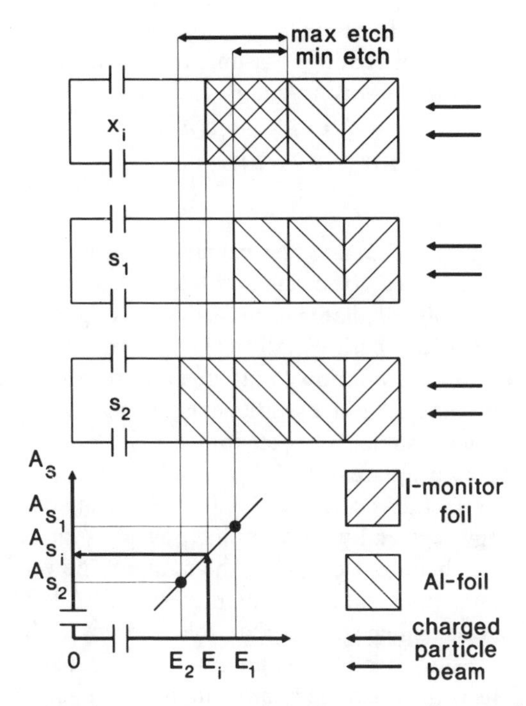

Fig. 7. Chemical etch and determination of corresponding standard activity.

dards (at least two: s_1 and s_2) are irradiated, for which additional aluminum foils (1 and 2) are inserted between the recoil foil and the standard. In this way standard activities (A_{s_1} and A_{s_2}) can be obtained for different incident energies (E_1 and E_2), which should cover the effective incident energies for the different samples. If the effective incident energy on a particular sample (after etch) is known (E_i), the standard activity (A_{s_i}) for a standard irradiated with exactly the same energy as the sample (E_i) can be obtained by linear interpolation. It is good practice to choose an incident energy just above the maximum cross-section (to obtain the highest activation of the analyte element or best detection limit and precision). As a consequence, the increase of the activity as a function of energy around E_i is quite low, and consequently linear interpolation introduces a negligible systematic error.

3.4 Energy reduction calculation

The effective incident energy on sample or standard depends on the initial incident energy, the different foils placed before sample or standard and the chemical etch process for the sample (if applicable). The energy reduction (from

E_1 to E_2) can be calculated from the nature and thickness of a foil (or the surface layer removed by chemical etching). Assume a foil with thickness L(cm) or X (g/cm^2), stopping power S[MeV/(g/cm^2)] and range R(g/cm^2), irradiated by charged particles of energy E_1. Using equations (1) and (3)

$$X = \int_0^L \rho \, dl = \int_{E_1}^{E_2} \frac{dE}{-S} = \int_0^{E_1} \frac{dE}{S} - \int_0^{E_2} \frac{dE}{S} = R(E_1) - R(E_2) \tag{31}$$

the emerging energy E_2 can now be calculated using energy–range tables. Similarly, the thickness X can be calculated to reduce the energy from E_1 to E_2.

As ranges are primarily calculated as mass thickness (g/cm^2), the thickness of a foil should be determined experimentally as g/cm^2. Therefore it is more convenient to weigh a foil of exactly known surface area, rather than to measure the thickness by a micrometer and use the mass density to calculate the mass thickness. In this way, sufficient precision is essentially attainable. A 5 μm aluminum foil is cut as a square of 10.0 ± 0.1 cm ($\pm$ precision) and weighed to be 135.0 ± 0.1 mg. The thickness calculated is 1.35 ± 0.03 mg/cm^2. The precision of the calculated energy is given by the stopping power multiplied by the precision of the thickness. For helium-4 particles, the stopping power is of the order of 10^2 MeV/(g/cm^2) for an incident energy of interest in CPAA. The precision of the energy is thus 3 keV, i.e. much smaller than the accuracy of the threshold energy (0.1 MeV). The precision is much better for helium-3 particles, deuterons or protons and for foils with an atomic number higher than that of aluminum (lower stopping power) or for foils with a mass density higher than aluminum or thicker than 5 μm (higher mass).

The thickness of the surface layer removed by chemical etching has to be determined in g/cm^2 in order to calculate the effective incident energy. Therefore the sample is weighed before etching (actually before irradiation) and after etching, the dimensions of the sample are measured and the surface area is calculated. The precision attainable for an etch (< 50 mg/cm^2) of a typical cylindrical sample (diameter 15.0 ± 0.1 mm, thickness 1000 ± 1 μm), weighed with 0.1 mg precision, is < 1 mg/cm^2. When this is multiplied by the stopping power [$< 10^2$ MeV/(g/cm^2)], the precision of the energy is < 0.1 MeV. The error in the final concentration is determined by the increase of activity with energy. This slope is known from the irradiation of different standards and in general is low (Section 3.3).

This procedure is to be preferred to measurement of the thickness by a micrometer before and after the etch and using the mass density to calculate the thickness in g/cm^2. Precautions have to be taken to ensure that the whole surface of the sample is etched uniformly. The procedure is only valid if the etching speeds for the irradiated side of the sample and the other side are identical. As there has been no proof to the contrary, the procedure proposed is expected to give accurate results. Alternative procedures [32, 33] have been proposed that are not subject to this possible systematic error.

3.5 Recoil nuclides

An approximate knowledge of the maximum energy and range of recoil nuclides is necessary to assess the minimum thickness of (1) the recoil foil (to stop recoil nuclides from the I-monitor foil, Section 3.2) or (2) the surface layer to be etched (to remove recoil nuclides formed from surface contamination, Section 3.3).

The maximum energy of the recoil nuclide B for a nuclear reaction A (a, b) B can be simplified to [34]

$$E_B = \frac{m_a m_B E_a}{(m_A + m_a)(m_B + m_b)} \left\{ 1 + \left(\frac{[m_A E_a + (m_A + m_a) Q] m_b}{m_a m_B E_a} \right)^{1/2} \right\}^2 \tag{32}$$

This energy is always lower than $E_a + Q$, which can be sufficient to estimate an upper limit for the range of the recoil nuclide. The range of the recoil nuclide in the target considered (aluminum foil or sample) can be found in the compilation of Northcliffe and Schilling [35] for all energies of interest in CPAA, for all recoil nuclides ($1 \leq Z \leq 103$), but only for 12 solid targets (Be, C, Al, Ti, Ni, Ge, Zr, Ag, Eu, Ta, Au and U). Correction for different mass numbers (p. 252) can be neglected. Table 3 summarizes a few range data of beryllium, nitrogen, zirconium and uranium ions (i.e. recoil nuclides) in beryllium, aluminum, zirconium and uranium targets at 5, 10 and 20 MeV. For comparison, the range of protons and helium-4 particles (high and low range respectively) in the same targets is given for an energy used in current practice.

TABLE 3.
Ranges in mg/cm² of 5, 10 and 20 MeV ions (beryllium-9, nitrogen-14, zirconium-90 and uranium-238) in different targets (beryllium, aluminum, zirconium and uranium). For comparison, the range of 15 MeV protons and 30 MeV helium-4 particles in the same targets

Ion	$E = 5$ MeV Target				$E = 10$ MeV Target				$E = 20$ MeV Target			
	Be	Al	Zr	U	Be	Al	Zr	U	Be	Al	Zr	U
^{9}Be	1.3	1.9	3.8	8.4	2.8	4.0	7.3	15	7.1	9.4	16	30
^{14}N	0.7	1.1	2.3	5.2	1.3	1.9	3.7	8.1	2.7	3.7	6.7	14
^{90}Zr	0.4	0.5	1.0	2.3	0.6	0.9	1.8	4.0	1.0	1.4	2.8	6.2
^{238}U	0.2	0.3	0.4	0.8	0.4	0.5	0.9	1.7	0.8	1.0	1.7	3.4

	Target			
Ion	Be	Al	Zr	U
p(15 MeV)	307	345	480	715
α(30 MeV)	89	103	150	234

3.6 Radiochemical separation

Instrumental analysis is possible if spectral or matrix interference can be avoided by an appropriate choice of the incident energy or the measuring conditions (Section 2.3). If not, the radionuclide B formed from the analyte element A has to be separated radiochemically from (an) interfering radionuclide(s) D formed from (an) interfering element(s) C. The latter technique is called radiochemical analysis. This section deals with some major differences between a radiochemical separation and a common chemical separation as used for non-nuclear methods of analysis.

For most charged particle-induced reactions, the atomic number of the radionuclide B is different from that of the analyte element A. This is the case e.g. for (p, n), (p, α), (d, n), (d, α), (^{3}He, n), (^{3}He, d), (α, n) and (α, d) reactions, but not for (p, d) and (^{3}He, α) reactions. The chemical separation to be developed for CPAA is thus different from that for all non-nuclear analytical methods and for some other methods based on activation analysis, such as thermal and fast neutron activation analysis using the (n, γ) or (n, 2n) reaction respectively and photon activation analysis using the (γ, n) reaction. It is also not necessary to separate the matrix, but rather the radionuclide(s) formed from the matrix element(s). Again, the atomic number of the radionuclide(s) is mostly different from that of the matrix element(s). Considering only the chemical separation involved, CPAA is observed to be an independent analytical method, not subject to the same systematic errors as other analytical methods.

A radiochemical separation has three important advantages over a conventional chemical separation. (1) Inactive carrier can be added for the elements to be separated (B and D). This avoids the difficulties of a chemical separation at the trace level. (2) Reagent impurities (or blanks) do not influence the detection limit capabilities of the analytical method. (3) Separations need not be quantitative nor even reproducible (see below).

For the choice and development of a radiochemical separation the following points have to be considered.

Selectivity of the separation As pointed out in Section 2.3.3, induced matrix activity (D) can be up to 10^6 times higher than the activity to be measured (B). It is not possible to resolve such a low activity in the presence of such a high activity. Therefore the decontamination factor (ratio of D before to that after separation) should be up to 10^4.

Quantitative nature of the separation Quantitative recovery of B is to be preferred (99% or above). If the separation is not quantitative and not even reproducible, determination of the yield for each individual separation is possible. Two approaches can be followed: addition before separation of an accurately known amount of an inactive or an active carrier, and measurement of the mass or the activity of the carrier after separation.

Speed of the separation To obtain an optimum detection limit, the time needed to perform the whole separation procedure should not exceed a few half-lives of the radionuclide to be measured.

Inactive carrier It is not only an advantage but mostly also a necessity to add inactive carriers of the elements to be separated. Indeed, if the atomic number of the radionuclide B is different from that of the analyte element A (which is usually the case; see above), this radionuclide B is produced 'carrier free'. The number of radionuclides (N) particles can be calculated from the activity (A) and the half-life ($t_{1/2}$):

$$N = At_{1/2}/\ln 2 \tag{33}$$

For $A = 1$ kBq and $t_{1/2} = 1$ d, the amount of substance is 10^{-16} mol. It is clear that, if there is no accidental addition of inactive element B (e.g. as impurity in the matrix or reagents), it is not possible to separate chemically such a low mass.

Detection efficiency of the measurement To obtain the optimum detection limit (or sensitivity), the activity should be measured after separation with the highest possible detection efficiency. Efficiency is also determined by the geometry (size and distance from the detector) of the source. Separation procedures ending with volumes of solution are less favorable than those ending with a precipitate.

Precision of the measurement The precision (i.e. reproducibility) of the measurement is determined by the counting statistics and also by the possibility of reproducing measurements with exactly the same detection efficiency. The former is expressed as the relative standard deviation calculated from counting statistics (i.e. $C^{-1/2}$, where C is the number of counts measured) and is decreased at higher detection efficiency. The reproducibility of measurement can be assessed as the relative standard deviation for the average of a number of measurements, and is decreased by lower detection efficiency conditions. Indeed, as the detection efficiency of a point source is proportional to r^2 (with r the distance from source to detector), the relative error in reproducing a measurement is $[(r + \Delta r)^2 - r^2]/r^2 \approx 2\Delta r/r$ (with Δr the error in the source–detector distance). Thus, a larger r (or a lower detection efficiency) will improve the reproducibility. Obtaining a good precision and a low detection limit are thus contradictory demands.

Self-absorption of the source Self-absorption of the source is in principle no problem, as long as the absorption is identical for all samples (and for standards, Section 3.7). Self-absorption can be important for low energy γ-rays or for sources containing components of high atomic number.

Separation of the matrix Although in principle not necessary, it may be convenient to separate the main matrix components and so reduce the mass of material involved in the separation procedure. The following separation process is then faster, leads to smaller volumes (of solution or precipitate) and allows further evaporation of a solution to a smaller volume if necessary. As a consequence, the detection efficiency is improved and the self-absorption is lowered substantially or becomes negligible.

Health physics The activity level required to perform precise measurements (kBq) provides no health physics risk (nSv/h). The induced matrix activity, however, may be much higher. It is good practice first to separate the main matrix activity, choosing a procedure that limits manual intervention to the strict minimum. Ion exchange chromatography, for instance, is a better choice than (manual) solvent extraction.

The experimental development of a radiochemical separation is supported by the use of radiotracers. The chemical separation is simulated step by step with the addition of radiotracers (and inactive tracers as well) for the element to be separated (B) and for the interfering element(s) (D), in order to check quantitativeness and selectivity. These radiotracers are preferably produced by thermal neutron activation (if such a facility is available) rather than by charged particle activation. Indeed, production by, e.g., a (p, n) reaction yields a radiotracer that has to be separated from the target material, whereas for production by an (n, γ) reaction the target material can be used as the inactive carrier. The tracer has to be brought to the same chemical form as the radionuclide to be separated. This is often not possible for the dissolution step preceding the actual separation procedure, as this should guarantee quantitative recovery of the radionuclide to be measured. Therefore alternative ways to assess the yield of a separation have been proposed [36, 37].

For each separation step, three measurements can be made: the initial activity, as a reference, and after separation the two fractions in which the major and the minor activity, respectively, is collected. From the measurement of the minor fraction the most precise yield can be calculated. However, it is not always easily possible to perform all measurements with the same detection efficiency. For a precipitation, for example, the reference measurement is made on the solution before precipitation. After precipitation and filtration, a fraction of the filtrate is taken (equal in volume to the reference) and the yield for the ion to be precipitated can be calculated in a very precise way. To check coprecipitation (e.g. of matrix activity) the precipitate is measured. As the detection efficiency is not comparable with that of the reference measurement, the result has to be considered as an estimate. This is largely sufficient for the purpose of evaluating the selectivity of this separation.

There is no need to apply the radiochemical separation to the standards, at least if monoelemental standards are used (Section 3.7). However, standards and samples have to be measured at identical detection efficiency. Therefore the standards are brought into exactly the same geometrical form as the samples after separation. If the last step of a separation is ion chromatography, the standards are dissolved and the total volume is brought exactly to the volume of the eluate. Possible differences in self-absorption have to be considered. The eluate may contain an acid other than the one used to dissolve the standard; the eluate may contain only the added inactive carrier (B), whereas the standard contains a larger amount of the analyte element (A).

3.7 Standards

The standards should fulfill the following conditions: (1) the concentration of the analyte element is known very accurately, (2) this concentration does not change during irradiation (e.g. due to heating or vacuum), (3) the radionuclide (or any compound of it) formed from the analyte element does not volatilize during or after irradiation, (4) the main matrix composition is known in order to calculate the F-factor, (5) no nuclear or spectral interferences occur, (6) they are solid. Moreover, it is very convenient if the following conditions are also fulfilled: (7) target thickness is slightly larger than the range, (8) the activity produced comes only from the analyte element(s).

'Monoelemental' standards can meet these requirements. For most metals, all conditions are fulfilled by the use of very pure foils of appropriate thickness. It is clear that, if only one element is present, the chance of interference is quite limited. For other elements compounds are chosen, pure and with well known stoichiometry (first and fourth conditions), stable when heated or under vacuum (second condition), solid (sixth condition), powdered (seventh condition) and with a limited number of components (fifth and eighth conditions). Also, the behavior of the radionuclide formed has to be considered. Boric acid as a boron standard, for instance, is a good choice for the $^{10}B(p, \alpha)^{7}Be$ reaction but the wong choice for the $^{10}B(d, n)^{11}C$ reaction, as carbon-11 can volatilize as carbon dioxide. The drawbacks of monoelemental standards are that for multielemental analysis different standards have to be used and that sample and standard are not identical. The latter implies different stopping power and self-absorption of the sample compared with the standard.

To avoid these drawbacks, a reference material with certified content for the analyte element(s) can be used. Reference materials are issued by the BCR (Community Bureau of Reference, Commission of the European Communities), NIST (National Institute for Standards and Technology, formerly NBS, USA) and so on for a number of industrial and environmental materials. The accuracy of the result(s) obtained, however, depends on the accuracy of the certified concentration(s). It is clear that such a standard cannot be used if CPAA has been applied to contribute to the certification analyses.

An alternative could be to simulate the sample matrix by mixing powdered components and to add an exactly known amount of the analyte element(s) or to apply the standard addition method (Section 2.4.6). However, the requirements of homogeneity are quite high considering the limited range of charged particles in matter. Monoelemental standards are to be preferred.

3.8 Activity measurement

Charged particle-induced reactions, such as (p, n), (p, α), (p, d) reactions, yield radionuclides that are over-rich in protons and consequently decay by positron (β^+) emission or electron capture (EC), mostly followed by gamma (γ) emission. Positrons are emitted with an energy between zero and a characteristic maximum energy. Electron capture involves measurement of X-rays or Auger electrons and is of no importance for CPAA. Gamma-rays are monoenergetic. A radionuclide (and the analyte element the radionuclide is produced from) can be identified by its characteristic γ-ray energy (qualitative analysis). The activity measured provides quantitative information. A limited number of radionuclides are pure positron emitters. Positron emitters are measured indirectly by the annihilation radiation: a positron loses its kinetic energy and annihilates with an electron, i.e. both electron masses are converted into energy, and two γ-rays are emitted in opposite directions; the energy of each γ-ray is 511 keV, i.e. the rest mass of an electron. It is clear that information about the characteristic maximum energy is lost during the annihilation process. A positron emitter has then to be identified by its half-life. It is clear that γ-ray spectrometry is the method of choice except for pure positron emitters.

3.8.1 Gamma spectrometry

To reduce spectral interferences, a detector with the highest energy resolution is required, i.e. a high purity germanium (HPGe) detector. The detector signal is amplified and fed into a multichannel analyzer (MCA): an analog-to-digital converter (ADC) measures the pulse height (i.e. the γ-energy) and a microprocessor (or computer) stores a digital energy spectrum. One channel represents the number of γ-rays measured in a small energy interval. Considering a typical energy resolution of 2 keV, the spectrometer is set at 0.5 keV per channel and a 4000-channel analyzer covers an energy range from zero to 2 MeV.

The characteristics of a HPGe detector are described by the performance towards the 1333 keV γ-ray of cobalt-60: (1) energy resolution, i.e. the FWHM (full width at half maximum intensity), typically 2 keV; (2) peak to Compton ratio, i.e. the ratio of photopeak height to Compton plateau height at 1 MeV, typically 40–60; (3) relative detection efficiency, i.e. relative to a 7.6 cm × 7.6 cm NaI detector at 25 cm source–detector distance, typically 10–50%. The absolute detection efficiency is of the order of 10^{-4} to 10^{-3}.

Pulses can be lost because of ADC dead time (the ADC is still busy processing the previous pulse) or because of pulse pile-up (two pulses are so close to each other in time that the ADC measures the sum of the two energies). It is clear that these phenomena are more probable at high count rates. Different correction methods have been proposed for ADC dead time, which is the most important source of counting losses: the life-time method [38], the actual-time method [39], the dead-time-stabilizer (DTS) method [40–42] and the Harms method [43]. Correction for counting losses due to pulse pile-up requires an amplifier equipped with pile-up rejection (PUR), i.e. a device capable of detecting and rejecting pile-up pulses. Each PUR communicates with the ADC, which applies correction using one of the methods mentioned above. Because the pulser method [44] corrects for both ADC dead time and pulse pile-up, it is the method of choice. It has been shown (both theoretically and experimentally) that all (systematic and random) errors are always below 2% if the following conditions are fulfilled: (1) the measuring time is > 1 min and less than one half-life of the radionuclide to be measured, and (2) the counting loss is < 50% [45].

Samples and standards have to be measured at the same detection efficiency, which is determined by the source–detector geometry and by self-absorption as well. For instrumental analysis of solid samples, it is mostly sufficient to measure samples and standards positioned at exactly the same distance from the detector with the irradiated side towards the detector. Because of the limited range of charged particles in matter, the self-absorption of sample and standard (and hence the difference in self-absorption) is very low or negligible except for very low γ-energies. For radiochemical analysis, however, special precautions have to be taken to ensure identical geometry and self-absorption for sample and standard (Section 3.6). Self-absorption (and differences in self-absorption) can be evaluated roughly assuming a point source absorbed by half the sample (or standard), using, e.g., the mass attenuation coefficients compiled by Hubbell [46]. If the difference in self-absorption is not negligible, it can be experimentally determined using tracer activity.

3.8.2 Positron counting

Radionuclides that are *pure* positron emitters have to be measured by their annihilation radiation (Section 3.8). Although this can be measured by a HPGe detector, it is more convenient to use a NaI scintillation detector, because (1) the detection efficiency is higher, (2) the cost and ease of operation are more favorable and (3) no profit can be made from the better energy resolution of a HPGe detector, as all positron emitters produce annihilation radiation of the same energy (511 keV). Therefore, spectrometry by a multichannel analyzer is not necessary; a single-channel analyzer is sufficient. The selectivity can be improved by the use of a γ–γ coincidence set-up. Two NaI detectors are used, the positron source being placed between them. By the annihilation event two 511 keV γ-rays are emitted simultaneously in opposite directions and can be detected by the two detectors. A coincidence circuit selects all simultaneous measurements. All γ-radiation detected from sources other than positron emitters is rejected, as well as that from all positron sources outside the gap between the detectors.

For qualitative measurements the half-life of the positron emitter has to be determined by decay curve analysis (Section 3.8.3). For quantitative measurements it is important to assess conditions of equal detection efficiency and to correct for possible counting losses. The former also implies that positron sources should annihilate completely.

Annihilation only occurs after slowing down of the positron. Therefore the positron source should be surrounded by enough material to assure that all positrons (even those at maximum energy) annihilate, i.e. more than the range of positrons with maximum energy. In practice much lower absorber thicknesses are sufficient to absorb most of the positrons, because of the continuous energy spectrum of a positron emitter and because, for a positron, the projected range (Section 2.1.3) is much smaller than the linear range. The linear range of 1 MeV to 1.9 MeV positrons in water is 5 to 9 mm. The projected range for a positron emitter with a maximum energy of 1 MeV is < 1.6 mm for 90% of the positrons and < 3 mm for 99% of them. For a positron emitter with a maximum energy of 1.9 MeV these figures are 3.7 and 5.5 mm [47]. The resulting annihilation radiation is only slightly absorbed (1% per mm of absorber) by such an absorber. Pure positron emitters used in CPAA are (maximum energy in MeV): carbon-11 (1.0), nitrogen-13 (1.2), oxygen-15 (1.7), fluorine-17 (1.7) and fluorine-18 (0.6). Other positron sources are sodium-22 (0.5), copper-64 (0.7) and gallium-68 (1.9), which are used to check stability of the measuring equipment and to evaluate differences in detection efficiency (see below).

There is a sharp decrease in detection efficiency when the sample is moved away from the center between the two detectors. Identical geometry for sample and standard is thus very important. If this is not possible, the ratio of detection efficiencies can be determined experimentally using copper-64 or germanium-68/gallium-68 tracer. Copper-64 can be produced by reactor irradiation of copper; after decay of copper-62 (β^+, $t_{1/2} = 9.73$ min), a nearly pure positron source is obtained with a half-life of 12.71 h. Gallium-68 (β^+, $t_{1/2} = 1.13$ h), a daughter of germanium-68 (EC, $t_{1/2} = 287$ d), provides a long-lived, nearly pure positron source. Germanium-68 is commercially available.

Correction for counting losses in a coincidence set-up can be applied using the dead time, i.e. the time the set-up needs to process one event.

$$R_c = R_m/(1 - R_m\tau) \tag{34}$$

where R_c is the corrected count rate, R_m is the measured count rate and τ is the dead time.

This equation is valid only for a non-extensible dead time, i.e. one that is constant as a function of the count rate. Although this is not the case for a γ–γ coincidence set-up, this correction can be applied for counting losses up to 10% (e.g. for $\tau = 10^{-6}$ min/count and $A < 10^5$ counts/min). The dead time can be experimentally determined by repeated measurements on a short-lived radionuclide, radioisotopically pure and with well-known half-life. Irradiation of pure graphite or polyethene with deuterons (< 12 MeV) produces the pure positron emitter nitrogen-13 with a half-life of 9.97 min. Measurements at low count rates (with negligible counting losses) and the half-life are used to extrapolate to higher count rates. From the extrapolated count rate (without counting losses), R_c and the measured count rate R_m, the dead time τ can be calculated from equation (34).

3.8.3 Decay curve analysis

As all positron emitters yield annihilation radiation of 511 keV, they are identified by their half-life. This is done by decay curve analysis (DCA), especially for mixtures of positron emitters of different half-life. DCA can also be applied for γ-ray spectrometry in the case of spectrometric interference.

The activity (mostly annihilation radiation) is measured repeatedly. A DCA computer program then calculates the count rate of the component(s). In its simplest form the program requires prior knowledge of the number of components and their half-lives. More complex DCA programs do not require any pre-knowledge of number and half-lives of the components but require more measurements. The DCA program according to Cumming [48] requires prior knowledge of the number of components and the approximative half-lives. Statistical parameters indicate whether erroneous suppositions have been made.

3.8.4 Decay correction

For equal detection efficiency, the ratio of the activities in the sample and the standard at the end of irradiation in equation (12) may be replaced by the ratio of count rates (measured activity). During a measuring time t_m, a number of counts C is measured after a waiting time t_w (i.e. between the end of irradiation and the start of measurement). Once the number of counts C has been corrected for counting losses (Sections 3.8.1 and 3.8.2), the count rate R can be calculated by

$$R = \frac{C\lambda}{e^{-\lambda t_w}(1 - e^{-\lambda t_m})} \tag{35}$$

3.9 Quantitation

This section gives an overview of the final calculation of the concentration in the sample unless the standard addition method (Section 2.4.6) has been applied. If nuclear or spectral interferences cannot be avoided (Section 2.3), a correction may be possible.

3.9.1 No interferences

A standard and a sample are supposed to be irradiated separately but with the same incident energy (Section 2.4.1). If this condition cannot be fulfilled exactly, because of etching the sample, a correction has to be applied (Section 3.3). The concentration of the analyte element in the sample can be calculated by equation (12) (Section 2.4.1) from:

- the concentration of the analyte element in the standard; monoelemental standards are to be preferred (Section 3.7);
- the ratio of the activity in the sample and the standard at the end of irradiation; if sample and standards are measured with equal detection efficiency (Sections 3.6, 3.8.1 and 3.8.2), the ratio of count rates can be calculated (Section 3.8.4);
- the ratio of the beam intensity for standard and sample; samples and standards may be irradiated with different beam intensities, correction being made by beam intensity monitor foils (Section 3.2), but the beam intensity should be constant during irradiation [Section 2.4.1, equation (9)], and is monitored by direct measurement of current (Section 3.1.3);
- the irradiation time, which should be measured automatically by an electronic clock;
- the decay constant $\lambda = \ln 2/t_{1/2}$, with $t_{1/2}$ = half-life, taken from Ref. 31;
- the F-factor [equation (13)]; sample and standard may have different stopping power (which is always the case for monoelemental standards), and correction is made by one of the standardization methods described in Section 2.4.

3.9.2 Correction for spectral interference

Such a correction (Section 2.3.2) is possible if two conditions are fulfilled: (1) the interfering radionuclide emits another kind of radiation (two γ-rays or a positron and one γ-ray), and (2) the second kind of radiation can be measured free from interference. A monoelemental standard for the element (C), i.e. the element the interfering radionuclide (D) is formed from, is irradiated and measured for both types of radiation. It is clear that the avoidance of a spectral interference is to be preferred to a significant correction because of the resulting unfavorable counting statistics. This method is, however, of practical interest in calculating an upper limit for the spectral interference.

3.9.3 Correction for nuclear interference

Correction for nuclear interference (Section 2.3.1) is possible if (1) the concentration of the interfering element is known, or (2) the concentration of the interfering element can be determined separately without nuclear interference from the analyte element, or (3) two irradiation conditions (i.e. two different energies of the same charged particle or two different particles) result in a sufficiently different interference factor (to be determined experimentally).

REFERENCES

1. U. Fano, *Annu. Rev. Nucl. Sci.* **13**, 1 (1963).
2. W. H. Bragg and R. Kleeman, *Philos. Mag.* **10**, 318 (1905).
3. H. Baumgart, W. Arnold, J. Günzl, E. Huttel, A. Hofmann, N. Kniest, E. Pfaff, G. Reiter, S. Tharraketta and G. Clausnitzer, *Nucl. Instrum. Methods Phys. Res., Sect. B* **B5**, 1 (1984).
4. H. H. Andersen and J. F. Ziegler, *Hydrogen Stopping Powers and Ranges in All Elements*, Pergamon Press, New York (1977).
5. J. F. Ziegler, *Helium Stopping Powers and Ranges in All Elemental Matter*, Pergamon Press, New York (1977).
6. K. A. Keller, J. Lange and H. Münzel, *Q-Values and Excitation Functions of Nuclear Reactions, Part a: Q-Values*, edited by H. Schopper, Springer-Verlag, Berlin (1973).
7. K. A. Keller, J. Lange, H. Münzel and G. Pfennig, *Q-Values and Excitation Functions of Nuclear Reactions, Part b: Excitation Functions for Charged-Particle Induced Nuclear Reactions*, edited by H. Schopper, Springer-Verlag, Berlin (1973).
8. F. K. McGowan and W. T. Milner, *Charged-Particle Reaction List, 1948–1971*, edited by Katharine Way, Academic Press, New York (1973).
9. K. Okamoto, (Ed.), *Handbook on Nuclear Activation Data*, Tech. Rep. Ser. No 273, International Atomic Energy Agency, Vienna (1987).
10. K. Okamoto (Ed.), *Proceedings of the IAEA Consultants' Meeting on Data Requirements for Medical Radioisotope Production*, INDC (NDS)-195/GZ, International Atomic Energy Agency, Vienna (1988).
11. D. Gandarias-Cruz and K. Okamoto, *Status on the Compilation of Nuclear Data for Medical Radioisotopes Produced by Accelerators*, INDC (NDS)-209/GZ, International Atomic Energy Agency, Vienna (1988).
12. O. Schwerer and K. Okamoto, *Status Report on Cross-sections of Monitor Reactions for Radioisotope Production*, INDC (NDS)-218/GZ, International Atomic Energy Agency, Vienna (1989).
13. K. A. Keller, J. Lange and H. Münzel, *Q-Values and Excitation Functions of Nuclear Reactions, Part c: Estimation of Unknown Excitation Functions and Thick Target Yields for p, d, ^{3}He and α Reactions*, edited by H. Schopper, Springer-Verlag, Berlin (1974).
14. M. Blann, in *Proceedings of the IAEA Consultants' Meeting on Data Requirements for Medical Radioisotope Production*, edited by K. Okamoto, INDC (NDS)-195/GZ, International Atomic Energy Agency, Vienna (1988), pp. 115–124.
15. A. Pavlik, in *Proceedings of the IAEA Consultants' Meeting on Data Requirements for Medical Radioisotope Production*, edited by K. Okamoto, INDC (NDS)-195/GZ, International Atomic Energy Agency, Vienna (1988), pp. 124–131.
16. K. Hata and H. Baba, in *Proceedings of the IAEA Consultants' Meeting on Data Requirements for Medical Radioisotope Production*, edited by K. Okamoto, INDC (NDS)-195/GZ, International Atomic Energy Agency, Vienna (1988), pp. 131–142.
17. W. Seelmann-Eggebert, G. Pfennig, H. Münzel and H. Klewe-Nebenius, *Chart of Nuclides*, 5th. Ed., Kernforschungszentrum Karlsruhe (1981).
18. F. W. Walker, J. R. Parrington and F. Feiner, *Nuclides and Isotopes*, 14th Ed., General Electric Co. (1989).
19. C. Vandecasteele and K. Strijckmans, *J. Radioanal. Chem.*, **57**, 121 (1980).
20. K. Casteleyn, K. Strijckmans and R. Dams, *Nucl. Instrum. Methods Phys. Res., Sect. B* **B68**, 161 (1992).

21. E. Ricci and R. L. Hahn, *Anal. Chem.* **37**, 742 (1965).
22. E. Ricci and R. L. Hahn, *Anal. Chem.* **39**, 794 (1967).
23. M. A. Chaudhri, G. Burns, E. Reen, J. L. Rouse and B. M. Spicer, *J. Radioanal. Chem.* **37**, 243 (1977).
24. K. Ishii, M. Valladon and J.-L. Debrun, *Nucl. Instrum. Methods* **150**, 213 (1978).
25. K. Ishii, M. Valladon, C. S. Sastri and J.-L. Debrun, *Nucl. Instrum. Methods* **153**, 503 (1978).
26. K. Ishii, C. S. Sastri, M. Valladon, B. Borderie and J.-L. Debrun, *Nucl. Instrum Methods* **153**, 507 (1978).
27. K. Masumoto and M. Yagi, *J. Radioanal. Nucl. Chem., Articles* **109**, 449 (1987).
28. M. Yagi and K. Masumoto, *J. Radioanal. Nucl. Chem., Articles* **111**, 359 (1987).
29. G. Wauters, C. Vandecasteele and J. Hoste, *J. Radioanal. Nucl. Chem.*, Articles **98**, 345 (1986).
30. N. De Brucker, J. Dewaele, K. Strijckmans and C. Vandecasteele, *Anal. Chim. Acta* **220**, 93 (1989).
31. G. Erdtmann and W. Soyka, *The Gamma Rays of the Radionuclides*, Verlag Chemie, Weinheim (1979).
32. M. Valladon and J.-L. Debrun, *J. Radioanal. Chem.* **39**, 385 (1977).
33. M. Valladon, G. Blondiaux, A. Giovagnoli, C. Koemmerer and J.-L. Debrun, *Anal. Chim. Acta* **116**, 25 (1980).
34. J. B. Marion and F.C. Young, *Nuclear Reaction Analysis, Graphs and Tables*, North-Holland, Amsterdam (1968).
35. L. C. Northcliffe and R. F. Schilling, *Nuclear Data Tables* **A7**, 233 (1970).
36. R. Mortier, C. Vandecasteele, K. Strijckmans and J. Hoste, *Anal. Chem.*, **56**, 2166 (1984).
37. K. Strijckmans, J. Dewaele and R. Dams, *Anal. Chim. Acta* **262**, 193 (1992).
38. D. F. Covell, *Anal. Chem.* **32**, 1086 (1960).
39. E. Junod, *J. Radioanal. Chem.* **20**, 113 (1974).
40. J. Bartošek, F. Adams and J. Hoste, *Nucl. Instrum. Methods* **103**, 45 (1972).
41. F. Adams, J. Hoste, J. Bartošek and J. Mašek, *J. Radioanal. Chem.* **15**, 479 (1973).
42. J. Bartošek, G. Windels and J. Hoste, *Nucl. Instrum. Methods* **103**, 43 (1972).
43. J. Harms, *Nucl. Instrum. Methods* **53**, 192 (1967).
44. K. Strijckmans, C. Vandecasteele and J. Hoste, *Anal. Chim. Acta* **89**, 255 (1977).
45. K. Strijckmans, *De bepaling van stikstof in Ti, Ni, Zr, Nb, Mo, Ta en W door aktiveringsanalyse met geladen deeltjes* Ph.D. thesis, Rijksuniversiteit Gent (1980), pp. 67–83.
46. J. H. Hubbell, *Int. J. Appl. Radiat. Isot.* **33**, 1269 (1982).
47. S. E. Derenzo, in *Positron Annihilation*, edited by R. R. Hasiguti and K. Fujiwara, Japan Institute of Metals, Sendai (1979), pp. 819–823.
48. J. B. Cumming, CLSQ, the Brookhaven Decay Curve Analysis Program, *Brookhaven Natl. Lab., [Rep.] BNL* (6470) (1962) (unpublished).

Chapter 11

ION BACKSCATTERING AND ELASTIC RECOIL DETECTION

Eero Rauhala

Accelerator Laboratory, Department of Physics, University of Helsinki, SF-00170 Helsinki, Finland

1 INTRODUCTION

Elastic backscattering and elastic recoil spectrometries are based on the elastic collision of charged particles. The energy distribution of charged particles is observed in both, which makes the experimental set-up for each similar and relatively uncomplicated. The differences arise from the different particles detected: ion backscattering consists in probing the heavier target atoms by lighter projectile ions, whereas in the elastic recoil method heavier projectiles are used and lighter recoiled particles, the atomic particles from the target, enter the detector. Backscattering readily reveals microscopic amounts of heavy elements in light samples, while light recoils from heavy samples are detected with good sensitivity. Both methods yield information on the depth distribution of many analyte elements in one experiment. The elastic collision between incident particles of mass M_1 and target particles of mass M_2 is schematically illustrated in Fig. 1 for the ion backscattering and elastic recoil events.

Ion backscattering spectrometry has been used for analytical purposes for more than three decades. A traditional Rutherford backscattering experiment is performed with ^{4}He ion projectiles in the Coulomb scattering energy region around 2 MeV. New methodological developments by the end of the 1980s include, e.g., the use of Rutherford scattering of heavier ions and helium and hydrogen ion scattering above the standard Rutherford energy region. Experimental developments are exemplified by microbeam and elaborate detection techniques.

An overwhelming number of studies on the application of ion backscattering in

Chemical Analysis by Nuclear Methods Edited by Z. B. Alfassi

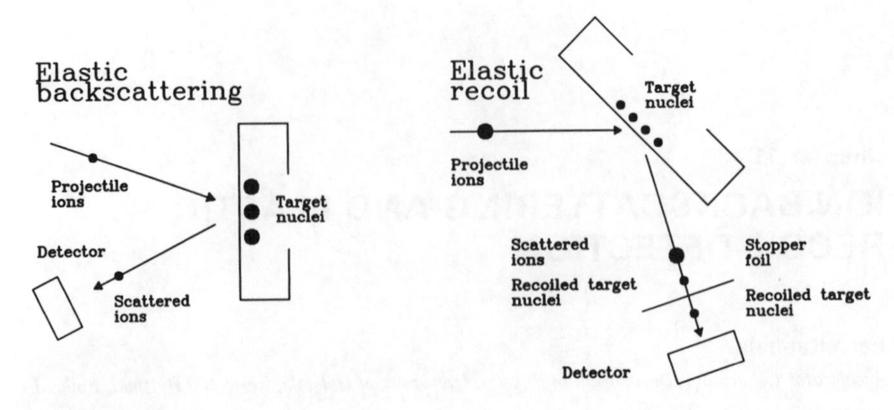

Fig. 1. Schematic diagrams of the experimental set-ups in ion backscattering and in elastic recoil spectrometry.

various fields of technology, metallurgy, biological, medical and environmental investigation has been published. The examples range from surface and bulk analysis of composition, crystalline structure and impurity content of multi-layered multi-element samples to studies of such processes as diffusion, solubility, corrosion, sputtering, oxidation, gettering and blistering. Textbooks and many extensive reviews of the method have been published (e.g. [1–5]). The method has been referred to, *inter alia*, as Rutherford backscattering spectrometry (RBS or BS), elastic backscattering spectrometry (EBS), proton elastic scattering (PES) and heavy ion Rutherford backscattering (HIRBS).

Elastic recoil spectrometry has a more recent origin. In 1971, Cohen, Fink and Degnan [6] reported the results of hydrogen profiling in self-supporting foils by using a proton–proton elastic collision *symmetric coincidence technique* in transmission geometry: the sample was bombarded by 17 MeV protons and the scattered and recoiled protons were detected behind the foil in coincidence at two detectors each positioned at 45° angle with respect to the incident beam. The symmetry of the collision event required that protons detected simultaneously at the two detectors must have originated from the same (p, p) elastic scattering event. Further experiments exploiting the symmetric H–H coincidence method have been presented in Refs. 7 and 8. Similar He–He coincidence experiments with high energy He ions were carried out by Smidt and Pieper to profile He in vanadium [9], and in [10]. An analogous *asymmetric coincidence technique* has been used by Moore *et al.* [11] and by Klein [12] for recoils with masses different from those of the incident ions: He, O and Cl ions at 10–40 MeV were employed to detect various elements in the range from ^{12}C to ^{63}Cu.

The elastic recoil detection method in its uncomplicated present form was introducced by L'Ecuyer *et al.* in 1976 [13]. A 35 MeV ^{35}Cl beam incident on

LiF/Cu/LiF samples was used. The scattered ^{35}Cl beam was stopped by 5–7 μm nickel foils in front of the detector.

Doyle and Peercy [14] adapted the method to hydrogen profiling in 1979. Other early papers in the 1970s and early 1980s include, e.g. L'Ecuyer, Brassard and Cardinal [15], Terreault *et al.* [16], Ross *et al.* [17], Wielunski, Benenson and Lanford [18], Groleau, Gujrathi and Martin [19], Nölscher *et al.* [20], Moreau *et al.* [21], Thomas *et al.* [22] and Turos and Meyer [23].

To date, hundreds of articles have been published on the application of the elastic recoil detection method. The applications range from studies of hydrogen and other light element analysis in various materials from semiconductor and superconductor technology, metallurgy, geology, polymer chemistry and biology to hydrogen adsorption and embrittlement in steels, archeological dating, space research and plasma–wall interaction in thermonuclear reactors.

Recent reviews of the elastic recoil detection method are found in Refs. 24–29. The terminology is more variable than in the case of ion backscattering: titles such as elastic recoil detection (ERD), elastic recoil spectrometry (ERS), forward recoil spectrometry (FRS), high energy ion recoil (HEIR), hydrogen recoil spectrometry (HRS) and others have been used.

2 THEORETICAL PRINCIPLES

Ion backscattering involves the observation of the lighter incident particle ($M_1 < M_2$) at a backscattering angle $\Theta > 90°$. In elastic recoil, the lighter recoiling target particle ($M_2 < M_1$) is detected at a forward recoil angle $\Theta < 90°$. The angle of incidence and the exit angle are denoted by Θ_1 and Θ_2, respectively. The collision geometry for the scattered and recoiled particles is outlined in Fig. 2.

The main analytical characteristics of the backscattering and elastic recoil

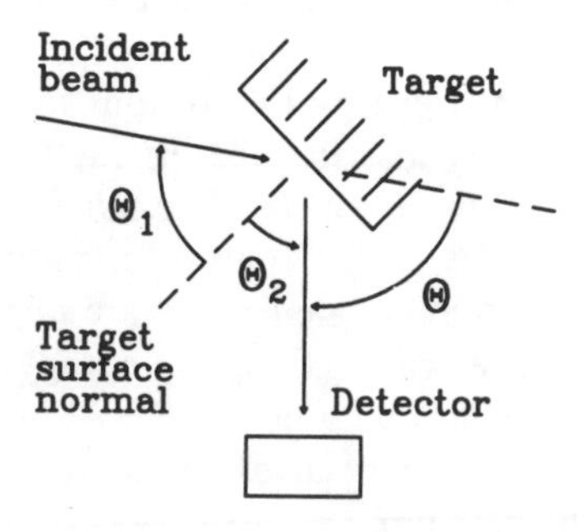

Fig. 2. Definition of angles and collision geometry for elastically scattered and recoiled particles.

methods—the capabilities of mass, depth and quantitative analysis—are theoretically treated in terms of the *collision kinematics, stopping power* and the *scattering cross-section,* respectively.

2.1 Mass perception—energetics of the elastic collision

Mass perception is based on the energy exchange between the elastically colliding particles. The energetics of the elastic collision may be calculated precisely when the masses of the colliding partners and the collision geometry are known. Conservation of energy and momentum in the interaction gives rise to constant ratios K of the incident and exit energies, E_0 and E_1. These ratios are called the *kinematic factors* and are expressed for the backscattered ions as:

$$K_{\mathrm{BS}} = \frac{E_1}{E_0} = \left[\frac{(M_2^2 - M_1^2 \sin^2\Theta)^{1/2} + M_1 \cos\Theta}{M_1 + M_2}\right]^2 \tag{1}$$

and for the elastic recoils as:

$$K_{\mathrm{ER}} = \frac{E_1}{E_0} = \frac{4M_1 M_2}{(M_1 + M_2)^2} \cos^2\Theta \tag{2}$$

The energy E_1 refers to the scattered particle and the recoiling target particle in equations (1) and (2), respectively.

The dependence of the kinematic factor on target mass is shown in Fig. 3 for several incident ions at typical detection angles. For $M_1 > M_2$ the kinematic factor K_{ER} increases with M_2 and has a maximum at $M_1 = M_2$. In the conventional set-up the stopper foil eliminates recoils for $M_2 > M_1$. Although K_{ER} is always single valued, the stopper foil gives rise to double values of detected energies [see Section 3.1.2].

2.2 Depth analysis—energy loss of particles in matter

Depth analysis is based on the energy loss of ions in the target material. The physical mechanisms of energy loss involve interactions between the incident particle, target electrons and target nuclei. The nuclear contribution is usually negligible from the point of view of the analysis method.

The main quantities needed in the calculations are *stopping power* and *stopping cross-section.* By definition, the ratio of the ion energy loss ΔE to the distance traversed Δx approaches the stopping power $S = \delta E/\delta x$ as Δx diminishes. The stopping cross-section ε is obtained by dividing the stopping power by the atomic density (atoms/cm^3) of the medium:

$$\varepsilon(E) = S(E)/N \tag{3}$$

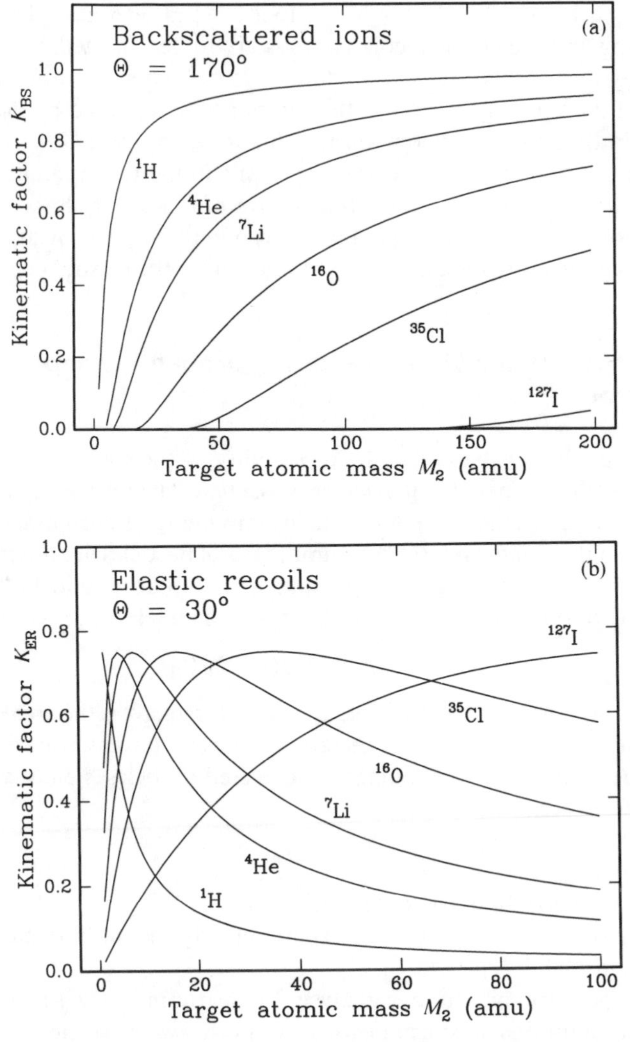

Fig. 3. Kinematic factors for elastically (a) scattered and (b) recoiled particles.

The basic equation for calculating the electronic stopping power in the high and in the intermediate energy region is the Bethe–Bloch equation [30, 31]. Recent semiempirical parametrizations for calculating the energy loss of ions have been presented by Ziegler and co-workers. The latest of these empirical models [32], based on the Brandt–Kitagawa theory [33], was published in 1985. It relates proton stopping powers by scaling to the stopping powers of heavier ions in all elemental matter. Accuracies of the order of 2–5% for proton and helium ions [34] and 10–15% for heavy ions [35] at energies relevant to ion backscattering and the elastic recoil methods are to be expected.

The energy loss in mixtures or compounds is to a good accuracy given by a linear sum of the stopping powers of the components, weighted by their relative abundance in the mixture or compound. This is the *Bragg additivity rule* [36]. Deviations from this rule are observed, especially in light compounds [37].

Energy straggling—the statistical fluctuation of ion energy loss—affects energy resolution for particles emerging from deep in the sample. Straggling in the stopper foil significantly impairs energy resolution in the elastic recoil method.

2.3 Quantitative analysis—scattering probability and cross-section

Quantitative analysis is based on the probability of the elastic collision event, characterized in terms of the *scattering cross-section*. The relevant quantities are related as follows. Assume incident particles impinging at normal incidence on the surface of a thin uniform target. A number dQ of particles are observed at the detector, positioned at an angle Θ expanding an acceptance solid angle $d\Omega$. The differential scattering cross-section $d\sigma/d\Omega$ is then defined as:

$$d\sigma/d\Omega = (1/Nt)\,[(dQ/d\Omega)/Q] \tag{4}$$

where Q is the number of particles incident on the target, Nt is the areal density of the target (in e.g. atoms/cm^2), N is the atomic density (atoms/cm^3) and t is the target thickness (nm). The total number of detected particles A emitted from the target may be written as:

$$A = d\sigma/d\Omega\,\Omega\,Q\,Nt \tag{5}$$

This equation is the basis of quantitative analysis because, when reserved, equation(5) gives Nt as a function of experimentally measurable data and the cross-section $d\sigma/d\Omega$.

In order to perform quantitative analysis, the scattering cross-section in given experimental conditions must be known accurately. Two separate cases must be examined: lower energy scattering from the screened Coulomb potential, referred to as *Rutherford scattering*, and *non-Rutherford scattering* taking place at higher energies when the nuclear potential and nuclear resonance scattering interfere. In collisions with large collision distances b, the inner electron shells provide screening of the nuclear charge and slight deviations from the Rutherford cross-section result. In close collisions, on ther other hand, when b becomes comparable with the sum of the nuclear radii, the nuclear force interactions produce significant deviations from the Rutherford cross-section.

Rutherford scattering and the usually small electronic screening effect may be calculated theoretically; experimental cross-section data are needed in the high energy region, since non-Rutherford cross-sections cannot be predicted accurately.

A complication may arise in the elastic recoil method owing to the *multiple scattering* contribution. This is a consequence of the small detection angle and the increase of the cross-section with decreasing detection angle. The effects and corrections are outlined in [25] according to the theory presented in Refs. 38 and 39. Multiple scattering is usually of little concern in ion backscattering [40].

2.3.1 Rutherford scattering

The Rutherford model of ion scattering assumes an interaction of two bare nuclear point charges and a repulsive Coulomb potential of the form $V \approx q/r$. The Rutherford formulae for the backscattering (subscript BS) and elastic recoil (ER) cross-sections in the laboratory frame of reference may be expressed as:

$$d\sigma/d\Omega_{BS} = \left(\frac{Z_1 Z_2 e^2}{16\pi\varepsilon_0 E}\right)^2 \frac{4}{\sin^4\Theta} \frac{[(1 - x^2\sin^2\Theta)^{1/2} + \cos\Theta]^2}{(1 - x^2\sin^2\Theta)^{1/2}} \tag{6}$$

$$d\sigma/d\Omega_{ER} = \left(\frac{Z_1 Z_2 e^2}{8\pi\varepsilon_0 E}\right)^2 \frac{1}{\cos^3\Theta}(1 + x)^2 \tag{7}$$

where $x = M_1/M_2$, e is the electron charge, and E is the collision energy. The most important functional dependences are the proportionalities of the cross-sections to the squares of Z_1, Z_2 and $1/E$.

2.3.2 Non-Rutherford scattering

As the incident ion energy approaches the Coulomb barrier height of the target nucleus, new scattering mechanisms enter to compete with the Rutherford scattering. Scattering from the nuclear potential tends to gradually enhance or diminish the cross-section relative to the Rutherford value, and abrupt changes in the cross-section are encountered at energies coinciding with the excited states of the compound nucleus formed. In a general case, the total scattering is thus a result of three interacting phenomena: *Rutherford scattering*, *nuclear potential scattering* and *nuclear resonance scattering*. The three processes take place simultaneously and add up coherently, making the analytical treatment of the net effect very complicated [41].

The high non-Rutherford energies are now used routinely to enhance the detection sensitivity of light elements. A considerable amount of experimental non-Rutherford backscattering cross-section data can be found in the literature. However, some of the older data measured in connection with the studies of nuclear level structure in the past decades are inaccurate. Recent measurements are found, e.g., for ^{1}H ions in Refs. 42–48 and for ^{4}He ions in [49–52]. Analytical fits for the proton non-Rutherford cross-sections are presented in Refs. 53 and 54.

The non-Rutherford backscattering method and data on cross-sections are reviewed for proton backscattering in Refs. 45, 5 and 54 and for ^{4}He ions in [55] and [56].

Figure 4 presents recent data on ^{4}He backscattering cross-section measurements for a ^{11}B target [57]. A deviation of the cross-section from the Rutherford prediction is indicated above 1.3 MeV and a resonance dip is shown at 2.06 MeV. The original data spans the energy region $E_{He} = 1.0–3.3$ MeV. When probing carbon, nitrogen or oxygen by ^{4}He ions, for example, non-Rutherford effects must be taken into account above 2.2 MeV.

For the elastic recoil detection, new experimental cross-section data (solid dots), a polynomial fit (solid line) [58] and a comparison of other recent data [59–64] on the ^{1}H(^{4}He, ^{1}H)^{4}He hydrogen elastic recoil process are presented in Fig. 5. The dotted line represents the calculated curve from Ref. 59. Many experimental problems inherent in the other measurements were overcome in the data of Ref. 58. The data show that non-Rutherford effects must be considered in elastic recoil analysis of hydrogen above $E_{He} \approx 1.3$ MeV. At 2.5 MeV, e.g., the cross-section exceeds the Rutherford value by a factor of ≈ 2.

3 SPECTRUM EVALUATION

A few examples of how to apply the theoretical principles outlined above to correlate the spectrum parameters to those of the sample are presented below.

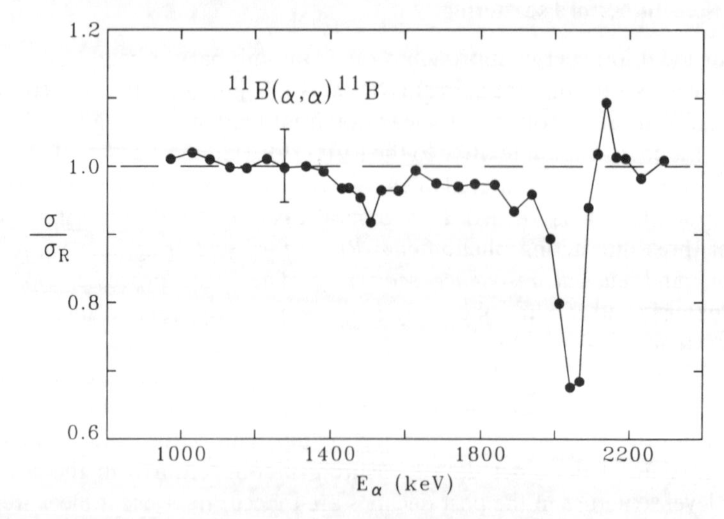

Fig. 4. Elastic scattering cross-sections for scattering of ^{4}He by ^{11}B through the scattering angle $\Theta = 170.5°$. (From Ref. 57).

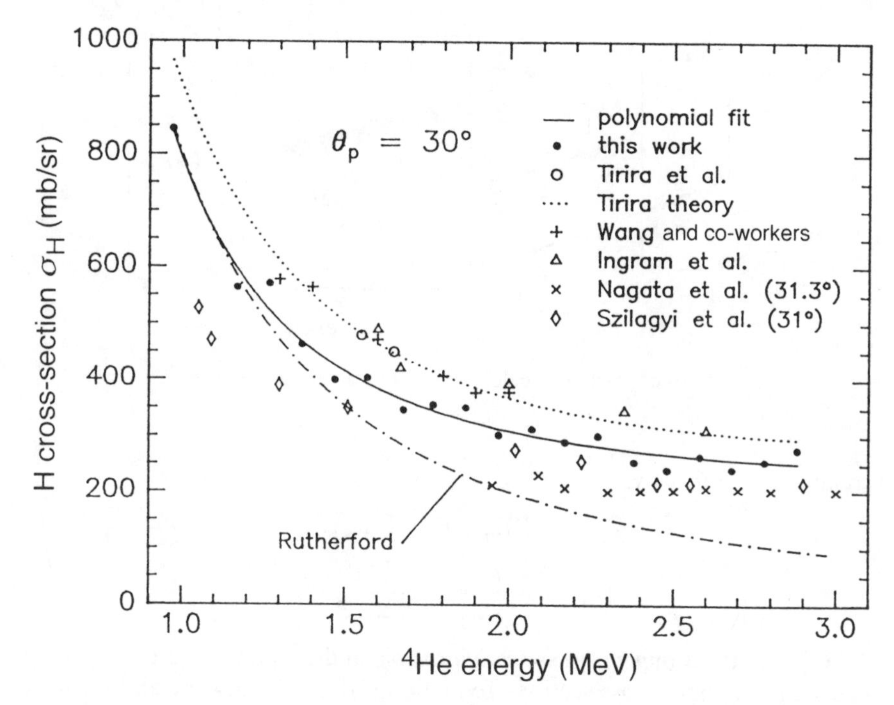

Fig. 5. Comparison of cross-sections for the $^{1}H(^{4}He, ^{1}H)^{4}He$ elastic recoil reaction of Ref. [58] at detection angle $\Theta = 30°$ with other measurements (see references in [58]). (From Ref. 58.))

Ion backscattering and elastic recoil are treated separately for clarity. Reviews of the formalism are presented, e.g, in Refs. 1 (RBS) and 25 and 65 (ERD).

3.1 Mass analysis and depth scale

3.1.1 Backscattering spectrum

A schematic illustration of a case of backscattering from a thick elemental sample and a corresponding signal of a backscattering spectrum are illustrated in Fig. 6. Assume an elemental sample bombarded by ions of energy E_0. The energy of the emerging ions after scattering from a depth x is the detected energy E_1. The energy of the ions scattered from the sample surface is KE_0, where K is the kinematic factor [equation (1)].

The energy difference observed for ions scattered from the surface and from depth x is $\Delta E = KE_0 - E_1$. The energy difference ΔE observed from the spectrum

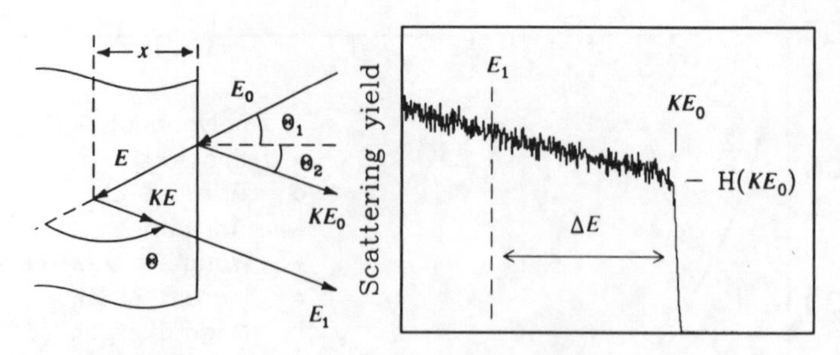

Fig. 6. Energy–depth scale in ion backscattering spectrometry.

is related to depth x by:

$$\Delta E = KE_0 - E_1 = [\varepsilon]_{\mathrm{BS}} Nx \tag{8}$$

$$[\varepsilon]_{\mathrm{BS}} = \frac{K}{\cos\Theta_1}\varepsilon_{\mathrm{in}} + \frac{1}{\cos\Theta_2}\varepsilon_{\mathrm{out}}$$

where $[\varepsilon]$ is the *stopping cross-section factor,* and $\varepsilon_{\mathrm{in}}$ and $\varepsilon_{\mathrm{out}}$ are the energy dependent stopping cross-sections [equation (3)] on the inward and outward paths of the ions. Near the sample surface, constant values $\varepsilon_{\mathrm{in}} = \varepsilon(E_0)$ and $\varepsilon_{\mathrm{out}} = \varepsilon(KE_0)$ are commonly used. This approximation is referred to as the *surface energy approximation,* and the corresponding $[\varepsilon]$ factor is denoted $[\varepsilon_0]$.

For a multi-elemental target, the energies and the energy losses after scattering depend on the atomic species of the collision partner.

3.1.2 Elastic recoil spectrum

In the elastic recoil method, the incident particle is different from the one detected. Usually a stopper foil is used to block the scattering of the incident ions to the detector. The treatment is analogous to that above with the effect of the stopper foil added. The energies for a simple elemental sample are illustrated in Fig. 7.

The stopping cross-section factors now include the energy loss contributions of both incident ions and recoiled target particles:

$$[\varepsilon]_{\mathrm{ER}} = \frac{K}{\cos\Theta_1}\varepsilon_{\mathrm{ions}} + \frac{1}{\cos\Theta_2}\varepsilon_{\mathrm{recoils}} \tag{9}$$

The energy difference ΔE_{T} of the recoils emitting from the surface of the target and from depth x, observed from the spectrum, is then written:

$$\Delta E_{\mathrm{T}} = KE_0 - E_1 = [\varepsilon]_{\mathrm{ER}} Nx \tag{10}$$

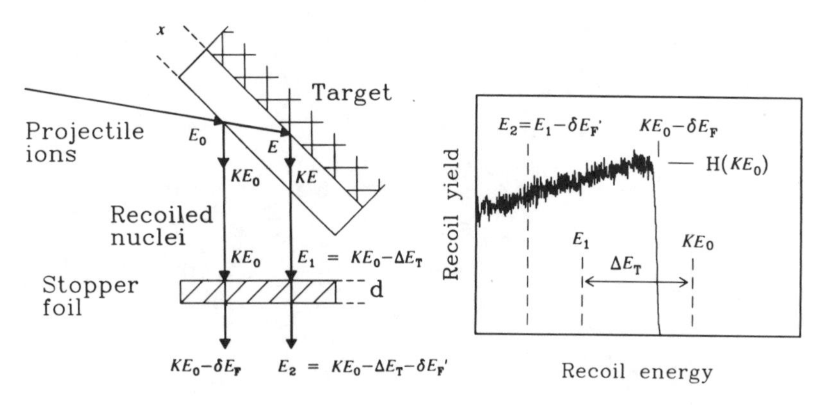

Fig. 7. Energy–depth scale in elastic recoil spectrometry.

Besides stopping the scattered incident ions, the stopper foil also reduces the energy of the recoils by an amount δE_F. This energy loss depends on the recoil energy and varies from δE_F to $\delta E_F'$ for recoils emitted from depth 0 to depth x in the target. In a multi-element sample, the recoil energy—and thus also δE_F—depends on target mass. As a consequence, a small distortion in the spectrum arises if the energy dependence of δE_F is ignored [24]. To avoid these difficulties, a correction to spectrum height [24] or numerical integration of expressions of the form $\varepsilon N x$ may then be applied.

Figure 8 illustrates an example of the effect of the stopper foil on detected recoil energies. Whereas the energy of the recoils increases with target mass before the foil (for $M_1 > M_2$), heavier recoils lose more energy in the foil and may thus have less energy at the detector than the lighter particles. The experimental stopping power data used in the calculations are from Ref. 66, and the other abbreviations refer to various semi-empirical stopping calculations.

The extension to multi-element, multi-layer structures or to layers deeper in the sample is in principle straightforward, but the notation in the calculations become complex as the number of elements and layers increases.

3.2 Quantitative analysis

The formalism for conversion of the spectrum signal heights and areas to amount of material in the sample is similar for the two methods discussed. In the following, stopping cross-sections and scattering cross-sections relevant for either scattered ions or recoiled target particles for $[\varepsilon]$ and $\sigma(E)$ should be applied.

According to equation (5), the total number of detected particles may be written as $\sigma\Omega Q N t$. Consider a thin layer of thickness $t = \tau_i$, corresponding to an

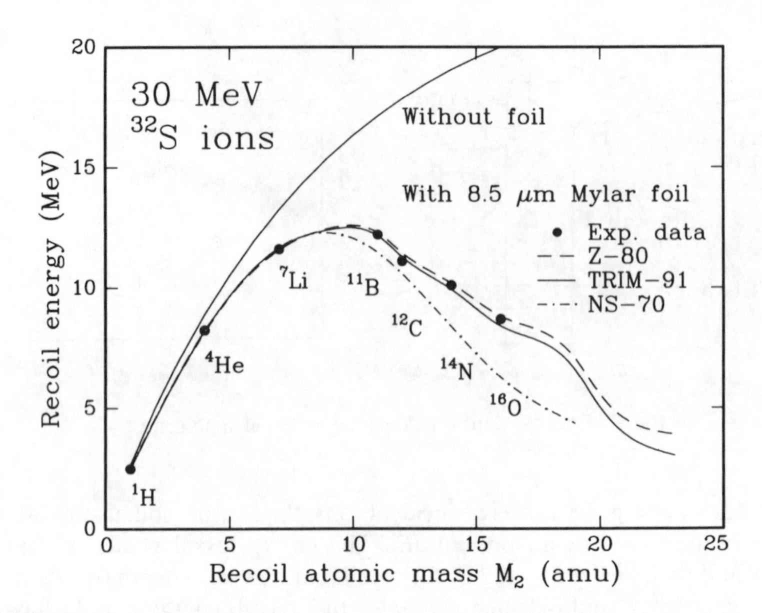

Fig. 8. The effect of stopper foil on the recoil energies detected. 30 MeV ^{32}S ions, 8.5 μm Mylar stopper foil and a 30° detection angle are assumed in the calculations.

energy width δE of channel i in the spectrum. The number of particles detected in that channel is thus $H_i = \sigma(E_i)\Omega Q N\tau_i$, where the scattering cross-section $\sigma(E_i)$ is evaluated at the collision energy E_i. From equations (8) and (10) it then follows that $N\tau_i$ may be written as $N\tau_i = \delta E/[\varepsilon]$. If the angle of incidence departs from normal, a geometrical factor $1/\cos\Theta_1$ must be added. In all, the height of a signal from an elemental sample surface becomes:

$$H_0 = \sigma(E_0)\Omega\, Q\, \delta E/[\varepsilon_0]\cos\Theta_1 \tag{11}$$

where the subscript zero refers to the sample surface. For a two-element sample A_mB_n, for example, the atomic ratio m/n of the elements may be written as:

$$\frac{m}{n} = \frac{H_{A,0}/\sigma_A(E_0)}{H_{B,0}/\sigma_B(E_0)} \cdot \frac{[\varepsilon_0]_A^{AB}}{[\varepsilon_0]_B^{AB}} \tag{12}$$

where the $[\varepsilon]$ factors depend on both the compound of the mixture and the scatterer.

When there is no signal overlapping or the signals are otherwise well separable, the total number of counts A contained in a signal from a thin film is often measurable. A is the sum of H_i over all channels i of the signal, i.e., the signal area $A = \Sigma_i H_i$. Assuming the surface energy approximation, and thus a constant scattering cross-section $\sigma = \sigma(E_0)$, it follows almost immediately that:

$$A_0 = \sigma(E_0)\, \Omega\, Q\, Nt / \cos \Theta_1 \tag{13}$$

providing the number of atoms per unit area Nt without knowledge of the energy loss of the ions in the film.

The fundamental problem of depth profiling is to find a relationship between the amount of material and the sample depth. Equations (8)–(13) relate signal height H and the amount of material in terms of areal density Nt (atoms/cm^2). Once the atomic density (atoms/cm^3) or the mass and specific gravity (g/cm^3) are known, the depth scale in units of length may be calculated according to the previous section.

4 EXPERIMENTAL TECHNIQUES

4.1 Ion backscattering

The requirements for the experimental apparatus in an accelerator facility to perform ion backscattering experiments are relatively uncomplicated. A simple scattering chamber equipped with a particle detector and amplifier/analyzer electronics are the basic needs. A schematic diagram of a standard target–detector set-up was shown in Fig. 1. A scattering angle near 180°, e.g. 170°, is a typical choice. Unconventional experimental set-ups are used, e.g., in glancing angle experiments, in experiments with absorber foils or for non-vacuum systems. Channeling experiments, elaborate electrostatic, magnetic and time-of-flight detector systems and various microbeam applications are examples of more complex experimental techniques.

4.1.1 Ion channeling

The ion backscattering/channeling measurements provide information on the crystalline structure of the sample. When the beam is aligned along axial or planar directions of a crystalline sample, a reduction in the number of backscattered ions is detected as compared with random orientation.

Figure 9 shows three-dimensional plots of backscattering spectra for 2 MeV ^{4}He ions incident on a Si/SiGe superlattice, with the beam aligned close to a crystal axis and planes. In this recent study [67], the implantation damage in a strained-layer superlattice was investigated. The spectra (yield–energy plane) are displayed as a function of alignment angle. The reduction in the backscattering yield is obvious near zero angles in parts (a) and (b). The Ge signals around 1500 keV and the Si signals around 1100 keV from the three 275 Å SiGe layers, alternated by 275 Å Si layers, are separated.

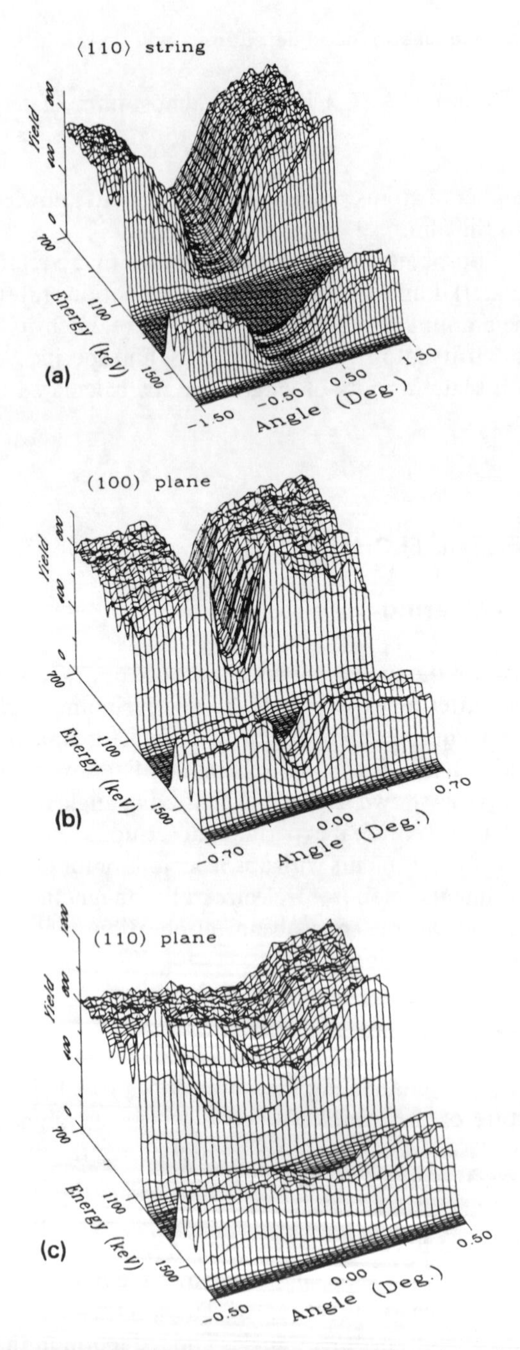

Fig. 9. The dependence of backscatter yield on sample orientation and energy for a Si/SiGe superlattice. In (a) the beam alignment is close to the ⟨110⟩ axis, in (b) and (c) close to the (100) and (110) planes, respectively. The three peaks around 1.6 MeV correspond to the Ge in the SiGe layers, and the Si edge is at 1.1 MeV. (From Ref. 67.)

The reduction is due to the steering effect of the atom rows, forcing the ions away from the scattering centers to the 'channels' in a crystal. Because of its insensitivity to atoms in a perfect crystal, the channeled beam acts as a sensitive probe to detect crystal imperfection.

Figure 10 presents another example of how ion channeling is applied to investigate the recovery of ion implantation damage by laser annealing treatment [68]. A random alignment ^{4}He ion backscattering spectrum of an Sb-implanted Si sample is shown in (a). Spectra labeled as (b), (c), and (d) illustrate the <001>-aligned, the laser annealed and an unimplanted <001>-aligned reference spectrum from a virgin Si crystal, respectively. The distributions around channel 270 show the depth profiles of the implanted Sb atoms for random, untreated aligned and laser-treated aligned orientations. The similarity of spectra (c) and (d) indicates an almost complete recrystallization.

4.1.2 Special arrangements

A major factor limiting the resolution in ion backscattering is the detector used. A standard choice is a silicon surface barrier (SSB) detector, which has an energy

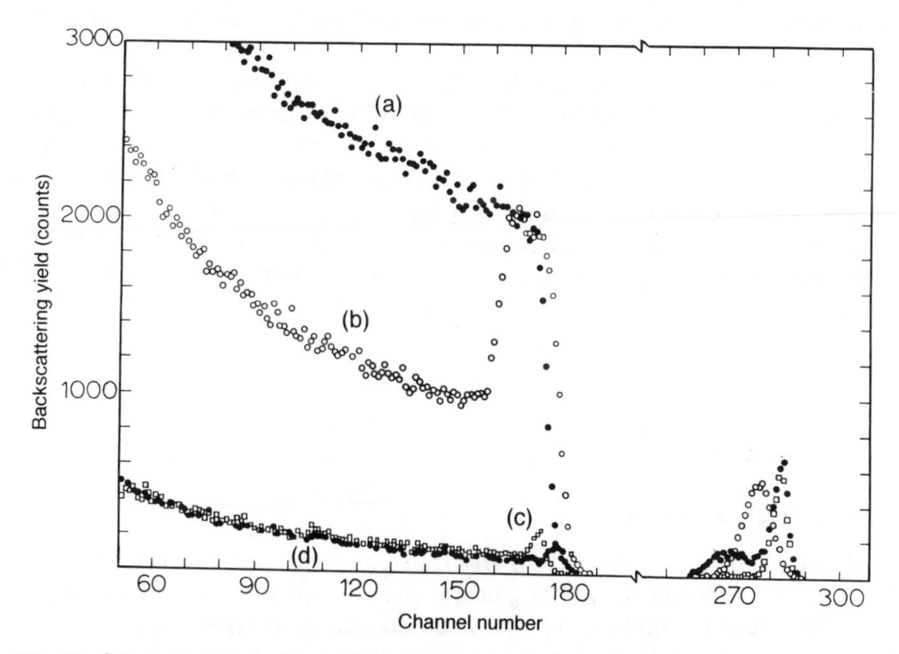

Fig. 10. ^{4}He ion backscattering/channeling spectra of a Si sample: (a) random, (b) ⟨001⟩-aligned incidence on 5×10^{15} Sb ions/cm^2 implanted sample, and (c) after 40 MW/cm^2 pulsed CO_2 laser annealing. The ⟨001⟩-aligned spectrum of the unimplanted crystal is shown in (d). (From Ref. 68.)

resolution of the order of 15 keV for ^{4}He ions but much worse for high energy heavier ions. These detectors are also short lived under heavy ion irradiation. In an attempt to overcome these disadvantages, various other methods of charged particle detection have been investigated. Recent studies include the use of, e.g., electrostatic energy analyzers (EEA) [69,70], magnetic spectrometers [71], time-of-flight (TOF) systems [72, 73], gas detectors [74], PIN photodiodes [75], passivated implanted planar silicon detectors (PIPS) [76], bolometers [77], and calorimeters [78].

The geometrical effect of path length enhancement may be employed by tilting the target. The target is tilted with respect to the beam and detector directions to make either the incident angle Θ_1 or exit angle Θ_2 (or both) close to 90°. This *glancing angle geometry* has been investigated in detail in Ref. 79. An almost tenfold improvement in depth resolution can be accomplished: from 10–30 nm for low MeV ^{4}He ions in standard conditions to better than 5 nm. As the beam hits and exits the target surface at small angles, a smooth surface is crucial.

Absorber foils [80, 81] and electronic data processing [82, 83] have been utilized to eliminate unwanted background signals from low energy ions scattered from the substrate material. In this way, a significantly improved sensitivity for the detection of heavier surface constituents or implanted atoms in deep range profiling is achieved. A high pulse rate and pile-up handling system is described in [84].

For target materials which do not tolerate a vaccum, external experimental set-ups have been developed [85, 86]. As with the more commonly used particle induced X-ray emission external beam systems, the beam is taken outside the vacuum through a thin exit foil or a differentially pumped small aperture. The surface barrier detector is placed either in vacuum or in a He gas atmosphere.

During the last decade, there has been an increasing number of experiments in which microbeams have been applied to materials analysis and to ion backscattering studies in particular. Recent reviews of microbeam facilities and applications are found in e.g. Refs. 87 and 88.

4.2 Elastic recoil

4.2.1 Conventional set-ups

Two modifications of experimental arrangements exist, used for thick and thin film samples, sometimes called the *glancing* and the *transmission* modes, respectively. The glancing mode set-up was depicted schematically in Figs. 1 and 7. Owing to the large inclination of the beam direction to the target surface, similar conditions for target smoothness as with glancing angle backscattering are essential. For the same reason the kinematic broadening of the beam energy width critically depends on beam collimation and defining apertures. It may be shown [89] that a curved slit instead of a rectangular one in front of the detector improves energy and depth resolutions by $\approx 50\%$.

The transmission geometry is illustrated schematically in Fig. 11 [90]. The transmission mode clearly requires a film thin enough to allow the recoiled particles to transmit through the film. No stopper foil is needed, however, if the range of the incident heavier ions does not extend through the sample film. In this particular experiment, depth resolutions of 38 nm and 29 nm at the surface and at a depth of 6 μm were achieved and depths as great as to 6.2 μm were analyzed.

4.2.2 Special arrangements

The stopper foil impairs energy resolution due to straggling. The foil may be omitted by using electromagnetic [91, 92] fields as particle filters in front of the detector. In the $E \times B$ *technique* (Refs. 93 and 94 and references therein) the scattered projectile ions and any unwanted charge states of recoiling particles are eliminated by using crossed magnetic and electric fields. Figure 12 shows a schematic diagram of the technique.

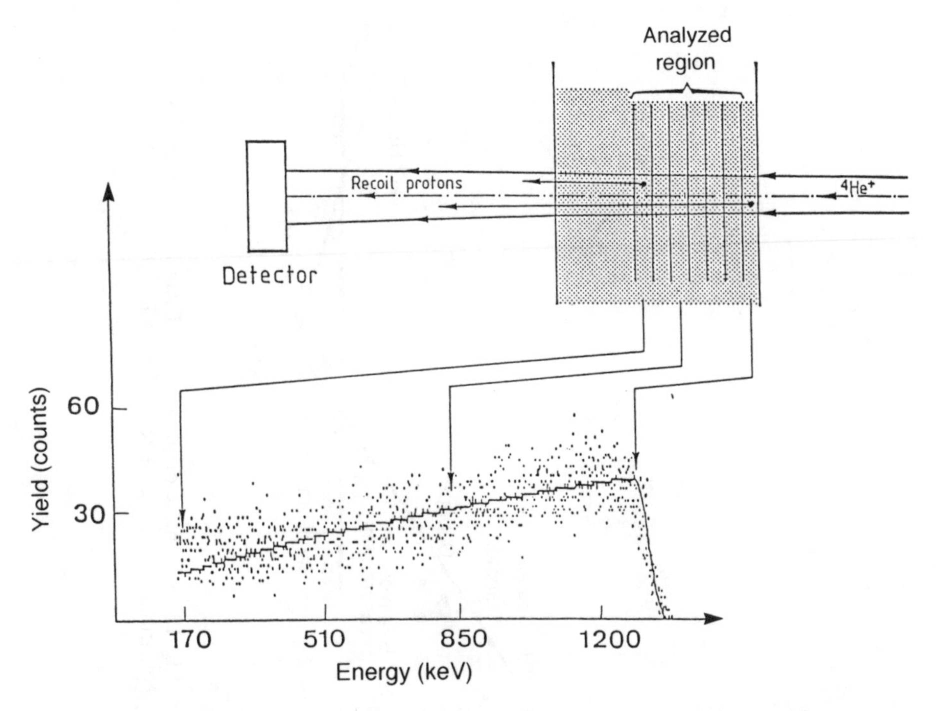

Fig. 11. Schematic illustration of the experimental arrangement and the resulting proton recoil spectrum in transmission mode recoil spectrometry. A 3 MeV ^4He microbeam incident on 25 μm Kapton was used. (From Ref. 90.)

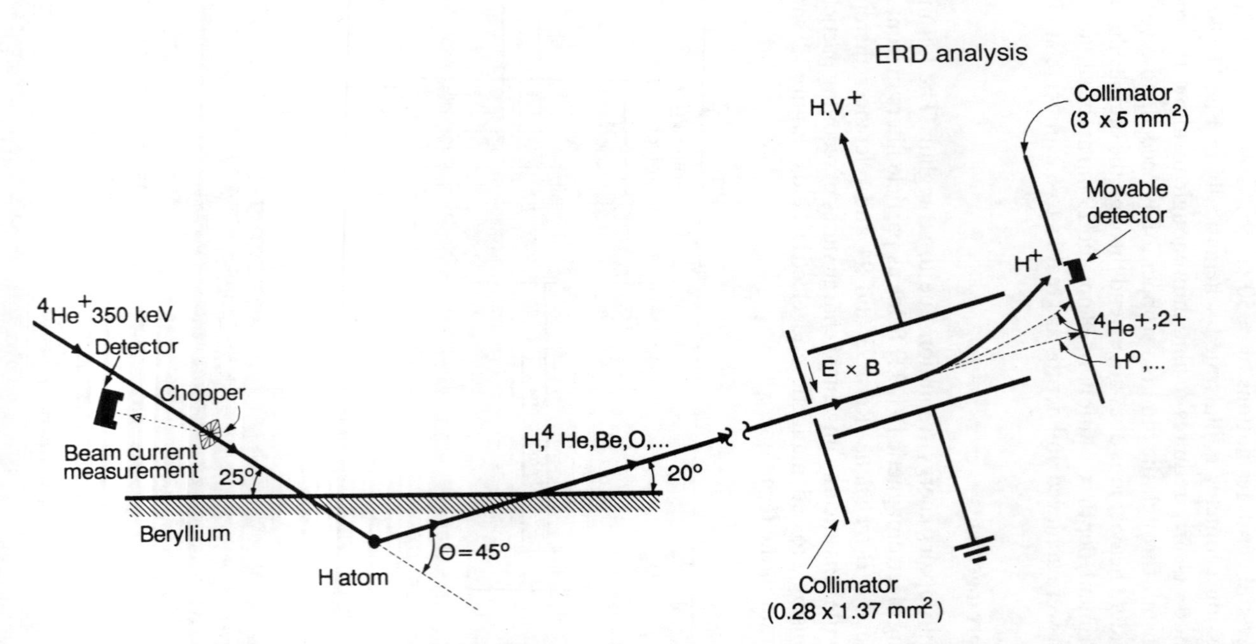

Fig. 12. Schematic diagram of the ERD $E \times B$ method showing the interaction geometry, the $E \times B$ filter, collimator positioning and the detector. (From Ref. 93.)

A simple way of obviating the need for a foil is the *kinematic separation method.* As there is a maximum scattering angle of heavy projectiles incident on lighter target atoms, one may place the detector at a sufficiently large detection angle [95].

In a standard measurement, the beam current density is limited because of the degradation of the sample by radiation damage. Suppressing the beam intensity and increasing the soild angle of the detector would solve the problem but also impair the depth resolution. To overcome the difficutly, a method called *scattering recoil coincidence spectrometry* (SRCS) or *elastic recoil coincidence spectroscopy* (ERCS) has been proposed [96]. By this method, detector solid angles of the order of steradians, instead of the usual millisteradians, could be employed. Many variations in geometry are possible. Figure 13 [97] illustrates one experimental set-up and a spectrum of a polycarbonate sample. The scattered ions and recoiled target particles were measured in time coincidence in transmission geometry. Depth information was derived from the measured energies of both scattered and recoiled particles. The method has been further developed, e.g., in Refs. 12 and 98–102. The idea of coincidence recoil measurements was applied earlier in the 1970s [6, 103] to reduce noise and aid in impurity identification.

The special characteristic of the elastic recoil detection method, of providing information on both energy and mass (or atomic number) of ions of the very material being analyzed, is exploited in *mass and energy dispersive recoil spectrometry.* The method has been elaborated by Swedish groups [104, 105] (and earlier references therein) and, e.g., in Refs. 19, 22 and 106–112. A typical experimental arrangement consists of a TOF or an energy loss ($\Delta E-E$) detector telescope system, where both the energy and the mass (atomic number) of the recoils can be measured.

Very recently a method of pulse shape discrimination of scattered incident ions and recoiled target particles was proposed [113]. Experiments were performed both in transmission and glancing angle geometry with 12.1 MeV alpha-particles incident on thin carbon films. It was demonstrated that pulses from longer range alpha-particles could be effectively suppressed using pulse shape discrimination and detector depletion layer tuning.

Other TOF systems have been described recently (e.g. [73, 114, 115]), ($\Delta E-E$) telescope systems in [116, 117] and magnetic spectrographs in [118, 119].

5 DATA ANALYSIS

The formalism outlined in Section 2.3 is rarely applied as such. More often nowadays, the basic theoretical principles are incorporated into computer programs in which the stopping powers according to semi-empirical models and theoretical scattering cross-sections are calculated and experimental data are

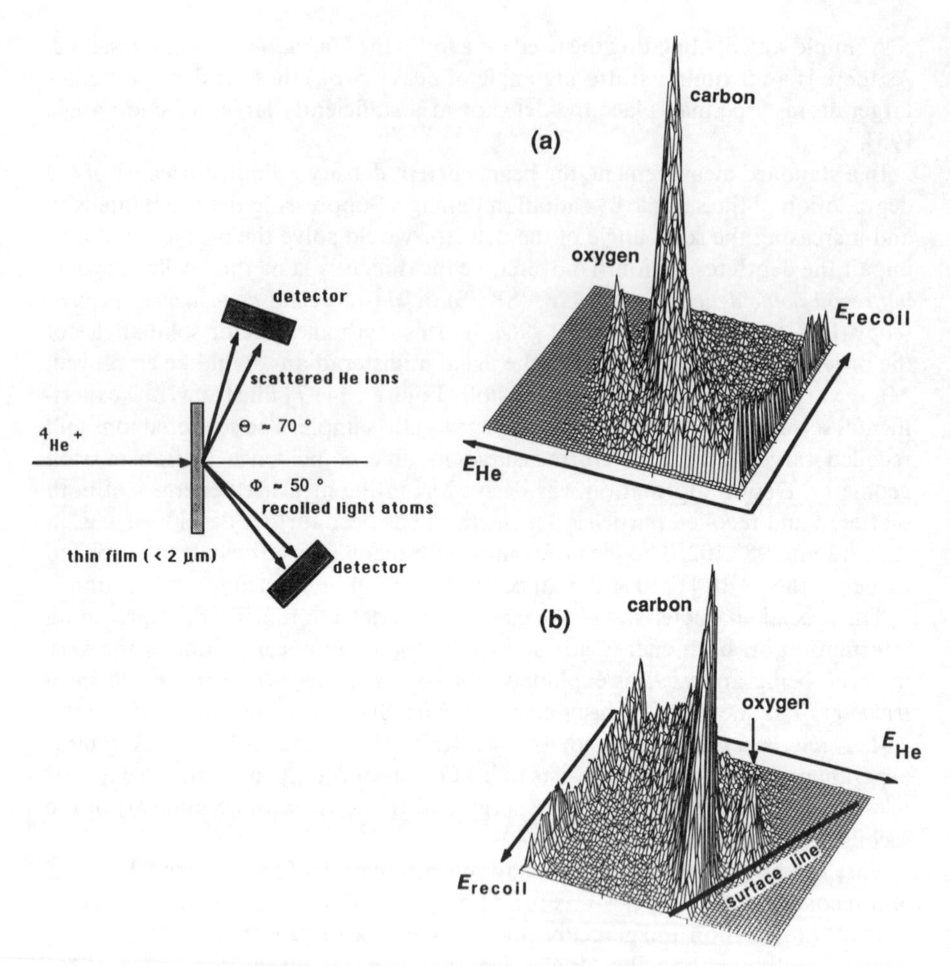

Fig. 13. Experimental setup for the scattering–recoil coincidence measurements. The incident ^4He beam is transmitted through the thin film sample. Scattered ions and recoiled light elements are detected. (a) The spetrum of a polycarbonate sample; (b) the same spectrum rotated by 180°. The coincidence counts are plotted as a function of the energy of the scattered ^4He and the recoiled elements. (From Ref. 97.)

implemented. Various corrections and fine details are readily taken into account. The often necessary iteration steps may be executed automatically.

The *simulation method* involves synthesizing a theoretical spectrum based on an assumed sample composition. Comparisons between the experimental and the synthesized spectra make it possible to modify the initial assumption about the sample composition and, through successive iterations, finally to reach a theroretical simulation of the experimental spectrum.

The simulation method for treating backscattering spectra was introduced by Ziegler, Lever and Hirvonen in 1976 [120]. Early work on computer data analysis in the 1980s includes, e.g., Refs. 121–125. Because of the similarity of the methods, only slight modifications are needed in a backscattering data analysis program to enable it to handle elastic recoil data as well. Programs capable of both kinds of calculations are described e.g. in Refs. 126 and 127. Other recent publications on data analysis include, e.g., [128, 129] (iterative computer analysis of ERD experiments), [130] (algorithms for the computation of screened cross-sections), [131] (pocket calculator RBS computations), [132] (RBS simulation by a 'retrograde' method), [133] (high speed data processing in micro RBS), [134] (analysis of RBS spectra from samples implanted to high dose), [135] (RBS element distributions from considerable depths), [136–138] (computer automated ion beam channeling systems) and [139] (Monte Carlo calculations of plural and multiple scattering in RBS).

An example of recent application of a computer program to backscattering data treatment is depicted in Fig. 14 [140]. The elastic scattering $^{12}C(^{4}He, {}^{4}He)^{12}C$ resonance at $E_{He} = 4.265$ MeV was utilized to detect minor carbon impurities at the surface and at the interface of a SrS–glass sample. In the treatment of spectra in the non-Rutherford energy region a computer program is imperative, especially when dealing with rapidly changing resonant cross-sections.

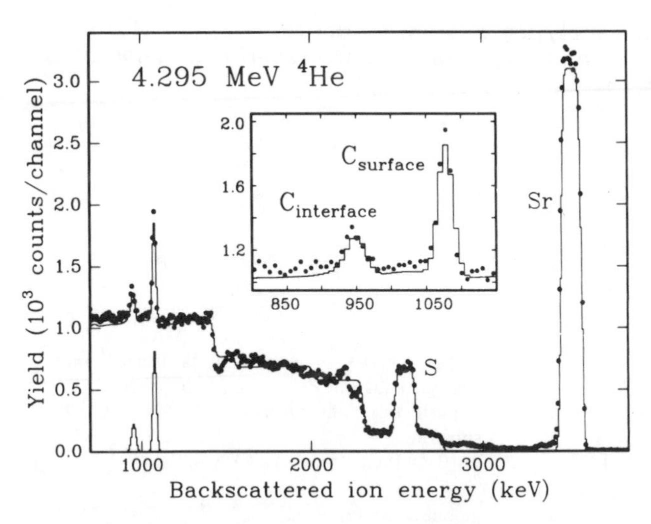

Fig. 14. An ion backscattering spectrum for 4.265 MeV ^{4}He ions incident on a sample consisting of a 1440×10^{15} atoms/cm^2 SrS film on a glass substrate. A carbon impurity (see insert) at the surface (30×10^{15} C/cm^2) and at the interface (18×10^{15} C/cm^2) are observed. (From Ref. 140.)

6 ANALYTICAL CHARACTERISTICS

The analytical qualities of both of the methods described vary considerably with the experimental conditions adopted. The ion type, energy and experimental set-up may be chosen to meet the demands of a particular analysis problem. Recent reviews of the analytical characteristics may be found in Refs. 5, 24 and 25 for the backscattering and elastic recoil methods. Typical parameters for ion backscattering and elastic recoil methods for low MeV energy ^{4}He ions are summarized in Table 1.

6.1 Ion backscattering

Increasing the ion energy beyond the Rutherford energy region typically enhances the sensitivity for light element detection in ^{1}H and ^{4}He ion backscattering. Figure 15 [141] illustrates the non-Rutherford enhancement effect of oxygen detection sensitivity for ^{1}H and ^{4}He ions. An almost 20-fold improvement for 8.8 MeV ^{4}He and a 5-fold improvement for 2.5 MeV ^{1}H ions relative to 2 MeV ^{4}He ions is observed. The ^{12}C(^{4}He, ^{4}He)^{12}C resonance at 4.265 MeV produces an

TABLE 1.
Typical main analytical characteristics for 1–2 MeV ^{4}He ion backscattering and elastic recoil analysis in standard experimental conditions

RBS and ERD	
beam current	1–50 nA
beam diameter	≈ 1 mm (in microbeam applications ≈ μm)
lateral resolution	≈ 1–4 mm (≈ μm)
analysis depth	≈ 1–3 μm
overall accuracy	2–5%
quantitative	
non-destructive	
(radiation damage induced)	
10–20 min/sample run	
RBS	
mass resolution	≥ 2 amu at target mass 50 amu, ≥ 20 at 200 amu
detection sensitivity	strongly substrate dependent, ≥ 10^{-2} atomic layers for heavy impurities on light substrates, tens of at% for light element in heavy substrates
depth resolution	≥ 10 nm (in glancing angle geometry ≥ 1 nm)
no standards needed	
ERD	
specificity	unique selection of ^{1}H, ^{2}H
detection sensitivity	5×10^{13} atoms/cm^{2}, 0.5–0.01 at%, little variation with substrate atomic number
depth resolution	varies with depth, 30–60 nm
standards	usually needed, e.g., H-implanted Si, polymer film

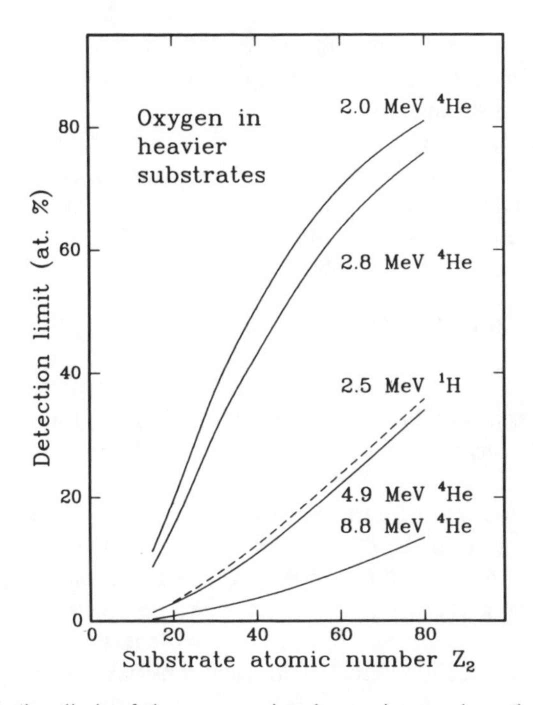

Fig. 15. The detection limit of the oxygen signal superimposed on the yield of heavier elements from a thick two-component mixture or compound. (From Ref. 141.)

increment greater than 100-fold over the conventional Rutherford cross-section for carbon [142].

Increasing the ion mass and energy improves the mass resolution for heavy targets. For example, 10 MeV ^{4}He and 20 MeV ^{12}C ions will produce a mass resolution of about 4 amu at target mass 200, a value 5 times better than the value of 20 amu for 2 MeV ^{4}He. Higher incident ion energies also allow greater depths to be analyzed. With 8 MeV ^{4}He ions, for example, depths of the order of 5–50 μm, depending on the target atomic number, may be reached; 3 MeV ^{1}H ions facilitate backscattering analysis to depths of about twice of that reached by 8 MeV ^{4}He ions.

Low energies and heavy ions have been used to increase the detection sensitivity for heavy elements in light substrates, recently in Refs. 80 and 143. The former study utilized a 400 keV ^{12}C ion beam and an absorber foil (40 μg/cm^2 carbon) to stop the ions scattered from the Si substrate. Sensitivities for heavy impurities of the order of 10^{11} atoms/cm^2 were achieved. At this low energy, and because of the absorber, good mass resolution should not be expected.

The use of more elaborate detection systems, TOF techniques, electrostatic or magnetic energy analyzers makes the resolutions even better. For example, TOF system resolution of 30, 75 and 100 keV for ^{35}Cl ions at 2, 10 and 13 MeV have been reported [144], values that could be compared with the 170, 280 and 300 keV reported for SSB detectors [145]. Mass resolutions for 25 MeV ^{35}Cl ions ranging from 1 amu at target mass 50 amu to 2.5 at 200 amu with an energy resolution of 100 keV were quoted in Ref. 144. Lower energy (< 500 keV) ^{4}He, ^{7}Li and ^{12}C ion backscattering characteristics of a TOF system have been investigated in Refs. 146, 147 and 148.

6.2 Elastic recoil detection

The dependence of energy resolution as a function of sample depth on various experimental factors in standard elastic recoil spectrometry is illustrated in Fig. 16 [63]. The calculations were carried out for 3.0 MeV ^{4}He ions incident on an Al(H) sample, assuming a 1 mm × 3 mm^2 slit at a distance of 100 mm and a 24 μm Mylar stopper foil. The dominating terms close to the surface are the geometrical effects and energy straggling in the stopper foil. Deeper in the sample, the multiple scattering and energy straggling contributions become more important.

The use of heavier incident ions and higher energies makes possible the analysis of light target elements heavier than hydrogen. Sensitivities similar to those of hydrogen detection with ^{4}He incident ions have been reported. As with backscattering, higher ^{4}He energies reach deeper depths. For 25–33 MeV ^{4}He ions, for example, depths of the order of 100–300 μm may be examined [25]. Using more elaborate coincidence or electromagnetic particle filtering techniques, time-of-flight or energy loss telescopes [see Section 4.2.2], resolutions and sensitivities are improved and particle identification and data interpretation are facilitated. With coincidence techniques and 25–33 MeV ^{4}He ion energies, for example, detection sensitivities as low as 10^{-5} at % may be attained and the lithium isotopes can be accurately assayed [25]. By combining TOF and coincidence techniques, depth resolutions better than 10 nm at the sample surface can be expected when probing light elements in heavy matrices by ≈ 1 MeV/amu ^{12}C incident ions [102]. A depth resolution of 1 nm was obtained when investigating carbon/boron multi-layers with a magnetic spectrograph using a 120 MeV ^{197}Au beam [118].

7 APPLICATIONS TO ELEMENTAL ANALYSIS

7.1 Ion backscattering

A typical application of standard RBS is shown in Fig. 17 [140], where a 10 min 2.0 MeV ^{4}He ion spectrum representing the diffusion of a small cobalt surface impurity into silicon is plotted. The inset shows the Co depth distribution

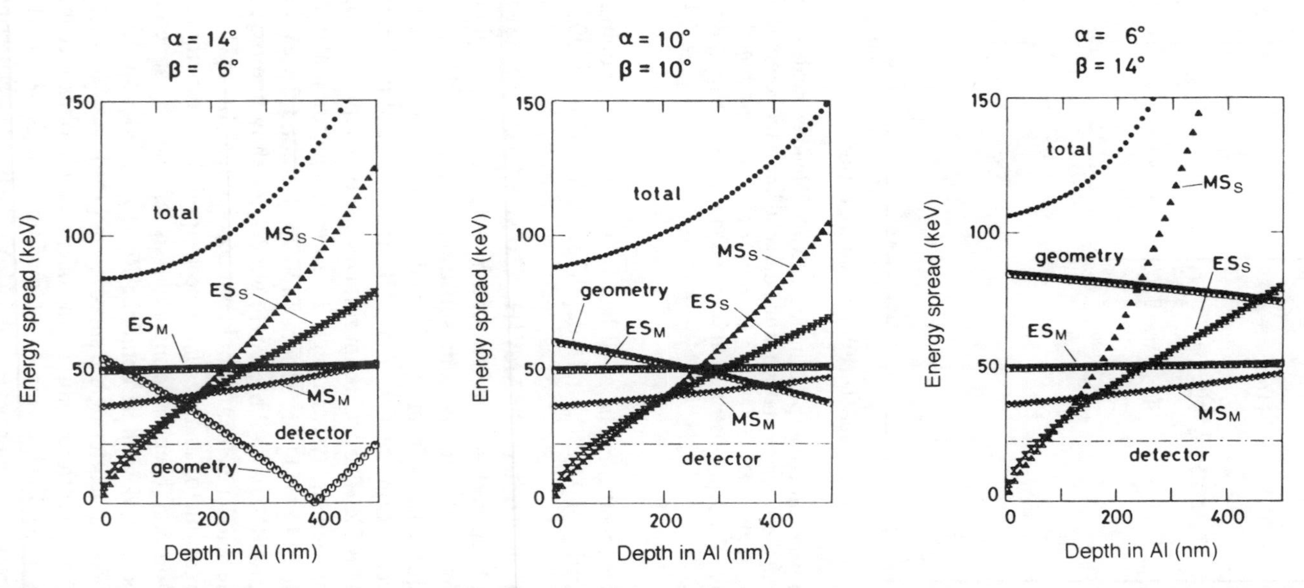

Fig. 16. Dependence of energy spreads on depth in Al for $\alpha = 6°$, $10°$ and $14°$ with $\alpha + \beta = 20°$ (α and β correspond to the angles $\alpha = 90° - \Theta_1$ and $\beta = 90° - \Theta_2$ of Fig. 2). Calculated curves are for detector resolution, geometrical resolution, multiple scattering (MS) and energy straggling (ES). Subscripts S and M indicate the sample and the Mylar stopper foil, respectively. (From Ref. 63.)

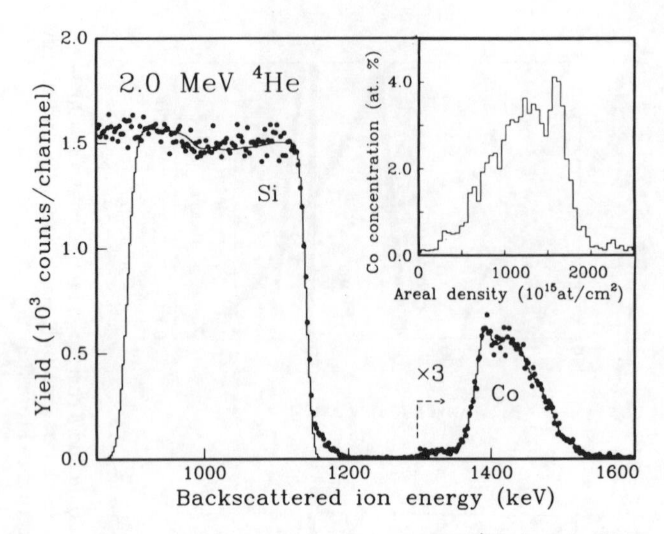

Fig. 17. Rutherford backscattering spectrum for 2.0 MeV ^{4}He ions incident on a Si(Co) sample. The dots form the experimental spectrum, the solid line is the computer simulated spectrum. Scattering angle is $\Theta = 170°$, with $\Theta_1 = \theta_2 = 5°$. The migration of cobalt into silicon is obvious. The Co depth distribution is shown in the insert. (From Ref. 140.)

extracted from the spectrum. The solid line is the result of a computer fit. A sensitivity of the order of 0.1 at % and a depth resolution of about 30 nm can be estimated.

Figure 18 [5] illustrates 2.5 MeV ^{1}H, 2.0 MeV ^{4}He, 8.0 MeV ^{7}Li and 22.0 MeV ^{12}C ion backscattering spectra of a Y_1 Ba_2 Cu_3 O_{7-x} high-T_C superconductor film on an aluminum oxide substrate. From the ^{1}H spectrum the oxygen may be quantified. From the ^{7}Li and ^{12}C ion spectra, the film areal density and concentrations of Cu, Y and Ba may be readily derived to an accuracy better than 5%. The ^{12}C beam resolves the two copper isotopes ^{63}Cu and ^{65}Cu. For oxygen analysis in high-T_C superconductor materials, either higher energy ^{4}He [149–151] or ^{1}H ions [150, 152] around 2 MeV have been employed. Standard RBS with 2 MeV ^{4}He would not show the oxygen signal adequately and the heavier elements would not be separated. In [153], ^{16}O beams for high-T_C superconductor analysis have been applied. The methods and the recent literature on ion backscattering analysis of high-T_C superconductor materials have been reviewed in detail in Refs. 141, 154 and 155. Oxygen measurements by elastic scattering are surveyed in [156].

Current proton backscattering studies include, e.g., microbeam element mapping applications [157, 158], a survey of medium energy ($E_p = 0.1-1$ MeV) backscattering [159], high energy ($E_p = 4.6$ MeV) elastic backscattering experiments for oxygen stoichiometry determination of reactively evaporated ZrO

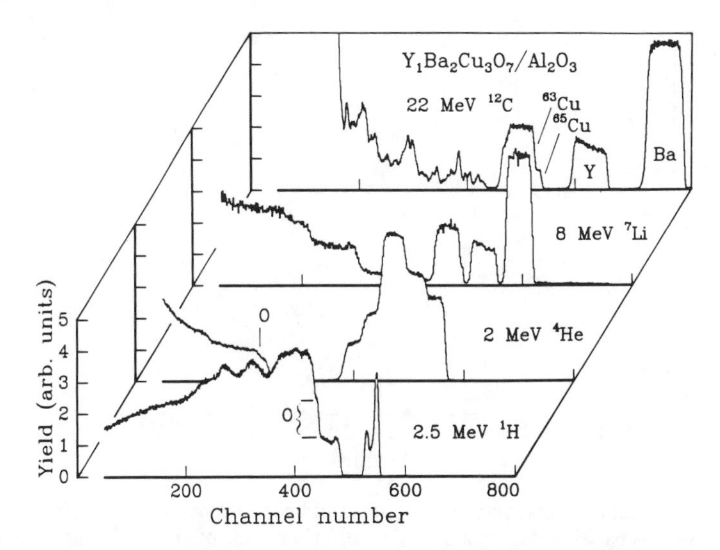

Fig. 18. Comparison of backscattering spectra for 2.5 MeV ^{1}H, 2.0 MeV ^{4}He, 8.0 MeV ^{7}Li and 22.0 MeV ^{12}C ions incident on a 3.20×10^{18} atoms/cm^2 $Y_1Ba_2Cu_3O_{7-x}$ high-T_C sample on an Al_2O_3 substrate. The scattering angle is $\Theta = 170°$, with $\Theta_1 = \Theta_2 = 5°$. (From Ref. 5.)

films [160], investigations of resonance effects in thin film backscattering [47], trace element analysis in biological samples [161] and characterization of silicon nitride and oxide layers on silicon [162, 163].

A recent publication [164] demonstrates the capabilities of high energy ^{4}He ion backscattering for nitrogen profiling in nitride films and nitrogen-implanted samples. Figure 19 presents a spectrum of a TiN film deposited on a Zr substrate for 6 MeV incident ^{4}He ions. The nitrogen concentration was analyzed as about 23% by mass and the film thickness as 1.3 μm. From an implanted sample, a sensitivity limit of the order of 10^{16} N/cm^2 was estimated, more than an order of magnitude lower than N concentrations used in many tribological applications. The detection limit for standard 2.0 MeV ^{4}He backscattering [5] is only ≈ 60 at % of N in bulk Zr.

Other current publications on ^{4}He ion backscattering applications are, e.g., Si, Al and Fe thin film analysis ($E_{He} = 3.576$ MeV) [165], study of ^{64}Ni implantation in ^{57}Fe thin films ($E_{He} = 5.7$ MeV) [166], investigation of oxidation of yttrium-implanted Co, Mo and Ta ($E_{He} = 8.8$ MeV) [167] and alloyed Ni/Au/Te/Ni/GaAs ohmic contacts [168]. The mutual diffusion of polystyrene and poly(xylenyl ether) [169], the recrystallization of neodymium-implanted GaP, GaAs and AlGaAs [170] and the precipitation and segregation of Sb at Si–SiO_2 interface during thermal oxidation [171] have been studied.

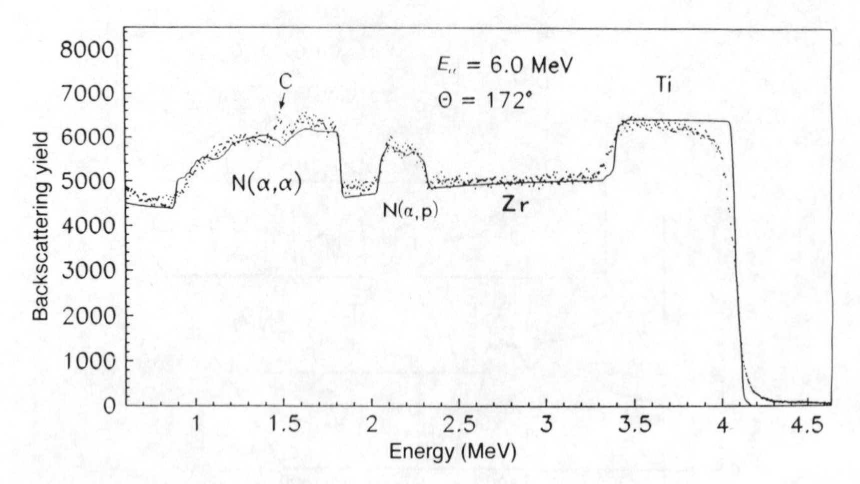

Fig. 19. Experimental (dots) and calculated (solid line) backscattering spectra of a 1.3 μm TiN/Zr device. A 6.0 MeV ^{4}He beam and a scattering angle of Θ = 172° were used. (From Ref. 164).

The 3.045 MeV ^{16}O(^{4}He, ^{4}He)^{16}O resonance has been utilized for oxygen depth distribution measurements in SiO and MoO films [172] and in depth profiling of low-*Z* elements in high-*Z* substrates [173]. Channeling/backscattering studies of SOI (silicon-on-insulator) structures by SIMOX (separation by implantation of oxygen) using 3.5 MeV ^{4}He [174], precipitation studies of antimony delta-doping layers in Si [175] (E_{He} = 2.0 MeV), analysis of III–V semiconductor hetero-epitaxial layered structures [176] and high energy Si-implanted InP [177] and structural characterization of CdS epilayers [178], for example, have been published.

Figure 20 [144] illustrates the improved depth resolution of heavy ion backscattering accomplished by using a TOF detector system. A spectrum for 10 MeV ^{12}C ions incident on a ten-layer $In_{0.18}$ $Ga_{0.82}$ As/GaAs superlattice on GaAs(Si) is shown. The InGaAs and GaAs layer thicknesses are 8 and 20 nm, respectively. The first In signals are well separated. The gradual decrease of depth resolution with increasing depth is due to beam energy straggling, which is shown by the Gaussian curves (representing the energy resolutions) overlaid at the low and high ends of the In signals. The detector flight path was 40 cm. With 25 MeV ^{12}C ions, a flight path of 83 cm and a sample tilt of 75°, a resolution of 2 nm was obtained.

Other current heavy ion applications are, e.g., indium depth profile measurements in InGaAs with 15 MeV ^{35}Cl ions [179], annealing behavior of near-surface stoichiometry profiles of GaAs wafers using 33 MeV ^{16}O ions [180] and microanalysis by 0.5–4 MeV heavy ion microprobe [181]. The application of

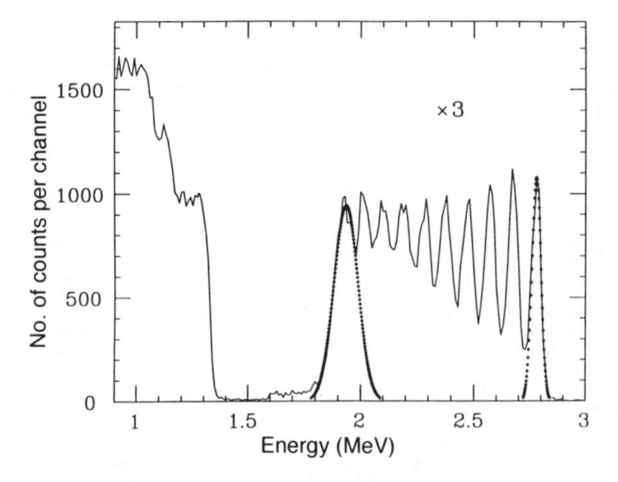

Fig. 20. 10 MeV ^{35}Cl beam backscattered from a ten-layer $In_{0.18}$ $Ga_{0.82}$ As/GaAs superlattice on GaAs(Si). (From Ref. 144.)

high resolution medium energy ion backscattering to SiO optical coatings exposed to a space environment was presented in [182]. The surface sensitivity of low energy nitrogen ions [183] and depth and mass resolutions with heavy ions [184] have been studied.

7.2 Elastic recoil detection

A typical case of elastic recoil spectrometry for hydrogen profiling is shown in Fig. 21 [185]. The spectrum, taken by a 2.55 MeV ^{4}He beam, is from a ZnS:Mn/$Al_xTi_yO_z$/Si structure of a thin film electroluminescent device. Kapton and Mylar standards and a detection angle of 30° were used. The probed depth was 0.6 µm. The conspicuous surface contamination peak around channel 350, typical of many samples, is caused by the adsorption of water and hydrocarbons at the surface. The hydrogen concentration was shown to fall well below 1 at % in ZnS:Mn and to increase to ≈ 3.5 at % at the interface.

Heavy ion elastic recoil studies of wear behavior of steel after nitrogen implantation are exemplified in Ref. 186. Figure 22 shows a recoil spectrum for a 30 MeV ^{35}Cl beam incident on a sandwich structure produced by deposition of SiN(H) and a-Si:H layers of 20 nm thickness. The ^{12}C, ^{14}N and ^{1}H recoils are shown separately. Even better particle separation, with a depth resolution of 5–10 nm, was attained by omitting the 9.2 µm Mylar stopper foil and using a Bragg (ΔE–E) ionization detector chamber [116]. Experiments with very heavy ions and high energies (^{129}Xe and ^{127}I beams of 260–420 MeV) are described in Ref. 112.

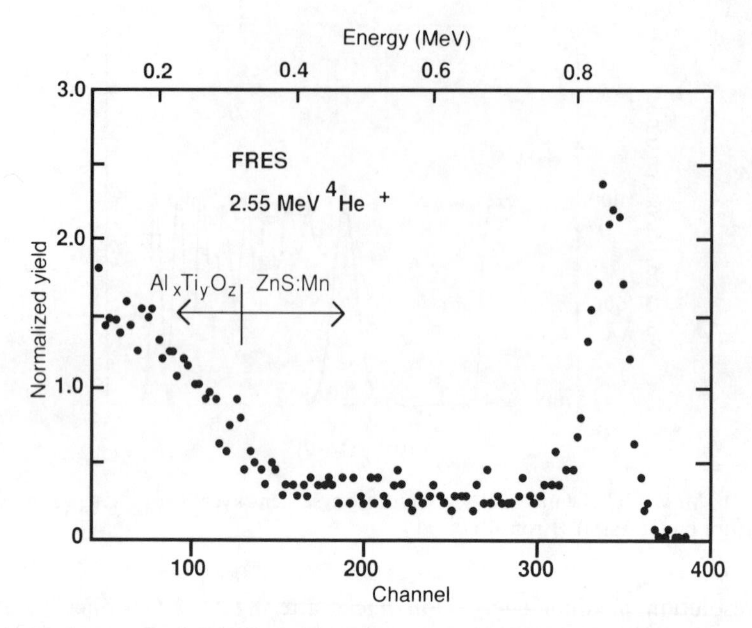

Fig. 21. Hydrogen elastic recoil spectrum from ZnS:Mn/$Al_xTi_yO_z$ sample. The angle of incidence of the 2.55 MeV ^4He ion beam is $\Theta_1 = 75°$. (From Ref. 185.)

The results obtainable with energy loss telescopes using mass and energy dispersive ERD are described e.g. in Refs. 104, 117 and 187. Figure 23 shows a plot of a mass–recoil energy matrix for an 1300 °C annealed N- and O-implanted Si sample, obtained with 48 MeV ^{81}Br ions incident at 67.5° to the sample surface normal. A time-of-flight energy (TOF–E) detector telescope was used. The Si, O and N signals are well resolved in mass. The dip in the Si signal corresponds to the buried oxide layer and the two peaks for a mass of 14 amu indicate nitrogen segregation to both of the buried Si/SiO_2 interfaces. In another study [188], the nitrogen detection sensitivity at 0.1–0.2 μm in the overlying single crystal Si was evaluated as 2×10^{19} at/cm^3, corresponding to 0.04 at %. A depth resolution of ≈ 80 nm at a depth of 550 nm for an Al depth profile in a $Ga_{(1-x)}Al_xAs$ quantum well heterostructure and a mass resolution better than 1 amu for recoils lighter than ^{27}Al were reported [189].

Other recent applications of the elastic recoil detection method include, e.g., studies of the surface chemistry of leached minerals [190], carbon and oxygen incorporation in reactively sputtered Cr–Si–Al films [191], the analysis of light adsorbate positions on crystal surfaces [192], and hydrogen and other light elements in gases [193]. Oil additive-associated compositional changes in sliding metal surfaces are investigated in [105]. Hydrogen depth profiles in bevelled proton-implanted semiconductors [194], in PVD silicon nitride films [195] and

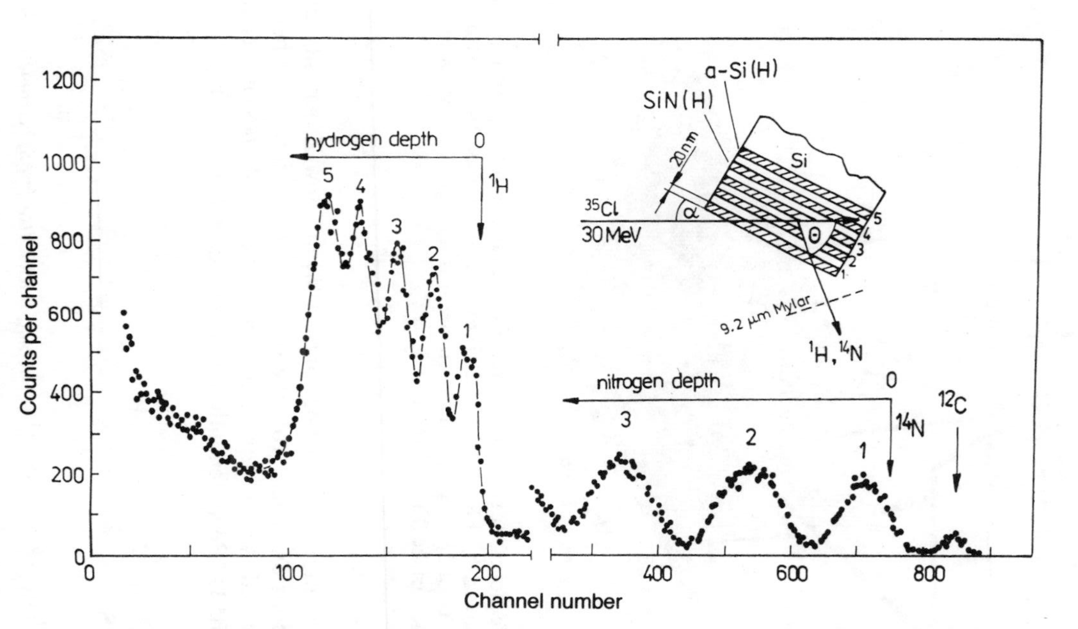

Fig. 22. An elastic recoil detection spectrum of a sandwich structure of five 20 nm SiN(H) and a-Si:H layers. The detection angle is $\Theta = 30^{\circ}$, with $\alpha = 90^{\circ} - \Theta_1 = 20^{\circ}$. (From Ref. 186.)

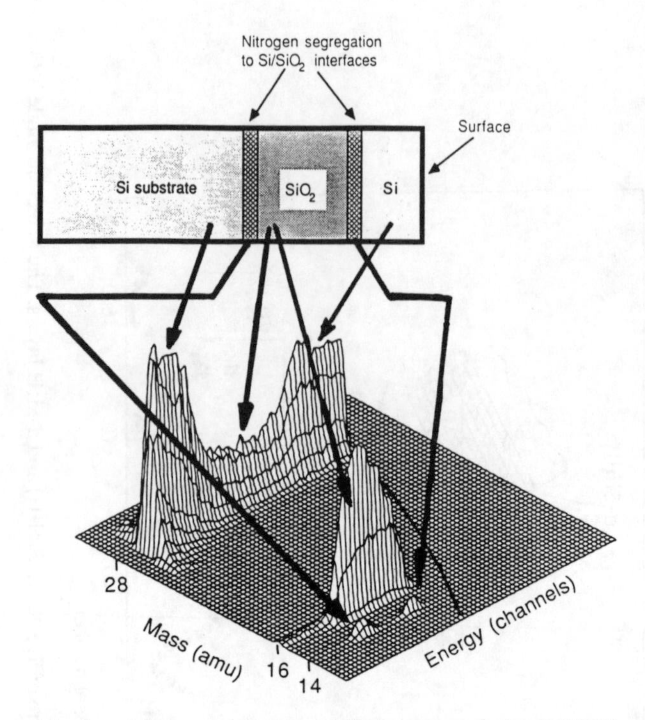

Fig. 23. Isometric plot of the sorted mass–recoil energy matrix for Si (100) implanted with 1×10^{17} N^+ followed by 1.8×10^{18} O^+ 200 keV ions/cm^2, respectively, at 600 °C and annealed at 1300 °C for 5 h. The curved line on the mass–energy plane indicates the energy of recoils from the surface (From Ref. 187.)

in diamond-like amorphous carbon films [196] and low-Z deposits and their depth distributions on probes from the vessel walls of fusion experiments [197] have been measured. A monochromatic neutron beam (14 MeV) is used in [198] to investigate the D–Pd system interphase boundary behavior in palladium.

REFERENCES

1. W.-K. Chu, J. W. Mayer and M. A. Nicolet, *Backscattering Spectrometry*, Academic Press, New York (1978).
2. S. U. Campisano, W. K. Chu, G. Foti, J. W. Mayer, M.-A. Nicolet, and E. Rimini, in *Ion Beam Handbook for Material Analysis*, edited by J. W. Mayer and E. Rimini, Academic Press, New York (1977) pp. 22–65.
3. G. Deconninck, *Introduction to Radioanalytical Physics*, Elsevier, New York (1978) pp. 76–111.

4. L. C. Feldman and J. W. Mayer, *Fundamentals of Surface and Thin Film Analysis*, North-Holland, New York (1986), pp. 13–68.
5. E. Rauhala, in *Elemental Analysis by Particle Accelerators*, edited by Z. B. Alfassi and M. Peisach, CRC Press, Boca Raton (1992) pp. 179–241.
6. B. L. Cohen, C. L. Fink and J. B. Degnan, *J. Appl. Phys.* **43**, 19 (1972).
7. P. Paduschek and P. Eichinger, *Appl. Phys. Lett.* **36**, 62 (1980).
8. M. F. C. Willemsen, A. M. L. Theunissen and A. E. T. Kuiper, *Nucl. Instrum. Methods Phys. Res., Sect. B* **B15**, 492 (1986).
9. F. A. Smidt and A. G. Pieper, *J. Nucl. Mater.* **51**, 361 (1974).
10. L. Shabason and W. J. Choyke, *Nucl. Instrum. Methods* **138**, 533 (1976).
11. J. A. Moore, I. V. Mitchell, M. J. Hollis, J. A. Davies and L. M. Howe, *J. Appl. Phys.* **46**, 52 (1975).
12. S. S. Klein, *Nucl. Instrum. Methods Phys. Res., Sect. B* **B15**, 464 (1986).
13. J. L'Ecuyer, C. Brassard, C. Cardinal, C. Chabbal, L. Deschenes and J. P. Labrie, B. Terreault, J. G. Martel and R. St.-Jaques, *J. Appl. Phys.* **47**, 381 (1976).
14. B. L. Doyle and P. S. Peercy, *Appl. Phys. Lett.* **34**, 811 (1979).
15. J. L'Ecuyer, C. Brassard and C. Cardinal, *Nucl. Instrum. Methods* **149**, 271 (1978).
16. B. Terreault, J. G. Martel, R. G. St.-Jaques and J. L' Ecuyer, *J. Vac. Sci. Technol.* **14**, 192 (1977).
17. G. G. Ross, B. Terreault, G. Gobeil, G. Abel, C. Boucher and G. Veilleus, *J. Nucl. Mater.* **128/129**, 730 (1984).
18. L. S. Wielunski, R. E. Benenson and W. A. Lanford, *Nucl. Instrum. Methods* **218**, 120 (1983).
19. A. Groleau, S. C. Gujrathi and J. P. Martin, *Nucl. Instrum. Methods* **218**, 11 (1983).
20. C. Nölscher, K. Brenner, R. Knauf and W. Schmidt, *Nucl. Instrum. Methods* **218**, 116 (1983).
21. C. Moreau, E. J. Knystautas, R. S. Timsit and R. Groleau, *Nucl. Instrum. and Methods* **218**, 111 (1983).
22. J. P. Thomas, M. Fallavier, D. Ramdane, N. Chevarier and A. Chevarier, *Nucl. Instrum. Methods* **218**, 125 (1983).
23. A. Turos and O. Meyer, *Nucl. Instrum. Methods Phys. Res., Sect. B* **B4**, 92 (1984).
24. J. E. E. Baglin, in *Encyclopaedia of Materials Analysis*, edited by C. R. Brundle, C. A. Evans, Jr. and S. W. Wilson, Butterworth-Heinemann, Stoneham (1992).
25. J. B. A. England, in *Elemental Analysis by Particle Accelerators*, edited by Z. B. Alfassi and M. Peisach, CRC Press, Boca Raton, (1992) pp. 243–278.
26. W. A. Lanford, *Nucl. Instrum. Methods Phys. Res., Sect. B* **B66**, 65 (1992).
27. J. E. E Baglin and J. S. Williams, in *Ion Beams for Materials Analysis*, edited by J. R. Bird and J. S. Williams, Academic Press Australia, Marrickwille, (1989), Chapter 3.
28. F. Pászti, *Nucl. Instrum. Methods Phys. Res., Sect. B* **B66**, 83 (1992).
29. F. H. P. M. Habraken, *Nucl. Instrum. Methods Phys. Res., Sect. B* **B68**, 181 (1992).
30. H. A. Bethe, *Ann. Phys.* **5**, 325 (1930).
31. U. Fano, *Ann. Rev. Nucl. Sci.* **13**, 1 (1963)
32. J. F. Ziegler, J. P. Biersack and U. Littmark, *The Stopping and Range of Ions in Solids*, Vol. 1, Pergamon Press, New York (1985).
33. W. Brandt and M. Kitagawa, *Phys. Rev. B* **B25**, 5631 (1982).
34. H. Paul, D. Semrad and A. Seilinger, *Nucl. Instrum. Methods Phys. Res., Sect. B* **B61**, 261 (1991).
35. E. Rauhala, in *Proc. High Energy and Heavy Ion Beams in Materials Analysis*, edited by J. R. Tesmer, C. J. Maggiore, M. Nastasi, J. C. Barbour, and J. M. Mayer, Materials Research Society, Pittsburgh (1990) pp. 61–71.
36. W. H. Bragg and R. Kleeman, *Philos. Mag.* **10**, 318 (1905).

37. D. I. Thwaites, *Nucl. Instrum. Methods Phys. Res., Sect. B* **B69**, 53 (1992).
38. B. P. Nigam, M. K. Sundaresan and T.-Y. Wu, *Phys. Rev.* **115**, 491 (1959).
39. J. B. Marion and B. A. Zimmermann, *Nucl. Instrum. Methods* **51**, 93 (1967).
40. J. P. Biersack, E. Steinbauer and P. Bauer, *Nucl. Instrum. Methods Phys. Res., Sect. B* **B61**, 77 (1991).
41. R. D. Evans, The Atomic Nucleus, McGraw-Hill, New York (1955).
42. M. Luomajärvi, E. Rauhala, and M. Hautala, *Nucl. Instrum. Methods Phys. Res., Sect. B* **B9**, 255 (1985).
43. E. Rauhala, *Nucl. Instrum. Methods Phys. Res., Sect. B* **B12**, 447 (1986).
44. E. Rauhala and M. Luomajärvi, *Nucl. Instrum. Methods Phys. Res., Sect. B* **B33**, 628 (1988).
45. E. Rauhala, *Nucl. Instrum. Methods Phys. Res., Sect. B* **B40/41**, 790 (1989).
46. J. M. Knox and J. F. Harmon, *Nucl. Instrum. Methods Phys. Res., Sect. B* **B44**, 40 (1989).
47. J. Liu, T. Xie and H. J. Fishbeck, *Nucl. Instrum. Methods Phys. Res., Sect. B* **B56/57**, 744 (1991).
48. G. H. Yang, D. Z. Zhu, H. J. Xu and H. C. Pan, *Nucl. Instrum. Methods Phys. Res., Sect. B* **B61**, 175 (1991).
49. J. A. Leavitt, L. C. McIntyre, Jr., P. Stoss, J. G. Oder, M. D. Ashbaugh, B. Dezfouly-Arjomandy, Z.-M. Yang and Z. Lin, *Nucl. Instrum. Methods Phys. Res., Sect. B* **B40/41**, 776 (1989).
50. J. A. Leavitt, L. C. McIntyre, Jr., M. D. Ashbaugh, J. G. Oder, Z. Lin and B. Dezfouly-Arjomandy, *Nucl. Instrum. Methods Phys. Res., Sect. B* **B44**, 260 (1990).
51. K. E. Hubbard, J. E. Martin, R. E. Muenchausen, J. R. Tesmer and M. Nastasi, in *Proc. High Energy and Heavy Ion Beams in Materials Analysis*, edited by J. R. Tesmer, C. J. Maggiore, M. Nastasi, J. C. Barbour, and J. M. Mayer, Materials Research Society, Pittsburgh (1990) pp. 165–173.
52. H. S. Cheng, X. Y. Lee and F. J. Yang, *Nucl. Instrum. Methods Phys. Res., Sect. B* **B56/57**, 749 (1991).
53. J. M. Knox, R. J. McLeod, D. R. Mayo and X. Qian, *Nucl. Instrum. Methods Phys. Res., Sect. B* **B45**, 26 (1990).
54. J. M. Knox, *Nucl. Instrum. Methods Phys. Res., Sect. B* **B66**, 31 (1992).
55. J. A. Leavitt and L. C. McIntyre Jr., in *Proc. High Energy and Heavy Ion Beams in Materials Analysis*, edited by J. R. Tesmer, C. J. Maggiore, M. Nastasi, J. C. Barbour, and J. M. Mayer, Materials Research Society, Pittsburgh (1990) pp. 129–138.
56. J. A. Leavitt and L. C. McIntyre, Jr., *Nucl. Instrum. Methods Phys. Res., Sect. B* **B56/57**, 734 (1991).
57. L. C. McIntyre Jr., J. A. Leavitt, M. D. Ashbaugh, Z. Lin and J. O. Stoner J. R., *Nucl. Instrum. Methods Phys. Res., Sect. B* **B64**, 457 (1992).
58. J. E. E. Baglin, A. J. Kellock, M. A. Crockett and A. H. Shih, *Nucl. Instrum. Methods Phys. Res., Sect. B* **B64**, 469 (1992).
59. J. Tirira, P. Trocellier, J. P. Frontier and P. Trouslard, *Nucl. Instrum. Methods Phys. Res., Sect. B* **B45**, 203 (1990).
60. H. Wang and G. Q. Zhou, *Nucl. Instrum. Methods Phys. Res., Sect. B* **B34**, 145 (1988).
61. Y. Wang, J. Chen and F. Huang, *Nucl. Instrum. Methods Phys. Res., Sect. B* **B17**, 11 (1986).
62. D. C. Ingram, A. W. McCormick, P. P. Pronko, J. D. Carlson and J. A. Woollam, *Nucl. Instrum. Methods Phys. Res., Sect. B* **B6**, 430 (1985).
63. S. Nagata, Y. Yamaguchi, Y. Fujino, Y. Hori, N. Sugiyama and K. Kamada, *Nucl. Instrum. Methods Phys. Res., Sect. B* **B6**, 533 (1985).

64. E. Szilagyi, F. Pászti, A. Manuaba, C. Hajdu and E. Kotai, *Nucl. Instrum. Methods Phys. Res., Sect. B* **B43**, 502 (1989).
65. B. L. Doyle and D. K. Brice, *Nucl. Instrum. Methods Phys. Res., Sect. B* **B35**, 301 (1988).
66. J. Räisänen and E. Rauhala, *Phys. Rev. B* **B41**, 3951 (1990).
67. M. Vos, C. Wu and I. V. Mitchell, T. E. Jackman, J.-M. Baribeau and J. P. McCaffrey, *Nucl. Instrum. Methods Phys. Res., Sect. B* **B66**, 361 (1992).
68. K. Naukkarinen, T. Tuomi and M. Blomberg, M. Luomajärvi and E. Rauhala, *J. Appl. Phys.* **53**, 5634 (1982).
69. D. J. O'Connor and T. Chunyu, *Nucl. Instrum. Methods Phys. Res., Sect. B* **B36**, 178 (1989).
70. Th. Enders, M. Rilli and H. D. Carstanjen, *Nucl. Instrum. Methods Phys. Res., Sect. B* **B64**, 817 (1992).
71. D. O. Boerma, F. Labohm, and J. A. Reinders, *Nucl. Instrum. Methods Phys. Res., Sect. B* **B50**, 291 (1990).
72. M. Döbeli, P. C. Haubert, R. P. Livi, S. J. Spicklemire, D. L. Weathers and T. A. Tombrello, *Nucl. Instrum. Methods Phys. Res., Sect. B* **B56/57**, 764 (1991).
73. E. Arai, H. Funaki, M. Katayama and K. Shimizu, *Nucl. Instrum. Methods Phys. Res., Sect. B* **B68**, 202 (1992).
74. M. H. Mendenhall, *Nucl. Instrum. Methods Phys. Res., Sect. B* **B10/11**, 596 (1985).
75. S. C. Gujrathi, D. W. Hetherington, P. F. Hinrichsen and M. Bentourkia, *Nucl. Instrum. Methods Phys. Res., Sect. B* **B45**, 260 (1990).
76. W. De Coster, B. Brijs, W. Vandervorst and P. Burger, *Nucl. Instrum. Methods Phys. Res., Sect. B* **B64**, 287 (1992).
77. S. Woiwod, R. M. Mueller, B. Stritzker and S. H. Moseley, *Nucl. Instrum. Methods Phys. Res., Sect. B* **B50**, 91 (1990).
78. H. H. Andersen, *Nucl. Instrum. Methods Phys. Res., Sect. B* **B12**, 437 (1985).
79. J. S. Williams and W. Möller, *Nucl. Instrum. Methods* **157**, 213 (1978).
80. J. A. Knapp and B. L. Doyle, *Nucl. Instrum. Methods Phys. Res., Sect. B* **B45**, 143 (1990).
81. H. Abel, B. Agius and J. Gyulai, *Nucl. Instrum. Methods Phys. Res., Sect. B* **B21**, 77 (1987).
82. Th. Maisch, V. Schüle, R. Günzler, P. Oberschachtsiek, M. Weiser, S. Jans, K. Izsak, and S. Kalbitzer, *Nucl. Instrum. Methods Phys. Res., Sect. B* **B50**, 1 (1990).
83. P. Oberschachtsiek, V. Schüle, R. Günzler, M. Weiser, and S. Kalbitzer, *Nucl. Instrum. Methods Phys. Res., Sect. B* **B45**, 20 (1990).
84. G. Amsel, E. Girard, G. Vizkelethy, G. Battistig, Y. Girard and E. Szilágyi, *Nucl. Instrum. Methods Phys. Res., Sect. B* **B64**, 811 (1992).
85. B. L. Doyle, *Nucl. Instrum. Methods* **218**, 20 (1983).
86. E. Rauhala, J. Räisänen and M. Luomajärvi, *Nucl. Instrum. Methods Phys. Res., Sect. B* **B6**, 543 (1985).
87. J. S. C. McKee and G. R. Smith, *Adv. Electron. Electron Phys.* **73**, 93 (1989).
88. F. Watt and G. W. Grime (Eds.), *Principles and Applications of High-Energy Ion Microbeams*, Adam Hilger, Bristol (1987).
89. D. K. Brice and B. L. Doyle, *Nucl. Instrum. Methods Phys. Res., Sect. B* **B45**, 265 (1990).
90. J. Tirira, P. Trocellier and J. P. Frontier, *Nucl. Instrum. Methods Phys. Res., Sect. B* **B45**, 147 (1990).
91. G. G. Ross, B. Terreault, G. Gobeil, G. Abel, C. Boucher and G. Veilleux, *J. Nucl. Mater.* **128/129**, 730 (1984).

92. C. R. Gossett, *Nucl. Instrum. Methods Phys. Res., Sect. B* **B15**, 481 (1986).
93. G. G. Ross and L. Leblanc, *Nucl. Instrum. Methods Phys. Res., Sect. B* **B62**, 484 (1992).
94. G. G. Ross, L. Leblanc, B. Terreault, J. F. Pageau and P. A. Gollier, *Nucl. Instrum. Methods Phys. Res., Sect. B* **B66**, 17 (1992).
95. G. G. Ross and B. Terreault, *J. Appl. Phys.* **51**, 1259 (1980).
96. W.-K. Chu and D. T. Wu, *Nucl. Instrum. Methods Phys. Res., Sect. B* **B35**, 518 (1988).
97. H. C. Hofsäss, N. R. Parikh, M. L. Swanson and W.-K. Chu, *Nucl. Instrum. Methods Phys. Res., Sect. B* **B45**, 151 (1990).
98. S. S. Klein, P. H. A. Mutsaers and B. E. Fischer, *Nucl. Instrum. Methods Phys. Res., Sect. B* **B50**, 150 (1990).
99. J. S. Forster, J. R. Leslie and T. Laursen, *Nucl. Instrum. Methods Phys. Res., Sect. B* **B45**, 176 (1990).
100. B. Gebauer, D. Fink, P. Goppelt, M. Wilpert and T. Wilpert, in *Proc. High Energy and Heavy Ion Beams in Materials Analysis*, edited by J. R. Tesmer, C. J. Maggiore, M. Nastasi, J. C. Barbour and J. M. Mayer, Materials Research Society, Pittsburgh (1990), pp. 257–267.
101. H. C. Hofsäss, N. R. Parikh, M. L. Swanson and W.-K Chu, *Nucl. Instrum. Methods Phys. Res., Sect. B* **B58**, 49 (1991).
102. H. A. Rijken, S. S. Klein and M. J. A. de Voigt, *Nucl. Instrum. Methods Phys. Res., Sect. B* **B64**, 395 (1992).
103. J. A. Moore, I. V. Mitchell, M. J. Bollis, J. A. Davies and L. M. Howe, *J. Appl. Phys.* **46**, 52 (1975).
104. H. J. Whitlow, in *Proc. High Energy and Heavy Ion Beams in Materials Analysis*, Edited by J. R. Tesmer, C. J. Maggiore, M. Nastasi, J. C. Barbour and J. M. Mayer, Materials Research Society, Pittsburgh (1990), pp. 73–85 and 243–256.
105. H. J. Whitlow, E. Johansson, P. A. Ingemarsson and S. Hogmark, *Nucl. Instrum. Methods Phys. Res., Sect. B* **B63**, 445 (1992).
106. L. E. Seiberling, *Nucl. Instrum. Methods Phys. Res., Sect. B* **B24/25**, 526 (1987).
107. A. M. Behrooz, R. L. Hedrick, L. E. Seiberling and R. W. Zurmühle, *Nucl. Instrum. Methods Phys. Res., Sect. B* **B28**, 108 (1987).
108. J. P. Stoquert, G. Guillaume, M. Hage-Ali, J. J. Grob, C. Canter and P. Siffert, *Nucl. Instrum. Methods Phys. Res., Sect. B* **B44**, 184 (1989).
109. S. C. Gujrathi , P. Aubry, L. Lemay and J.-P. Martin, *Can. J. Phys.* **65**, 950 (1987).
110. A. Houdayer, P. F. Hinrichsen, S. C. Gujrathi, J. P. Martin, S. Monaro, L. Lessard, K. Oxorn, C. Janicki, J. Brebner, A. Belhadfa and Y. Yelon, *Nucl. Instrum. Methods Phys. Res., Sect. B* **B24/25**, 643 (1987).
111. J. P. Thomas, M. Fallavier and A. Ziani, *Nucl. Instrum. Methods Phys. Res., Sect. B* **B15**, 443 (1986).
112. P. Goppelt, B. Gebaucer, D. Fink, M. Wilpert, Th. Wilpert and W. Bohne, *Nucl. Instrum. Methods Phys. Res., Sect. B* **B68**, 235 (1992).
113. S. S. Klein and H. A. Rijken, *Nucl. Instrum. Methods Phys. Res., Sect. B* **B66**, 393 (1992).
114. E. Arai, H. Funaki, M. Katayama, Y. Oguri and K. Shimizu, *Nucl. Instrum. Methods Phys. Res., Sect. B* **B64**, 296 (1992).
115. S. C. Gujrathi and S. Bultena, *Nucl. Instrum. Methods Phys. Res., Sect. B* **B64**, 789 (1992).
116. E. Hentschel, R. Kotte, H. G. Ortlepp, F. Stary and D. Wohlfarth, *Nucl. Instrum. Methods Phys. Res., Sect. B* **B66**, 242 (1992).
117. W. M. A. Bik, C. T. A. M. de Laat and F. H. P. M. Habraken, *Nucl. Instrum. Methods Phys. Res., Sect. B* **B64**, 832 (1992).

118. G. Dollinger, T. Faestermann and P. Maier-Komor, *Nucl. Instrum. Methods Phys. Res., Sect. B* **B64**, 422 (1992).
119. G. J. Sandker, P. Eeken, W. M. A. Bik, K. van der Borg and F. H. P. M. Habraken, *Nucl. Instrum. Methods Phys. Res., Sect. B* **B64**, 292 (1992).
120. J. F. Ziegler, R. F. Lever, and J. K. Hirvonen, in *Ion Beam Surface Layer Analysis*, Vol. 1, Edited by O. Meyer, G. Linker and F. Käppeler, Plenum Press, New York (1976) p. 163.
121. E. Rauhala, *Univ. Helsinki Rep. Ser. Phys.* (HU-P-227) (1983).
122. P. Borgesen, R. Behrisch, and B. M. U. Scherzer, *Appl. Phys. A* **A27**, 183 (1982).
123. P. A. Saunders and J. F. Ziegler, *Nucl. Instrum. Methods* **218**, 67 (1983).
124. E. Rauhala, *J. Appl. Phys.* **56**, 3324 (1984).
125. L. R. Doolittle, *Nucl. Instrum. Methods Phys. Res., Sect. B* **B9**, 334 (1985); **B15**, 27 (1986).
126. RUMP Computer Program Manual, Computer Graphics Service, Lansing, New York (1989).
127. G. Vizkelethy, *Nucl. Instrum. Methods Phys. Res., Sect. B* **B45**, 1 (1990).
128. J. Tirira, J. P. Frontier, P. Trocellier and P. Trouslard, *Nucl. Instrum. Methods Phys. Res., Sect. B* **B54**, 328 (1991).
129. K. Oxorn, S. C. Gujrathi, S. Bultena, L. Cliche and J. Miskin, *Nucl. Instrum. Methods Phys. Res., Sect. B* **B45**, 166 (1990).
130. M. H. Mendenhall and R. A. Weller, *Nucl. Instrum. Methods Phys. Res., Sect. B* **B58**, 11 (1991).
131. A. Climent-Font, *Nucl. Instrum. Methods Phys. Res., Sect. B* **B61**, 541 (1991).
132. Y. Serruys, *Nucl. Instrum. Methods Phys. Res., Sect. B* **B61**, 221 (1991).
133. M. Takai, Y. Katayama, A. Kinomura, T. Lohner and S. Namba, *Nucl. Instrum. Methods Phys. Res., Sect. B* **B64**, 277 (1992).
134. K. Jensen and G. C. Farlow, *Nucl. Instrum. Methods Phys. Res., Sect. B* **B59/60**, 643 (1991).
135. R. D. Edge, *Nucl. Instrum. Methods Phys. Res., Sect. B* **B35**, 309 (1988).
136. P. K. van Staagen, J. R. Williams, P. A. Barnes and R. D. Gilchrist, *Nucl. Instrum. Methods Phys. Res., Sect. B* **B56/57**, 785 (1991).
137. D. J. Diskett, A. J. Avery and R. E. T. Marshall, *Nucl. Instrum. Methods Phys. Res., Sect. B* **B64**, 836 (1992).
138. U. S. Fischer, M. Döbeli, M. Suter, M. Alurralde, M. Victoria, R. Gotthardt, R. Schäublin, H. Rühl, W. Wölfli and S. Schwyn, *Nucl. Instrum. Methods Phys. Res., Sect. B* **B64**, 249 (1992).
139. P. Bauer, E. Steinbauer and J. P. Biersack, *Nucl. Instrum. Methods Phys. Res., Sect. B* **B64**, 711 (1992); see also E. Steinbauer, P. Bauer and J. P. Biersack, *Nucl. Instrum. Methods Phys. Res., Sect. B* **B45**, 171 (1990); J. P. Biersack, E. Steinbauer and P. Bauer, *Nucl. Instrum. Methods Phys. Res., Sect. B* **B61** 77 (1991).
140. J. Saarilahti and E. Rauhala, *Nucl. Instrum. Methods Phys. Res., Sect. B* **B64**, 734 (1992).
141. E. Rauhala, J. Saarilahti and N. Nath, *Nucl. Instrum. Methods. Phys. Res., Sect. B* **B61**, 83 (1991).
142. P. Revesz, J. Li, Gy. Vizkelethy and J. W. Mayer, L. J. Matienzo and F. Emmi, *Nucl. Instrum. Methods Phys. Res., Sect. B* **B58**, 132 (1991).
143. H. Abel, B. Agius and J. Gyulai, *Nucl. Instrum. Methods Phys. Res., Sect. B* **B21**, 77 (1987).
144. M. Döbeli, P. C. Haubert, R. P. Livi, S. J. Spicklemire, D. L. Weathers, and T. A. Tombrello, *Nucl. Instrum. Methods Phys. Res., Sect. B* **B47**, 148 (1990).
145. P. F. Hinrichsen, D. W. Hetherington, S. C. Gujrathi, and L. Cliche, *Nucl. Instrum. Methods Phys. Res., Sect. B* **B45**, 275 (1990).

146. M. H. Mendenhall and R. A. Weller, *Nucl. Instrum. Methods Phys. Res., Sect. B* **B47**, 193 (1990); **B59/60**, 120 (1991).
147. J. W. Rabalais, *J. Vac. Sci. Technol., A* **A9**, 1293 (1991).
148. H. Bu, M. Shi, K. Boyd, and J. W. Rabalais, *J. Chem. Phys.* **95**, 2882 (1991).
149. J. A. Martin, M. Nastasi, J. R. Tesmer, and C. J. Maggiore, *Appl. Phys. Lett.* **52**, 2177 (1988).
150. J. C. Barbour, B. L. Doyle, and S. M. Myers, *Phys. Rev. B* **B38**, 7005 (1988).
151. B. Blanpain, P. Revesz, L. R. Doolittle, K. H. Purser, and J. W. Mayer *Nucl. Instrum. Methods Phys. Res., Sect. B* **B34**, 459 (1988).
152. E. Rauhala, J. Keinonen and R. Järvinen, *Appl. Phys. Lett.* **52**, 1520 (1988).
153. M. Döbeli, U. S. Fischer, M. Suter and W. Wölfli, *Nucl. Instrum. Methods Phys. Res., Sect. B* **B63**, 68 (1992).
154. J. Keinonen, J. Räisänen and E. Rauhala, in *Studies of High Temperature Superconductors,* Vol. 4, edited by A. V. Narlikar, Nova Science, USA/Canada (1989), pp. 239–262.
155. M. Nastasi, in *Structure–Property Relationships in Surface Modified Ceramics*, edited by C. J. McHargue *et al.*, Kluwer Academic (1989) pp. 447–501.
156. K. K. Bourdelle, *Nucl. Instrum. Methods Phys. Res., Sect. B* **B66**, 247 (1992).
157. F. Watt, G. W. Grime, A. J. Brook, G. M. Gadd, C. C. Perry, R. B. Pearce, K. Turnau and S. C. Watkinson, *Nucl. Instrum. Methods Phys. Res., Sect. B* **B54**, 123 (1991).
158. G. W. Grime, F. Watt, A. R. Duval and M. Menu, *Nucl. Instrum. Methods Phys. Res., Sect. B* **B54**, 353 (1991).
159. N. Matsunami, K. Kitoh, J. Kanasaki and N. Itoh, *Nucl. Instrum. Methods Phys. Res., Sect. B* **B45**, 412 (1990).
160. A. Caridi, E. Cereda, S. Fazinic, M. Jaksic, G. M. Braga Marcazzan, M. Scagliotti and V. Valkovic, *Nucl. Instrum. Methods Phys. Res., Sect. B* **B64**, 774 (1992).
161. W. M. Kwiatek, J. Lekki, C. Paluszkiewicz and N. Preikschas, *Nucl. Instrum. Methods Phys. Res., Sect. B* **B64**, 512 (1992).
162. W. Jiang, P. Zhu, A. Donag and S. Yin, *J. Appl. Phys.* **70**, 2610 (1991).
163. V. Havranek, V. Hnatovic, J. Kvitek, I. Obrusnik, V. Rybka and V. Svorcik, *Nucl. Instrum. Methods Phys. Res., Sect. B* **B68**, 223 (1992).
164. H. Artigalas, A. Chevarier, N. Chevarier, M. El Bouanani, E. Gerlic, N. Moncoffre, B. Roux, M. Stern and J. Tousset, *Nucl. Instrum. Methods Phys. Res., Sect. B* **B66**, 237 (1992).
165. F. C. Stedile, R. Hübler, I. J. R. Baumvol and W. H. Schreiner, *Nucl. Instrum. Methods Phys. Res., Sect. B* **B64**, 756 (1992).
166. G. Marest and M. A. El Khakani, *Nucl. Instrum. Methods Phys. Res., Sect. B* **B59/60**, 833 (1991).
167. K. M. Kramer, J. R. Tesmer and M. Nastasi, *Nucl. Instrum. Methods Phys. Res., Sect. B* **B59/60**, 865 (1991).
168. K. Wuyts, J. Watte, R. E. Silverans, H. Bender, M. Van Hove and M. Van Rossum, *J. Vac. Sci. Technol.*, B **B9**, 228 (1991).
169. R. J. Composto and E. J. Kramer, *J. Mater. Sci.* **26**, 2815 (1991).
170. A. Kozanecki and R. Groetzschel, *J. Appl. Phys.* **69**, 1300 (1991).
171. J. S. Williams, M. Petravic, Y. H. Li, J. A. Davies and G. R. Palmer, *Nucl. Instrum. Methods Phys. Res., Sect. B* **B64**, 156 (1992).
172. V. Hnatovicz, H. Macholdt and F.-W. Richter, *Nucl. Instrum. Methods Phys. Res., Sect. B* **B62**, 247 (1991).
173. W. De Coster, B. Brijs, J. Goemans and W. Vandervorst, *Nucl. Instrum. Methods Phys. Res., Sect. B* **B64**, 417 (1992).
174. K. Touhouche, Y. Tao, A. Yelon, G. Kajrys, Y. Trudeau, K. Oxorn, S. Bultena and G. Gagnon, *Nucl. Instrum. Methods Phys. Res., Sect. B* **B59/60**, 676 (1991).

175. L. J. van Ijzerndoorn, C. W. Fredriksz, C. van Opdorp, D. J. Gravesteijn, D. E. W. Vandenhoudt, G. F. A. van de Walle and C. W. T. Bulle-Lieuwma, *Nucl. Instrum. Methods. Phys. Res., Sect. B* **64**, 120 (1992).
176. R. Flagmeyer, *Nucl. Instrum. Methods Phys. Res., Sect. B* **B68**, 190 (1992).
177. S. M. Gulwadi, R. K. Nadella, O. W. Holland and M. V. Rao, *J. Electrochem. Mater.* **20**, 615 (1991).
178. G. Leo, A. V. Drigo, N. Lovergine and A. M. Mancini, *J. Appl. Phys.* **70**, 2041 (1991).
179. M. Döbeli, P. C. Haubert T. A. Tombrello, J.-I. Chyi D. Huang and H. Morkoc, *Nucl. Instrum. Methods Phys. Res., Sect. B* **B52**, 72 (1990).
180. G. Gagnon, A. Houdayer, J. F. Currie and A. Azelmad, *J. Appl. Phys.* **70**, 1036 (1991).
181. Y. Horino, Y. Mokuno, A. Chayahara, M. Kiuchi, K. Fujii, M. Satou and M. Takai, *Nucl. Instrum. Methods Phys. Res., Sect. B* **B64**, 358 (1992).
182. M. H. Mendenhall and R. W. Weller, *Opt. Lett.* **16**, 1466 (1991).
183. W. Lang and J. Weidhass, *Nucl. Instrum. Methods Phys. Res., Sect. B* **B64**, 796 (1992).
184. Q. Yang and D. J. O'Connor, *Nucl. Instrum. Methods Phys. Res., Sect. B* **B67**, 98 (1992).
185. R. Lappalainen, J.-P. Hirvonen, P. J. Pokela and J. Alanen, *Thin Solid Films* **181**, 259 (1989).
186. C. Neelmeijer, R. Grötzschel, E. Hentschel, R. Klabes, A. Kolitsch and E. Richter, *Nucl. Instrum. Methods Phys. Res., Sect. B* **B66**, 242 (1992).
187. H. J. Whitlow, C. S. Petersson, K. J. Reeson and P. L. F Hemment, *Appl. Phys. Lett.* **52**, 1871 (1988).
188. H. J. Whitlow, K. J. Reeson, P. L. J. Hemment and C. S. Petersson, in *Selected Topics in Electronics Materials*, edited by B. R. Appleton, D. K. Biegelsen, W. L. Brawn and J. A. Knapp, Mater. Res. Soc. Extended Abstract EA18, Pittsburgh, PA (1988) pp. 149–152.
189. H. J. Whitlow, G. Possnert and C. S. Petersson, *Nucl. Instrum. Methods Phys. Res., Sect. B* **B27**, 448 (1987).
190. G. W. Arnold, H. R. Westrich and W. H. Casey, *Nucl. Instrum. Methods Phys. Res., Sect. B* **B64**, 542 (1992).
191. C. Neelmeijer, R. Grötzschel, R. Klabes, U. Kreissig and G. Sobe, *Nucl. Instrum. Methods Phys. Res., Sect. B* **B64**, 461 (1992).
192. J. Schultz and E. Taglauer, P. Feulner and D. Menzel, *Nucl. Instrum. Methods Phys. Res., Sect. B* **B64**, 588 (1992).
193. R. S. Sokhi, J. B. A. England and G. M. Field, *Nucl. Instrum. Methods Phys. Res., Sect. B* **B40/41**, 809 (1989).
194. C. Ascheron, D. Lehmann, C. Neelmeijer, A. Schindler and F. Bigh, *Nucl. Instrum. Methods Phys. Res., Sect. B* **B63**, 412 (1992).
195. A. Markwitz, M. Bachmann, H. Baumann, K. Bethge and E. Krimmel, *Nucl. Instrum. Methods Phys. Res., Sect. B* **B68**, 218 (1992).
196. X. Long, X. Peng, F. He, M. Liu and X. Lin, *Nucl. Instrum. Methods Phys. Res., Sect. B* **B68**, 266 (1992).
197. R. Behrisch, R. Grötzschel, E. Hentschel and W. Assmann, *Nucl. Instrum. Methods Phys. Res., Sect. B* **B68**, 245 (1992).
198. B. G. Skorodumov and I. O. Yatsevich, *Nucl. Instrum. Methods Phys. Res., Sect. B* **B64**, 388 (1992).

Chapter 12

NUCLEAR REACTION ANALYSIS OR MORE GENERALLY CHARGED PARTICLE (ACTIVATION) ANALYSIS

Friedel Sellschop
Schonland Research Centre for Nuclear Sciences
and
Department of Physics, University of the Witwatersrand, Johannesburg, South Africa

1 INTRODUCTION—THE THIRD GENERATION OF NUCLEAR ANALYSIS

As the contents list of this book indicates, the term 'nuclear analysis' embraces many techniques. It is interesting to reflect on the evolution of the field. Undoubtedly the first generation was that of *neutron* activation analysis, initially purely instrumental. Within that era the impact of the large-volume semiconductor photon detector, with its greatly superior energy resolution, was seminal in realizing the potential of the technique. We saw, too, the addition of radiochemical methods to enhance the analytical power and applicability. Although *charged particle* (activation methods) (NRA or nuclear reaction analysis) enjoyed early recognition, their more general use evolved later, in what I like to regard as a second generation. This was so for several reasons, including the advent of reliable, easy-to-operate accelerators, the surface barrier semiconductor detector, and the appreciation that the apparent limitation of the extent of sample probed did not disqualify the method; indeed, the surface and near-surface regions were important regions of analysis. This second generation saw the advent of nuclear analysis directed to single crystal materials, notably silicon, and the recognition of the ion channeling effect with features which went beyond simply elemental analysis.

Chemical Analysis by Nuclear Methods Edited by Z. B. Alfassi

The third (and current) generation of nuclear analysis is characterized both by the proven acceptance evident in the widespread *routine* use with automated systems and the many advantages taken of on-line computer control and analysis, and by the progression of more sophisticated procedures which give highly specific information on the host compound not accessible to other analytical methods.

A reasonable ideal for any analytical method is to be multi-elemental, in the sense of simultaneous rather than sequential determination of elements. Only two accelerator-based methods can to a substantial extent claim this advantage. These are (Rutherford) backscattering and the ion-induced X-ray methods, but even these are not simply so, with effective low (Z, A) cut-offs, significant dispersion and sensitivity features and matrix effects being important.

By analogy with neutron activation analysis (NAA), it might be expected that charged particle *activation* analysis would be multi-elemental in character. However, nuclear reaction and structural characteristics vary widely throughout the Periodic Table—there is no simple correlation of nuclear excitation response with A or Z. This is of course true also for neutron activation analysis, but is exacerbated in the case of NRA: in distinction to NAA, many more reaction channels are generally open (Fig. 1), each with a characteristic Q-value, and there is also the effect of the Coulomb barrier. Consequently, it is more realistic to appreciate NRA as element or elements specific. Even though this may appear as a limitation if our quest was for multi-elemental capability, in fact NRA can claim that there is a nuclear reaction appropriate for chemical analysis for virtually every element of the Periodic Table. In reality, NRA plays a role of particular importance in the analysis of the lightest nuclides, which are inaccessible to NAA, since typically neutron irradiation produces stable nuclides in this region of the Periodic Table.

2 PRINCIPLES

Charged particles have a finite range, and this is further characterized by the range straggling effect. The consequence is that, unlike the neutron activation case, an effective volume is sampled and, within this volume, analysis is further complicated by the rapid change in energy of the incident particle and the consequent changes in the cross-sections of the nuclear reactions relevant to the analysis. Furthermore, if the reaction has a negative Q-value, the cross-section drops to zero before the end of the actual range. Although this would appear to make an analysis more complicated, advantage can be taken of these features by, for example, varying the energy of the incident particle; the data can then be deconvoluted to give depth dependent data for the analysed element. If there is a (sharp) nuclear resonance in the excitation function, this can be readily exploited to give depth dependence with well defined depth resolution.

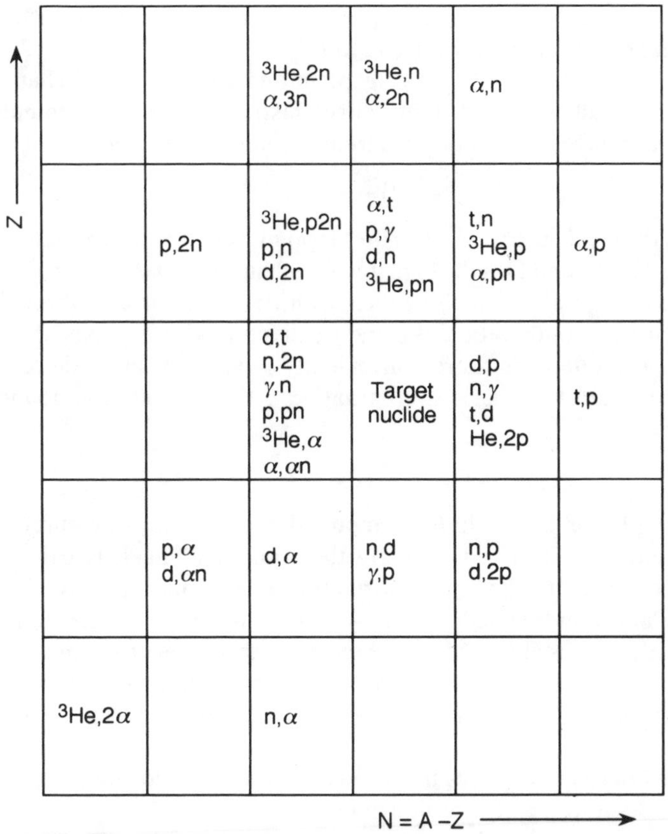

Z ↑

		^{3}He,2n α,3n	^{3}He,n α,2n	α,n	
	p,2n	^{3}He,p2n p,n d,2n	α,t p,γ d,n ^{3}He,pn	t,n ^{3}He,p α,pn	α,p
		d,t n,2n γ,n p,pn ^{3}He,α α,αn	Target nuclide	d,p n,γ t,d He,2p	t,p
	p,α d,αn	d,α	n,d γ,p	n,p d,2p	
^{3}He,2α		n,α			

N = A − Z →

Fig. 1. Diagrammatic presentation of nuclear reaction-induced transformations caused by the bombardment of a specific target nuclide by some of the lighter charged particles (with neutron- and photon-induced reactions included for comparison).

Basic data on energy loss and straggling are thus essential in the case of NRA. The standard reference is that of Northcliffe and Schilling [1], with useful supplementary references for specific commonly used ions being those of Andersen and Ziegler [2] and Ziegler and Chu [3]. Betz [4] has given a prescription for scaling from known ion data to other ions. Data on energy straggling can be found in the work of Lindhard and Scharff [5].

At all times it is well to remember that, unlike the situation for NAA, for the particles and energies most typically deployed in NRA, the actual sample volume interrogated, as prescribed by the range and the charged particle beam diameter, is in fact small. Hence questions must arise as to how meaningful NRA is when dealing with bulk samples of unknown homogeneity. These legitimate questions should be confronted as a statistical problem, as has been conveniently presented by Watterson, Sellschop and Zucchiatti [6].

If a nuclear reaction is energetically permitted, the reaction mechanism(s) that prevail are broadly subdivided into direct (fast) or delayed (compound nucleus). In the former the reaction proceeds from the initial state to the final state, as

$$\mathrm{a} + \mathrm{B} \longrightarrow \mathrm{c} + \mathrm{D}$$

where a is the incident particle, B is the target nucleus, c is the emitted (sometimes detected) particle, and D is the final (recoiling) nucleus. D can be produced in one or more excited states, which in their turn de-excite directly through photon emission or by a combination of β and γ emission. Such photons are well suited for NRA, and often are more conveniently used than direct detection of the charged or neutral particles c. Alternatively, the delayed compound nucleus reaction sequence

$$\mathrm{a} + \mathrm{B} \longrightarrow \mathrm{C} \longrightarrow \mathrm{c} + \mathrm{D}$$

can lead to photon decays in the compound nucleus (direct capture) so that the final nucleus D = C, or additionally the compound nucleus can decay in a manner independent of its mode of formation, after a time typically of the order of 10^{-16} s, to one or more final nuclei D. Again the final nuclei D may be unstable to β decay with the possibility of associated γ-ray emission, or the nuclei D may be formed each in one or more excited states with the resulting possibilities of γ or of β and γ emissions. These distinctive scenarios are presented in detail by Sellschop and Annegarn [7].

Preferred nuclear reactions in various energy ranges for medium weight and heavier nuclei respectively are shown in Table 1 [7].

The parameterization of the rate of such nuclear reactions has been presented by Sellschop and Annegarn [7]. For a simple two-body reaction

$$\mathrm{A} + \mathrm{X} \longrightarrow \mathrm{B} + \mathrm{Y}$$

we may reasonably expect that the production rate can be simply expressed as

$$\mathrm{d}N_{\mathrm{B}}/\mathrm{d}t = \Phi \sigma N_{\mathrm{A}}$$

where Φ is the flux density of the projectiles (x), expressed in $\mathrm{cm}^{-2}\mathrm{s}^{-1}$, σ is the cross-section for this reaction, expressed in cm^2, and N_{A} is the number of atoms of target nuclide A in the volume sampled by the beam.

For a situation in which the product nucleus B is itself unstable, with a decay constant λ, the population B will follow the radioactive decay law

$$-\mathrm{d}N_{\mathrm{B}}/\mathrm{d}t = \lambda N_{\mathrm{B}}$$

so that the net production rate must be

$$\mathrm{d}N_{\mathrm{B}}/\mathrm{d}t = \Phi \sigma N_{\mathrm{A}} - \lambda N_{\mathrm{B}}$$

TABLE 1.
Preferred nuclear reactions in various energy ranges

Energy range of the incident particle	Medium nuclides ($25 < A < 80$)				Heavy nuclides ($80 < A < 250$)			
	n	p	d	α	n	p	d	α
0–1 keV	n, γ	–	–	–	n, γ	–	–	–
1–500 keV	n, γ	p, n p, γ p, α	d, p d, n	α, n α, γ α, p	n, γ	–	–	–
0.5–10 meV	n, α n, p	p, n p, α	d, p d, n d, pn d, 2n	α, n α, p	n, p n, γ	p, n p, γ	d, p d, n d, pn d, 2n	α, n α, p α, γ
10–50 MeV	n, 2n n, p n, np n, 2p n, α	p, 2n p, n p, np p, 2p p, α	d, p d, 2n d, pn d, 3n d, t	α, 2n α, n α, p α, np α, 2p	n, 2n n, p n, pn n, 2p n, α	p, 2n p, n p, pn p, 2p p, α	d ,p d, 2n d, np d, 3n d, t	α, 2n α, n α, p α, np α, 2p

From K. H. Lieser, *Treatise on Analytical Chemistry,* 2nd Ed., Part 1, Vol. 14, *Theory and Practice*, edited by L. M. Kolthoff, P. J. Elving and V. Krivan, Wiley, New York (1986), Chapter 1.

Integration of this differential equation gives

$$N_B = \Phi\sigma N_A/\lambda(1 - e^{-\lambda t})$$

Hence the activity A of B as a function of irradiation time is

$$A = -dN_B/dt = \lambda N_B$$
$$= \Phi\,\sigma\,N_A(1 - e^{-\lambda t})$$

The term $(1 - e^{-\lambda t})$ is known as the saturation factor, and characterizes the shape of the activation curve.

Since our objective in CP(A)A is to determine N_A, it is useful to express the activation equation in more practical form as

$$A = 0.622\,\Phi\,\sigma f m^*/M(1 - e^{-\lambda t})$$

where f is the isotopic abundance of the nuclide A in the element, M is the atomic mass of the element, and m^* is the mass of the element containing the nuclide A. Careful selection of irradiation and decay times can optimize our analysis or lead to the best compromise for a specific analysis.

Extension of these activation principles is necessary to account for the more typical situation in which one seeks to analyse for the element of interest (often in trace quantities) in a matrix of different composition. The volume of material interrogated by the beam is determined by the stopping power and range of the mixture of sample(s) cum matrix (m), not of the pure sample material. If c is the concentration of the sample in the matrix, by weight rather than atomic, then

$$n = \text{number of nuclides of A per g of s+m}$$
$$= c\, N_{\text{Avog}} f/M_{\text{s+m}}$$

Then, for a target s + m of thickness greater than the range of the incident charged particle, we can develop the earlier expressions [7] to give

$$A = icf\,(1 - \mathrm{e}^{-\lambda t(\text{irradiation})}) \cdot \mathrm{e}^{-\lambda t(\text{decay})} \rho_{\text{s+m}} \frac{N_{\text{Avog}}}{M_{\text{s+m}}} \cdot \int_{E_i}^{0} \sigma(E)\, \mathrm{d}E \frac{\mathrm{d}x}{\mathrm{d}E}$$

where i = beam intensity (s^{-1}), and ρ = density.

Clearly, to use this activity expression, we need to know adequately the excitation function $\sigma(E)$ over the energy range of relevance, and in absolute units. Relative excitation functions are common, but reliable information in absolute terms is not as common as should be the case. Hence it is usual to resort to the use of standards or reference materials in which the concentration of the nuclide of interest is well specified and the sample + matrix composition is well known. Calibration using standards is most rigorously achieved if a range of standards is used, thereby establishing a calibration curve covering a range of values which includes that of the unknown sample.

Assuming that the standard amounts of the nuclide of interest are in a matrix different to that in which the nuclide of unknown concentration is embedded, we can compare the two activities as follows on the basis of the same irradiation and decay times and the same beam currents (i):

$$\frac{A_{\text{sample}}}{A_{\text{standard}}} = \frac{c_{\text{sample}}\, M_{\text{stand+matrix}}\, \rho_{\text{sample+matrix}} \displaystyle\int_{E_i}^{0} \sigma(E)\, \mathrm{d}E \left[\frac{\mathrm{d}x}{\mathrm{d}E}\right]_{\text{sample+matrix}}}{c_{\text{standard}}\, M_{\text{sample+matrix}}\, \rho_{\text{stand+matrix}} \displaystyle\int_{E_i}^{0} \sigma(E)\, \mathrm{d}E \left[\frac{\mathrm{d}x}{\mathrm{d}E}\right]_{\text{stand+matrix}}}$$

The integrals can be addressed using the Bethe–Bloch energy loss rate expression

$$-\frac{\mathrm{d}E}{\mathrm{d}x} = \frac{2\pi e^4 z^2 M}{m}\frac{}{E} NZ \ln\left(\frac{4mE}{IM}\right)$$

where z = atomic number of the projectile, Z = atomic number of the target atom, v = particle/projectile velocity, m = mass of the electron, e = charge of the electron, N = number of atoms per unit volume, and I = average ionization potential of target atoms. Hence we can proceed to reduce the integrals to

$$\int_{E_i}^{0} \sigma(E)\, \mathrm{d}E \frac{\mathrm{d}x}{\mathrm{d}E} = \frac{K}{NZ} \int_{E_i}^{0} \sigma(E)\, E \left(\frac{4mE}{IM}\right)^{-1} \mathrm{d}E$$

where $K = m/2\pi e^4 z^2 M$.

Since the logarithmic term varies slowly with E and can thus be taken as a constant, we may write

$$\int_{E_i}^{0} \sigma(E)\,\mathrm{d}E\,\frac{\mathrm{d}x}{\mathrm{d}E} = \frac{K^\star}{NZ}\int_{E_i}^{0} \sigma(E)\,E\,\mathrm{d}E$$

Ricci and Hahn [8] first pointed out that this integrated differential cross-section function defined an area in parameter space [$\sigma(E)$ vs. x] which extended on the x axis to the range R appropriate to the stopping medium. Dividing this area by R gave a constant effective cross-section, or in their terminology an 'average' cross-section

$$\sigma_{\mathrm{ave}} = \frac{\int_0^R \sigma(E)\,\mathrm{d}x}{R} = \frac{\int_{E_i}^{0} \sigma(E)\frac{\mathrm{d}x}{\mathrm{d}E}\,\mathrm{d}E}{\int_{E_i}^{0}\frac{\mathrm{d}x}{\mathrm{d}E}\,dE}$$

$$= \frac{\int_{E_i}^{0} \sigma(E)\,\mathrm{d}E}{\int_{E_i}^{0} E\,\mathrm{d}E}$$

This formulation is apparently independent of any matrix properties, and the simplification enables us to replace

$$\int_0^R \sigma(E)\,\mathrm{d}x$$

by $R\,\sigma_{\mathrm{ave}}$, so that we can determine the concentration of the sample from

$$c_{\mathrm{sample}} = \frac{A_{\mathrm{sample}}\,c_{\mathrm{stand}}\,M_{\mathrm{sample+matrix}}\,\rho_{\mathrm{stand+matrix}}\,R_{\mathrm{stand+matrix}}}{A_{\mathrm{stand}}\,M_{\mathrm{stand+matrix}}\,\rho_{\mathrm{sample+matrix}}\,R_{\mathrm{sample+matrix}}}$$

Hence CP(A)A can be rendered a quantitative analytical tool and, as we shall see, enjoys widespread use, having carved a niche, or perhaps a set of niches, where it is of particular value.

3 APPLICATIONS OF ELEMENTAL ANALYSIS

An exhaustive survey of the literature has been conducted to assemble an inventory of case histories for CP(A)A. These data are presented in summary form as a set of Periodic Tables showing elements reported as analysed by protons (Fig. 2). Similar Periodic Tables for incident deuterons, tritons, helium-3 and helium-4 ions are given in Ref. 7. The major applications represented in these

CP(A)A applications: incident protons

H																	He
Li	Be											B	C	N	O	F	Ne
Na	Mg											Al	Si	P	S	Cl	Ar
K	Ca	Sc	Ti	V	Cr	Mn	Fe	Co	Ni	Cu	Zn	Ga	Ge	As	Se	Br	Kr
Rb	Sr	Y	Zr	Nb	Mo	Tc	Ru	Rh	Pd	Ag	Cd	In	Sn	Sb	Te	I	Xe
Cs	Ba		Hf	Ta	W	Re	Os	Ir	Pt	Au	Hg	Tl	Pb	Bi	Po	At	Rn
Fr	Ra																

La	Ce	Pr	Nd	Pm	Sm	Eu	Gd	Tb	Dy	Ho	Er	Tm	Yb	Lu
Ac	Th	Pa	U	Np	Pu	Am	Cm	Bk	Cf	Es	Fm	Md	No	Lw

Fig. 2. Elements reported as analysed by protons

figures are in the fields of metallurgy, semiconductors, geology, the environment, biology, medicine, art and archaeology—a subdivision into these fields is attempted in Ref. 7.

An indication of the sensitivity for elemental analysis is presented for protons. (Fig. 3). Similar figures of sensitivity for incident deuterons, tritons, helium-3 ions, alpha-particles and lithium-7 ions are given in Ref. 7.

4 BEYOND PURE ELEMENTAL ANALYSIS

The late second generation and the third generation of charged particle analysis are characterized by the evolution of the general technique to levels of specificity and detail that transcend merely elemental analysis, important as that is. A number of examples will be presented.

4.1 Ion channelling

This refinement will be shown to have the effect of improving the depth resolution in impurity analysis, of improving the sensitivity of light element analysis in a heavier matrix and, most importantly, of being capable of giving information on the sites occupied in the matrix by the analysed impurities (lattice location).

The reference just made to 'lattice location' gives the essential clue that in ion channelling one is dealing with a matrix in single crystal form. This should not be interpreted as being relevant only to the esoteric: single crystals and their intrinsic

CP(A)A applications: sensitivity indicators
Protons: 10 MeV; 1 μA; up to 1 h irradiation;
assume no interferences

Limits of detection

0.1–1.0 ng/g

1–10 ng/g

10–100 ng/g

H																	He
Li	Be											B	C	N	O	F	Ne
Na	Mg											Al	Si	P	S	Cl	Ar
K	Ca	Sc	Ti	V	Cr	Mn	Fe	Co	Ni	Cu	Zn	Ga	Ge	As	Se	Br	Kr
Rb	Sr	Y	Zr	Nb	Mo	Tc	Ru	Rh	Pd	Ag	Cd	In	Sn	Sb	Te	I	Xe
Cs	Ba		Hf	Ta	W	Re	Os	Ir	Pt	Au	Hg	Tl	Pb	Bi	Po	At	Rn
Fr	Ra																

La	Ce	Pr	Nd	Pm	Sm	Eu	Gd	Tb	Dy	Ho	Er	Tm	Yb	Lu
Ac	Th	Pa	U	Np	Pu	Am	Cm	Bk	Cf	Es	Fm	Md	No	Lw

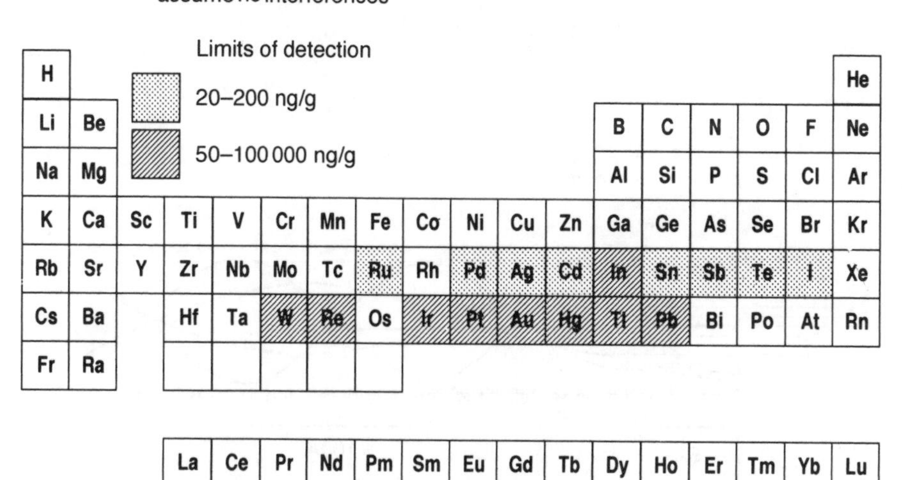

Fig. 3(a) and (b). Sensitivity for elemental analysis using 10 MeV protons.

and induced/implanted impurities are of widespread relevance in modern semiconductor technology.

Channelling occurs when a well-collimated ion beam is directed close to a major symmetry direction, plane or axis, of a single crystal [9]. This is illustrated in exaggerated form in Fig. 4. Correlated small-angle scatterings of the charged particle have a consequence that the particle is prevented from reaching sufficiently small impact parameters with the atomic nuclei to undergo large-angle Rutherford scattering, which would remove it from the channelling trajectory. As a result, the contribution to the backscattering spectrum from the matrix is substantially reduced, by a factor of about two orders of magnitude, enhancing the sensitivity for analysis of the lighter impurities, particularly those on or near the surface. There is always some surface disorder of course, hence there is an ion–surface interaction that is greater than that deeper in the crystal, within the ordered planar or axial structure.

Figure 4 is admittedly exaggerated, but one may well appreciate that the gentle steering of the channelled charged particle involves so many crystal atoms that one may invoke a continuum model with the nuclear charge of the atoms in a row or plane uniformly averaged or smeared along the row or plane. The essence of the Lindhard approach is just this, that the interaction of the channelled particle with an atomic row can be faithfully described in terms of a continuum potential $U_a(r)$, where r is the perpendicular distance from the row.

Following the presentation of Feldman and Mayer [10], $U_a(r)$ is then the value of the atomic potential averaged along the atomic row with atomic spacing d. Hence, for the axial case

$$U_a(r) = \frac{1}{d}\int_{-\infty}^{+\infty} V(\sqrt{z^2 + r^2})\,dz$$

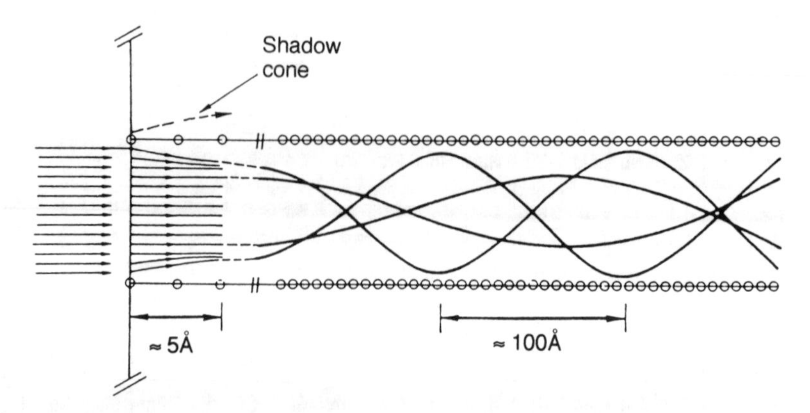

Fig. 4. Exaggerated presentation of trajectories of particles undergoing correlated scattering at the surface and within a crystal.

where $V(\bar{r})$ is the screened Coulomb potential and $\bar{r}$ is the spherical radial coordinate, with $\bar{r}^2 = (z^2 + r^2)$.

This so-called standard potential is then

$$V(\bar{r}) = Z_1 Z_2 e^2 \left(\frac{1}{\bar{r}} - \frac{1}{\sqrt{\bar{r}^2 + C^2 a^2}} \right)$$

where C^2 is usually taken as equal to 3 and a is the Thomas–Fermi screening distance. Hence, for the axial continuum potential

$$U_a(r) = \frac{Z_1 Z_2 e^2}{d} \ln \left[\left(\frac{Ca}{r} \right)^2 + 1 \right]$$

where d is the average distance between the atoms in the rows.

The magnitude of this potential is of the same order as for atomic potentials [10], that is, 223 eV at $r = 0.1$ Å for He along the ⟨110⟩ rows of Si ($d = 3.84$ Å).

The total energy E of a particle inside the crystal is then

$$E = \frac{p_{\parallel}^2}{2M} + \frac{p_{\perp}^2}{2M} + U_a(r)$$

where $p_{\parallel}$ and $p_{\perp}$ are the parallel and perpendicular components of the momentum with respect to the string direction. Accordingly

$$p_{\parallel} = p \cos\psi$$

$$p_{\perp} = p \sin\psi$$

and

$$E = \frac{p^2 \cos^2 \psi}{2M} + \frac{p^2 \sin^2 \psi}{2M} + U_a(r)$$

Since channelling angles are small, we may use the small angle approximation, and consequently we can equate the last two terms with the transverse energy, so that

$$E_{\perp} = \frac{p^2 \psi^2}{2M} + U_a(r)$$

The kinetic energy and the potential energy contribution are clearly distinguished. The total energy is conserved, and in this approximation it is important to appreciate that the transverse energy is conserved, which may be considered as a definition of the channelling condition.

Accordingly we can now establish the critical angle for channelling by equating the transverse energy at the turning point $U(r_{\min})$ to the transverse energy at the midpoint, viz.

$$E \psi_c^2 = U(r_{\min})$$

so that

$$\psi_c = \left[\frac{U(r_{min})}{E}\right]^{1/2}$$

The thermal vibration of the atoms causes a smearing of the atom positions, thereby setting a lower limit to the minimum distance for which a row can provide the necessary correlated sequence of scatterings required for the channelling condition. A reasonable first approximation to the critical angle can be obtained by substituting $r_{min} = \rho$, where ρ^2 is 2/3 of the mean square thermal vibration amplitude; hence

$$\psi_c(\rho) = \frac{\psi_1}{\sqrt{2}}\left|\ln\left[\left(\frac{Ca}{\rho}\right)^2 + 1\right]\right|^{1/2}$$

where

$$\psi_1 = \left(\frac{2Z_1Z_2e^2}{Ed}\right)^{1/2}$$

The calculated values of $\psi_c(\rho)$ compare well with experimental measurements, and track the temperature dependence.

Another characteristic parameter associated with channelling, in addition to the critical angle, is the minimum yield, in other words the remaining yield when in optimal channelling alignment. For head-on incidence with $\psi = 0$, we can picture that around each string of atoms is an area πr_{min}^2 within which particles cannot channel, whereas particles incident at $r > r_{min}$ can channel. Then the minimum fraction of particles that can never be channelled is simply

$$\frac{\pi r_{min}^2}{\pi r_0^2}$$

where r_0 is the radius associated with each string. But

$$\pi r_0^2 = \frac{1}{Nd}$$

where N is the atomic concentration of atoms and d is the atomic spacing along the string. This minimum yield χ_{min} in a backscattering measurement is the yield of close encounter events in a precisely oriented channelling configuration. With $r_{min} \approx 0.1$ Å, we may expect some 99% of the particles to be channelled, that is to say, a minimum yield of order 1%. In practice the measured minimum yields are typically some few percent, indicating the role of other parameters such as defects.

The continuum approach can be applied equally effectively to planar channelling, leading to a critical angle defined as

$$\psi_p = \left[\frac{U_p(y_{min})}{E}\right]^{1/2}$$

where

$$y_{\min} \cong \rho \Big/ \sqrt{2}$$

is the one-dimensional vibrational amplitude. The characteristic angle for planar channelling is then

$$\psi_2 = \left(\frac{2\pi Z_1 Z_2 e^2 a N d_p}{E}\right)^{1/2}$$

which is of the order of the critical angle for planar channelling.

Experimentally we find that critical angles for planar channelling are 2 to 4-fold smaller than critical angles for axial channelling.

From a geometrical projection for planar channelling, the minimum yield may be expected to be

$$\chi_{\min}(\text{planar}) = \frac{2y_{\min}}{d_p}$$

which is substantially larger than the corresponding value for axial channelling. The value of the minimum yield for good planar channelling directions is typically of the order of 10–20%.

The thrust of our interest in this chapter is CP(A)A, and the introduction of channelling concepts is important in that this opens up opportunities that go well beyond mere elemental analysis, and indeed make possible detailed analysis not accessible by other techniques. One such opportunity of particular importance is the prospect of specific *lattice location* of impurities. The angular dependence of the yield in a nuclear reaction, when scanned through a major axis or plane, will, in terms of the theoretical background presented above, vary with characteristics reflecting the critical angle and the minimum yield. Evidently, if an impurity atom occupies a *substitutional site*, its particular nuclear signal will reflect the same angular dependence as the host lattice. At the other extreme, if impurity atoms are randomly distributed in the host, unrelated to the host crystal structure, the yield scan will be independent of angle. These extremes are presented, following Feldman and Mayer [10], in Fig. 5. Between these two extremes lie a myriad of detailed possibilities, and it is evident that, by angular scanning of reaction yields for different major axial and/or planar directions, the positions of occupied sites may be at least delimited if not precisely established. Furthermore, one can appreciate that here is a tool that can track the dynamic behaviour of impurities such as dopants, for example, as a function of temperature.

The detailed interpretation of such angular scans in relation to major crystal axial and planar directions requires an equally detailed knowledge of the flux distribution of channelled particles.

A channelled particle is confined within equipotential contours, U_T as for example the continuum contours shown in Fig. 6 [10]. We should interpret that $U_T = \Sigma U_i$, the sum of the individual potentials U_i of the nearby rows or planes.

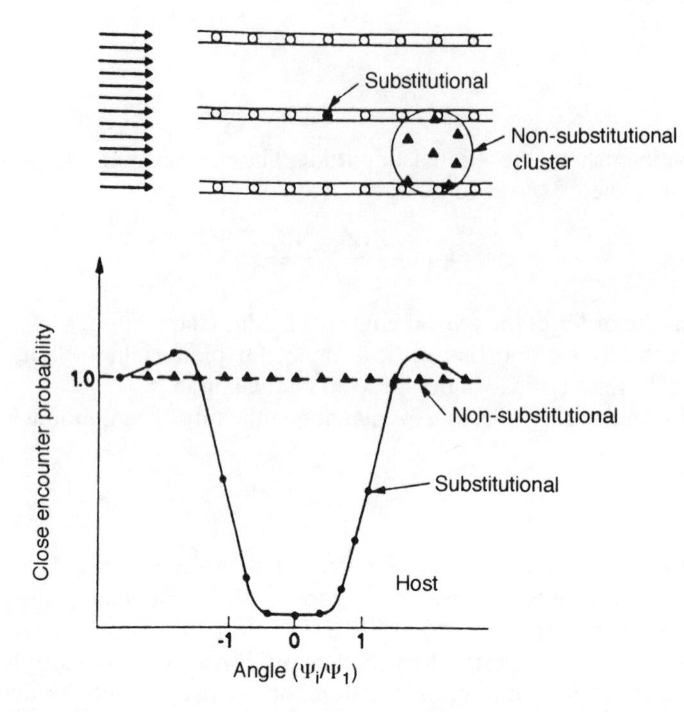

Fig. 5. Schematic representation of the close encounter probability curve expected for impurities in substitutional and randomly located sites in a single crystal.

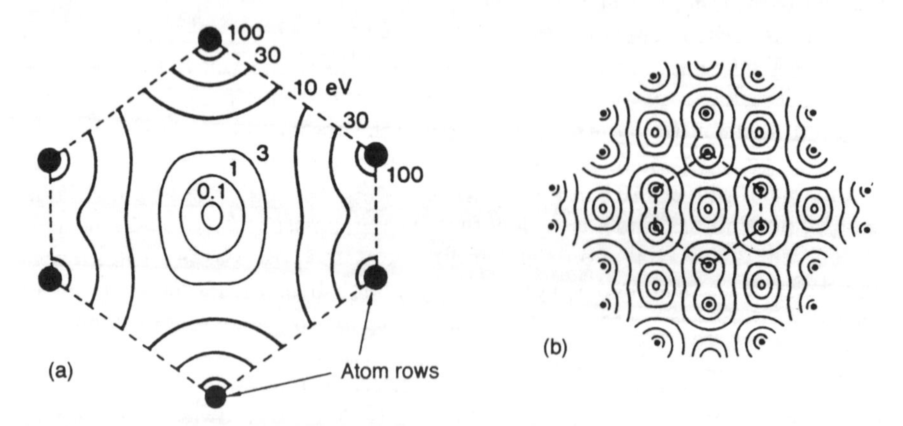

Fig. 6. (a) Equipotential contours for the axial continuum potential for the case of He in the ⟨110⟩ direction of Si, showing the change in shape of the potential contours corresponding to the geometry of the channel. (b) The potential contours for an array of channels of the type shown in (a).

Hence a particle with a given transverse energy $E_\perp$ must always lie within a region given by

$$U_T(r) \le E_\perp$$

As can be seen from Fig. 6, for silicon the 3 eV contour closes within the centre of the channel, implying that particles with transverse energy < 3 eV have their trajectories confined to a particular channel. For channelled particles with higher transverse energies, for example $E_t \ge 10$ eV, the confinement is not within one channel, but the particles are guided by the cylindrically symmetric potentials around the axial rows.

Normalizing to unit probability to find a particle somewhere in its allowed area $A(E_\perp)$, the probability of finding a particle of transverse energy $E_\perp$ at any point r is

$$P(E_\perp, r) = \frac{1}{A(E_\perp)}, \qquad E_\perp \ge U_T(r)$$

$$P(E_\perp, r) = 0 \qquad E_\perp < U_T(r)$$

The allowed area is defined by an equipotential contour as shown in Fig. 6. So, for example [10], a 1 MeV particle entering the centre of the channel at an angle of 0.18° will have $E_\perp = 10$ eV, and thus has an equal probability of being found at any point within the area defined by the equipotential contour $U_T = 10$ eV. Calculation of the flux distribution of ions for a channelled beam of $\psi = 0$ is based on the following premises:

- conservation of transverse energy: a particle that enters at r_{in} cannot get closer to the row than r_{in}
- cylindrical symmetry
- for two-dimensional axial channelling a particle has a uniform probability of being found in its allowed area, viz. πr_{in}^2, where r_{in} is the initial distance from the string at the crystal surface, i.e.

$$P(r_{in}, r) = \frac{1}{\pi r_0^2 - \pi r_{in}^2}, \qquad r > r_{in}$$

$$= 0, \qquad r < r_{in}$$

The flux distribution $f(r)$ inside the crystal is obtained by integrating over all initial impact parameters:

$$f(r) = \int_0^{r_0} P(r_{in}, r)\, 2\pi r_{in}\, dr_{in}$$

$$= \int_r^{r_0} \frac{1}{\pi r_0^2 - \pi r_{in}^2}\, 2\pi r_{in}\, dr_{in}$$

$$= \ln \frac{r_0^2}{r_0^2 - r^2}$$

The effect of channelling is seen as transforming an originally spatially uniform distribution into a peaked distribution.

Such flux distributions clearly display the most prominent feature of channelling, namely that the flux intensity and hence the close–encounter probability approaches zero near the atomic rows, i.e. as $r \to 0$.

Expanding $f(r)$ for small r near the atom rows gives

$$f(r) \approx \frac{r^2}{r_0^2}$$

This simple expression for the flux distribution is useful in CP(A)A for estimating the scattering intensity for substitutional impurities which lie at $r = 0$.

Another feature which is important for CP(A)A analysis is that, at the centre of the channel where $r = r_0$, besides the evident peaking of the flux, the intensity far exceeds unity, which is the value that would correspond to the normal (non-channelled) particle density. This has the important consequence that the scattered yield from *interstitial impurities* located near the centre of the channels will be significantly greater than in the normal, non-channelled case—we have in channelling a sensitivity amplifier for interstitial and near-interstitial impurities! This was elegantly demonstrated by Andersen *et al.* [11] for Yb impurities in Si.

Pursuing the importance of channelling in CP(A)A, we now consider the powerful role it plays in the analysis of *surfaces*. In order to interpret such studies, one needs to consider in detail the interaction of an incident beam with the single first surface layer of atoms and, for simplicity, the second or nearest next layer.

One needs to calculate the flux distribution $f(r_2)$ at the second atom as a result of scattering interactions with the first (surface) atom. Using the small angle approximation to pure Coulomb scattering [10], it can be shown that the lateral distance r_2 to the second atom has a minimum, a distance of closest approach R_c which is the Coulomb shadow cone radius

$$R_c = 2\left(\frac{Z_1 Z_2 e^2 d}{E}\right)^{1/2}$$

The flux $f(r_2)$ then becomes

$$\begin{aligned} f(r_2) &= 0 \quad \text{for} \quad r_2 < R_c \\ &= \frac{1}{2}\left(\frac{1}{\sqrt{1 - R_c^2/r_2^2}} + \sqrt{1 - R_c^2/r_2^2}\right) \quad \text{for} \quad r_2 > R_c \end{aligned}$$

The change in the flux distribution is so sharp at $r_2 = R_c$ that the curvature occurs within a distance that is small compared with the thermal vibration amplitude of the atoms in the crystal. Consequently it is reasonable to use a delta function as an approximation for $f(r_2)$. We need now to calculate the intensity (I_2) of the scattering from the second atom—this is given by the overlap of the calculated flux distribution $f(r_2)$ with the Gaussian positional distribution of the second atom. This has been evaluated [10] as

$$I_2 = \left(1 + \frac{R_c^2}{\rho^2}\right) e^{-R_c^2/\rho^2}$$

so that the total intensity (I) ascribable to the surface peak is given by the unit contribution due to the first atom and to I_2:

$$I = 1 + I_2$$

This total intensity for the surface peak is dependent on the parameter ρ/R_c only, in other words on the ratio of the thermal vibration amplitude to the shadow cone. For values of $\rho < R_c$, the topmost surface atom does effectively shadow the underlying atoms from direct close encounters with the beam. If a crystal of excellent quality is used, de-channelling is low, and the energy spectrum for backscattering is distinguished by the surface peak, with a low continuum towards ever lower energies. Channelling presents us therefore with a fine analytical tool for the study of surfaces and, specifically, of the *arrangement of surface atoms*.

This sensitivity of channelling to the surface structure is exemplified in simplified form in Fig. 7. Four simple cases [10] are presented in this figure; also represented are the respective backscattering spectra showing the surface peak. The spectra shown by the dashed curves represent the scattering yield from a crystal with an ideal surface for the case that the thermal vibration amplitude ρ is much less than the shadow cone radius R_c. This ensures that the surface peak intensity corresponds to one atom per row in this *ideal* case (Fig. 7(a)). The crystal with a *reconstructed surface*, where the surface atoms are displaced in the plane of the surface, represents a situation where the second atom is not shadowed (Fig. 7(b)). In this case the surface peak intensity is twice that of the ideal crystal. The surface peak is sensitive also to the surface atoms being *relaxed*, in that they are displaced normal to the surface plane, but in this event one must use non-normal incidence, so that the shadow cone established by the surface atoms is not aligned with the atomic rows in the bulk (Fig. 7(c)). Normal incidence in this case would yield a surface peak intensity equivalent to one monolayer. The combination of measurements at normal and oblique incidence unambiguously reveals the presence of relaxation. The final example (Fig. 7(d)) is that of a surface adsorbate of higher Z than the substrate atoms, where the surface adsorbate atom can shadow the atoms of the substrate if $R_{\text{adsorbate}} > \rho_{\text{substrate}}$. The Z dependence of Rutherford scattering permits of clear separation between substra-te and adsorbate. However, as in the example given in Fig. 7(d) the adsorbate is shown as positioned exactly over the surface atoms, we see in addition to the well resolved adsorbate peak that the substrate surface peak is reduced in intensity.

The fourth example, that of the adsorbate, represents a class of problem that has widespread application in CP(A)A. Take, for example, the *epitaxial* growth of a layer of gold on a single crystal substrate of silver. If it is truly epitaxial and thus

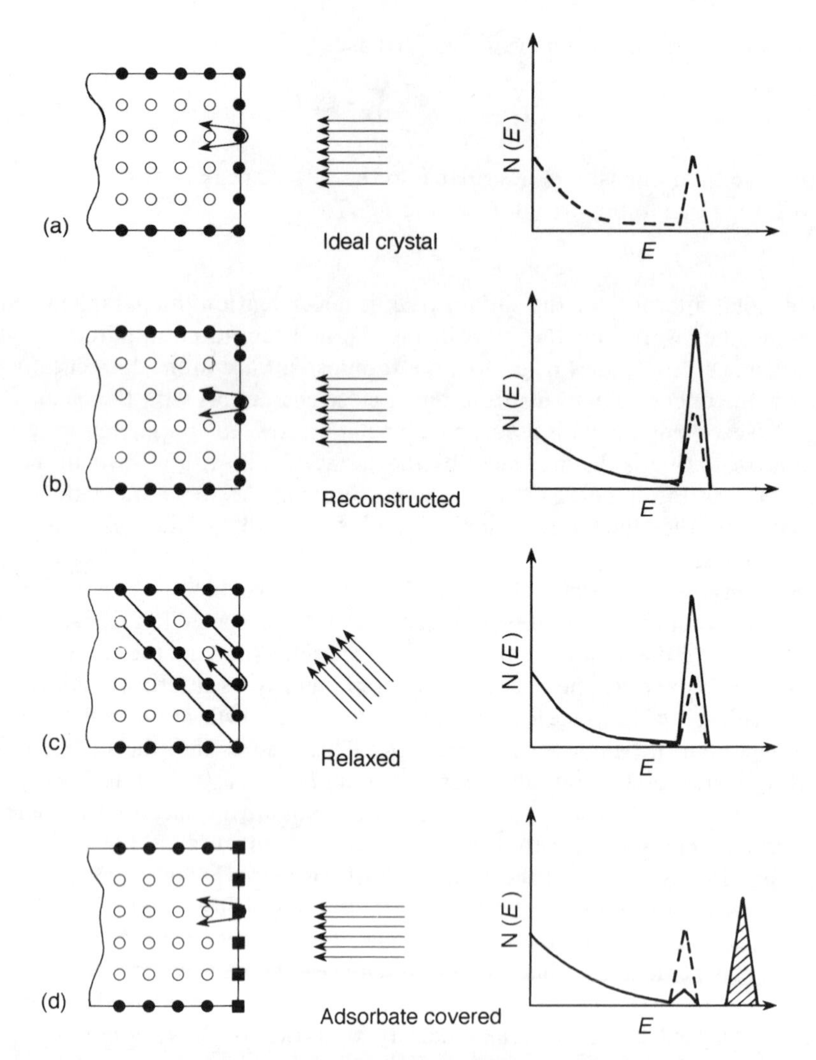

Fig. 7. Representations of different surfaces on a simple cubic crystal. The backscattered spectra shown on the right hand side represent the expected signal from the different structures. The dashed line in each case represents the signal from the bulk-like crystal.

in perfect register, atom for atom, then the shadowing of the substrate for adsorbate layers of a fraction of a monolayer is but partial; once the adsorbate thickness takes on dimensions of one monolayer or greater, the shadowing becomes effective. By exploiting the channelling effect in this way, one is able quantitatively to determine the extent of true epitaxial registration or true lattice constant match, and also the thickness of the covering layer.

Even when epitaxy is not at issue, channelling holds singular advantages in CP(A)A for *thin film analysis.* This arises through the suppression of the single crystal substrate contribution to the measured (backscattering) spectrum, consequently giving an increased sensitivity for the analysis of the amorphous overlayer. This is particularly significant for light impurities in the overlayer, which in the normal (non-channelled) situation would be a minor perturbation on the dominant substrate contribution, extracted inaccurately at best. In addition, from a careful analysis of the substrate the surface peak, the interface strain in the crystal substrate can be assessed.

4.2 Time differential perturbed angular distributions (TDPAD): microscopic CP(A)A

TDPAD is a variant of CP(A)A which takes advantage of the de-excitation of an excited nucleus through observation of the de-excitation γ-rays. The normal angular distribution of these de-excitation photons is perturbed through the hyperfine interaction at the excited nucleus by the electron configuration at its immediate site. This causes a precession of the nuclear quadrupole moment in the hyperfine field, which is observed as a rotation of the angular distribution pattern of the nuclear de-excitation γ-rays. The parameterization of the observed time differential rotation allows one to attribute characteristic binding conditions and residence sites to the excited nucleus. The quantities that can be extracted from the spectra are [12]

- f_i the fraction of ions occupying site i in the crystal
- ω_{0_i} the quadrupole precession frequency for fraction f_i
- ν_Q the quadrupole coupling constant $= eQV^{i}_{zz}/h = 10/3\pi \times \omega_0$ for the nuclear spin 5/2
- V^{i}_{zz} the principal component of the electric field gradient
- η_i the asymmetry coefficient of the field gradient for each configuration f_i, given by $\eta_i = \dfrac{(V^{i}_{xx} - V^{i}_{yy})}{V^{i}_{zz}}$

and

- δ_i the (assumed gaussian) distribution of the electric field gradient, where small values reflect a sharply defined field gradient and a well defined binding condition.

To measure this perturbed or rotated angular distribution of the decay γ-rays, two detectors are set up at 0° and 70.5° for the geometry of a diamond structure cubic lattice [these are commonly BaF_2 detectors, which have better time characteristics than NaI(Tl)]. The charged particle beam is pulsed (typically, pulse width 3 ns and repetition rate 500 ns). In the between-pulses interval the two detectors record the decay of the relevant nuclear level. So, for example, it is common to study fluorine as a dopant. The reaction $^{19}F(p, p')\,^{19}F$ populates the $(5/2)^+$ level at 197 keV: this second excited state has an electric quadrupole moment $Q = 0.70$ b and a lifetime $\tau = 128$ ns.

The time spectra from the two detectors are used to generate a spin rotation ratio

$$R_{exp}(t) = 2\frac{\rho(0^o) - \rho(70.5^\circ)}{\rho(0^o) + 2\rho(70.5^\circ)}$$

where $\rho(\theta) = N(\Theta)\,N(\Theta)$ and $N_{1,2}(\Theta)$ are the normalized counting rates in the detectors 1 and 2 at angle Θ.

The *theoretical* spin rotation function is

$$R_{th}(t) = \frac{W(\theta, \varphi, 0^o, t) - W(\theta, \varphi, 70.5^\circ, t)}{W(\theta, \varphi, 0^o, t) + 2W(\theta, \varphi, 70.5^\circ, t)}$$

with

$$W(\Theta, t) = 1 + A_2(1)A_2(2)\sum_{n=o}^{3} \hat{S}_n^{eff}\cos[g_n(\eta)\omega_o t]$$

The coefficients S_n^{eff} contain all information on the angles θ and φ, i.e. on the geometry of the electric field gradients and the detector arrangement, and also on Θ, the angle between the detectors (here Θ has been chosen as 70.5° and typically along $\langle 111 \rangle$ for the diamond type cubic lattice). The coefficients S_n^{eff} also depend on the asymmetry parameter η. Each coefficient S_n^{eff} is an average value over all equivalent electric field gradient orientations. The values have been computed and tabulated [13].

The term $g_n(\eta)$ takes care of the frequency ratio of the quasi-harmonics of the nuclear state with g_n taking on integral numbers for $\eta = 0$.

The anisotropy coefficient, somtimes denoted as $A_{k_1}\,A_{k_2}$, is presented above as $A_2(1)\,A_2(2)$ and has been tabulated [13]. These coefficients may be corrected for contributions from $A_4(1)$ and/or $A_4(2)$, if necessary, by an appropriate multiplicative factor to S_n^{eff}.

For a *single crystal* (sc) host matrix the R ratio becomes

$$R_{th}^{sc} = \frac{2}{3}\sum_i f_i[A_{k_1k_2}^{eff} f(\sigma, \delta_i) G_{k_1k_2}(\theta, \varphi, \eta_i, \omega_{0_i}, t)]$$

The factor f_i takes into account that more than one unique residence site exists for the ^{19}F ions, each site with a population fraction f_i. The quantity σ is the experimental time resolution and δ is the frequency distribution factor for the fraction f_i.

For a *polycrystalline* (*pc*) *host* the theoretical spin rotation function, because of the averaged angles θ and φ becomes

$$R_{th}^{pc} = f_i[SA_{kk}^{eff}\,f(\sigma, \delta_i)\,G_{kk}^{eff}(\eta_i, \omega_{0_i}, t)]$$

where A_{kk}^{eff} and G_{kk}^{eff} are the effective anisotropy and the perturbation coefficient respectively. They are corrected with respect to a finite solid angle. G_{kk} differs from $G_{k_1k_2}$ insofar as the angles θ and φ have been averaged.

From the observed and normalized time spectra, the ratios $R_{exp}(t)$ can be formed for the pair of detectors and for each set of measurements. Such spectra are then fitted to several different combinations $R_{th}^{sc}(t)$ and $R_{th}^{pc}(t)$ to determine the number of lattice sites taken up and to evaluate the relevant parameters f_i, ω_0, η_i and σ_i.

From such *microscopic* CP(A)A experiments one may therefore expect the answers to the following physics questions to be revealed:

(1) for specific dopants implanted at well defined temperatures, what site or sites with distinctive electric field gradients are populated within the time window of a few hundred nanoseconds, for particular combinations of probe and matrix?
(2) for site(s) so indicated, how sharp is the identification and to what extent can these sites, in terms of symmetry and location, be assigned in the structure of the matrix?
(3) microscopic CP(A)A through TDPAD studies is a dynamic tool, and the lattice sites with their unique characteristics can be studied in quest of answers as to whether there is any probe dependence, temperature dependence or dose dependence.

In the case of diamond [12, 14, 15] and the (recoil) implantation of fluorine therein, fascinating data are revealed through TDPAD microscopic CP(A)A! A typical $R_{exp}(t)$ is shown in Fig. 8. This figure shows clearly that no fewer than three frequencies are required to fit the data adequately: two of these are well matched by single crystal theory with the orientation of the electric field gradient along the $\langle 111 \rangle$ axis. They match calculations based on molecular orbital and cluster models [16, 17] and are concluded to be, respectively, the *intra-bond* or *bond-centered* site ($\approx$ 60.6 MHz) and a *distorted substitutional* site ($\approx$ 55.1 MHz). The third frequency ($\approx$ 31.8 MHz) is interesting: it cannot be fitted to any single crystal site, but is well matched to a polycrystalline or amorphous model; the frequency itself is consistent with that for a H–F bond. This is obviously an important component, as it is measured as being the most highly populated of all the three sites and challenges one with the question of the origin of this hydrogen; in addition, this polycrystalline fraction shows a strong dose dependence. There is strong independent evidence [18] that all natural diamonds in their crystallization genesis in the upper mantle of the Earth include microscopically small droplets of the parental magma in an essentially homogeneous distribution throughout their bulk. These magma droplets are each within their minute volume amorphous, rich in water and gas (possibly CH_4), so that the rationale is logically developed that within these is the hydrogen source. Emphatically, however, the power is demonstrated of TDPAD as a *dynamic microscopic diagnostic CP(A)A probe* with few, if any, peers. We are able to exploit the

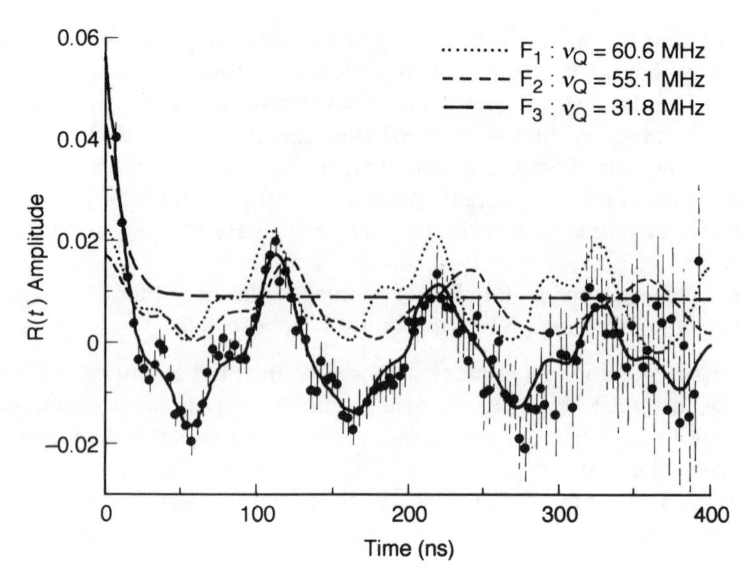

Fig. 8. Reduced TDPAD time spectrum for the recoil implantation of ^{19}F in type IIb diamond at 150 °C.

extremely high degree of locality of TDPAD with its r^{-3} dependence, arising from the conflation of the gradient with the coulombic interaction. We exploit also the short time window ('snapshot') of interrogation, before secondary effects such as diffusion can complex or distort the primary physical realities. The facility of studying what may be described as 'hot atom chemistry' is evident. It is also evident that the wealth of information arising from the use of the fluorine example justifies the extension of the TDPAD method to a large selection of pulsed incident ions: this is most effectively achieved, minimizing radiation damage, by using (primary) pulsed heavy ions themselves rather than recoil-implanted heavy ions produced by the use of pulsed protons.

A variant of TDPAD is the time differential perturbed angular correlation (TDPAC) technique. This is justifiably a version of CP(A)A in that selected (radioactive) charged particle beams are implanted into matrices of interest in pursuit of an evolving understanding of dopant behaviour. The TDPAC method differs from TDPAD in that it uses a cascade of two γ-rays in a nuclear decay through a specific excited state which has to have a quadrupole moment of sufficient strength and be of a convenient lifetime. The most commonly used radioactive nuclide is indium-111, which decays by emission of a $\beta-$particle with a half-life of 2.8 days to cadmium-111. In the de-excitation of cadmium-111 to its ground state, the transitions go through an intermediate state which satisfies the criteria, with the relevant γ-rays having energies of 173 and 247 keV respectively. Similar studies have been made with implanted hafnium-181. Such experiments, at a very early stage of the methodology and instrumentation, reveal that in natural diamond at least two sites are populated. A valuable variant of the TDPAC technique is to measure not the γ-rays from the de-excitation of the final nucleus but the conversion electrons themselves: these will be sensitive to crystal structure in their emerging transit through the crystal lattice through the blocking and channelling effects, giving direct information complementary to that from the TDPAC data on lattice location.

Emphasis has been laid, in relation to the TDPAD and TDPAC methods, on the intrinsic microscopic nature of the CP(A)A analytical capability for dopant lattice location; *macroscopic* CP(A)A in dopant distribution studies is readily realized, for example, in the implantation of stable isotopes into matrices and the depth distribution analysis thereof through the use of resonant nuclear reactions. Examples are the study of carbon in diamond (through the implantation of ^{13}C) and of lithium, boron and fluorine in diamond [19]. Such studies reveal also the relevance of specific chemistry in the details of depth distributions of dopants.

4.3 Muon spin rotation and muonium

The muon is ever more an important charged particle for unique analysis, an important contributor to CP(A)A. This is true for the muon in both positive and negative states. With the now ready availability of intense fluxes of muons, at 'meson factories', there has been rapid application of muons as probes in the study of materials. That this is so should not surprise us. The muon has for some time been anticipated as a unique charged particle probe, as the interaction of the muon with its immediate environment can be tracked through its spin polarization [20]. Furthermore, the muon (in its negative form ostensibly a heavy electron) interacts through weak nuclear and electromagnetic forces, so that radiation damage effects are minimal in the material under investigation. The relevant parameters for the muon are:

mass	$206.7684\,m_e$
charge	$\pm e$
spin	$\frac{1}{2}$
magnetic moment μ_μ	$3.1833448\,\mu_p$
lifetime τ_μ	$2.1994\,\mu s$

Muons are produced through the reaction

$$\pi^+ \rightarrow \mu^+ + \bar{v}_\mu$$

with
$$\tau_\pi = 2.6 \times 10^{-8}\,s$$

In this reaction parity is violated, so that the muons are polarized along the axis of the beam. In their turn the muons decay as

$$\mu^+ \rightarrow e^+ + \bar{v}_e + v_\mu$$

Parity is again violated and the positrons have an asymmetrical distribution with respect to the muon spin. When such muons are stopped in a sample of analytical interest, they will precess at a frequency determined by the magnetic moment of the muon and the magnetic field which it experiences. A set of positron detectors located in the precession plane will measure the asymmetric decay distribution precessing at the same frequency and giving rise to an oscillatory counting rate. The frequency (or frequencies) thus determined experimentally allows one to deduce the magnetic properties of the site which the stopped muons populated before their decay. The theoretical description of the process must include a damping function which describes how the muons are depolarized in the sample. The depolarization is rich in information, and muon spin rotation (μSR) studies can reveal muon diffusion in metals, phase transitions and defect structures, *inter alia*.

When, however, a positive muon is stopped in an insulator or semiconductor, it may bind with an electron to form the *atom* 'muonium' ($\mu^+e^- = \mathrm{M}\mu$). This atom is of considerable interest. Although the muonium mass is substantially less than that of a hydrogen atom, the reduced mass for the two is closely similar. We may therefore think of muonium as a light isotope of hydrogen with similar (chemical) properties. The are obvious advantages in studying the behaviour of hydrogen in solids through muonium, particularly as the majority of the hydrogen in the important materials silicon and diamond is present in a form (or forms) which is neither optically nor electrically active. Muonium can readily be traced as described above for the muon, and there is no radiation damage to the solid as would have been the case if, for example, hydrogen had been implanted for study. Just as the muon or proton in metals can be regarded as the simplest impurity problem in conductors, muonium as a substitute for hydrogen is a prototype of an impurity centre in a non-metal, and it is the simplest dopant in a semiconductor.

The comparative properties of hydrogen and muonium are given in Table 2. The formation of the paramagnetic muon state, muonium, can be observed as

TABLE 2.

	H	Mμ
Mass (m_e)	1837.15	207.769
Reduced mass (m_e)	0.999456	0.995187
Ground state radium (nm)	0.0529465	0.0531736
Ground state energy (eV)	– 13.5984	– 13.5403

before for the bare muon, that is, by the characteristic precession pattern of the muon spin polarization, but now under the influence of the bound electron and the applied field. A similar μSR experiment is configured, taking advantage of the fact that positron emission is most probable in the direction of the muon spin, so that the positron counting rate is modulated with the muon precession frequencies.

What may we expect to observe? For a muonium atom in its 1 s ground state in an external field $\bar{B}$, the spin Hamiltonian function is

$$H_{M\mu} = A\bar{S}_\mu \cdot \bar{S}_e + g_e\mu_e\bar{S}_e \cdot \bar{B} + g_\mu\mu_\mu\bar{S}_\mu \cdot \bar{B}$$

where $\bar{S}_\mu$ and $\bar{S}_e$ are the spin operators for the muon and the electron respectively, and the vacuum values of the g factors and the magnetic moments are known. The hyperfine constant

$$A = h\nu_0 = h\omega_0 = -8\pi/3\,|\psi(0)|^2\, g_e\mu_e g_\mu\mu_\mu$$

where $|\psi(0)|^2$ is the electron density at the muon. A precise value for ν_0 in vacuum is 4463.30235(52) MHz.

The Hamiltonian function can be solved analytically to give the eigenvalues

$$\frac{E_{1,3}}{h} = v_{1,3} = \frac{v_0}{4} \pm \frac{\Gamma B}{2} \mp \gamma_\mu B$$

and

$$\frac{E_{2,4}}{h} = v_{2,4} = -\frac{v_0}{4} = \pm\frac{(v_0^2 + \Gamma^2 B^2)^{\frac{1}{2}}}{2}$$

where

$$\gamma_\mu = \frac{g_\mu\mu_\mu}{h} = 13.55 \text{ MHz/kG}$$

and

$$\Gamma = \frac{(g_e\mu_e - g_\mu\mu_\mu)}{h} = 2.82 \text{ MHz/G}$$

The energy level diagram for muonium is predicted as given in Fig. 9. In a longitudinal external magnetic field, where the initial μ^+ polarization is parallel with the applied field, only transitions obeying the selection rule $\Delta M = 0$ may occur. This is ν_{24} only. Recently, in longitudinal field studies, the quenching of the polarization in diamond showed strong temperature dependence and also a strong dependence on diamond type—this augurs well for defect studies. In an external magnetic field transverse to the initial polarization, the transitions $\Delta M = \pm 1$ are allowed. These correspond to ν_{12}, ν_{23}, ν_{34} and ν_{14}. Instrumental limitations generally exclude observation of the latter two frequencies, but the two lower frequencies are clearly observed.

However, other frequencies are also observed, and to give interpretation to these additional frequencies it is deduced that two different muonium states exist.

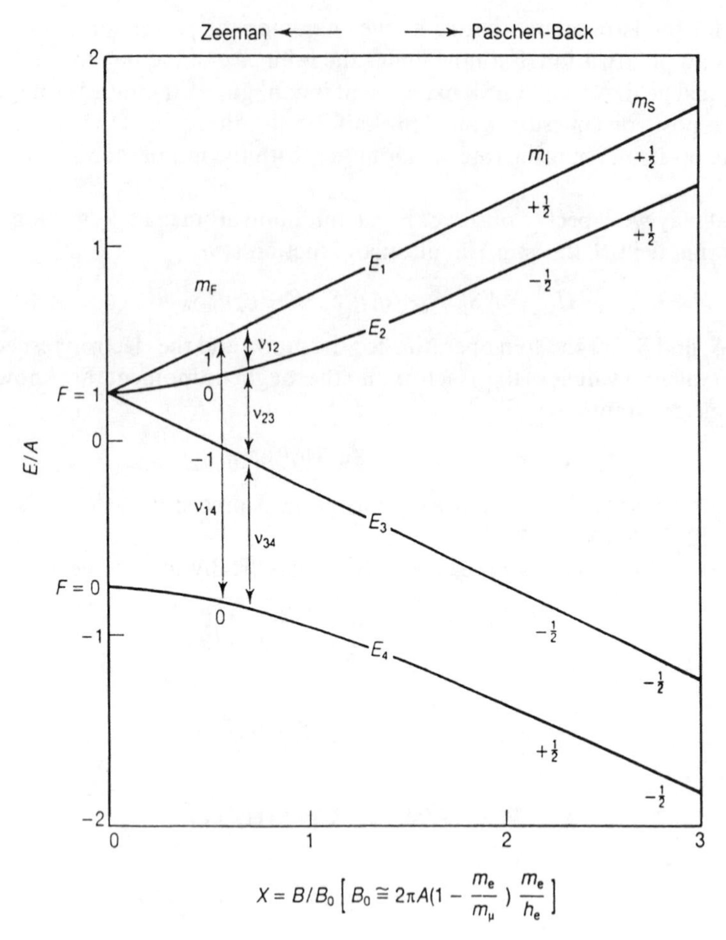

Fig. 9. Breit–Rabi diagram for normal muonium. The magnetic field X is measured in units of B_0. In a vacuum, B_0(vac) = 0.15868 T. The transitions indicated correspond to μSR frequencies in a transverse field.

The muonium state considered above is coded as normal muonium (Mμ) and the additional muonium state is called anomalous muonium (M$\mu^{\star}$). The M$\mu^{\star}$ frequencies are well reproduced by an anisotropic spin Hamiltonian function, the hyperfine interaction of which has an axial symmetry about one of the four equivalent $\langle 111 \rangle$ directions z, the $\langle 111 \rangle$ axes in the crystal. The Hamiltonian can be constructed and analytical solutions found for $\theta = 0°$ and $\theta = 90°$. A Breit–Rabi diagram for anomalous muonium can be determined, from which predictions can be made. For zero external field, three M$\mu^{\star}$ frequencies should be observed, given by

$$\nu_{12} = \nu_{32} = \frac{(A_{\parallel} - A_{\perp})}{2h}$$

$$\nu_{14} = \nu_{34} = \frac{(A_{\parallel} + A_{\perp})}{2h}$$

$$\nu_{24} = \frac{A_{\perp}}{h}$$

which satisfy the relationship

$$\nu_{12} + \nu_{14} = \nu_{24}$$

To illustrate the analytical power of muon and muonium spin rotation in CP(A)A, some results in the study of diamond are given [21]. Both normal and anomalous muonium are observed in diamond. This formation is shown to be an intrinsic property of Group IV semiconductors. $M\mu$ is formed in less than 10^{-10} s after the μ^+ enters the target. From the detailed temperature dependence of the $M\mu^\star$ frequencies, it is concluded that $M\mu^\star$ forms from the precursor $M\mu$ with the conservation of both muon and electron polarization. With increasing temperature we observe the transformation $M\mu \rightarrow M\mu^\star$.

From a study of the transition from normal to anomalous muonium in diamond, it is established that the isotropic part of the $M\mu^\star$ hyperfine interaction is negative. This implies that $M\mu^\star$ is not just a distorted version of isotropic $M\mu$, as this would require a positive electron spin density at the muon. An alternative model of $M\mu^\star$ is a paramagnetic complex similar to a molecular radical, with the unpaired electron removed from the muon. The weak negative spin density at the muon is then the result of an exchange polarization mechanism.

Another result is that the thermally activated conversion of $M\mu$ to $M\mu^\star$ in diamond proceeds without the loss of the electron polarization. At room temperature $M\mu$ and $M\mu^\star$ are observed simultaneously in the μSR spectra; both are formed promptly upon μ^+ implantation. The thermally activated conversion of $M\mu$ to $M\mu^\star$, together with the fact that $M\mu^\star$ is observed in diamond up to at least 1100 K, establishes that $M\mu^\star$ is the more stable state. There is evidence that only a single jump is required by $M\mu$ to form $M\mu^\star$.

Normal muonium is generally interpreted to be at a *tetrahedral interstitial* site. The anisotropic hyperfine interaction of $M\mu^\star$ implies that it is immobile in its site. Modelling of this site is consistent with $M\mu^\star$ being situated near the centre of the host bond. Cluster calculations of the potential energy surface traversed by muonium in moving from the tetrahedral interstitial site to the centre of a strongly stretched C—C bond in diamond reproduces the observation that the $M\mu$ state is metastable, but apparently overestimates the true barrier height. This *bond-centred site* for $M\mu^\star$ is attractive also, as it was one of the strongly favoured sites found in the TDPAD studies described earlier.

It is evident that muon spin rotation has entered the CP(A)A family as a method of unique value, and much can be anticipated from it in the future in the nuclear analysis field.

4.4 The special role of μ^- beams in CP(A)A: μ^-SR

If a negative muon is used in addressing a solid of interest, it can be captured in the outermost Bohr orbits of an atom. In this event then, within a time interval of $\approx 10^{-14}$ s, the negative muon cascades from an orbit with quantum number $n > -7$ to the 1 s ground state orbital. The negative muon behaves as a heavy electron, the final 1 s orbital being some 207 times smaller than for the 1 s electron case. Therefore, during the final stage of the implanted negative muon thermalization, it is in fact inside the host nucleus.

In the final orbital transitions, characteristic photons are emitted, but these are MeV in energy for the muon, whereas they might have had keV energy for the electron. These are interesting in their own right as diagnostic indicators—a special case of CP(A)A!

During this cascading process of the negative muon, most of its polarization is lost, but sufficient remains for μ^-SR experiments. In the 1 s orbital the negative muon screens one nuclear charge! Hence negative muon capture transforms an atom of charge number Z into one with charge number $Z-1$. In a semiconductor material, the negative muon capture by a matrix nucleus on a regular lattice site is electrically equivalent to the sudden (and damage-free) introduction of an acceptor atom. We may therefore investigate the electronic process associated with the sudden introduction of an acceptor atom by the use of the μ^-SR technique, in a time window not accessible to other techniques of 1 ns – 1 μs after muon implantation. The damping rate in this time window is immeasurably small, hence the measured asymmetry of the electron emission is a measure of the muon spin polarization.

In silicon, the spin depolarization of muons captured in 1 s orbitals is thought to be caused by the interaction of the muon magnetic moment with the magnetic moments of the unpaired electrons of quasi-aluminium where the q-Al is in interstitial sites as a result of the recoil of the cascade gamma-quanta. For silicon, only below 30 K are the Frenkel pairs sufficiently long lived to give rise to measurable muon depolarization. An experimental μ^-SR measurement in silicon as a function of temperature shows the persistence of 4% asymmetry from 300 K to ≈ 40 K, and then a steady reduction of the asymmetry in going from 40 K to ≈ 4 K!

The mechanism should be tested for other materials such as Ge and diamond. The comparative parameters of relevance are shown in Table 3.

For diamond, the ratio of the gamma recoil energy to the displacement threshold energy is so small that the muonic atoms (quasi-boron) should remain

TABLE 3.

Material	E_{2-1} (keV)	$E_{\gamma-\text{recoil}}$ (eV)	E_d (eV)	A (%)
Si	403	3.11	12.9	4.0
Ge	1772	22.8	14.5	< 0.1
Diamond	75.6	0.26	55	3.0

on their lattice sites, forming a low lying acceptor level that is occupied even at low temperatures. Then all electrons are paired, so that there should be no further muon depolarization. This is precisely as observed experimentally, with the measured asymmetry of $\approx 3\%$ remaining constant from 300 K to 4 K!

By contrast, in germanium, the q-gallium interstitials are paramagnetic, and muon depolarization should be so fast that no muon polarization should be observed at all; this remains to be confirmed experimentally.

The excellence of the results for silicon and diamond gives confidence that at least a qualitative understanding of the behaviour of negative muons in elemental semiconductors has been successfully achieved. We are indeed able to study in this way and get insights into electronic processes on a time scale not accessible to standard semiconductor physics—a powerful example of CP(A)A.

5 THE IMMEDIATE FUTURE

At this time we therefore conclude a review on nuclear reaction analysis or, more broadly, charged particle (activation) analysis, with the knowledge that the standard techniques are firmly established and becoming ever more routine, and with clear evidence that the niche that has been established for these standard techniques will remain relevant and significant. In addition, however, we see that CP(A)A has moved beyond mere elemental analysis and has already developed into a number of more sophisticated methods, some of which have been addressed in this article, of singular and unique value in a number of fields of materials science. One is left with no doubt, however, that there is yet more to come!

6 ACKNOWLEDGEMENTS

It is a pleasure to acknowledge the wide span of collaborators and students over many years who have been vital to the execution of our own contributions to CP(A)A—though it is invidious to identify any particular individual, I am happy to record the close collaboration I have enjoyed in recent years with my erstwhile student and now valued colleague, Simon Connell. Notable also has been the role of key technical staff without whom little would have been possible, again it is with diffidence that I identify individuals, but it is appropriate that I record the enormous contributions that Mick Rebak has made in our work in the skilful preparation of diamond specimens for use as 'nuclear' targets. The ongoing support and encouragement of the Foundation for Research Development, the De Beers Industrial Diamond Division (Pty) Ltd and, in particular, their seniors officers the late Henry Dyer (to whom this article is dedicated), Corrie Phaal and Rob Caveney, and the University of the Witwatersrand is recorded with great appreciation.

REFERENCES

1. L. C. Northcliffe and R. F. Schilling, Range and stopping power for heavy ions, *Nucl. Data Tables* **A7**, 233 (1970).
2. H. H. Andersen and J. F. Ziegler, *Hydrogen stopping Powers and Ranges in All Elements*, Plenum Press, New York (1977).
3. J. F. Ziegler and W. K. Chu, Stopping cross-sections and backscattering factors for the ^{4}He ions in matter, *At. Data Nucl. Data Tables* **13**, 463 (1974).
4. H. D. Betz, *Rev. Mod. Phys.* **44**, 465 (1972).
5. J. Lindhard and M. Scharff, *Mat. Fys. Medd. Dan. Vid. Selsk.* **27**, (15) (1953).
6. J. I. W. Watterson, J. P. F. Sellschop and A. Zucchiatti, Sampling adequacy and measurement significance in ion beam analysis, *Nucl. Instrum. Methods Phys. Res., Sect. B* **B28**, 554 (1987).
7. J. P. F. Sellschop, Charged particle (activation) analysis, in *Elemental Analysis by Particle Accelerators*, edited by Z. B. Alfassi and M. Peisach, CRC press, Boca Raton (1992), pp. 75–149.
8. E. Ricci and R. L. Hahn, *Anal. Chem.* **37**, 742 (1965); **39**, 794 (1967).
9. J. Lindhard, *Mat. Fys. Dan. Vid. Selsk.* **34** (14) 1 (1965).
10. L. C. Feldman and J. W. Mayer, *Fundamentals of Surface and Thin Film Analysis*, North-Holland, Amsterdam (1986).
11. J. U. Andersen, O. Andreason, J. A. Davies and E. Uggerhoj, *Radiat. Eff.* **7**, 25 (1971).
12. J. P. F. Sellschop, S. C. Connell, K. Bharuth-Ram, H. Appel, E. Sideras-Haddad and M. C. Stemmet, *Mater. Sci. Eng.* **B11**, 227 (1992).
13. D. Wegner, *Hyperfine Interactions* **23**, 179 (1985).
14. J. P. F. Sellschop, in *The Properties of Natural and Synthetic Diamond*, edited by J. E. Field, Academic Press, New York (1992), pp. 81–179.
15. E. Sideras-Haddad, S. H. Connell, J. P. F. Sellschop, K. Bharuth-Ram, M. C. Stemmet S. Naidoo and H. Appel, *Nucl. Instrum. Methods Phys. Res., Sect. B* **B64**, 237 (1992).
16. S. Connell K. Bharuth-Ram, H. Appel, J. P. F. Sellschop, M. C. Stemmet and J. E. Lowther, *Solid State Commun.* **68**, 587 (1988).
17. W. Verwoerd, *Nucl. Instrum. Methods Phys. Res., Sect. B* **B35**, 509 (1988).
18. H. W. Fesq, D. M. Bibby, C. S. Erasmus, E. J. D. Kable and J. P. F. Sellschop, *Phys. Chem. Earth* **9**, 817 (1975).
19. T. D. Derry, R. A. Spits and J. P. F. Sellschop, *Mater. Sci. Eng.* **B11**, 249 (1992).
20. A. M. Stoneham, *Helv. Phys. Acta* **56**, 449 (1983).
21. E. Holzschuh, W. Kundig, P. F. Meier, B. D. Patterson, J. P. F. Sellschop, M. C. Stemmet and H. Appel, *Phys. Rev. A* **A25**, 1272 (1982).

Chapter 13

PARTICLE-INDUCED X-RAY EMISSION

U. A. S. Tapper*

Ion Beam Analysis Division, National Accelerator Centre, Faure 7131, South Africa

W. J. Przybyłowicz† and H. J. Annegarn

Schonland Research Centre for Nuclear Sciences, University of the Witwatersrand, Private Bag 3, WITS 2050, South Africa

1 INTRODUCTION

Particle-induced X-ray emission (PIXE) is a method in which X-ray emission is used for elemental analysis. Vacancies in inner atomic shells are normally created by high energy hydrogen or helium ions, with kinetic energies of millions of electron volts (MeV). At these energies, X-ray excitation cross-sections are similar to those of electron beams in the 10–50 keV energy range used for electron microprobe microanalysis (EPMA). Compared with the electron microprobe, the PIXE technique has some distinct advantages, resulting in its rapid development over the last 20 years. Major features of PIXE are its multi-elemental character (in principle, all elements from B to U can be measured), high sensitivity (absolute detection limits down to 1 pg and relative detection limits down to 0.1 mg/kg), smooth variation of relative detection limits with atomic number, the ability to analyse small amounts of material (1 mg or less), speed of analysis (1–10 min per specimen) and the possibility for automation. PIXE is generally considered as a non-destructive technique. Sample preparation, if any, is relatively simple. In

* Present address: De Beers Diamond Research Laboratory, P.O. Box 916, Johannesburg 2000, South Africa

† On leave from the Faculty of Physics and Nuclear Techniques, The Academy of Mining and Metallurgy, 30-059 Cracow, Poland

Chemical Analysis by Nuclear Methods Edited by Z. B. Alfassi

principle, any dry sample which can be fitted into a vacuum sample chamber can be analysed 'as is'. It is also possible to extract an ion beam out of a chamber (under vacuum) into atmosphere—the external beam mode. This mode enables analyses of bulky samples and is especially useful when analysing delicate and valuable specimens, like old publications (e.g. the Gutenberg Bible), paintings, stamps and papyrus documents.

The first demonstration of the PIXE method as a powerful analytical technique was in 1970 [1]. The original application, and still one of the most important, was the analysis of aerosol samples—air-borne particles deposited onto a thin filter substrate. For this sample type, no existing technique can compete with PIXE in terms of detection limits, speed and economy. Other particulate sample types are also favourable for PIXE analysis, e.g. particles extracted from water samples or blood, because competing wet chemistry techniques such as inductively coupled plasma mass spectrometry (ICP–MS) require that the samples be completely dissolved prior to analysis.

The use of a focussed ion beam as an excitation source provides a microanalytical tool with unique potential. Standard PIXE typically uses a beam size of the order of 10 mm diameter. Micro-PIXE is performed using a beam probe which

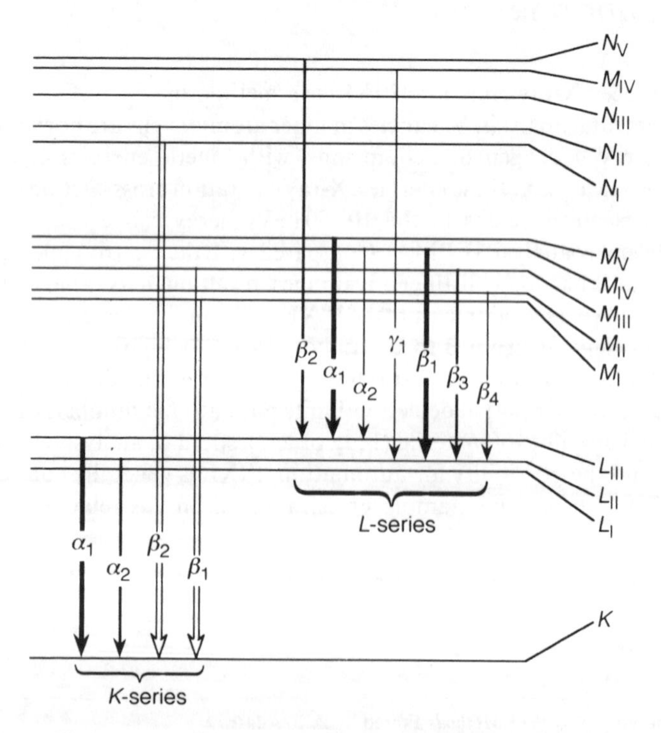

Fig. 1. Electron transitions responsible for *K*- and *L*-series X-rays [2].

can be as small as 0.5 μm. By attaining detection limits of the order of 1 ppm, extremely high absolute sensitivity can be obtained. For instance, if a 1 μm thick tissue section is probed by a 1 μm beam, and detection limits of 10 ppm are reached, trace elements with a mass of 10^{-17} g will be detected. However, the microprobe mode is thoroughly described in the next chapter, and here we shall instead concentrate on more general aspects of PIXE.

The range of an ion beam incident on a solid target depends on its charge and velocity. For a typical situation of hydrogen ion beams of a few MeV energy, the range is a few tens of μm. PIXE is insensitive to the surface condition, such as thin oxide layers. Hence, although the technique is surface oriented, the ion probe penetrates deeply enough to reach the bulk of most sample materials.

While slowing down in a sample, the beam induces several physical processes, some of them carrying element-specific information. Among them, creation of atomic inner shell vacancies has the largest cross-sections. An excited atom constitutes an unstable system—a rearrangement of atomic electrons will take place until the atom again reaches the minimum potential energy allowed by the Pauli exclusion principle. Released energy appears either as X-ray photons or Auger electrons; either can be used for analytical purposes. X-ray energies are a function of the atomic number Z of the target atom. Filling of vacancies by outer electrons gives rise to either K_α or K_β emission lines, if the vacancies were in the K shell, or L_α, L_β or L_γ emission lines, if the vacancies were in the L shell (Fig. 1). The following approximate expressions can be used to calculate X-ray energies:

$$E(K_\alpha) = Z^2/100\,\text{keV}$$

$$E(L_\beta) = Z^2/750\,\text{keV}$$

$$E(M_\gamma) = Z^2/3000\,\text{keV}$$

Emitted X-rays are normally detected by an energy dispersive Si(Li) detector. This detector type is sensitive to X-rays in the 1–30 keV range. From the expressions above it can be concluded that K X-rays from elements of atomic number 11 (Na) to ≈ 55 (Cs) can effectively be detected. For the heavier elements, L and M X-rays have to be used. L X-rays are useful for analysis of elements of atomic number 70 (Yb) to 82 (Pb). The M X-rays are usually not used for analysis but may be present in the spectrum. Elements in the in-between region, the rare earth elements, are not well suited for standard PIXE analysis owing to severe spectral interferences between their L X-rays—of relatively low fluorescence yield—and K X-rays of lighter elements. Despite this, the broad range of elements available in a single analysis is very attractive in several fields of application.

The instrument used for PIXE analysis consists of a small accelerator, accelerating ions to MeV energies, and a vacuum chamber with a sample changer (Fig. 2). An X-ray detector views the sample at a short distance. The detector output signal is fed to a multichannel analyser (MCA), hosted by a computer, via pulse processing electronics.

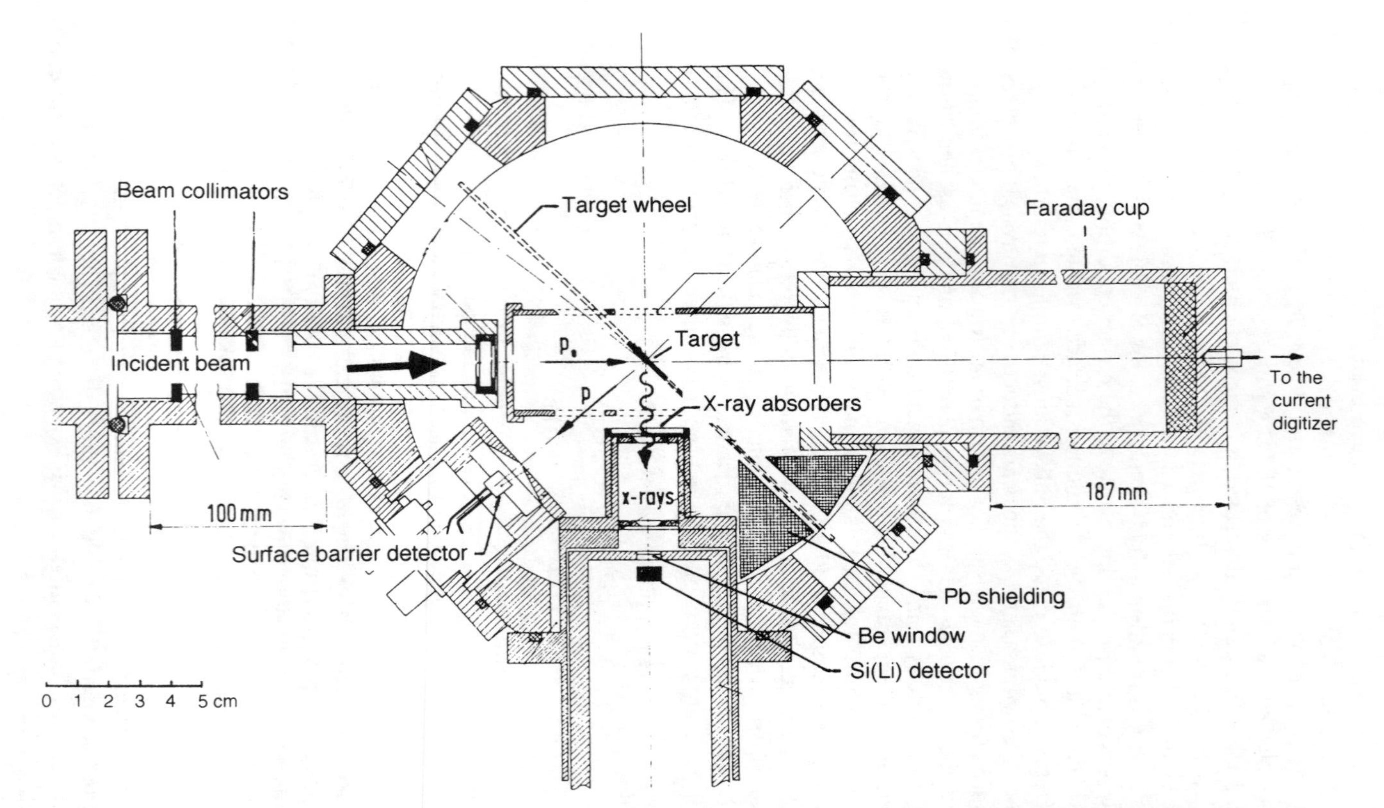

Fig. 2. Design details of the typical PIXE chamber, shown in a horizontal cross-section through the chamber at the beam axis (top view). The proton beam enters from the left and strikes a target mounted on the target wheel, which is located in the centre of the chamber at 135° to the beam axis. Proton induced X-rays pass through the absorber and strike the X-ray detector crystal through a Be window. Backscattered protons strike the surface barrier particle detector. Protons passing through the target and those scattered in the target are collected by the Faraday cup and Faraday chamber, situated in the middle of the PIXE chamber [3].

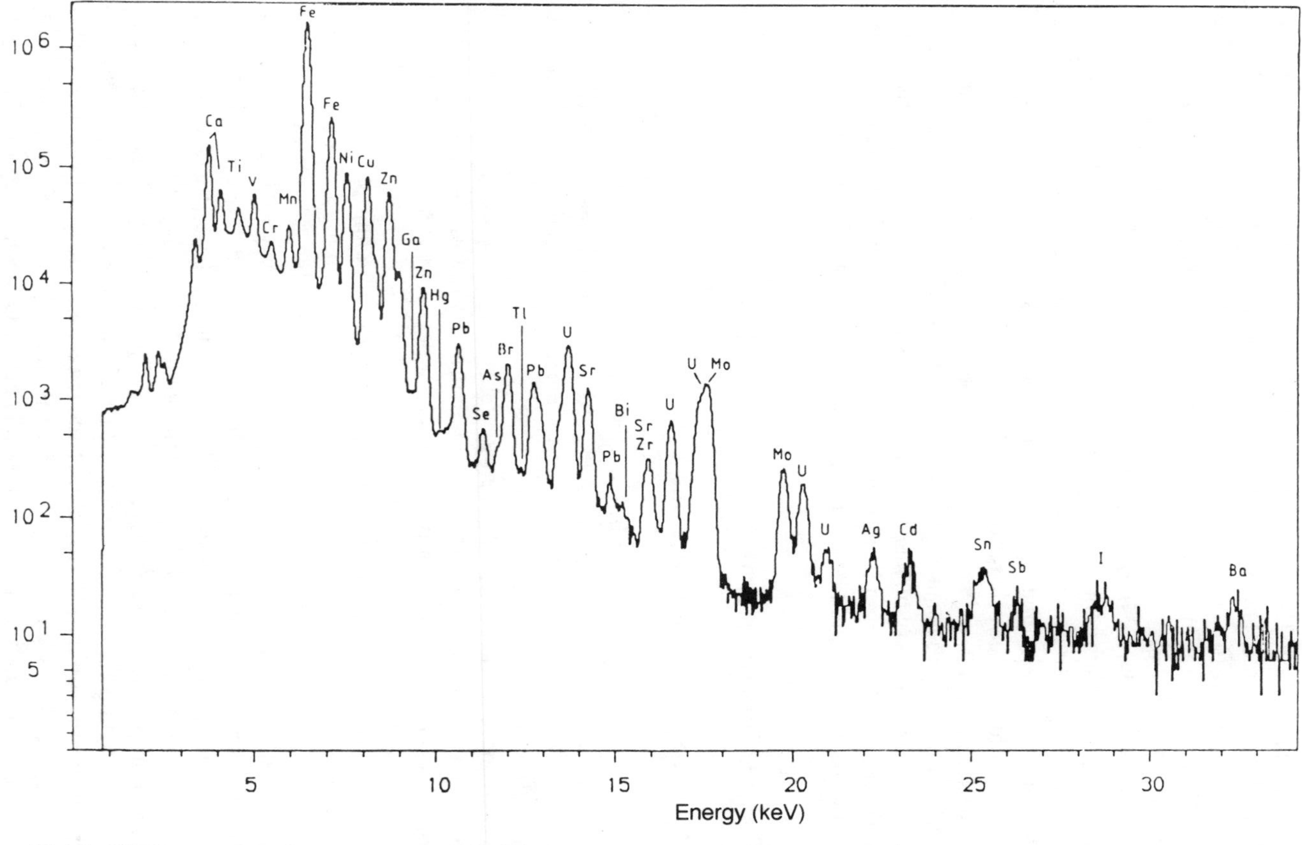

Fig. 3. PIXE spectrum of a seawater sample. The water was preconcentrated and spotted on to a thin plastic foil [4].

An example of a PIXE spectrum from a water sample evaporated onto a thin film is given in Fig. 3. The continuous background under the peaks is due mainly to bremsstrahlung from electrons accelerated in collisions with the incident high energy ions. As the atomic processes in the sample are well known, it is possible to calculate the elemental concentrations of the target from the spectrum without reference to standards. In practice, the set-up is calibrated using certified standard samples. Because the physics of the underlying processes is known, the risk of unknown interferences in the relatively complicated spectrum is small, giving a safe analytical technique. Thus, from peak areas in the spectra, elemental concentrations can be calculated.

An excellent textbook which thoroughly describes the PIXE technique and its applications was published in 1988 by Johansson and Campbell [5], and gives the reader the opportunity further to penetrate this field. To date, six international conferences on PIXE and its applications have been held. The two most recent sets of proceedings could be used to locate existing PIXE facilities [6, 7].

2 X-RAY PRODUCTION

2.1 Ionization cross-sections

The desired result of bombarding a specimen with particles is to eject bound electrons from the *K* or *L* atomic orbits, with subsequent emission of characteristic X-rays. It is thus important to know the correct values of the cross-sections for these processes.

There have been several theoretical approaches describing the ionization process, with the object of deriving analytical expressions for the ionization cross-section [8]. The currently most used model in the context of PIXE is the ECPSSR model [9]. It is a refinement of a Plane Wave Born Approximation (PWBA) model, using perturbation theory for the handling of the interaction. The initial state is a plane wave (describing the incoming ion) and a bound electron, and the final state is the wave function for the scattered ion and the released electron. With corrections for energy loss during interaction (E), electrostatic coulomb deflection from the atomic nucleus (C), perturbation of the atomic stationary states (PSS) and relativistic effects (R), agreement is close to experimental data. Figure 4 shows the theoretical ECPSSR predictions of *K* and *L* shell ionization cross-sections as functions of proton energy *E* and target atom. The *K* shell data are now well established, with theoretical predictions matching experiment to within 3–5% for all but the lowest ion energies [11]. The *L* shell data are more complex, with three subshells and many more X-ray peaks. X-ray line production cross-sections are typically accurate to within 5–15% for a wide range of light ion energies and target combinations [11]. The most frequently used ionization cross-sections for *K* and *L* subshells are those of Cohen and

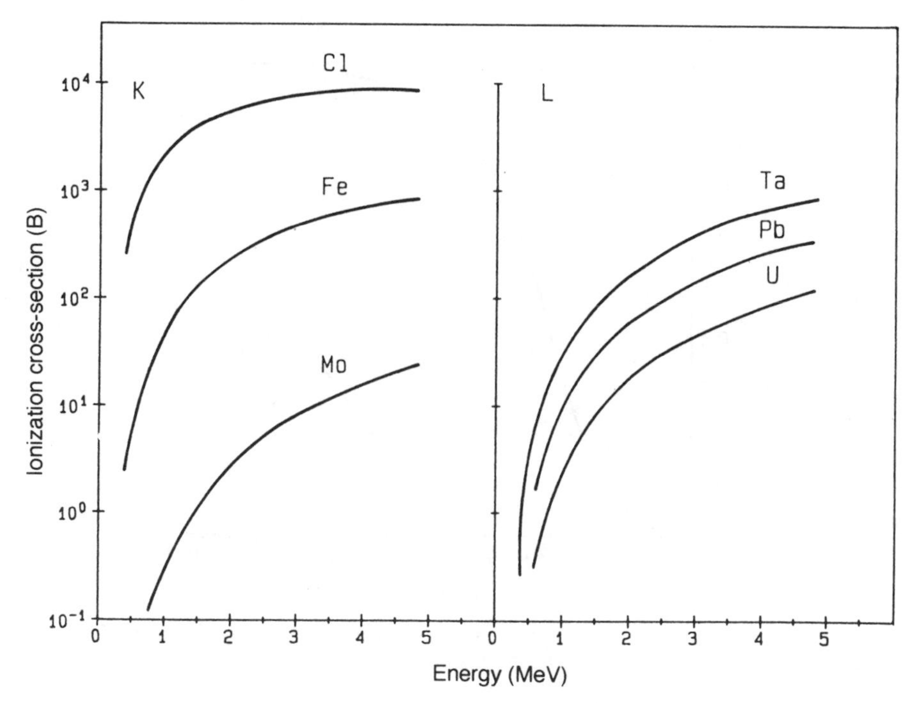

Fig. 4. The *K* and *L* shell ionization cross-sections as functions of proton energy *E* and target atom. The values are theoretical ECPSSR predictions [10]. From Johansson and Campbell [5].

Harrigan [10], calculated for elements from $Z = 6–95$ and for proton and α-particle beam energies between 100 keV and 10 MeV.

Cross-sections for alpha beams can be estimated as 4 times the corresponding proton cross-section for the same ion velocity, i.e. energy per nucleon. For example, 2 MeV protons have an Fe *K* shell ionization cross-section of 251 barn, whereas the value for 8 MeV alphas is 922 barn.

2.2 Characteristic radiation

Once a vacancy has been produced, irrespective of excitation mode, electrons in the excited atom will rearrange and release excess energy as either X-ray photons or Auger electrons. The ratio between the two competing processes, called the fluorescence yield, is shown in Fig. 5, as a function of atomic number *Z*, for both *K* and *L* shell and in Table 1. Figure 1 explains the origin of the principal *K* and *L* X-ray lines (*K* and *L* series) from electron transitions from higher levels to levels

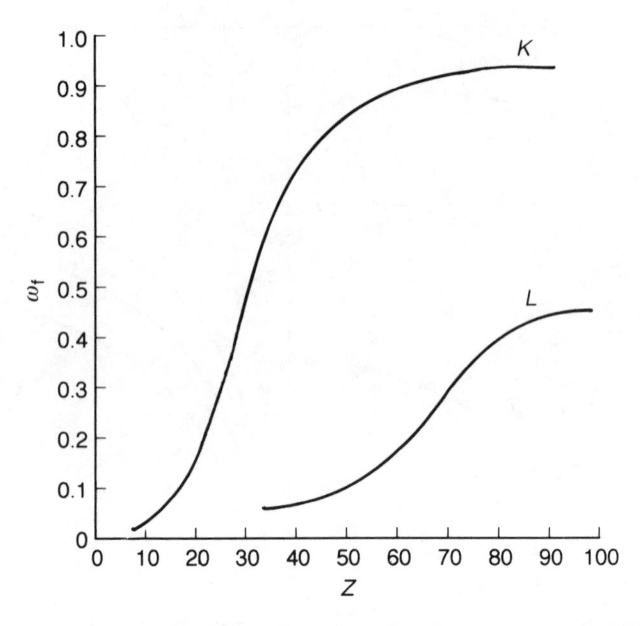

Fig. 5. Fluorescence yields (K- and L-shell) as functions of atomic number Z [2].

TABLE 1.
Data related to the atomic K shell and to K X-ray emission. U_K is the K electron binding energy [12]. The five K X-ray energies listed [12] are those of the principal lines; at lower Z values, where Si(Li) detectors have insufficient resolution to justify considering the members of the K_α and K_β doublets as separate, a weighted mean energy is given. The K fluorescence yield ω_K is from a semi-empirical fit to selected experimental data [13]. The intensity ratio of the K_α and K_β groups was interpolated graphically using theoretical values [14] except for the values between atomic numbers 20 and 30, which are means of selected experimental values [15]. Reproduced from Johansson and Campbell [5] with permission.

		K X-ray energies (keV)			
Z	U_K	K_α	K_β	ω_K	K_β/K_α
11	1.072	1.041		0.0213	
12	1.303	1.254		0.0291	
13	1.558	1.485	1.557	0.0387	0.0134
14	1.839	1.740	1.836	0.0504	0.0294
15	2.149	2.014	2.139	0.0642	0.0472
16	2.472	2.308	2.464	0.0804	0.0659
17	2.823	2.622	2.816	0.0989	0.0862
18	3.206	2.957	3.190	0.1199	0.1088
19	3.608	3.313	3.587	0.1432	0.1211
20	4.039	3.691	4.014	0.1687	0.1230
21	4.490	4.089	4.461	0.1962	0.1272
22	4.966	4.509	4.933	0.2256	0.1304

TABLE 1 *(continued)*

Z	U_K	K X-ray energies (keV)					ω_K	K_β/K_α
		K_α		K_β				
23	5.466	4.950		5.429			0.2564	0.1328
24	5.991	5.412		5.948			0.2885	0.1336
25	6.538	5.895		6.491			0.3213	0.1348
26	7.111	6.399		7.058			0.3546	0.1355
27	7.711	6.925		7.652			0.3880	0.1365
28	8.332	7.461	7.478	8.266			0.4212	0.1377
29	8.981	8.028	8.048	8.905			0.4538	0.1388
30	9.661	8.616	8.639	9.571		9.660	0.4857	0.1409
31	10.367	9.224	9.250	10.267		10.370	0.5166	0.1440
32	11.104	9.855	9.887	10.980		11.100	0.5464	0.1493
33	11.867	10.508	10.544	11.724		11.860	0.5748	0.1560
34	12.658	11.182	11.222	12.494		12.650	0.6019	0.1624
35	13.474	11.878	11.924	13.290		13.470	0.6275	0.1683
36	14.326	12.595	12.648	14.109		14.312	0.6517	0.1727
37	15.200	13.335	13.395	14.958		15.186	0.6744	0.1780
38	16.105	14.098	14.165	15.832		16.085	0.6956	0.1831
39	17.039	14.884	14.959	16.734		17.013	0.7155	0.1876
40	18.000	15.691	15.775	17.662		17.969	0.7340	0.1913
41	18.983	16.521	16.615	18.618		18.952	0.7512	0.1950
42	20.000	17.375	17.480	19.601		19.965	0.7672	0.1981
43	21.044	18.251	18.367	20.612		21.005	0.7821	0.2010
44	22.117	19.150	19.279	21.649		22.074	0.7958	0.2040
45	23.220	20.253	20.216	22.699	22.724	23.172	0.8086	0.2070
46	24.350	21.019	21.176	23.791	23.819	24.299	0.8204	0.2100
47	25.514	21.990	22.162	24.912	24.941	25.455	0.8313	0.2130
48	26.711	22.984	23.173	26.060	26.094	26.644	0.8415	0.2160
49	27.940	24.002	24.210	27.238	27.276	27.893	0.8508	0.2195
50	29.200	25.044	25.271	28.443	28.485	29.111	0.8595	0.2230
51	30.491	26.110	26.359	29.679	29.725	30.392	0.8676	0.2266
52	31.814	27.202	27.473	30.944	30.995	31.704	0.8750	0.2305
53	33.170	28.318	28.613	32.239	32.295	33.047	0.8819	0.2340
54	34.566	29.459	29.779	33.564	33.625	34.421	0.8883	0.2368
55	35.985	30.625	30.973	34.920	34.987	35.824	0.8942	0.2405
56	37.441	31.817	32.194	36.304	36.378	37.262	0.8997	0.2433
57	38.925	33.034	33.442	37.720	37.801	38.730	0.9049	0.2455
58	40.444	34.279	34.720	39.171	39.258	40.238	0.9096	0.2470
59	41.991	35.550	36.026	40.653	40.748	41.774	0.9140	0.2490
60	43.569	36.847	37.361	42.166	42.271	43.340	0.9181	0.2504

K and *L*. Bearden [16] compiled a comprehensive tabulation of energies for all series.

For elements lighter than Cl, the K_α and K_β lines cannot be resolved using a standard Si(Li) detector and they are observed as a single peak. This occurs also owing to the low intensity of the K_β line for low atomic number elements. The intensity ratios of the K_α and K_β groups (branching ratios) are given in Table 1. For *K* X-rays these ratios are known to a relatively high precision, though there still exists a slight difference between measured and calculated ratios in the region of atomic numbers $Z = 21$–32 [17]. It is generally accepted that the branching ratios for L X-rays (alpha-, beta- and gamma-) are not known with high precision. Here the picture is complicated by the dependence of the branching ratio on the mode of excitation, i.e. the ratios differ slightly for different incident ion energies. A useful source of relative X-ray intensities in PIXE spectra is that of Cohen and Harrigan [18].

2.3 Continuous background

The continuous background underlying the characteristic X-rays is bremsstrahlung radiation, produced by scattering of charged particles. The cross-sections for inner electron ionization, which give rise to the element-characteristic X-ray photons, are of similar size for electrons of some tens of keV and for few MeV heavy ions (p, He^+). At these energies the respective incident particles have a velocity of the same order as the inner shell electrons in the target. This condition is not chosen arbitrarily but because it is known that the ionization cross-section reaches a peak when a resonance between these velocities is attained. Comparing X-ray spectra obtained by electron and ion beam excitation, the advantage for ion beams stems from the lower bremsstrahlung background, rather than enhanced X-ray production. The signal-to-noise ratio is enhanced 2–3 orders of magnitude in the ion case. Heavy ions in the beam are not deflected much and therefore do not produce much bremsstrahlung. The dominant contribution to bremsstrahlung in a PIXE spectrum comes from electrons accelerated by the incident ions. This is referred to as secondary electron bremsstrahlung.

The maximum energy transfer from an ion with mass M_p and energy E_p to a free electron m_e, modelling electrons in the outer shells, can be approximated to $4m_e E_p/M_p$, which for 2.5 MeV protons (a common beam energy used in PIXE) is 5.4 keV. Secondary electron bremsstrahlung only contributes to X-ray energies below this value. Background above this limit is due to bremsstrahlung from ejected bound electrons and Compton scattered gamma-ray photons. Gamma-rays arise from nuclear reactions induced by the beam impinging on collimators and the target. Compton scattered photons reaching the X-ray detector crystal result in a continuous background extending to high energies. The two main background components are shown in Fig. 6. Compton background may become very noticeable in some instances, e.g. while irradiating Teflon filter samples, owing to prolific gamma-rays from the $^{19}F + p$ reactions. The bremsstrahlung background shows angular dependence, peaking at 90° and having a minimum at 180° relative to the beam, whereas the X-rays do not. The X-ray detector should accordingly be mounted at a backward angle, typically 135°. More detailed discussion on continuous background in PIXE is given by Ishii and Morita [20].

Apart from the continuous background produced in ion–atom collisions,

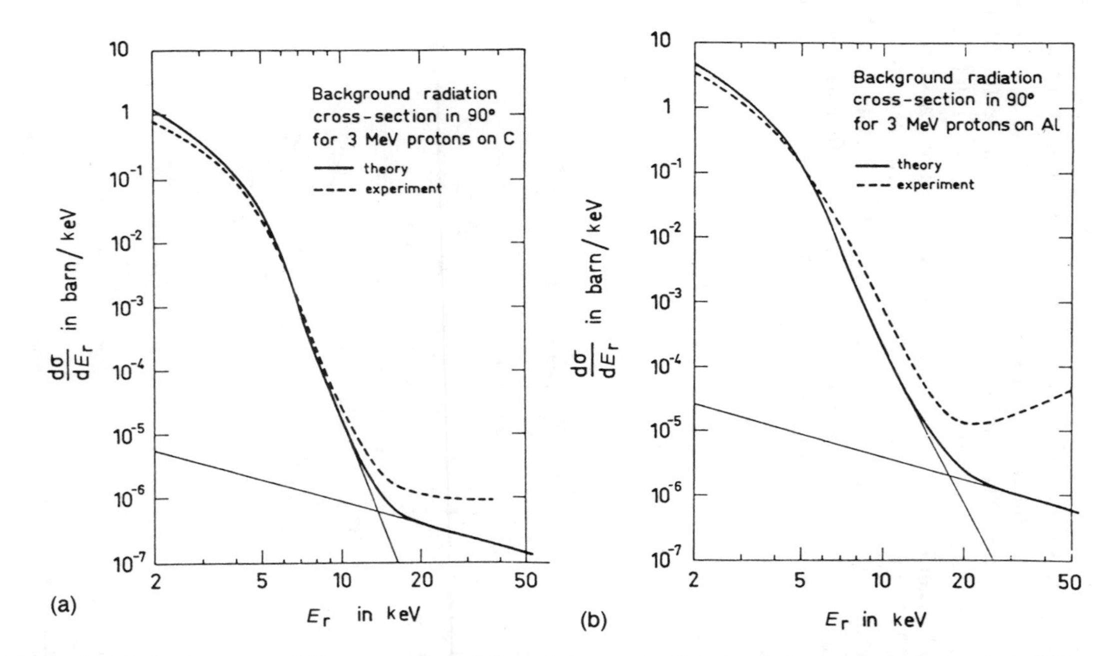

Fig. 6. Background radiation spectra at 90° expressed as differential cross-section dependence on photon energy E_γ for a thin elemental targets (a) carbon, and (b) aluminium, with proton energy 3 MeV. The full curves are calculated and the dashed curves measured [19].

other sources of background may seriously affect PIXE spectra. Incident protons can be elastically scattered by heavy elements in a sample and enter the active region of the detector. The signal from these scattered protons is large in comparison to the characteristic X-rays and contributes strongly to the continuous background and system dead time. It can, however, be easily eliminated by putting a proton absorber in front of the detector. For example, a 100 μm Mylar foil is thick enough to stop 3 MeV protons. Another potentially large contribution to the continuous background arises from the charging of an insulating sample by the ion beam. The resulting intermittent discharge to the nearest conductor produces bursts of bremsstrahlung X-rays, which result in a huge background. This can be eliminated by several ways. Coating a sample with a conducting carbon film is a common solution, adpated from the electron microprobe technique, when analysing e.g. mineralogical samples. Another possibility is to spray a sample with an electron current sufficient to maintain neutrality. In some cases, introducing a gas such as helium into the sample chamber is preferred.

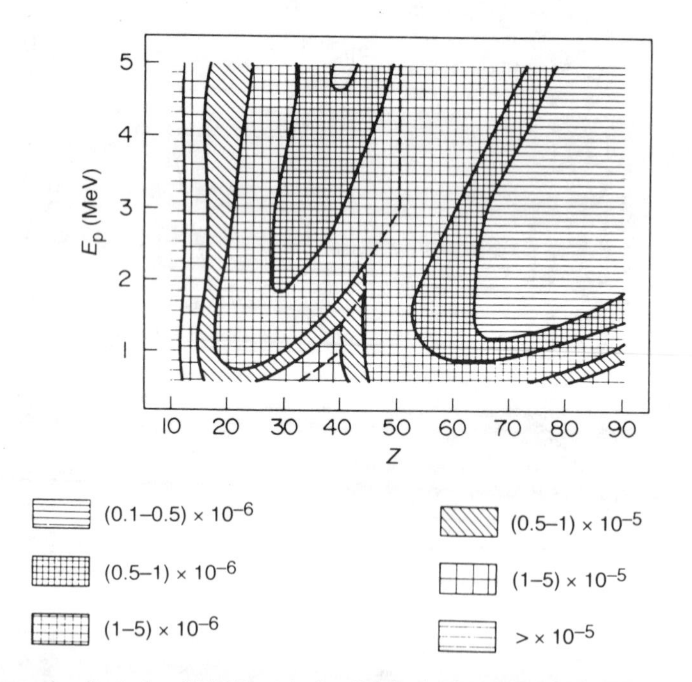

Fig. 7. Detection limits in of PIXE analysis of biological samples. The left hand side of the dotted line is for K X-ray detection and the right side for L_α X-ray detection [20].

2.4 Detection limits

The limit of detection (LOD) is determined by the relationship $N_p \geqslant 3\sqrt{N_B}$, where N_P is the number of counts in a characteristic X-ray peak and N_B is the number of background counts in the corresponding interval. Thus, detection limits depend crucially on the continuous background. The latter depends on the sample, its matrix components and the specific measuring conditions. Furthermore, detection limits can change dramatically if an interfering X-ray peak from another element coincides with a trace element peak of interest. In several fields of application the sample matrix consists of light elements whose X-rays have energies too low to reach the detector, e.g. the organic matrix of soft tissue samples. In such cases detection limits will be determined by the secondary electron background.

Figure 7 shows an example of detection limits achievable for biological samples [20], calculated for typical experimental conditions: incident beam of protons of energy between 0.6 and 5.0 MeV; integrated beam current 1 μC. The Si(Li) detector was at 135° detection angle. Pure plastic films are often used to simulate this type of sample. An example of measured LODs using 5 MeV alpha-particles is shown in Fig. 8, using a selection of plastic films as targets [21].

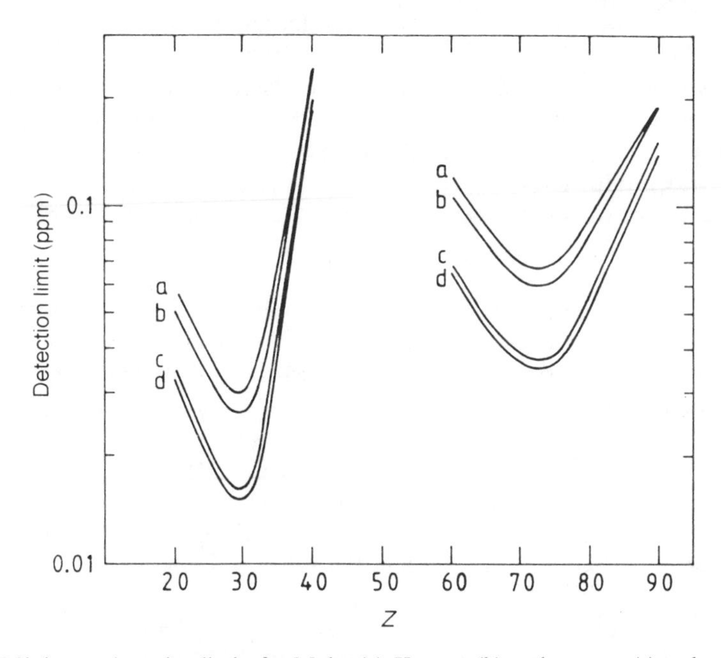

Fig. 8. Minimum detection limits for Mylar (a), Kapton (b), polystyrene (c) and polyethylene (d) bombarded with 5 MeV α-particles, for 5 μC beam charge [21].

2.5 Intensity of fluorescence radiation

2.5.1 Basic formalism

For a given characteristic X-ray emitted by an element, its yield can be calculated from the formula:

$$Y_p = (CN_0\Omega Q\varepsilon)/(4\pi We)\int_{E_0}^{E_f} \frac{\sigma_p^X(E)T_Z(E)\,dE}{S(E)} \tag{7}$$

where the photon attenuation

$$T_Z(E) = \exp\left[-\mu\int_{E_0}^{E} \frac{\cos(\theta)_i\,dE}{\cos(\theta)_0 S(E)}\right] \tag{8}$$

and Y_p = X-ray yield of element p, C = relative concentration by weight of element p, N_0 = Avogadro's number, Ω = solid angle subtended by the detector, Q = total beam charge hitting the target, ε = detector efficiency, including filter effects, W = atomic weight of the element, e = ion charge, E_0 = incident proton energy, E_f = final proton energy, $\sigma_p^X(E)$ = X-ray production cross-section for a peak p, μ = mass attenuation coefficient, $S(E)$ = matrix stopping power calculated using Bragg's rule, and θ_i and θ_0 = angle of the incoming beam and the outgoing X-ray with respect to the target surface normal.

Equation (7) is valid for light ions hitting heavy targets and requires that ions slow down along straight paths and that they have a unique charge state as a function of incident energy. Contributions to the X-ray yield from target recoil and ion straggling effects are neglected [11].

The X-ray production cross-sections σ_p^X for a peak p for the K and L subshell vacancies can be calculated from different formulae. For the K subshell:

$$\sigma_{Kp}^X = \omega_K \sigma_K^i (\Gamma_p/\Gamma_K) \tag{9}$$

where ω_K is the K shell fluorescence yield, σ_K^i is the ionization cross-section and (Γ_p/Γ_K) is the ratio of the width of the p transition relative to the total K shell width. The equation is simple, as the K shell has no subshells. For the L shell it is more complicated:

$$\sigma_{Lp}^X = (\sigma_{L1}^i(f_{13} + f_{12}f_{23}) + \sigma_{L2}^i f_{23} + \sigma_{L3}^i)\omega_{Li}(\Gamma_{Lp}/\Gamma_{Li}) \tag{10}$$

where the peak p originates from an initial vacancy in the L_i subshell. Here σ_{L1}^i, σ_{L2}^i and σ_{L3}^i are ionization cross-sections for each subshell; the f_{ij} and the ω_{Li} are the Coster-Kronig transition rates and the subshell fluorescence yields respectively.

For $i = 1$ we put $\sigma_{L2} = \sigma_{L3} = f_{13} = 0$ and $f_{12} = f_{23} = 1$; for $i = 2$ we put $\sigma_{L3} = f_{13} = 0$, $f_{23} = 1$; and for $i = 3$ all three L subshell ionization cross-sections are used [11].

The integral in equation (7) is the equivalent of the so-called ZAF (atomic number, absorption, fluorescent enhancement) correction in electron microprobe analysis, except that it neglects any secondary fluorescence contribution, which will be explained later. In order to calculate it correctly, the major sample components must be known and preferably also minor components, as they could introduce analytical errors in unfavourable situations. It does not mean that PIXE analysis is impossible without additional analysis of the major matrix elements. Several schemes of analyses with known and unknown matrices are discussed by Johansson and Campbell [5] and also by Pineda [22]. However, analytical results are not very sensitive to the matrix composition and these schemes are not much used. Stopping powers needed to solve the integral (8) are known precisely. Specifically, uncertainties in the stopping power database in TRIM, the Monte Carlo program for calculation of ion penetration into solids [23], are only 1–2%. Necessary attenuation coefficients are known less precisely. The mass absorption coefficients database for energies 1–40 keV by Leroux and Thinh [24] is accurate to about 5% in general. Thus the overall precision is normally stated as 5–10%.

It is important to keep in mind that the rapid decrease in proton ionization cross-section with depth in the specimen (i.e. with decreasing energy) results in X-ray production occurring mostly in the near-surface portion of the proton track. This effect is accentuated further by attenuation of the created X-rays in the

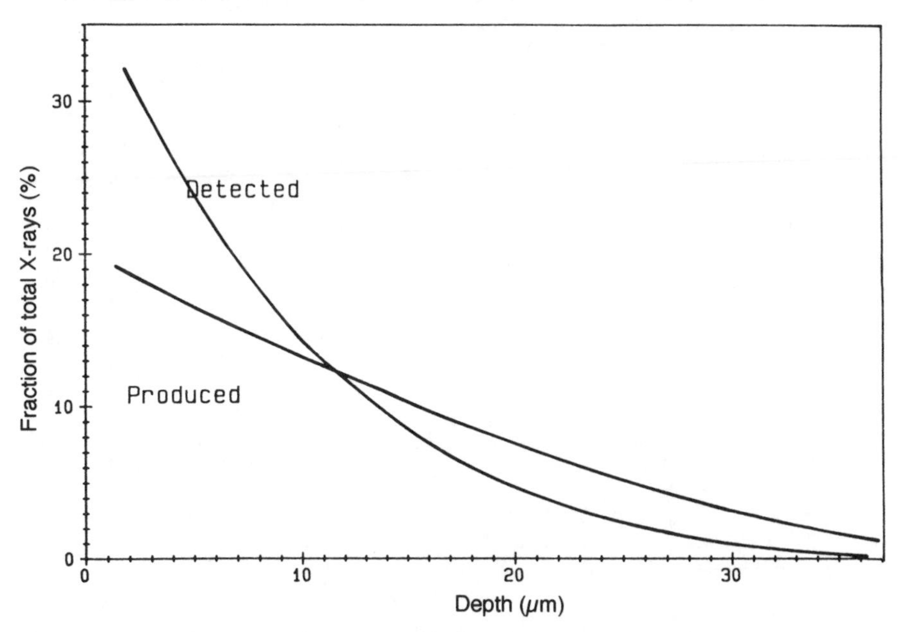

Fig. 9. Depth distribution of X-ray production and origin of detected X-rays for trace selenium in a pyrite (FeS_2) matrix. The proton energy is 3 MeV, incident beam normal to target surface, X-rays detected at 45° to target surface [25].

matrix. Johansson and Campbell [25] illustrate this for the case of selenium homogeneously distributed in a mineral matrix (pyrite). Half of the Se *K* X-ray production occurs in the first 9.5 μm, and half of the X-rays actually detected originate in the first 5.7 μm, whereas the estimated proton range is more than 30 μm (Fig. 9).

2.5.2 Thin target approximation

Equation (7) is a general expression, allowing for calculation of the yield of a characteristic X-ray emitted by an element of interest in all situations, i.e. for thin, intermediate or thick targets. Intermediate thickness targets are those for which beam particles pass through the sample, but their energy loss cannot be neglected. For two extreme cases, simplifications are possible. For very thin targets, proton energy losses in the sample are negligible and the energy dependent terms in the integral over dE in equation (7) are constant. The X-ray absorption for the emerging X-ray may also be neglected. Hence, the dependence of yield of a characteristic X-ray is a linear function of the element concentration and no matrix information is necessary. This is the easiest case of PIXE quantification,

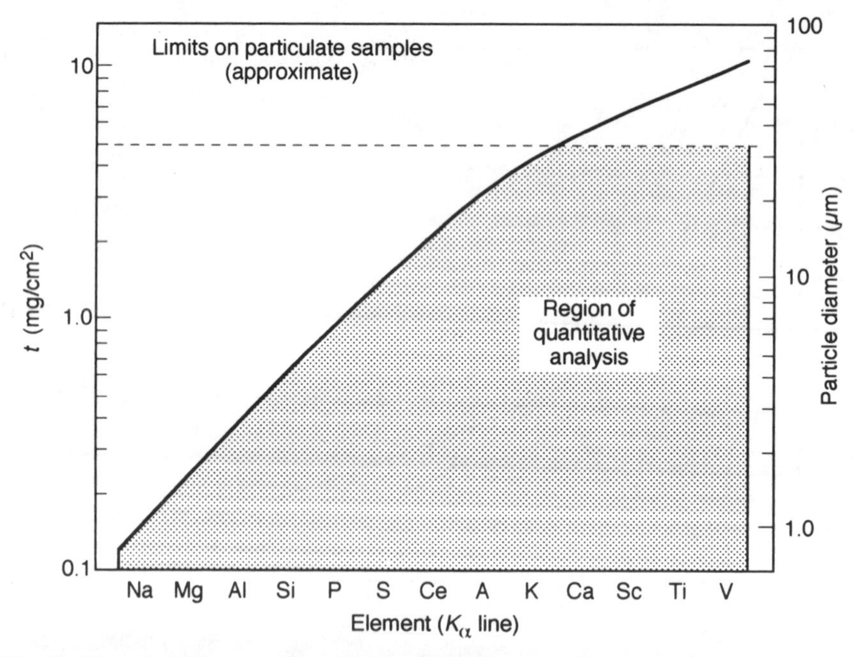

Fig. 10. Deposit thickness and particle size limitations for 30% or lower corrections to the number of X-rays N_x for aerosol particle samples. The 5 mg/cm^2 upper limit is associated with ion beam energy loss corrections [26].

and most analyses, especially when the method was first introduced, were done using this assumption.

However, as discussed later, preparation of thin samples is generally not trivial and in many cases is impossible. Many apparently thin samples violate the thin specimen criterion, a fact which sometimes was ignored in the early stages of PIXE development. Thicknesses up to a few mg/cm^2 are not infrequent, for which proton energy losses should be taken into account. Another important point is that the sample thickness giving rise to significant corrections depends strongly on the X-ray energy line used for quantification, e.g. K X-rays from yttrium (15 keV) only lose 2% for a light biological matrix of thickness corresponding to 600 keV ion energy loss, whereas aluminium (1.5 keV) K X-rays lose 84% for only 100 keV ion energy loss [22]. This phenomenon is important specifically in the multi-elemental analysis of aerosol samples, as shown by Cahill [26] (Fig. 10). For aerosol particles of 10 μm diameter, S may be determined with less than 30% absorption correction. For Al, particles of only 2 μm or greater (equivalent to a thickness of 0.4 mg/cm^2) will have absorption corrections of 30%, limiting accuracy to $\pm 10\%$ or above. Therefore modern spectrum evaluation software calculates X-ray yield using equation (7) without any simplifications. Necessary matrix information is obtained, for example, from simultaneous Rutherford backscattering (RBS) measurements. Johansson and Campbell review methods of sample thickness determinations in a related chapter [5]. The most straightforward method for determining sample thickness is detection of Coulomb scattered beam particles during analysis, as this quantity, for a given matrix, is proportional to the target thickness.

2.5.3 Thick targets

In the second extreme case of the X-ray yield equation, the proton beam is completely stopped in the sample, i.e. the final proton energy in equation (7) becomes zero. In comparison with samples of intermediate thickness, the quantification for thick targets is also easier because it is not necessary to know the target thickness and hence the final proton energy. Thus it is easier to calculate correctly the integral in equation (7). Measurement of sample thickness is generally not a trivial task. Sophisticated sample preparation also requires a certain know-how and experience to ensure that elemental losses and contamination are avoided. Besides, for some materials, e.g. valuable archaeological specimens, alloys, rocks or mineralized tissues, preparation of thin targets is impossible and thick target analysis is the only option. However, the possibility of analysing untreated samples is one of the major advantages of PIXE.

Detection limits in thick target PIXE (TTPIXE) are generally slightly higher in comparison with those in thin target analysis, typically in the few ppm range, but very dependent on sample matrix and element analysed. The only two requirements for thick target preparation are a flat surface and sample uniformity. The required smoothness depends on the X-rays to be measured, the degree of their absorption in the matrix and the accuracy to be achieved [27]. However, apart from the obvious convenience in sample preparation, some experimental problems are almost exclusive to thick targets. One is the possibility of beam charge buildup on insulating specimens and resulting background increase in the spectra, as discussed earlier. Another problem is heating due to deposition of the full beam energy in the specimen, which limits permissible beam currents and thus reduces sensitivity. Beam charge measurements, discussed later, are more difficult.

A common situation in thick target analysis is that one of the major components has X-ray lines in the spectrum. The range of peak heights will differ dramatically between major element peaks, e.g. several percent of iron (with high X-ray production cross-section), and elements present at a few parts per million, which can result in a factor of 10^5 in peak height difference. The main PIXE applications are trace element analyses. For these, high count rates from matrix components are an undesirable feature, resulting in longer dead time and complicated spectra with many interferences. A possible solution is selective absorption of the X-rays from these matrix elements, using thin layers of elements with absorption edges just below the X-rays of concern, the so-called critical absorbers. Generally it is an element one or two Z numbers down from the matrix element. A simpler practice is to use Al layers of appropriate thickness, which enables one to achieve detection limits of a few ppm in 3–5 min for the elements in the most desired regions ($26 < Z < 42$ and $80 < Z < 90$). This is done at the expense of lower Z elements, which are generally measurable by other techniques. Using X-ray absorption filters with holes (funny filters) in order to transmit only 5–25% of the low energy portion of X-rays is less popular owing to a more difficult quantification.

In some cases an intense X-ray line from one of the sample components may induce secondary X-ray emission within the sample. The secondary X-ray will then be counted as if generated by the bombarding beam and will thus constitute a source of error—the calculated concentration will be too high. Secondary X-rays may be either K X-rays of lighter elements or L X-rays of lower energy. The intensity of secondary X-ray generation depends on the sample composition. Secondary fluorescence corrections in thick target PIXE are discussed by Johansson and Campbell [5]. Secondary fluorescence effects are included in spectrum software packages, like GUPIX [28] or GEO-PIXE [29].

3 X-RAY SPECTROSCOPY

3.1 X-ray detectors

Elemental characteristic K X-rays range in energy from a few hundred eV for the light elements to 100 keV for the heaviest. It is in this region that a detector used in

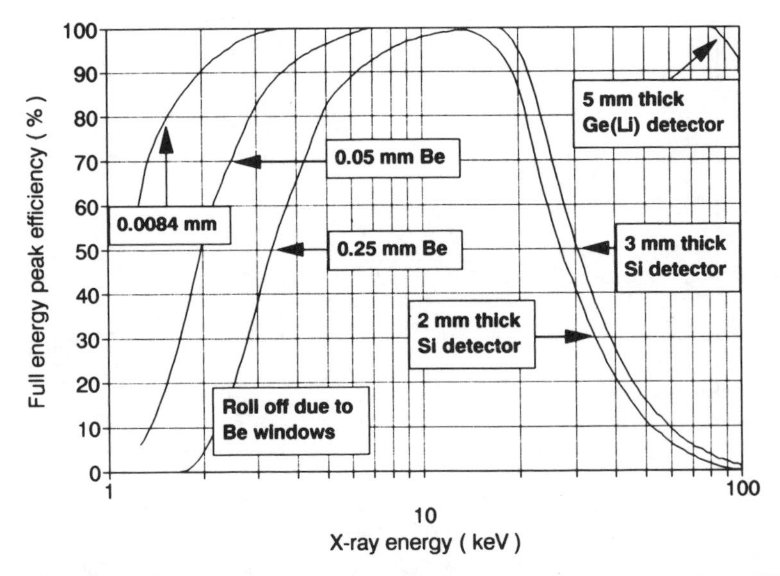

Fig. 11. Spectral response of Si(Li) detectors of two different thicknesses and two different Be windows. The efficiency of a germanium detector is also included for comparison.

a PIXE set-up should be efficient. Though there are several detector types which can be used in this region, by far the most appropriate are semiconductor detectors. In fact, the development of the PIXE technique was triggered by the invention of the energy dispersive Si(Li) detector. Such detectors are sensitive to X-rays with energies between 1 and 30 keV. Spectral responses for two different thicknesses of a typical Si(Li) detector are shown in Fig. 11, which also illustrates the effect of different Be window thicknesses in front of the detector. Above 30 keV the X-ray cross-section falls steeply and it is then not efficient to measure *K* X-rays from the heavier elements. Instead, *L* X-rays are used for the analysis of heavy elements, as they are more intense and are more efficiently detected in the Si(Li) detector.

Ideally a detector should give a response with a pulse amplitude linearly related to the energy deposited in the crystal and with infinitely sharp peak resolution. Si(Li) X-ray detectors give a full absorption Gaussian peak of finite width, accompanied also by a Si escape peak and a low energy tail originating from incomplete charge collection. A spectrum collected from an ^{55}Fe radioactive source, producing manganese K_α (5.898 keV) and K_β (6.490 keV) X-rays, is shown in Fig. 12. Compared with other radiation detectors, the Si(Li) X-ray detectors have a relatively favourable response function; note that Fig. 12 has a logarithmic scale on the ordinate. The relative amount of pulses falling in the Gaussian shaped peak is of the order of 99%, leaving 1% of distorted and unwanted pulses due to the other physical processes taking place in the crystal.

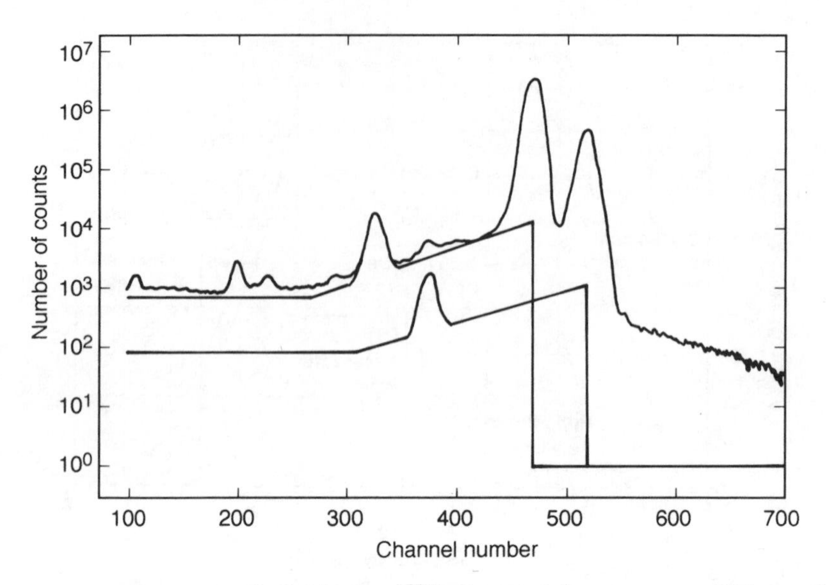

Fig. 12. X-ray spectrum collected from an ^{55}Fe source, giving manganese K X-rays. The Gaussian full absorption K_a and K_b peaks can be seen. The calculated contributions for the related Si escape peaks and low energy tails are also shown [30].

The silicon escape peak arises from induced Si K_α X-rays in the crystal. If an induced Si K_α X-ray photon escapes out of the crystal the output pulse will be lowered by the corresponding energy, i.e. 1.740 keV. Hence each peak will be accompanied by an escape peak 1.740 keV below the full absorption Gaussian peak. The tailing from X-ray detectors has a more complicated origin and has been extensively studied [31, 32]. The distorted pulses giving the tailing appear because of incomplete charge collection near the sruface of the crystal, but tailing differs from detector to detector for the same geometry.

Above the full energy peak, nothing but the slope of the Gaussian should be present. There is, however, the possibility that two photons may impinge at the detector within the pulse processing time, i.e. a pulse pile-up event. This gives rise to overlapping output pulses with variable amplitudes up to the sum of the two photons. The pulse height depends on the degree of overlap between the interfering pulses. For partial overlap, overlapping pulses can be detected electronically, and the distorted output pulses can both be rejected by the acquisition system. Pulses separated by less than a critical time (pulse pair resolving time), of the order of 300 ns, cannot be distinguished. An output pulse will appear as a correctly shaped pulse of amplitude equivalent to the sum of the two contributing events. This results in several extra peaks in the spectrum, situated at energies corresponding to the sum of dominant peaks. The most likely pile-up event is summation of two photons from the major spectrum peak.

The upper limit of X-ray energies detected is set by the likelihood of photon transmission through the semiconductor crystal. The X-ray cross-section is approximately proportional to Z^3 for X-rays. At 35 keV the absorption in a 3 mm thick Si(Li) detector is 50%, as can be seen in Fig. 11. For the detection of higher energy X-ray photons another semiconductor material, germanium, is chosen. High purity germanium detectors (HPGe) can be used up to and above 100 keV, which enables detection of characteristic K X-rays for all elements. Although HPGe detectors potentially offer improved resolution, in practice noise from reverse leakage current in the detector results in poorer resolution at energies below 6 keV [33]. For Si(Li) detectors escape peaks are much less pronounced than for Ge detectors, reducing peak interferences. The sensitivity for the X-ray energies could also be a disadvantage, as the Compton scattering from gamma-rays increases with the higher Z materials—Compton scattered gamma-rays comprise the dominant spectrum background at energies above the secondary electron bremsstrahlung background.

3.2 Spectrum fitting

To determine elemental concentrations quantitatively from X-ray spectra, the peak area of the X-ray line used for quantification must be measured. In addition, for a thick target, the integrated X-ray production cross-sections, X-ray attenuation and secondary fluorescence corrections should be calculated. Several computer packages exist which fit a function to the measured spectrum for the extraction of the peak area, and some of these also provide the thick target corrections. Three recently developed programs are mentioned below.

The spectrum contains a Gaussian full absorption peak followed by its tailing and escape peak, as well as a continuous spectrum background. The spectrum usually includes X-ray lines from different elements which cannot be resolved by the X-ray detector, and the interfering X-ray lines have to be separated according to their branching ratios. Also, the pulse pile-up peaks have to be analytically described and stripped off the spectrum. As the spectrum background is built up of several processes, which are highly dependent on the sample matrix, most software uses a semi-empirical approach for the fit. In 1986, an intercomparison of the programs available by that time was published, showing excellent internal agreement [34].

For thin targets, the HEX program [35] has been used extensively, and a modernized version, WITS-HEX, has recently become available [36, 37]. The program is written in PASCAL and operates on an IBM-PC or compatible computer, and has a menu-based user-friendly interface. WITS-HEX is distributed as public domain software and is presently under test by several users. The package is best for thin samples with a light matrix, e.g. aerosol, water or biological samples on a plastic backing. For thicker samples the software must include a thick target correction, i.e. must be able to completely solve equation (7). Among software packages satisfying this requirement, two are widely used, GUPIX [28] and GEO-PIXE [29]. Both use extensive data libraries and have proved reliable and accurate. GEO-PIXE uses the method of least squares to fit the background parameters, which is the usual approach in the PIXE programs, whereas GUPIX removes background by digital filtering. This relies on the fact that a Si(Li) spectrum, viewed as though the energy axis were a time axis, contains three frequency bands: the continuum, which has low frequency; the peaks, which have intermediate frequencies, and high frequency channel-to-channel fluctuations due to counting statistics. The advantage of this approach is that no assumption needs to be made regarding the analytical form of the background, and the number of parameters involved in the least squares fitting is reduced, thus increasing the deconvolution speed. GUPIX can handle thin, intermediate and thick samples as well as heavy element matrices. It can be run either in manual or in batch mode and is a convenient choice for off-line (after measurements) deconvolutions. One small disadvantage of the digital filtering approach is that the user cannot be sure if all the peaks present in the spectrum have been taken into account in deconvolution.

GEO-PIXE has been developed for more specific applications, mainly for trace element analysis of mineral grains using a PIXE microprobe, but can also be used for normal macro-beam analysis. In samples where any of the major, or minor, elements are visible in the spectrum there is a large dynamic range of peak heights, imposing a certain demand on the fitting procedure. This is a common situation for mineral samples, e.g. the ubiquitous iron peaks, and the program has been optimized in order to handle accurately small peaks with poor statistics. Thick target corrections are calculated once for each matrix. This enables fitting and trace element determination of large batches of samples without the need for the extensive and time consuming thick target calculations for each sample. In particular, the secondary fluorescence corrections require long computation times, and can therefore be switched off for matrices where the effect is negligible. The spectrum fitting typically takes ≈ 10 s at a VAX work station. Finally, multiplication of the fitted peak areas by calibration factors gives the trace element concentrations. Attenuation in X-ray absorption filters is also taken into account at the final stage, to allow the use of the same thick target corrections independent of the filter.

A remaining problem for reaching the ultimate precision of analysis is inclusion of a correct peak line shape. Specifically, the tailing properties are variable, and may be inherent to each individual detector, even from the same manufacturer. The line shape influence on the results becomes most important when a small peak is situated on the tail of a much larger peak of higher energy. A visual inspection of the spectrum fitting may help, however, in tracing this problem.

Some changes in the peak shape, almost exclusively for light elements, may be due to the chemical effects of X-rays [38]. These effects are troublesome in conventional PIXE elemental analysis but understanding is gradually evolving of the underlying physics. Chemical effects show up in several forms: energy shift, emission probability change, satellite formation as a result of multiple electron vacancies or of electron pick-up of counter atoms, and peak profile change (e.g. broadening).

3.3 Calibration factor

For the particle energies used in PIXE analysis, the ionization cross-section is largest for the lighter elements and decreases monotonically towards higher Z values. This is partly counteracted by the fluorescence yield, which increases with the element number. Sensitivities are cut off at the low energy end by X-ray absorption in filters and the detector window. A typical sensitivity curve $K(Z)$ is shown in Fig. 13 [22].

All parameters in equation (8) are known and can be found in tabulations. They are, however, only known to limited precision, so standardless analysis is limited to a precision of $\approx 10\%$. In particular, X-ray filter thicknesses have to be known

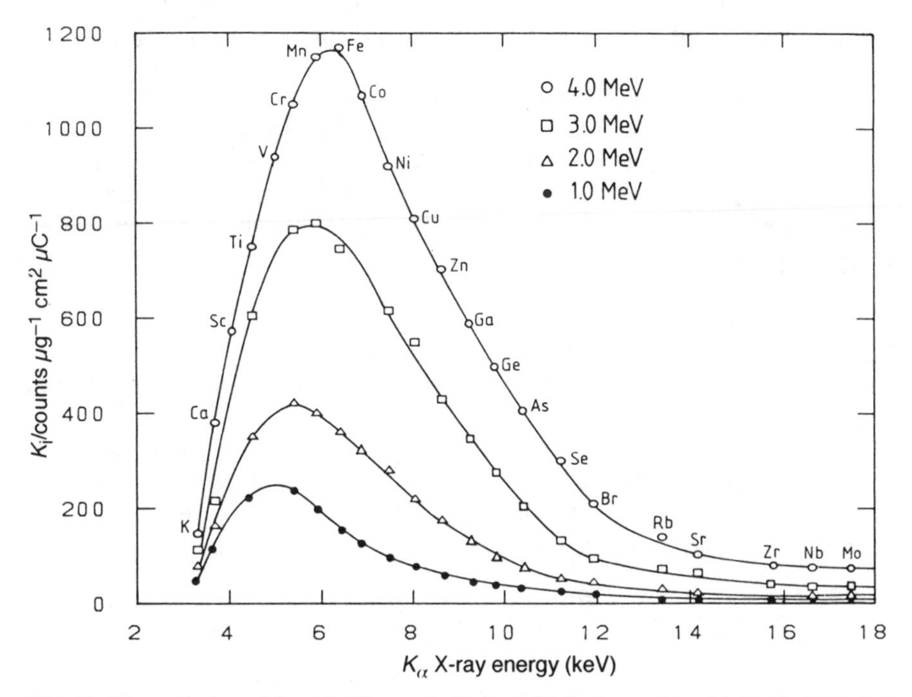

Fig. 13. The variation of the thin film analytical sensitivity factor K_z with K_a X-ray energy for a range of incident proton energies [22].

with great precision for heavily suppressed low energy X-ray lines, giving limited accuracy for these elements. To enhance the precision, the sensitivity factor for an element can be determined directly by the calibration of the set-up using thin film standards with certified areal densities $M_a(Z)$. Such standards are available with values guaranteed within a few percent [39]. The geometrical factor (detector solid angle) responsible for the height of the calibration curve for the heavy elements (high X-ray transmission) is then determined accurately by fitting a calibration curve to a number of analyses of standard samples. Secondly, the low energy cut-off is determined by adjusting the total X-ray attenuation (filter and detector window thickness) to the light elements, giving an overall precision better than 5% for all elements. Hence it is not necessary to calibrate for every single element because of the fixed shape of the curve (Fig. 13), determined by the smoothly varying Z-dependence of the X-ray production cross-section and the X-ray transmission to the detection system.

In conclusion, the most accurate measurement of detector solid angle and system X-ray transmission is obtained by fitting calculated to measured concentrations of standard samples. Following this calibration procedure, standardless analysis can be performed with a precision of $\approx \pm 5\%$. The calibration of the instrument is maintained as long as the geometric parameters remain unchanged. The normal procedure is to intermittently re-analyse selected standards, such as gallium phosphate (GaP), which has both low and intermediate energy peaks, to confirm that the calibration is still valid.

To check the calibration for thick targets (see Section 2.5.3) single element standards can be used. They have the advantage of being much less sensitive to radiation damage; some elements are known to be volatile, which changes the areal density of thin film standards. The detector solid angle and the system X-ray transmission must, of course, retain their values, and the precision will be determined by the quality of the thick target corrections. Our experience is that the above mentioned software packages give the same precision as in the case of thin target calibration ($\pm 5\%$). The corrections in most cases are relatively insensitive to the matrix composition, and also complex samples give good accuracy, as has been demonstrated in the analysis of thick targets of certified reference materials.

As thin and thick target analyses differ in the sense that the beam current is measured differently, using a Faraday cup behind the sample in the former case and collecting the charge introduced into the sample itself in the latter, both calibrations need to be carried out.

3.4 Sample chamber

Figure 2 shows a typical PIXE chamber for measurements in vacuum. The beam coming from the accelerator is guided to the sample and is integrated for purposes of calibration. One requirement is that the minimum possible X-ray and gamma-ray background should be generated near the Si(Li) detector. Therefore the choice of materials for collimators, specimen holders and other chamber components is critical. In the case of a thick specimen the beam is stopped completely in the target, whereas when traversing a thin specimen the beam may diverge and strike places in the chamber other than the Faraday cup used to integrate the charge. The usual positions of the Si(Li) detector are at 90° or 135° to the beam direction. The 135° position has the advantage of electron bremsstrahlung reduction (see background considerations), but the 90° angle allows for shorter sample–detector distance, and this sometimes gives better results in terms of limits of detection. Usually the detector is outside the chamber and the X-rays have to pass through a thin window in the chamber. X-ray absorption in this window should be minimized. For this, beryllium or Mylar of thickness 0.025–0.125 mm is the preferred window material. For the analyses of light elements (below $Z = 20$) the detector has to be inserted directly into the chamber. However, a minimum thickness absorber, preferably Be, is mandatory to prevent backscattering protons from entering the detector, since otherwise the resulting large pulses would cause severe spectral degradation. It is also useful to have a selection of different filters mounted on a wheel or ladder, with the possibility of fast replacement according to experimental requirements. The advantage of these filters is that an intense part of the spectrum can be suppressed, with relative enhancement of regions of interest.

3.5 Beam charge measurements

When analysing thin samples, the incident ions (protons) pass through the target and are collected in a beam stop designed as a Faraday cup. The charge deposited in the beam stop is used for quantification. Secondary electrons may also reach the beam stop, rendering the proton charge readings inaccurate. A solution is to couple the target chamber and beam stop to act as a single Faraday cup isolated from the rest of the beam line. In addition, a suppressor electrode is placed in front of the target and held at a negative potential (−50 to −200 V) to prevent secondary electrons from leaving the sample surface. Alternatively, such an electron suppressor may be placed at the entrance of an electrically isolated beam stop.

A different approach to quantify the beam intensity makes use of the signal from backscattered ions as the beam traverses a thin self-supporting foil before hitting a specimen. The disadvantage is the fragility of the foil and beam broadening due to multiple scattering. A similar approach, which avoids the fragility problem, is to measure the backscattered signal from a thin (0.02 – 0.03 μm) gold layer rotating regularly through the beam, thus intercepting 10–30% of the total fluence.

4 APPLICATIONS

4.1 Biology and medicine

PIXE applications in this field constitute about 25% of all scientific programmes involving this technique [40]. They also present many perspectives for the future, especially as far as ion microbeams are concerned. The lowest detection limits are obtained for matrices consisting of light elements ($Z < 10$), making biological samples with organic matrices favourable research subjects. There are a number of reasons for the low detection limits. Light elements do not contribute to detectable characteristic radiation which could obscure X-rays of heavier trace elements. Secondary electron bremsstrahlung radiation, the major background component in PIXE spectra, is roughly proportional to the average atomic number of the matrix and thus lowest for light element matrices [41]. Most of the trace elements essential for life, and some toxic elements, fall in the region of atomic number where PIXE has its lowest detection limits. The essential major and minor biological elements all have atomic numbers below 21 and thus their characteristic X-ray lines can be easily eliminated from the spectrum or reduced, when necessary, through the use of X-ray absorption filters. In PIXE, even for large infinitely thick specimens, the probed sample mass is at most a few mg. This fact makes the technique adequate in biological applications, where very often the sampled mass is small. High sensitivity in absolute terms allows for analyses of volumes down to 0.5 μl [42] and masses of <100 μg [43].

There are several recent reviews of PIXE applications in biology and medicine, e.g. those by Maenhaut [44–46]. Johansson and Campbell have devoted several chapters to this subject [47]. In this field of application, sample and specimen preparation are crucial and various techniques are employed. In certain cases the sample can be bombarded directly or after very limited preparation (e.g. after drying, cutting or polishing). This is possible with bone, teeth, hair, fingernails, skeletons of animals, mineralized structures in plants and also with various botanical samples, such as leaves, wood, bark or tree rings. Preparation of pellets requires mixing of sample material with a binder such as graphite. Using graphite has the advantage of making the pellet conductive, which eliminates specimen charging during irradiation in vacuum. On the other hand, adding a binder involves a danger of introducing impurities and dilutes the material under investigation. When preparing powdered targets, it is necessary to avoid particle size effects in the analysis; this is particularly important for the lighter elements (Na to Ca) [48].

Several procedures may be used to prepare thin specimens [49–51]. Homogeneous solid samples or organs are first dehydrated and then may be cut into thin slices (10–100 μm thick) and deposited onto a thin backing film [52]. Soft tissues may be embedded in paraffin [53, 54] or other organic material before microtome sectioning. However, a common problem related to this chemical fixing is that the redistribution of the more mobile elements may occur and impurities in the chemicals can distort the analyses. It is generally accepted that the best way of preparing samples for high resolution morphological studies (in micro-PIXE) and for meaningful measurements of elemental concentrations is cryofixation [55]. Only very fast cooling of the material ($>1000\,K/s$) does not disrupt the shape and elemental distribution of the cells and tissues at high spatial resolutions. Owing to the relatively low coefficient of thermal conductivity of biological tissue, freezing fresh tissue without significant damage can only be achieved in thin sections of no more than $\approx 100\,\mu m$. In bulk tissue, only the outside 50–100 μm layer to tissue will escape significant damage [55].

It was stated above that PIXE is generally a non-destructive technique. However, for delicate organic materials this is not valid. The energy deposited in the specimen by the ion beam may lead to radiation and heat-induced damage, i.e. to charring and partial volatilization of matrix elements (H, C, O) and even losses of some analyte elements (the halogens, S, As, Se, Hg, Pb). Thermal damage effects may sometimes become visible, especially in microprobe mode, for which the local current density may become extremely large. However, this does not necessarily mean that the results of elemental analysis are affected. The prevailing opinion is that elemental losses of those elements which it is possible to measure with PIXE are more associated with thermal damage (beam dose rate) than with ionization damage (total dose), although both play a role [45]. Limiting the dose rate is much more critical than the total dose in avoiding losses of inorganic minor or trace elements in macro-PIXE analysis of biological specimens. Ionization is responsible for most losses of light elements. Problems of the outgassing of organic and other samples during ion beam bombardment can be solved by using the non-vacuum approach. The merits and disadvantages of external beam PIXE have been reviewed by Williams [56] and recently by Räisänen [57].

Beam-induced specimen damage is much smaller for thin and semi-thick targets. The latter are especially attractive in this respect. For example, when a $5\,mg/cm^2$ thick specimen is analysed with a 2.5 MeV proton beam, only 30% of the beam energy is deposited in target, whereas the emerging X-ray intensity is 80% of that of an infinitely thick specimen [45].

It is also important to make the right choice of backing for biological thin specimens [47, 49, 58]. Commercially available plastic films are available, e.g. Mylar, Kapton and Hostaphan. These have suitable properties of heat and radiation resistance and low levels of impurities. Polyethylene is susceptible to heat damage and the Cl in polyvinyl chloride produces unacceptable interference. The minimum thickness of Mylar is $\approx 2\,\mu m$. The advantage of using thinner support foils (down to 0.1 μm) is the reduced continuous background. The general technique for producing such foils is to dissolve the desired plastic in a volatile hydrophobic solvent. A drop of the solution is put onto the surface of a beaker of water. As the drop spreads out and the solvent evaporates, a thin film of the plastic remains, which can be lifted onto a target frame. A recently reported recipe involves the dissolution of polyvinylformal (PVF) in tetrahydrofuran (THF) [59]. As THF is denser than water and would normally sink, the flotation was performed on the surface of a sucrose solution. Homogeneous films of only 0.1 μm thickness were obtained, $< 5\%$ as thick as the thinnest commercial Mylar films. Samples as small as 10–100 μg of biological material, e.g. a fish egg, could be analysed successfully on a PVF film.

In a general way, biological applications can be divided into a few groups: studies on different diseases, and on cancer, and trace element distribution in the body. Among these applications are numerous analyses of tissues of experimental animals, human blood serum, and animal and human brain. Hair analysis has also received special attention in the hope that trace element distribution in hair could serve as a monitor of elemental concentrations in the body. Biological materials—plants, algae etc.—can be used as indicators of environmental pollution. One interesting example is the analysis of tree rings. Stepping or scanning a millimetre-sized beam over a core bored from a living tree [60] or over a pine bark section [61] and measuring various elements in the rings from successive years allows one to trace the effects of environmental changes.

Probably the most intensively studied biological applications are investigations of elemental distributions within biological cells. Information so obtained relates to basic biochemical processes which sustain life. Such studies, however, require the use of nuclear microprobes, which is discussed in Chapter 14.

4.2 Aerosol studies

PIXE analyses of aerosol particles were among the very first applications of this method. In a pioneering article in 1970, Johansson, Akselsson and Johansson [1] reported on a 'dry deposition' aerosol experiment. Also in 1970, the PIXE group at Crocker Nuclear Laboratory, Davis, started size selected multi-elemental aerosol analyses [62]. Currently, aerosol related programmes form about 20% of all PIXE applications [40]. Comprehensive reviews of PIXE in aerosol research have been published recently by Koltay [63] and Maenhaut [64].

Suspended atmospheric particles, deposited on thin plastic membrane filters, form ideal targets for PIXE. In contrast to other applications, PIXE should be considered here as a macro-analytical technique, used to determine major and minor (% and mg/g) concentrations in very small samples, rather than as a trace analysis technique (μg/g amounts in large samples). The optimum mass loading of a filter or an impaction substrate is a monolayer of particles. This gives an areal mass loading between 100 $\mu g/cm^2$ and a few mg/cm^2, depending on the size of the particles. Thus the corrections for X-ray attenuation and ion energy loss are small; the air particulate sample can generally be treated as a thin target. The limits of mass loading/particle size which can be treated by the thin target approximation, taken as a less than 30% absorption correction, are also a function of the X-ray energy (element) of interest. Such limits are presented as a guide in Fig. 10. This figure illustrates that, for Na and Mg, absorptions tend to be high, even for smaller particles. For this reason, PIXE aerosol studies generally report quantitatively only for elements from Al upwards, even if Na peaks, for example, can be clearly resolved from the spectra.

The size and concentration of environmental aerosols are governed by processes occurring on time scales from a few seconds to months. An ideal particle sampling instrument should be capable of time-sequence sampling, to follow variable atmospheric phenomena on an appropriate time scale, and also capable of separation into at least coarse (2.5–15 μm) and fine (< 2.5 μm) mode fractions. Splitting of samples on a time or size basis often results in very low mass loadings, which limits the ability of conventional analytical techniques to handle such samples. As PIXE is ideally suited to analysis of such minute samples, a versatile range of samplers has evolved to match low sampling rates (0.5 to 15 l/min) in size- and time-resolved samplers to the requirements of PIXE [65].

Typical urban aerosol monitoring requirements are for samples to be collected over 24 h intervals. Such samples are typically well loaded, and are amenable to analysis by several techniques. However, the sampled material represents an

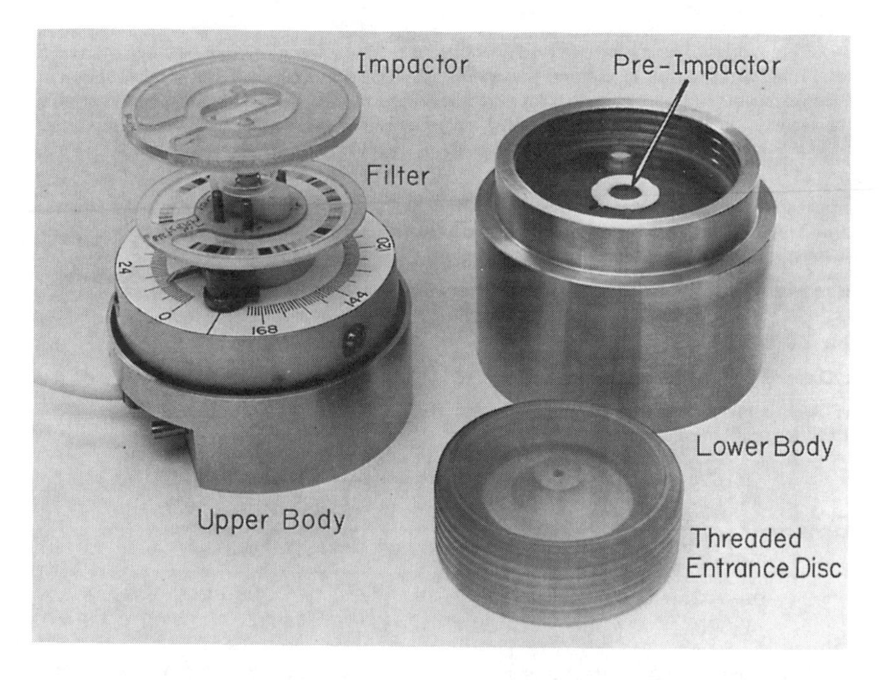

Fig. 14. Streaker aerosol particle sampler, developed to exploit the requirements and features of PIXE analysis. The filter disc can be analysed in up to 160 sample steps.

integration over possibly numerous atmospheric processes and source directions. A range of time-sequence samplers has been evolved [65], which sample for periods from 15 min to 4 h on continuously moving stepped filter media. In either case, the filter is exposed to PIXE analysis as a series of steps. One of the more widely used samplers of this type is the circular streaker sampler, which samples onto a polycarbonate filter membrane supported by a circular frame (Fig. 14) and may be operated as either a total filter or a two stage coarse and fine mode sampler, with both stages amenable to PIXE analysis.

An example of diurnal fluctuations of Mn concentrations in a highly industrialized region is shown in Fig. 15. Time resolution of streaker samples approaches the useful resolution inherent in the output from continuous gas analysers.

Integrated simultaneous time- and multi-stage size-resolved aerosol sampling has been achieved in a sampler known as the DRUM [66]. In this device, particles are impacted through rectangular orifices onto greased plastic foils wrapped around rotating drums. After exposure, the plastic foils are mounted on a frame for stepped analysis through a PIXE beam.

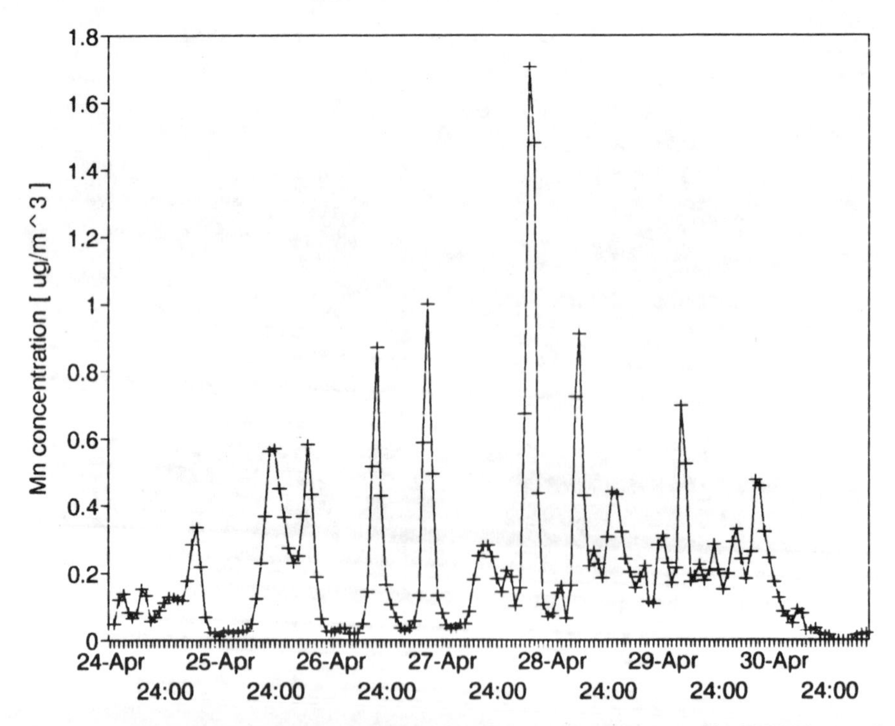

Fig. 15. Hourly Mn concentrations in an urban atmosphere, measured by PIXE from a 7 day circular streaker sample (Fig. 14). The sharp peaks are characteristic of a strong local source, in this case a large Mn smelter 20 km distant.

Elements that can be determined by PIXE constitute only about one third of the mass of a typical urban aerosol. The major part is composed of light elements, mainly H, C, N and O. For many of these aerosol samples, specifically also the time-sequence samples, it is possible to combine PIXE with other simultaneous nuclear analyses, or to perform other non-destructive off-line analyses. Particle elastic scattering (PESA) has been used to determine C, O and N [67–69], forward alpha scattering (FAST) for hydrogen (and, by calculation, organic carbon) [70], and proton-induced gamma emission (PIGE) for the lighter elements F, Na, Mg, Al and Si [71, 72]. Off-line techniques include β-particle absorption for total mass and optical absorption for inorganic (black) carbon. Using such combinations in association with PIXE, it is possible to obtain near-total elemental analyses of time- and size-resolved aerosols [73].

A major application of multi-elemental analyses of aerosol samples has been in the field of air pollution source apportionment receptor modelling. Concentrations of various airborne species (elements) are measured at ambient sampling sites (receptor sites). By comparing the receptor composition with a set of likely source elemental profiles, using the chemical mass balance (CMB) model, it is possible to estimate the individual source contributions [74]. It was specifically the development of the CMB model which helped to transform the use of multi-elemental aerosol data, such as provided by PIXE analysis, from being a powerful, but mainly academic, research technique into a routine and useful tool for regulatory purposes [75].

Other applications of PIXE in aerosol studies have ranged from underground mining environments to stratospheric aircraft sampling, industrial pollution, meso-scale pollution transport phenomena, effects of fine aerosols on global climate, nutrient recycling in biomass burning, and analysis of Arctic air particles, both from the air and historical analysis by recovery of particles from ice cores. The special features of small sample size and sensitivity to a broad range of elements will ensure that PIXE will remain the method of choice in many aerosol studies.

4.3 Geoscience

Geological applications of PIXE can be divided into the following groups:

- studies of the origin of the elements and the Solar system;
- exploration for precious materials;
- extension of established geological studies, like partitioning, zoning and fractionation, similar to those undertaken with electron probes, but to lower concentrations.

Most geological applications require the use of microbeams, because of the inherent heterogeneity of most minerals. However, the beam spot size requirements are not very exacting; a beam spot of 10–20 μm is already very useful for many applications. Several reviews on PIXE applications in the geosciences have been published [5, 76–85].

In contrast to biological and aerosol applications, sample preparation for geological applications of PIXE is simple and fast. The only important requirement when using solid bulk samples is a flat polished surface. For whole rock analyses, samples can be prepared by grinding and compressing the powder into pellets.

Despite promising pilot studies concerning the use of macro-beam PIXE for routine analyses of geological samples [86], replacement of small X-ray tubes and even radioisotope excitation by PIXE does not seem likely. Sample heterogeneity in either solid or pressed powder geological samples presents a fundamental barrier to PIXE and other ion beam analyses of bulk geological material [87]. The increased range of high energy protons in high energy PIXE and of higher energy X-rays effectively increases the volume sampled, but this is not a broadly applicable solution. For accurate trace analysis of bulk geological samples, some form of sample preparation involving fusion, dissolution or preconcentration is necessary. The purely instrumental approaches of PIXE in such cases have a limited role.

Methods of sample preconcentration are briefly reviewed by Annegarn and Bauman [76]. Elements of interest often occur concentrated in specific minerals, which can be separated from the matrix using chemical or physical techniques. When inter-element interference is the difficulty, selected elements may be concentrated by chemical dissolution of the entire sample. One method of preconcentration is the separation of heavy minerals from siliceous matrices using hydrofluoric acid. Fire assay is a standard technique for preconcentration of Au and platinum group elements (PGEs) in ores. For PGEs, fire assay results in a NiS button containing PGEs as sulphides. The NiS can be dissolved in concentrated HCl, leaving behind a few g of insoluble PGE sulphides, which can be filtered onto a Nuclepore membrane, forming a target suitable for prompt PIXE [88] or delayed X-ray emission (DEX) following proton activation [89].

Sometimes preconcentration takes place in nature. Gas diffusing through micro-fissures in the Earth's crust entrains particles representative of the bedrock through which it is passing. These particles can be sampled at the surface using Mylar films, held in inverted cups, and analysed by PIXE [90]. The phenomenon, called geogas analysis, has successfully located Ni, Cu, As and Pb ore bodies lying as far as 1000 m beneath the surface. In a similar way, one can measure the selective condensation of volcanic sublimates [91].

The rare earth elements (REEs) are of great interest to geologists, and there have been several trials to explore PIXE capabilities in this direction. For the generally used proton energies of 2–4 MeV, interferences caused by overlap of REE L lines and X-ray lines of other elements typically present in geological material make such analyses less convenient, although success in this direction has been reported [92]. Other approaches have been tested, such as analysing *K* X-rays, instead of *L* X-rays, using higher energy protons (40–60 MeV) [93–95]. Detection limits of 1 ppm have been obtained [95].

Archaeological applications are very similar to many geological applications from an analytical point of view. Sample preparation is usually very easy, if required at all. Precious archaeological materials tend to be analysed without any preparation. The materials (clays, glass, ceramics, metals [96, 97]) are almost always thick samples composed of heavy elements, giving rise to a large background and high counting rate from the matrix. The sensitivity for trace element analysis is then much poorer than for matrices of low atomic weight. Because of their large dimensions, the archaeological artifacts are seldom introduced into the small vacuum PIXE chambers and the external beam approach is often used.

5 COMPARISON WITH OTHER TECHNIQUES

From its inception PIXE has found many applications, many of them in the disciplines of biology and medicine, geology and archaeology, environmental (aerosol) and materials science. It is natural to specify the advantages of using PIXE instead of other methods of elemental analysis and to stress its use in such cases. The situation has changed somewhat during the past 20 years, owing to rapid developments in various analytical techniques. An extensive recent critical review of PIXE relative to its competitors was written by Maenhaut [98]. Comparisons should distinguish between microanalytical techniques and bulk analyses.

PIXE and XRF have several common features, since both techniques are based on X-ray emission. PIXE has by far the better absolute sensitivities: minimum detectable amounts are 10^{-15}–10^{-16} g, which means that for a millimetre-sized proton beam one can expect absolute detection limits in the picogram range, whereas the minimum detectable amount in XRF is in the nanogram range [5]. However, this may not be the only decisive factor, and for thick, homogeneous samples there is no real advantage in using PIXE instead of XRF [99–101]. For standard particle energies the ionization cross-sections of the elements in particle-induced excitation decrease with increasing atomic number, whereas in X-ray photon excitation the cross-sections increase with increasing atomic number. Background intensities also follow this trend and detection limits are largely controlled by it. Therefore, some authors state that in bulk analyses XRF is the better technique for light elements (Na through Ca) [101]. Matrix effects in PIXE are smaller, as slowing down of protons does not depend strongly on matrix composition and is partly compensated by the matrix effect caused by X-ray attenuation, whereas in XRF both the exciting primary radiation and the secondary X-rays are attenuated by the matrix and there is no such compensation. On the other hand, the much smaller volumes analysed by PIXE may not be representative of the bulk sample. This presents a fundamental barrier to all bulk analyses using ion beam techniques [76, 87].

Both techniques allow non-destructive, multi-element analyses of solids with little or no sample preparation. PIXE should be the method of choice as the only non-destructive technique available when only milligram amounts or less of solid sample are available or for analysis of thin superficial layers on a bulk sample.

A new variant of XRF—total reflection X-ray fluorescence—is most useful for aqueous samples with low concentrations of dissolved material or for the analysis of minute amounts of solid samples. However, a requirement that particulate matter has to be transferred to a total reflector is a serious drawback and restricts the range of possible applications [98].

Nuclear activation analysis (NAA) is an important technique of trace element analysis, mostly used for solid samples. The activating particles are thermalized neutrons from a nuclear reactor. Most applications tend to use instrumental activation analysis (INAA), but post-irradiation radiochemical separation (RNAA) is also possible. In favourable cases, detection limits down to ng/g are possible using INAA, and down to pg/g using RNAA [103]. However, sensitivities and detection limits vary in irregular ways as a function of atomic number.

For the analysis of liquids, optical spectrometric techniques and, particularly, inductively coupled plasma mass spectrometry (ICP–MS) are frequently more suitable than PIXE [98]. Moreover, optical spectrometric instruments and mass spectrometers can be directly interfaced to chromatographic columns, allowing for integrated instruments and one-step analysis. However, detection limits of the order of 1–10 ng/l reached for metals in sea water (after preconcentration) almost ten years ago [4] make PIXE still highly competitive with the above-mentioned techniques.

The position of PIXE versus other techniques changes dramatically when one refers to microanalysis. Here there exist fewer alternatives and one can easily show some decisive advantages for micro-PIXE. New nuclear microprobe facilities have recently been installed in several laboratories around the world, ensuring their greater accessibility [103–106].

REFERENCES

1. T. B. Johansson, R. Akselsson and S. A. E. Johansson, *Nucl. Instrum. Methods* **84**, 141 (1970).
2. B. Dziunikowski, Energy Dispersive X-ray Fluorescence Analysis, Elsevier, Amsterdam (1989).
3. M. Budnar, M. Kregar, U. Miklavžič, V. Ramšak, M. Ravnikar, Z. Rupnik and V. Valković, *Nucl. Instrum. Methods* **179**, 249 (1981).
4. E.-M. Johansson and S. A. E. Johansson, *Nucl. Instrum. Methods Phys. Res., Sect. B* **B3**, 154 (1984).

5. S. A. E. Johansson and J. L. Campbell, PIXE: A Novel Technique for Elemental Analysis, Wiley, New York (1988).
6. Proc. Fifth Int. Conf. on PIXE and its Analytical Applications, Amsterdam, August 21–25, 1989. *Nucl. Instrum. Methods, Phys. Res., Sect. B* **B49** (1990).
7. Proc. Sixth Int. Conf. on PIXE and its Analytical Applications, Tokyo, Japan, July 20–24, 1992. *Nucl. Instrum. Methods, Phys. Res., Sect. B* **B75** (1993).
8. T. Mukoyama, *Int. J. PIXE* **1**, 209 (1991).
9. W. Brandt and G. Lapicki, *Phys. Rev. A* **23**, 1717 (1981).
10. D. D. Cohen and M. Harrigan, *At. Data Nucl. Data Tables* **33**, 255 (1985).
11. D. D. Cohen, *Nucl. Instrum. Methods, Phys. Res., Sect. B* **B49**, 1 (1990).
12. C. M. Lederer and V. S. Shirley (Eds.), *Table of Isotopes*, 7th Ed., Wiley Interscience, New York (1978).
13. W. Bambynek, Personal communication of material presented verbally at the International Conference on X-ray and Inner Shell Process in Atoms, Molecules and Solids, Universtiy of Leipzig (1984).
14. J. H. Scofield, *Phys. Rev. A* **9**, 1041 (1974).
15. A. Perujo, J. A. Maxwell, W. J. Teesdale and J. L. Campbell, *J. Phys. B: At. Mol. Phys.* **20**, 4973 (1987).
16. J. A. Bearden, *Rev. Mod. Phys.* **39**, 78 (1967).
17. S. A. E. Johansson and J. L. Campbell, Ref. 5, p. 12.
18. D. D. Cohen and M. Harrigan, *At. Data Nucl. Data Tables* **34**, 393 (1986).
19. F. Folkmann, C. Gaarde, T. Huus and K. Kemp, *Nucl. Instrum. Methods* **116**, 487 (1974).
20. K. Ishii and S. Morita, *Int. J. PIXE* **1**, 1 (1990).
21. S. A. E. Johansson, *Int. J. PIXE* **2**, 33 (1992).
22. C. A., Pineda, Thick target particle-induced X-ray emission, in Elemental Analysis by Particle Accelerators, edited by Z. B. Alfassi and M. Peisach, CRC Press, Boca Raton (1992), pp. 279–305.
23. J. F. Ziegler, *TRIM—The Transport of Ions in Matter*, IBM-Research, Yorktown, NY (1992).
24. J. Leroux and T. P. Thinh, *Revised Tables of X-Ray Mass Attenuation Coefficients*, Report Corporation Scientifique Claisse Inc., Quebec (1977).
25. S. A. E. Johansson and J. L. Campbell Ref. 5, p. 98.
26. T. A. Cahill, *Nucl. Instrum. Methods* **181**, 473 (1981).
27. J. L. Campbell and J. A. Cookson, *Nucl. Instrum. Methods Phys. Res., Sect. B* **B3**, 185 (1984).
28. J. A. Maxwell, J. L. Campbell and W. J. Teesdale, *Nucl. Instrum. Methods Phys. Res., Sect. B* **B43**, 218 (1989).
29. C. G. Ryan, D. R. Cousens, S. H. Sie, W. L. Griffin, G. F. Suter and E. Clayton, *Nucl. Instrum. Methods Phys. Res., Sect. B* **B47**, 55 (1990).
30. N. P.-O. Larsson, U. A. S. Tapper and B. G. Martinsson, *Nucl. Instrum. Methods Phys. Res., Sect. B* **B43**, 574 (1989).
31. J. L. Campbell, *Nucl. Instrum. Methods Phys. Res., Sect. B* **B49**, 115 (1990).
32. J.-X. Wang and J. L. Campbell, *Nucl. Instrum. Methods Phys. Res., Sect. B* **B54**, 499 (1991).
33. P. J. Statham, Limitations and potential for EDX spectrometry in electron beam instruments. Inst. Phys. Conf. Ser. No. 119: Section 10, 425 (1991).
34. J. L. Campbell, W. Maenhaut, E. Bombelka, E. Clayton, K. Malmqvist, J. A. Maxwell, J. Pallon and J. Vandenhaute, *Nucl. Instrum. Methods Phys. Res., Sect. B* **B14**, 204 (1986).
35. G. I. Johansson, *X-Ray Spectrom.* **11**, 194 (1982).
36. A. D. Lipworth, H. J. Annegarn, S. Bauman, T. Molokomme and A. J. Walker, *Nucl. Instrum. Methods Phys. Res., Sect. B* **B49**, 173 (1990).

37. A. D. Lipworth, H. J. Annegarn and M. A. Kneen, *Nucl. Instrum. Methods Phys. Res., Sect. B* **B75**, 127 (1993).
38. K. Yoshihara and J. Iihara, *Int. J. PIXE* **2**, 93 (1992).
39. D. W. Mingay, *J. Radioanal. Chem.*, **78**, 127 (1983).
40. T. A. Cahill, J. Miranda and R. Morales, *Int. J. PIXE* **1**, 297 (1991).
41. Y. J. Uemura, Y. Kuno, H. Koyama, T. Yamazaki and P. Kienle, *Nucl. Instrum. Methods* **153**, 573 (1978).
42. F. Sommer and B. Massonnet, *Nucl. Instrum. Methods Phys. Res., Sect. B* **B22**, 201 (1987).
43. K. Kemp and G. Danscher, *Histochemistry* **59**, 167 (1979).
44. W. Maenhaut, *Nucl. Instrum. Methods Phys. Res., Sect. B* **B35**, 388 (1988).
45. W. Maenhaut, *Scanning Microsc.* **4**, 43 (1990).
46. W. Maenhaut, *Anal. Chim. Acta* **195**, 125 (1987).
47. S. A. E. Johansson and J. L. Campbell, Ref. 5, pp. 177–199.
48. D. G. Jex, M. W. Hill and N. F. Mangelson, *Nucl. Instrum. Methods Phys. Res., Sect. B* **B49**, 141 (1990).
49. N. F. Mangelson and M. W. Hill, *Scanning Microsc.* **4**, 63 (1990).
50. R. D. Vis, *The Proton Microprobe: Applications in the Biomedical Field*, CRC Press, Boca Raton (1985), 197 pp.
51. F. Watt and G. W. Grime (Eds.), *Principles and Applications of High-Energy Microbeams*, Adam Hilger, Bristol (1987), 400 pp.
52. M. Uda, K. Maeda, Y. Sasa, H. Kusuyama and Y. Yokode, *Nucl. Instrum. Methods Phys. Res., Sect. B* **B22**, 184 (1987).
53. S. C. Yeh, T. C. Chu, H. J. Lin and C. C. Hsu, *Nucl. Instrum. Methods Phys. Res., Sect. B* **B17**, 349 (1986).
54. H. J. Annegarn, A. E. Pillay, J. C. A. Davies, D. Faure and J. P. F. Sellschop, *Nucl. Instrum. Methods Phys. Res., Sect. B* **B35**, 415 (1988).
55. F. Watt and J. P. Landsberger, *Nucl. Instrum. Methods Phys. Res., Sect. B* **B77**, 249 (1993).
56. E. T. Williams, *Nucl. Instrum. Methods Phys. Res., Sect. B* **B3**, 211 (1984).
57. J. Räisänen, *Int. J. PIXE* **2**, 393 (1992).
58. S. B. Russell, C. W. Schulte, S. Faiq and J. L. Campbell, *Anal. Chem.* **53**, 571 (1981).
59. Y. Iwata, T. Fujiwara and N. Suzuki, *Int. J. PIXE* **2**, 381 (1992).
60. A. H. Legge, H. C. Kaufmann and J. W. Winchester, *Nucl. Instrum. Methods Phys. Res., Sect. B* **B3**, 507 (1984).
61. T. Raunemaa, M. Hannikainen, M. Kulmala and P. Hari, *Nucl. Instrum. Methods Phys. Res., Sect. B* **B22**, 473 (1987).
62. P. K. Mueller, A. Alcocer, T. A. Cahill, R. Sommerville and R. Flocchini, Elemental analysis by alpha excited X-ray fluorescence, *Proc. Am. Chem. Soc.* (1971).
63. E. Koltay, *Int. J. PIXE* **1**, 93 (1990).
64. W. Maenhaut, *Int. J. PIXE* **2**, 609 (1992).
65. H. J. Annegarn, T. A. Cahill, J. P. F. Sellschop and A. Zucchiatti, *Phys. Scr.* **37**, 282 (1988).
66. T. A. Cahill, C. Goodart, J. W. Nelson, R. A. Eldred, J. S. Nasstrum and P. J. Feeney, Proc. Int. Symp. Particulate and Multiphase procedures, Miami Beach, 22–26 April, 1985, Hemisphere Publishing, Washington, DC.
67. M. Bohgard and E.-M. Johansson, *Nucl. Instrum. Methods Phys. Res., Sect. B* **B3**, 268 (1984).
68. B. G. Martinsson, *Nucl. Instrum. Methods Phys. Res., Sect. B* **B15**, 636 (1986).
69. B. G. Martinsson, *Nucl. Instrum. Methods Phys. Res., Sect. B* **B22**, 356 (1987).
70. T. A. Cahill, Y. Matsuda, D. Shadoan, R. A. Eldred and B. H. Kusko, *Nucl. Instrum. Methods Phys. Res., Sect. B* **B3**, 263 (1984).

71. C. Boni, A. Caridi, E. Cereda and G. M. Braga Marcazzan, *Nucl. Instrum. Methods Phys. Res., Sect. B* **B47**, 133 (1990).
72. C. Boni, A. Caridi, E. Cereda, G. M. Braga Marcazzan and P. Redaelli, *Nucl. Instrum. Methods Phys. Res., Sect. B* **B49**, 106 (1990).
73. H. J. Annegarn and W. J. Przybyłowicz, *Nucl. Instrum. Methods Phys. Res., Sect. B* **B75**, 582 (1993).
74. P. K. Hopke, *Receptor Modeling in Environmental Chemistry*, Wiley, New York (1985).
75. J. G. Watson, J. C. Chow and C. V. Mathai, Receptor models in air resources management: a summary of the APCA International Specialty Conference, *J. Air Pollut. Control Assoc.* **39**, 419 (1989).
76. H. J. Annegarn and S. Bauman, *Nucl. Instrum. Methods Phys. Res., Sect. B* **B49**, 264 (1990).
77. G. Kullerud, R. M. Steffen, P. C. Simms and F. A. Rickey, *Chem. Geol.* **25**, 245 (1979).
78. C. G. Ryan, D. R. Cousens, S. H. Sie and W. L. Griffin, *Nucl. Instrum. Methods Phys. Res., Sect. B* **B49**, 271 (1990).
79. C. G. Ryan, D. R. Cousens, S. H. Sie, W. L. Griffin, G. F. Suter and E. Clayton, *Nucl. Instrum. Methods Phys. Res., Sect. B* **B47**, 55 (1990).
80. F. Watt, G. W. Grime and D. G. Fraser, Microbeam applications in the earth sciences, in Principles and Applications of High Energy Ion Microbeams, edited by F. Watt and G. W. Grime, Adam Hilger, Bristol (1987), pp. 239–272.
81. J. D. MacArthur and X-P. Ma, *Int. J. PIXE* **1**, 311 (1991).
82. C. G. Ryan and W. L. Griffin, *Nucl. Instrum. Methods Phys. Res., Sect. B* **B77**, 381 (1993).
83. S. H. Sie, *Nucl. Instrum. Methods Phys. Res., Sect. B* **B75**, 403 (1993).
84. L. J. Cabri, *Mineralogical Mag.* **56**, 289 (1992).
85. S. H. Sie, C. G. Ryan, D. R. Cousens and W. L. Griffin, *Nucl. Instrum. Methods Phys. Res., Sect. B* **B40/41**, 690 (1989).
86. K. G. Malmqvist, H. Bage, L.-E. Carlsson, K. Kristiansson and L. Malmqvist, *Nucl. Instrum. Methods Phys. Res., Sect. B* **B22**, 386 (1987).
87. J. I. W. Watterson, J. P. F. Sellschop and A. Zucchiatti, *Nucl. Instrum. Methods Phys. Res., Sect. B* **B28**, 554 (1987).
88. H. J. Annegarn, C. S. Erasmus and J. P. F. Sellschop, *Nucl. Instrum. Methods Phys. Res., Sect. B* **B3**, 181 (1984).
89. A. E. Pillay, C. S. Eramus, A. H. Andeweg, J. P. F. Sellschop, H. J. Annegarn and J. Dunn, *Nucl. Instrum. Methods Phys. Res., Sect. B* **B35**, 555 (1988).
90. K. Kristiansson and L. Malmqvist, *Geoexploration* **24**, 517 (1987).
91. P. Aloupogiannis, J. P. Toutain, G. Robaye, I. Roelandts, J. P. Quisefit and G. Weber, *Nucl. Instrum. Methods Phys. Res., Sect. B* **B49**, 277 (1990).
92. P. S. Z. Rogers, C. J. Duffy, T. M. Benjamin and C. J. Maggiore, *Nucl. Instrum. Methods Phys. Res., Sect. B* **B3**, 671 (1984).
93. J. J. G. Durocher, N. M. Halden, F. C. Hawthorne and J. S. C. McKee, *Nucl. Instrum. Methods Phys. Res., Sect. B* **B30**, 470 (1988).
94. M. Peisach and C. A. Pineda, *Nucl. Instrum. Methods Phys. Res., Sect. B* **B49**, 10 (1990).
95. J. S. C. McKee, G. R. Smith, Y. H. Yeo, K. Abdul-Retha, D. Gallop, J. J. G. Durocher, W. Mulholland and C. A. Smith, *Nucl. Instrum. Methods Phys. Res., Sect. B* **B40/41**, 680 (1989).
96. C. P. Swann and S. J. Fleming, *Nucl. Instrum. Methods Phys. Res., Sect. B* **B22**, 407 (1987).
97. S. J. Fleming and C. P. Swann, *Nucl. Instrum. Methods Phys. Res., Sect. B* **B22**, 411 (1987).

98. W. Maenhaut, *Nucl. Instrum. Methods Phys. Res., Sect. B* **B49**, 518 (1991).
99. R. Klockenkämper, B. Raith, S. Divoux, B. Gonsior, S. Brüggerhoff and E. Jackwerth, *Fresenius' Z. Anal. Chem.* **326**, 105 (1987).
100. L.-E. Carlsson and R. K. Akselsson, *Adv. X-ray Anal.* **24**, 313 (1981).
101. J. P. Willis, *Nucl. Instrum. Methods Phys. Res., Sect. B* **B35**, 378 (1988).
102. C. Erdtmann and H. Petri, in *Treatise on Analytical Chemistry*, 2nd Ed., Part I, Vol. 14, edited by I. M. Kolthoff, P. J. Elving and V. Krivan, Wiley, New York (1986), p. 419.
103. J. I. W. Watterson, R. W. Fearick, S. H. Connell, H. J. Annegarn, W. J. Przybyłowicz, A. H. Andeweg, I. McQueen and J. P. F. Sellschop, *Nucl. Instrum. Methods Phys. Res., Sect. B* **B77**, 79 (1993).
104. W. J. Przybyłowicz, J. I. W. Watterson, H. J. Annegarn, S. H. Connell, R.W. Fearick, A. H. Andeweg and J. P. F. Sellschop, *Nucl. Instrum. Methods Phys. Res., Sect. B* **B75**, 539 (1993).
105. M. Jakšić, L. Kukec and V. Valković, *Nucl. Instrum. Methods Phys. Res., Sect. B* **B77**, 49 (1993).
106. U. A. S. Tapper, W. R. McMurray, G. E. Ackermann, G. De Villiers, D. Fourie, P. J. Groenewald, J. Kritzinger, C. A. Pineda, H. Schmitt, K. A. Springhorn and T. Swart, *Nucl. Instrum. Methods Phys. Res., Sect. B* **B77**, 17 (1993).

Chapter 14

USE OF MICROPROBES

Ulf Lindh
Division of Physical Biology, Department of Radiation Sciences, Uppsala University, Uppsala, Sweden

1 INTRODUCTION

Since the invention of the nuclear microprobe in 1970 [1], new areas of application have rapidly been introduced. Applications include mineralogy, metallurgy, microelectronics, geology, archaeology, biology and medicine. The first International Conference of Nuclear Microprobe Technology and Applications was organized in Oxford in 1987. The meeting was attended by more than 70 scientists from 14 countries, and the proceedings include 47 papers on various themes of microprobe technology and applications [2]. The second conference in the series was held in Melbourne, Australia, in 1990. This conference was attended by 60 scientists from 13 countries. The proceedings include 71 papers well reflecting the rapid expansion of the technology and its applications [3]. The Department of Radiation Sciences organised the third conference in 1992 at Uppsala, where 70 scientists from 15 countries gathered and presented 80 papers, which are published in the proceedings [4]. The fourth conference is scheduled for 1994 and is to be held in Shanghai, China.

A few books have been published on the application of nuclear microprobes. The first, by Vis, appeared in 1985 [5]. Two years later, Watt and Grime published a second [6]. The field of nuclear microscopy, as the technique is often called, is only 20 years old, and it is remarkable that so many new areas of application have evolved during the two decades.

This overview of applications is concentrated on the biomedical field. For a complete account of the uses of nuclear microscopy the reader is referred to the conference proceedings mentioned above.

Chemical Analysis by Nuclear Methods Edited by Z. B. Alfassi

2 NUCLEAR MICROSCOPY OF CELLS AND TISSUES

Most of the samples under investigation in biology and medicine are principally composed of hydrogen, carbon, oxygen and nitrogen. Besides this organic matrix, there are minor amounts of such metals as sodium, potassium, magnesium and calcium. Furthermore, there are very small amounts of the so-called trace elements. A biologist probably defines the trace elements as follows. The elements H, C, N, O, Na, Mg, P, S, Cl, K and Ca form the bulk of living matter. These are the major and minor elements and they are therefore excluded from the 90 naturally occurring elements. From the remaining 79, we exclude the noble gases, as they do not seem to have any biological function. The remaining 73 elements are the trace elements. Of these, 17 are considered or suspected to be essential for life: Li, F, Si, V, Cr, Mn, Fe, Co, Ni, Cu, Zn, Se, As, Mo, Sn, I and Pb. There are some doubts about Li and Pb being essential. All of them are metals except Se and As.

The analytical problem posed to the biomedical researcher is that of finding heavier elements embedded in a light matrix. The widely used technique of particle induced X-ray emission with energy dispersive solid-state detectors provides multi-element capacity and minimum detection limits of parts per million or less [7]. To extract concentrations of trace elements, the mass under the probe has to be correctly assessed. This can be accomplished in several ways. In the early days of nuclear microscopy, it was common to use a method devised for the electron microscope [8]. This method relies on the fact that the bremsstrahlung caused by primary and secondary electrons is proportional mainly to the number of carbon atoms in the irradiated volume, i.e. proportional to the mass under the probe. This method was adapted for use with PIXE some years later [9]. Other, more accurate, methods have been devised for nuclear microscopy. They rely on nuclear scattering of the ions in the primary beam [10, 11] or on a nuclear reaction in an aluminium foil behind the specimen [12].

The important and hard-to-accomplish problem of specimen preparation will not be addressed here. The reader is directed toward the books by Roos and Morgan [13] and Robards and Sleytr [14].

There are several types of interaction which take place when a beam of nuclear projectiles impinges on a specimen. In this chapter we will discuss three interactions, all of which can occur simultaneously. They will prove to be very important for the nuclear microscopy of biological or medical specimens.

As was mentioned above, PIXE is a technique with mutli-element capacity and high sensitivity. The practical range of detectable elements, however, is from sodium and heavier elements up to uranium. The restriction to elements from sodium and heavier is imposed by a thin ($<7\,\mu m$) beryllium window in front of the detector crystal. The window effectively filters out most of the photons with energy lower than 1 keV. Besides PIXE, Rutherford backscattering spectroscopy (RBS) is utilized. In RBS, the energy of elastically scattered particles is measured, thus providing information on the matrix composition and density of the sample

[15]. Perhaps the most exciting method employed in nuclear microscopy is scanning transmission ion microscopy [16, 17]. Using this technique, nuclear particles transmitted through the specimen are detected, and measurement of their energy loss gives information on the structure and density of the sample. Because of the extended range of nuclear projectiles in organic matrices ($>100\ \mu m$ for 3 MeV protons), thick samples can be imaged, and because of the high rigidity of the nuclear projectiles, spatial resolution is reasonably well maintained throughout the sample.

The volume of biomedical work undertaken by nuclear microscopists is steadily increasing, although most of the work is still centred on only a few groups. A few examples of biomedical nuclear microscopy will be discussed in some detail here. One reason for limiting the account is that a complete review would mainly be repetition of recent work. The other reason is that in this way it is easier to estimate the role of nuclear microscopy in solving specific biomedical problems rather than compiling applications where nuclear microscopy could be used. The problems chosen are (i) the controversy of trace element involvement in the aetiology of Alzheimer's disease, (ii) dermatology, (iii) cardiovascular disease, (iv) AIDS research, (v) the involvement of heavy metals from implants in the aetiology of the chronic fatigue syndrome, and (vi) protective effects of essential trace elements. Merely a 'mini-review' of nuclear microscopy in biomedicine will be presented.

3 MINI-REVIEW OF BIOMEDICAL NUCLEAR MICROSCOPY

Modern medicine has become very dependent on laboratory tests and the interpretation of such results. Although electron microscopy has been used for decades, it is used only sparsely in medical diagnostics. X-ray microanalysis using electron microscopes has not been used extensively in medical diagnostics, although it can be used for intracellular detection and localization of electrolytes that are minor elements in tissues. However, the interest in trace elements has grown significantly during the last decades; although the determination of trace elements in humans has mostly been restricted to whole blood and plasma or serum. There is always an optimum range of concentration in the diet, below which deficiency symptoms occur and above which symptoms of toxicity appear. Our current knowledge of elements essential to mammals is imperfect and somewhat controversial.

Researchers in medicine and trace element biology have mostly concentrated on essential trace elements, especially copper, zinc and selenium in recent years. Detailed knowledge is abundant for the concentration interval where toxicity occurs. This is explained by the fact that the higher concentrations are easier to determine quantitatively. There is an increasing interest in those non-essential elements which have been accumulating in the environment because of man's activities. The elements attracting most of the interest have been cadmium, mercury and lead, but silver, arsenic, chromium, manganese, antimony, tin and vanadium also fall into this category. The list of trace elements under investigation will inevitably grow, parallelling the introduction of analytical methods and instruments with multielement capacity at very low concentrations.

Trace elements are often found as important cofactors in enzymes, the most prominent example being zinc. The importance of zinc as an essential trace element is evident from the fact that more than 200 enzymes are zinc dependent [18]. Trace elements are necessary for maintenance and regulation of cell function compartmentalization, gene regulation, and regulation of membrane functions, although some of the elements involved cannot, by definition, be regarded as trace elements. The development of life has provided elaborate, diverse and subtle mechanisms to control trace element homeostasis. The trace elements thus exert a profound influence on human health and disease states.

The homeostasis and interaction of trace elements with each other and with non-essential trace elements such as heavy metals in biological systems is very complex. In spite of the research effort expended, detailed knowledge of trace element functions is rather scarce. Much information has been gained with radioactive nuclides in cell cultures and in experimental animal models. With the arrival of positron emission tomography it has been possible to perform dynamic studies with short-lived radionuclides. The gross concentration of major, minor and trace elements in man is quite well characterized by various analytical methods [19]. The literature, however, lacks a good compilation of major, minor and trace element concentrations in animal tissues.

The Uppsala group has carried out a series of investigations to estimate the medical value of nuclear microscopy of isolated individual blood cells from healthy control individuals [20] and patients suffering from various diseases such as leukaemias [21], inflammatory connective tissue diseases [22], Down's syndrome [23], cystic fibrosis [24] etc.

The alteration of trace-element metabolism in association with chronic inflammatory diseases has been a field under study for many decades [25]. During recent years, clinical trace element research has come into focus, partly through the development of antirheumatic therapies which involve the manipulation of certain elements, but also because of the discovery of the central role certain elements have in free radical generation and lysosomal degranulation. The Uppsala group recently reported conspicuous accumulation of calcium and magnesium in granulocytes and erythrocytes from patients with rheumatoid arthritis and seronegative arthritides [26]. These results induced speculation about the consequences of cellular accumulation in the inflammatory process. However, later it was found that strontium [27] as well as manganese [28] also appeared in increased concentrations in granulocytes from patients with rheumatoid arthritis, while zinc appeared in reduced amounts [29].

Because of the genetic background, the inflammatory disease ankylosing spondylitis, or Bechterew's disease, has been studied thoroughly. Only individuals with the tissue antigen HLA-B27 develop the disease. Four groups were established, viz. healthy age- and sex-matched controls, patients, HLA-B27$^+$ healthy relatives and healthy HLA-B27$^-$ relatives. The HLA are histocompatibility antigens, the antigens on living cells that distinguish self from non-self. They are glucoproteins on the surface of all nucleated cells and are called *human leukocyte antigens* (HLA) because they were first identified on circulating human lymphocytes. The genes that code for histocompatibility antigens are located in the major histocompatibility complex on the short arm of chromosome 6. The genes are found in three different loci: A, B and C.

Three blood cell types were isolated and studied by nuclear microscopy. The elemental profiles comprised magnesium, calcium, manganese, iron, copper, zinc and strontium. First degree B27$^+$ relatives showed significant accumulation of cellular magnesium, calcium, manganese and iron compared with the healthy controls. Compared with the patients, relatives and controls showed significantly lower values of magnesium, calcium, manganese and iron [30].

The multielemental approach to individual cell analysis results in a large data set, and it does not seem to be adequate to compare results element by element. To interpret the interactions between major, minor and trace elements it is necessary to use multivariate statistics. Principal component analysis [31] was used to show that there was a similar lack of correlation between manganese and zinc in the control group and the group with HLA-B27$^-$ relatives. Zinc and manganese were similarly correlated in the patient group and the group with HLA-B27$^+$ relatives. In the same way, calcium and iron were correlated in the patient group and in the group with HLA-B27$^+$ relatives. Calcium and zinc correlated quite well in the control and in HLA-B27$^-$ relative groups. The use of multivariate statistical methods gave hints as to the direction of further studies.

The need to investigate the copper content of individual human fibroblast cells, as an adjunct to the investigations of Menkes disease, motivated Allan, Camarakis and Legge to exploit nuclear microscopy [32]. Menkes disease, which is also known as Menkes steely hair syndrome or the kinky hair syndrome, is an X-linked genetic disease characterized by a severe disturbance in copper metabolism. Unfortunately this inherited brain disease is progressive and fatal. The mutant gene associated with Menkes disease is expressed in cultured cells. Cell cultures could therefore be used as a controlled environment for the study of the disease.

The initial fibroblast cultures were sourced from skin samples that had been extracted by punch biopsy. The cells were grown on thin nylon foils and, after freeze-drying, the foil supported a monolayer of fibroblast at $\approx 60\%$ confluence; some cells were lost during washing and freeze-drying. Nuclear microscopy of the cells as well as healthy fibroblasts revealed an approximately sixfold increase in the copper level of Menkes cells. This level of increase of copper in cultured fibroblasts provides the nuclear microscope with a means of identifying individual Menkes cells by the examination of the copper content of the cells. The possibility of such individual cell identification will allow nuclear microscopy to play a role in the detection of sufferers from and carriers of Menkes disease.

4 SLIM-UP—AN EXAMPLE OF A NUCLEAR MICROSCOPE

The predecessor of the new *Slim-Up* (Scanning Light Ion Microprobe in Uppsala) was the Studsvik nuclear microprobe. The latter was developed and established by cooperation between the Division of Physical Biology, Uppsala University and Atomenergi AB, Studsvik. It became operational in 1975, and was the fourth nuclear microprobe in the world. It was designed to be a miniaturized copy of the Harwell microprobe, the first operational nuclear microprobe, designed by John Cookson [1], who is considered to be the father of nuclear microscopy. The Studsvik nuclear microprobe was continuously improved, and the eventual spatial resolution was $2.9 \times 2.9\ \mu m^2$ at 100 pA of proton current [33]. The nuclear microscopy community has agreed upon using a particle intensity corresponding to 100 pA as a standard for comparisons of spatial resolution.

The Studsvik nuclear microprobe was transferred from a single-ended accelerator at Studsvik to the tandem accelerator of the The Svedberg Laboratory, Uppsala in 1987, and was eventually dismantled in 1989. The Slim-Up is dedicated primarily to nuclear microscopy of environmental, biological and medical specimens. Recent development in lens design (Oxford Microbeams 1989) enables sub-μm elemental mapping to be carried out routinely. The major innovation in the design of the new lenses is that the magnet circuit is a single piece of stress-relieved iron profiled to μm accuracy. The novel design is free from higher order parasitic aberrations.

The beam of light ions, usually protons, is delivered by an EN-tandem accelerator fed by an off-axis direct extract duoplasmatron ion source. Two feedback-controlled magnetic dipoles align the beam with the optical axis of the probe-froming lenses. The magnet dipoles are placed before the object aperture (Fig. 1). Upstream and before the lenses, a magnetic beam scanning unit of current-controlled ferrite cores is installed. The scan area at the sample position comprises a two-dimensional, $2 \times 2\ mm^2$, matrix of $65\,536 \times 65\,536$ pixels. Arbitrarily shaped areas of the scan matrix can be irradiated repeatedly for multiples of 10^{-14} C or 1.0 μs. The target chamber is an octagonal construction containing the following parts: (i) a sample holder movable in two directions perpendicular to the beam, (ii) a microscope, for viewing from the back of the target, at maximum magnification of $\times$ 600, with the turret inside the chamber, (iii) in one turret position, a Faraday cup for monitoring of beam intensity, (iv) a $\times$ 150 stereomicroscope yielding an oblique view of the sample front, (v) a retractable

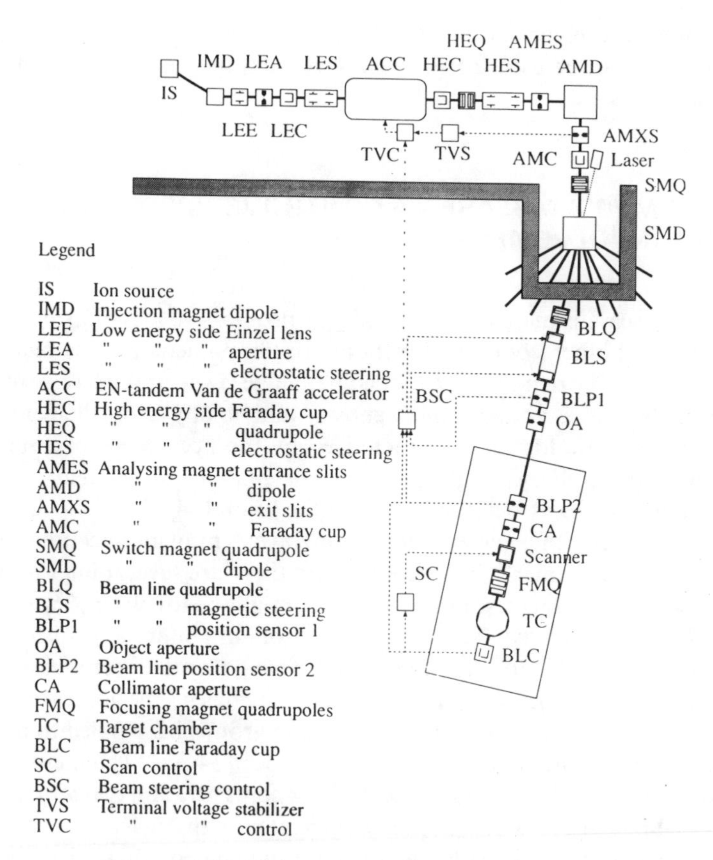

Fig. 1. The Slim-Up facility at the Svedberg Laboratory in Uppsala.

30 mm^2 Si(Li) detector (147 eV resolution at 5.9 keV) covered by a 7 μm beryllium window and mounted at 135° to the beam direction with a maximum solid angle of 980 msr, and (vi) a channeltron electron detector to yield a secondary electron image of the object.

Particle detectors will be installed to detect forward-scattered and back-scattered particles as well as transmitted ions. A Ge(Li) detector will be introduced to facilitate nuclear reaction analysis. Equipment for image analysis will also be integrated with the nuclear microscope. The objective of this is to simplify recognition of the regions of interest in the samples.

The beam control and data acquisition are performed by Intel 386/387 based personal computers equipped with dedicated expansion cards. Data are read simultaneously from eight channels with 16 bit resolution into a buffer and further to a mass storage memory device. One channel is devoted to multiplexed signals from up to eight detectors with ≤ 8192 channels resolution. The other seven channels can be used for the necessary sampling of accumulated charge, time, coordinates etc.

5 ALZHEIMER'S DISEASE—A CONTROVERSY OF TRACE ELEMENT INVOLVEMENT

Alzheimer's disease is manifested clinically as progressive loss of memory and a deterioration in other cognitive functions. Initially, memory loss and spatial disorientation are evident, followed at a later stage by symptoms of perital lobe dysfunction (aphasia, apraxia and agnosia) and finally gross dementia. The disease is neuropathologically characterized by the occurrence of a minimum density of neurofibrillary tangles and neuritic plaques distributed within the hippocampus and the association cortex of the brain [34, 35].

There is no convincing evidence for the involvement of an infectious agent in the aetiology of Alzhemer's disease. However, there are several lines of circumstantial evidence that suggest that aluminium might be involved [36, 37]. Some groups have demonstrated the co-localization of aluminium and silicon within plaque cores [38–41]. Although these results have already been questioned in some papers and have been disputed [42, 43], there has been a general belief that aluminium levels are increased in old age. Some groups have failed to detect an elevation of aluminium and silicon in the plaques [34, 35]. The controversy is complicated by several facts. Studies of the bulk concentration of major and trace elements in brain tissue are not readily correlated with nuclear microscopy or electron microscopy of senile plaque cores. Recent epidemiological studies have linked environmental aluminium with an increased prevalence of dementia [44, 45]. Also conspicuous is the finding by Crapper McLachlan *et al.* [46] that the rate of disease progression of clinically diagnosed Alzheimer's disease patients was decreased on administration of desferroxamine, a drug that preferentially binds iron, but also aluminium.

The Oxford group has undertaken a well designed study to shed some light upon the controversy over aluminium in Alzheimer's disease [47]. The suspicion that has been discussed by representatives of this group and various other authors is that aluminium and silicon are introduced during sample preparation. Five known cases of Alzheimer's disease were investigated by nuclear microscopy. The specimens were stained immunohistochemically to identify the senile plaques [48]. The group performed analyses of more than 100 plaque cores. Fewer than 10% of the plaque cores were found to contain aluminium. Where found, aluminium was co-localized with silicon and frequently also with a combination of calcium, titanium and iron. However, such deposits were also found in background scans. The latter finding suggests problems with contamination. With these results at hand, Watt and colleagues directed their research towards the reagents used in the staining procedure. Aluminosilicates have been shown to occur in air-borne dust [49]. The dust is composed of aluminosilicates up to $\approx 4\,\mu m$ in size, and these particles comprise a significant fraction of the air-borne dust. Tests carried out on the reagents used in the stains for plaque identification showed traces of contamination, even though the reagents had been filtered through 0.2 μm Millipore filters. Nuclear microscopy of the filtrate evaporated on clean Pioloform film showed accumulation of inorganic species including aluminosilicates. A hypothesis raised by the group is that the reagents contain aluminosilicates either as a very fine suspension of sub-μm particles or in solution, which crystallize out as the solvent evaporates. Furthermore, the contaminant colloid may preferentially locate on the plaques because the amyloid tends to be 'sticky'. The obvious solution to the problem is to carry out nuclear microscopy on unstained tissue specimens.

In this situation, the researchers found themselves confronted with a challenge. Unstained tissue sections usually provide very little contrast, making identification of structures almost impossible. With a primary beam of nuclear projectiles, various phenomena apart from PIXE can be employed in the analysis. Scanning transmission ion microscopy (STIM) has structural imaging capability. It has been shown that STIM is able to map senile plaques and cores in unstained tissue [50].

Plaques were identified using off-axis STIM by scanning quickly (≈ 4 min) over an area of $\approx 400 \times 400\,\mu m^2$. Once located, a $100 \times 100\,\mu m^2$ scan over the plaque was carried out using STIM, RBS and PIXE (taking ≈ 40 min). Specific analyses were then carried out using a small area scan over the plaque core ($5 \times 5\,\mu m^2$), and a larger scan ($100 \times 100\,\mu m^2$) for background control over a tissue region away from the plaque. Figure 2 shows a STIM map of a core-containing plaque identified in unstained, untreated tissue, and Fig. 3 shows the corresponding elemental maps of phosphorus, sulphur, carbon and nitrogen from the PIXE and RBS data. Table 1 shows the results of the analyses of over 80 unstained plaque cores from the hippocampus and the middle temporal gyrus of four cases of Alzheimer's disease and three age-matched controls. The results show no evidence of aluminium in the plaque cores at the detectable limit of analysis of 15 μg/g [47]. The implication from the work of the Oxford group is that the involvement of aluminium in the aetiology of Alzheimer's disease should be revised to consider the probable contamination of tissue by aluminosilicates present in most reagents.

Even though the controversy may not be solved entirely by these experiments, they show the potential of nuclear microscopy. It seems that there is no other technique available today with the capability to resolve both the morphology of tissue sections and the distribution of trace elements in the sections.

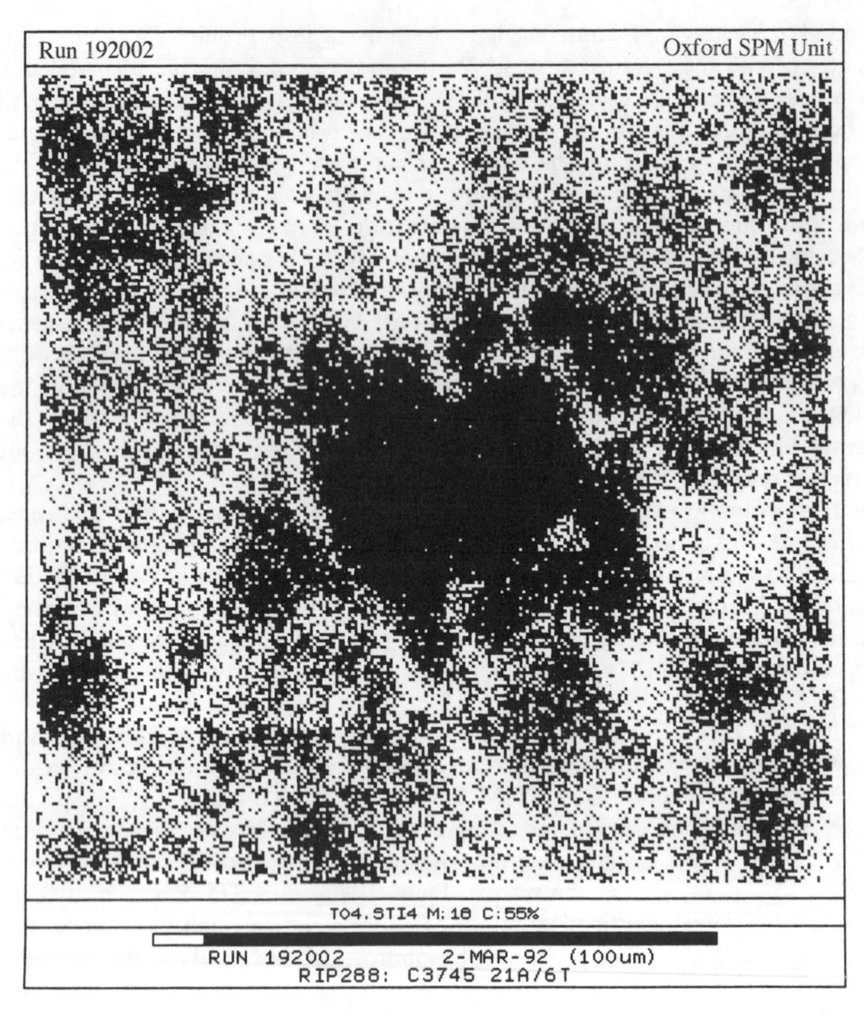

Fig. 2. STIM image of a senile plaque and core, found in unstained, untreated tissue. (From Ref. 47.)

6 APPLICATIONS IN DERMATOLOGY

Nuclear microprobe analysis of skin is an established method for the study of elemental distributions in normal and pathological situations [51]. The main focus of interest has been the distribution of physiologically important elements such as the electrolytes sodium, potassium, chloride, magnesium and calcium, the major elements phosphorus and sulphur and the trace elements iron, copper and

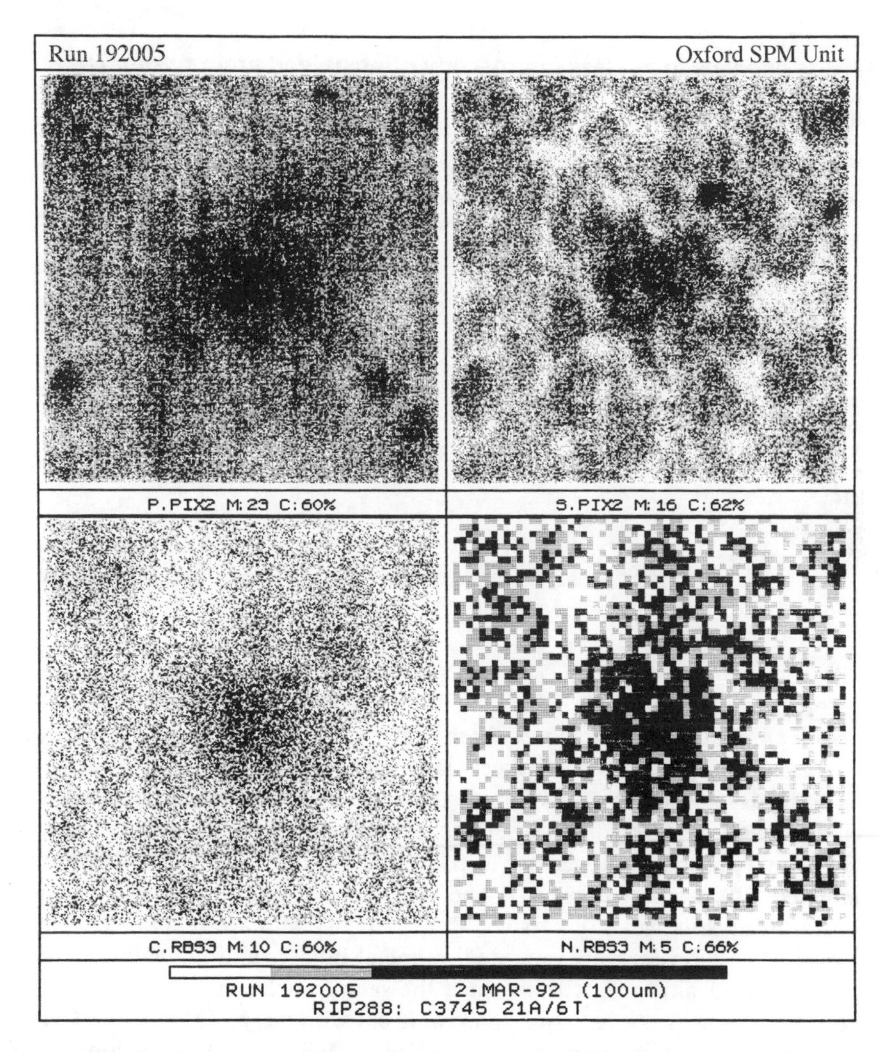

Fig. 3. PIXE and RBS maps of the senile plaque imaged using STIM in Fig. 2. Top right: phosphorous, top left: sulphur, bottom left: carbon, bottom right: nitrogen. (From Ref. 47.)

zinc. A change in the quotient of sodium to potassium has been used as an indicator of cell or tissue damage; normally the concentration of potassium is kept high in the cells and low outside. The quotient may also be a marker of regeneration activity in the cell [52]. Earlier data, from the group at Lund, about the trace element distribution have suggested that an abnormal iron distribution in apparently normal skin from a psoriatic patient implied abnormal metabolism of this element as one of the many dysfunctional features in the skin of psoriatics.

TABLE 1.
Results of scans in nuclear microscopy of unstained brain tissue. The cases CA, CB and CC are controls. The average area scanned was $100 \times 100\,\mu m^2$ [47]

Case and region (temporal cortex, hippocampus)	Total no. of of plaques scanned	Total no. of Al deposits in scans	Number of plaque cores scanned	Number of plaque cores with Al	Number of background scans	Total no. of Al deposits in backgrounds
B h	10	0	10	0	5	0
B tc	10	1	10	0	7	0
C h	9	0	9	0	13	0
C tc	9	2	9	0	3	0
D h	11	1	11	0	12	1
D tc	12	1	12	0	16	2
E h	6	0	6	0	6	1
E tc	11	0	11	0	13	0
CA tc					10	2
CA h					10	1
CB tc					10	1
CB h					13	0
CC tc					9	2
CC h					10	0

Atopic dermatitis, which is a constitutional condition in 10–20% of the Swedish population, has previously been shown to harbour an abnormal production for barrier lipids; this partially explains the increased transepidermal water loss in the dry non-eczematous skin of atopics [53]. Although immunologically well documented, the disorder is yet unexplained in many of its appearances.

The Lund group has carried out a recent investigation of pathological skin as compared with normal skin [54]. The skin samples were obtained by punch biopsies from three healthy persons, three patients with atopy and three with psoriasis. The biopsies were mounted on metal holders and immediately snap frozen in liquid nitrogen, within which the samples were subsequently stored before sectioning. Sections 16 μm thick were cut with a cryostat at $-20\,^{\circ}C$ and transferred to holders carrying a thin Kimfol foil. The skin sections were irradiated with a 1 nA beam of 2.5 MeV protons with a probe size of 5 μm. On selected samples, the epidermis was located and a rectangular beam scan was used to cover an area from the stratum corneum to the basal cell layer. Figures 4 and 5 show the results of the analyses.

In the normal skin, iron and zinc peak in the stratum germinatum and decrease towards the skin surface. The calcium level increases in the epidermis, which might reflect the influence of calcium on the differentiation process in the epidermis. Phosphorus peaks at about the same place as iron and zinc.

In the psoriatic skin, the most conspicuous finding is the dramatic increase of iron towards the skin surface. Sulphur shows nearly the same distribution as in normal skin, whereas phosphorus and calcium display double peaks. In normal skin chlorine increases inwards from the surface, whereas in psoriatic skin chlorine shows a broad peak. Data extracted from RBS spectra show, interestingly, that the mass of psoriatic skin increases with depth, whereas normal skin displays the largest mass closer to the surface.

The atopic skin shows distributions of sulphur, potassium and chlorine similar to those of the normal skin. However, the peaking of calcium is missing. The level of phosphorus is also a little higher and the peak less pronounced in the atopic case. Copper displays a sharp peak at the same location as the phosphorus peak. The concentration of iron is higher and the peak is broader, as with zinc. The mass profile is shifted into the sample as compared to the normal skin.

7 APPLICATIONS TO CARDIOVASCULAR DISEASE (ATHEROSCLEROSIS)

Arteries have walls constructed of three coats and a hollow core, called a *lumen*, through which the blood flows. The inner coat of an arterial wall, the *intima*, is composed of a lining of endothelium that is in contact with the blood and a layer of elastic tissue. The middle coat, or *tunica media*, is usually the thickest layer. It consists of elastic fibres and smooth muscle. The outer coat, the *adventitia*, is composed principally of elastic and collagenous fibres.

Because of the structure of the middle coat in particular, arteries have two major properties: elasticity and contractility. When the ventricles of the heart contract and eject blood into the large arteries, the arteries expand to accommodate the extra blood. Then, as the ventricles relax, the elastic recoil of the arteries forces the blood onwards. The contractility of an artery comes from its smooth muscle, which is arranged longitudinally and in rings around the lumen, somewhat like a doughnut, and is supplied by sympathetic branches of the autonomic nervous system. When there is sympathetic stimulation, the smooth muscle contracts, squeezes the wall around the lumen, and narrows the vessel.

Atherosclerosis is the deposit of lipid in the intima of large and medium sized arteries to form plaques. The lipid deposits contain cholesterol. The fatty material may later be replaced by dense connective tissue and calcium deposits. *Arteriosclerosis*, the hardening of the arteries, is characterized by a thickening of the intima, making the tunica media less elastic. Fat gradually accumulates between the elastic and collagenous fibres to produce lesions that protrude into the lumen. The calcification may lead to cell death, being responsible for the disruption of the arterial wall and subsequent plaque formation.

Very little is known about elemental involvement in atherosclerosis. There are indications of significant variations of potassium and calcium in blood serum. Vis *et al.* [55] showed by elemental mapping that there was a clearly visible accumulation of calcium in the affected area which was accompanied by an increase in zinc concentration.

The nuclear microprobe group at Lund has studied the elemental contents in arteries of atherosclerotic individuals, and has tried to relate this information to data about cholesterol, very low density lipids and high density lipids in blood serum [41]. Samples were collected during autopsy, immediately frozen with a Freon spray and mounted on a cork onto the microtome plate. Approximately 5–7 μm thick sections were produced at a temperature well below − 20 °C. The cryosections for nuclear microscopy were mounted on a Kimfol foil and freeze-dried.

The elemental contents seem to decrease with progressing atherosclerosis. In moderate atherosclerosis, the elemental contents were closer to that of advanced atherosclerosis than was expected from histopathology. In normal cases, or at very early stages of the disease, iron, copper and zinc were homogeneously distributed. All elements were found concentrated at the arterial border. In deeper regions, roughly 40 μm away from the arterial border, calcium, phosphorus and iron seemed to appear in granules, the diameter of which varied from

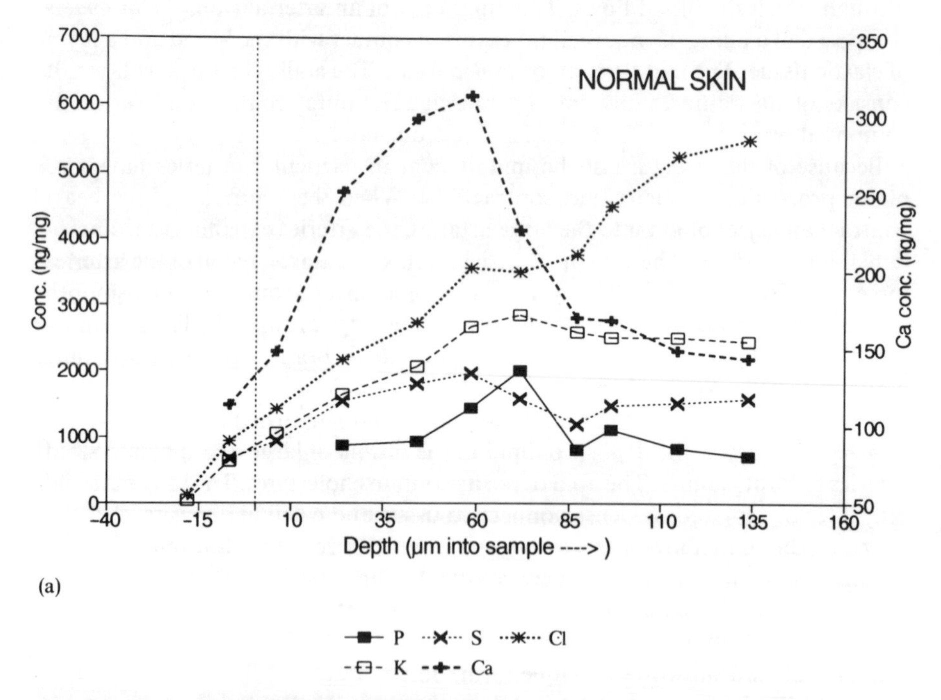

Fig. 4. Major and minor element concentrations in (a) normal skin, (b) psoriatic skin and (c) atopic skin. The skin surface is indicated by the vertical dotted line. Note the different scale for calcium concentration to the right. (From Ref. 54.)

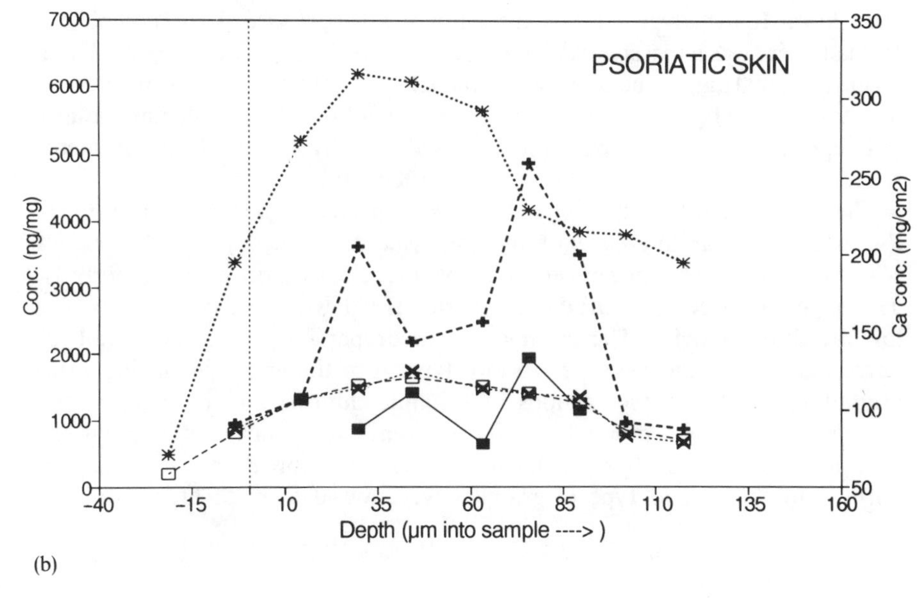
PSORIATIC SKIN
Conc. (ng/mg)
Ca conc. (mg/cm2)
Depth (μm into sample ---->)
7000
6000
5000
4000
3000
2000
1000
0
350
300
250
200
150
100
50
-40
-15
10
35
60
85
110
135
160

(b)

P
S
Cl
K
Ca

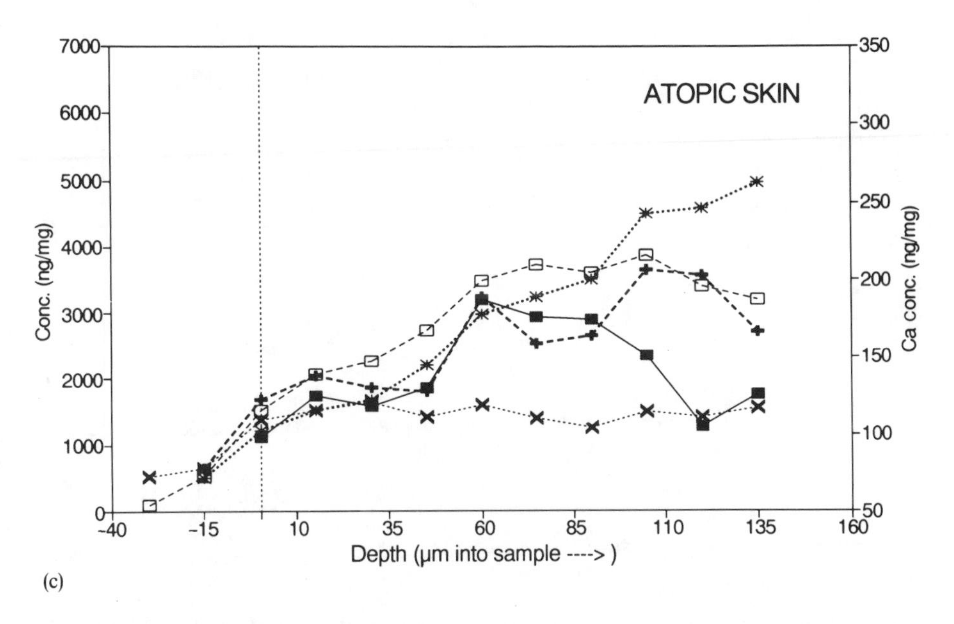
ATOPIC SKIN
Conc. (ng/mg)
Ca conc. (ng/mg)
Depth (μm into sample ---->)
7000
6000
5000
4000
3000
2000
1000
0
350
300
250
200
150
100
50
-40
-15
10
35
60
85
110
135
160
P
S
Cl
K
Ca

(c)

10 to 20 μm. In the advanced stage these granules disappeared almost completely. Instead, there was a massive calcified border, where the concentration of calcium reached 10 000 μg/g. The normal epithelial concentration was estimated at 500–900 μg/g. The phosphorus concentration followed the calcium pattern, although there was a tendency for increased P/Ca ratios in advanced atherosclerosis. This seemed also to be true for the K/Cl ratios.

The trace elements iron, copper and zinc displayed higher concentrations in the epithelium than in adjacent muscular tissue. In tissue damaged by arteriosclerosis, however, the concentrations of the trace elements dropped significantly. This might be expected, as the damaged arterial wall with dead cells display very low metabolic activity. The granules found deeper into the arterial wall for normal and moderate cases were of two types. For the first type, having a size around 10 μm, the concentrations of sulphur, chlorine and potassium were slightly increased as compared with the adjacent tissue. On the other hand, the concentrations of phosphorus, calcium and iron were more than twice as high as normal. In the second type of granules were found a much higher calcium

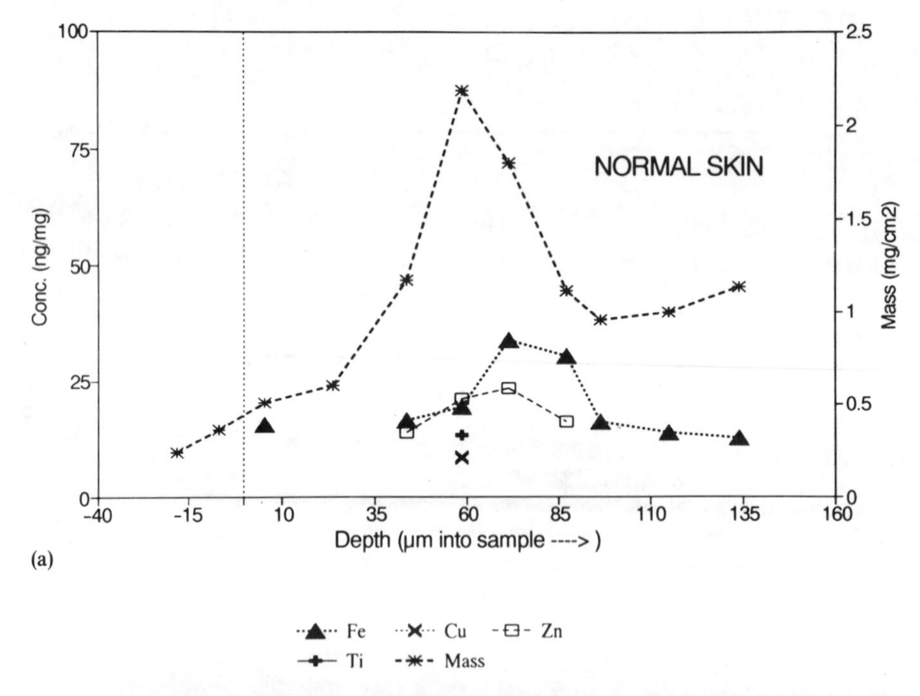

Fig. 5. Trace element concentrations in (a) normal skin, (b) psoriatic skin and (c) atopic skin. The skin surface is indicated by the vertical dotted line. Note the scale for the mass to the right. (From Ref. 54.)

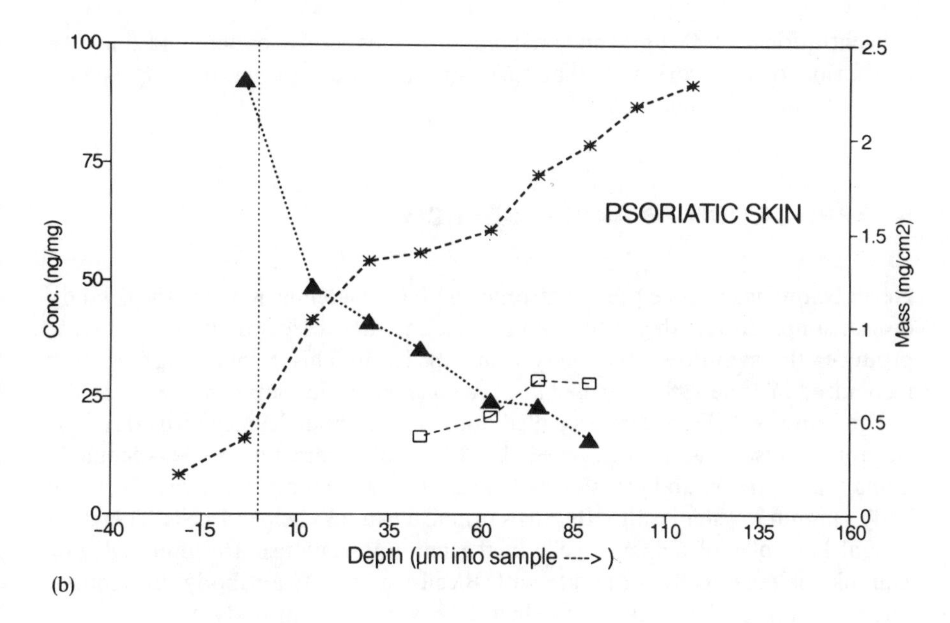

(b)

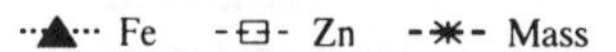

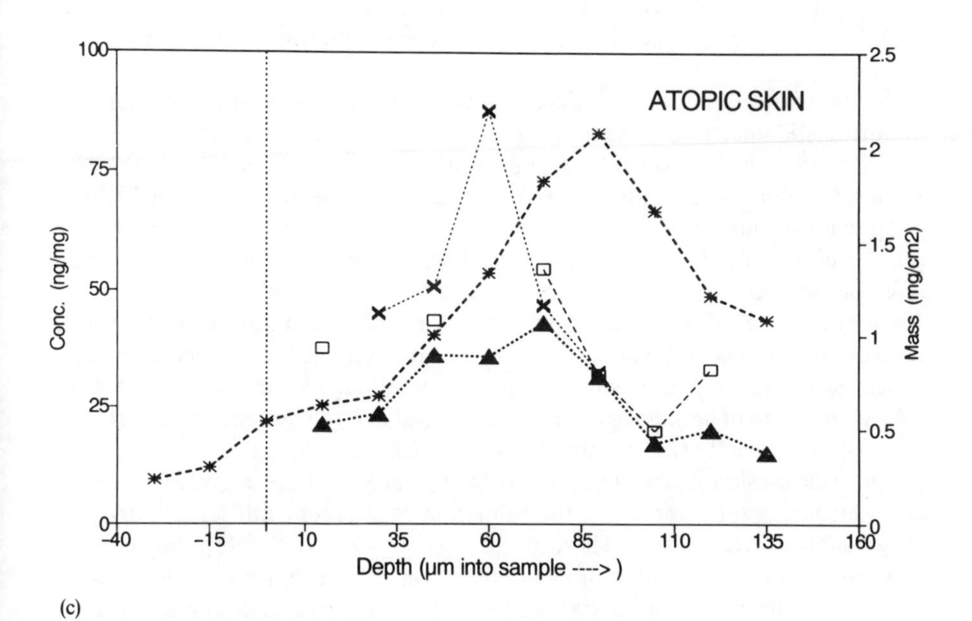

(c)

Fe Cu Zn Mass

concentration (> 5000 μg/g) and an increase of phosphorus, copper and zinc. The conclusion of the work was that these granules were probably plaques in a different stage of development [54].

8 APPLICATION TO AIDS RESEARCH

Never before has science been confronted with an epidemic in which the primary disease simply lowers the victim's immunity, and then a second unrelated disease produces the symptoms that may result in death. This primary disease, first recognized in June 1981, is called *acquired immune deficiency syndrome* (AIDS).

The cause of AIDS is a virus called human immunodeficiency virus (HIV). A variant of this virus, known as HTLV-I, causes a rare type of leukaemia in humans and attacks and transforms T cells. The reason the AIDS virus affects the body's immune system is that the virus primarily attacks helper T cells. Therefore several key roles of helper T cells in the immune response are inhibited. For example, helper T cells cooperate with B cells to amplify antibody production. The reduction of the number of helper T cells inhibits antibody production by descendants of B cells against the AIDS virus and other microbes. Also, helper T cells secrete interleukin 2, which stimulates proliferation of killer T cells. With the loss of helper T cells, fewer killer T cells are available to destroy various antigens. In AIDS, the ratio of helper T cells to suppressor T cells, which is normally 2:1, is reversed.

The deterioration of host defences allows for the development of cancer and opportunistic infections of various kinds. The two diseases that most often kill AIDS victims are Kaposi's sarcoma and *Pneumocystis carinii* pneumonia. Kaposi's sarcoma is a deadly form of skin cancer prevalent in equatorial Africa but previously almost unknown in the industrial world. It arises from endothelial cells of blood vessels and produces painless, purple or brownish lesions that resemble bruises on the skin or on the inside of the mouth, nose or rectum. *Pneumocystis carinii* pneumonia is a rare form of pneumonia caused by the protozoan *Pneumocystis carinii.* It results in shortness of breath, persistent dry cough, sharp chest pains, and difficulty in breathing. AIDS victims are also subject to a form of herpes that attacks the central nervous system and a bacterial infection that usually causes tuberculosis in chickens and pigs.

The nuclear microprobe at the Micro Analytical Research Centre, Melbourne, has been successfully applied in the validation of organometallic and inorganic drugs against AIDS [56]. There is an urgent need to develop satisfactory inhibitors of the human immunodeficiency virus. All currently used drugs have serious side effects. Any anitiviral drug is likely to be toxic also to normal cells. The aim in any drug development programme that is based on attacking the virus directly must be to maximize the ratio of toxicity to virus versus toxicity to normal cells and to optimize the delivery of the drug to the site of the virus.

A series of drugs has been designed by a Melbourne group and these have been extensively tested for toxicity and effectiveness [57]. However, evidence on the relative penetration of the drug into the cells was lacking as well as information on the composition within the cell. The drugs were designed to inhibit the replication of HIV in T-lymphocytes. They were tested for this activity and for their toxicity to healthy cells in a continuous human T-lymphocyte cell line (MT2 cells) and in peripheral blood lymphocytes from healthy donors. A characteristic of these drugs is that they contain heavy metal atoms which are not normally found in cells. These can be used for identification of the drugs. As the drugs are toxic to normal cells, the concentrations had to be kept below 5–200 μg/ml. This means that the concentration of heavy metals from the drug incorporated into the cells will probably be well below the level that can be measured by X-ray microanalysis in an electron microscope. The nuclear microscope, with its far better sensitivity, especially for the heavy metals, is well suited for the task. By elemental mapping, which is a technique common in nuclear microscopy, it should be within reach to describe the distribution of the drug within the cell and its nucleus. Such elemental mapping requires a probe resolution of ≈ 1 μm. Very few nuclear microscopes can reach this resolution; the Oxford and Melbourne nuclear microscopes are two that can.

Lymphocytes were incubated with growth medium containing the drug at various sub-toxic levels, and the medium was then removed by resuspending the cells in a buffer of ammonium acetate. Droplets of the cell suspension were placed on nylon foils, immediately snap frozen in isopentane cooled by liquid nitrogen and subsequently transferred at liquid nitrogen temperature to a high vacuum vessel in which they were freeze-dried. This buffer was previously shown to be well suited for such work by the group [58]. The ammonium salt appears to evaporate without leaving any traces during freeze-drying. The lymphocytes, however, do not tolerate long exposure to this buffer. Extracellular traces of the drug were consequently removed by repeated centrifugation in pure growth medium. Removal of the growth medium was finally accomplished by resuspension in ammonium acetate for only ≈ 10 min.

The features of nuclear microscopy were very nicely exploited by the Melbourne group, in that individual lymphocytes were scanned with 3 MeV protons at a resolution of 1 μm. To avoid radiation damage, the beam current was kept as low as 30 pA. The results were that the drug was incorporated into the cell and that the ratio of cobalt to tungsten remained unchanged. The intracellar distribution of phosphorus, cobalt and tungsten is depicted in Fig. 6. The higher concentration of phosphorus in the cell nucleus makes possible the elemental identification of this organelle, as can be seen in the figure. It is possible in this way to extract the concentration of the intranuclear component of the drug.

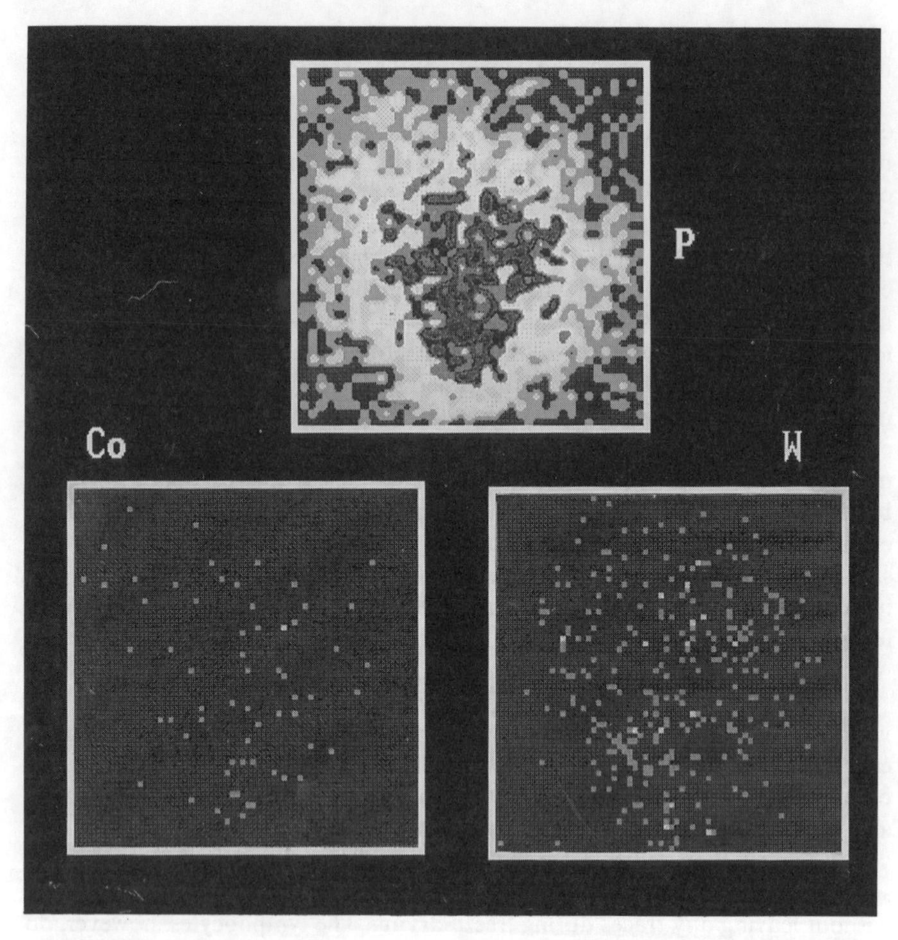

Fig. 6. PIXE maps of the lymphocyte intracellular distribution of phosphorus (P), cobalt (Co) and tungsten (W) after exposure to a subtoxic level of an anti-AIDS drug in the culture medium. (From Ref. 56.)

9 APPLICATIONS TO THE CHRONIC FATIGUE SYNDROME

The main advantage that can be extracted from our experience with nuclear microscopy of individual cells is that skew distributions of trace elements can be revealed. If a very small fraction of a cell type has a very different concentration of one or more trace elements than that in the greater part of such cells, a deviation could remain undetected using bulk analytical methods. This is especially obvious when studying cell types that occur in minor fractions. In such cases, bulk analytical methods may fail to be useful at all.

Anything that affects our environment is eventually going to affect man. Our primary concern here is exposure to heavy metals. There are several sources of exposure to metals. The problem of defining what is meant by heavy metals is, however, omitted here. As an example of a controversial question, we will use the internal exposure of man to mercury and to other metals with adverse effects emanating from dental silver amalgams.

The debate about the adverse effects of dental amalgams has been particularly animated, at least in Sweden and the Scandinavian countries. The benefits of using silver amalgam have been questioned since it was first introduced in modern times. The use of the name silver amalgams disguises the main component, viz. mercury. Dental or silver amalgams are composed of $\approx 50\%$ of mercury and various concentrations of other metals, such as silver, copper, tin, zinc and sometimes palladium and indium.

Dental amalgams are referred to as alloys of silver, tin, copper and mercury. The liquid mercury which is mixed with the alloy powder to make the amalgam is not stable in the prepared fillnig. It was discovered in Sweden in the early 1980s that many people suffered from a diffuse but handicapping syndrome, the cause of which was veiled in obscurity. A series of tentative investigations led a Swedish group to summarize the experience of the *metal syndrome* [59]. This report was later succeeded by another suggesting that the disease or diffuse syndrome was caused by the release of heavy metals from dental amalgams [60].

Lindh and Tveit [61] and Brune, Brunell and Lindh [62] performed investigations in the late 1970s and the early 1980s, the results of which pointed to the fact that the silver amalgams were not stable filling materials and that they were a serious hazard through the leakage of heavy metals to the human body, especially mercury. The routes of exposure comprise inhalation of mercury vapour, swallowing of fractured amalgam particles and ionic transport.

The obvious investigation to perform on patients suffering from the diffuse syndrome, which was named metal syndrome, was analysis of the heavy metal concentrations in blood plasma. Although the patient group displayed significantly higher plasma concentrations of mercury than the control group, it was not possible to identify individuals in the patient group solely by their plasma concentrations of mercury [63]. Analysis of individual blood cells by nuclear microscopy revealed much better discriminative power than plasma concentrations. More than 80% of the patients had measurable concentrations of mercury in one or a few blood cells whereas the controls did not display one cell with detectable mercury concentrations.

To prove the potential of high resolution nuclear microscopy, the beam size was decreased from $3 \times 3\,\mu m^2$ to $1.5 \times 1.5\,\mu m^2$. This meant that the proton intensity was reduced from 100 pA to 25 pA, a proton current that cannot be used for routine analytical purposes owing to the increased analysis time. In case of very fragile specimens, however, such beam currents may have to be used. The scan was collapsed to $15 \times 15\,\mu m^2$, centring one neutrophil granulocyte. The data collected during a continuous scan for slightly more than three hours were used to produce the elemental maps of Fig. 7. The elemental maps of zinc and mercury emanate from one neutrophil granulocyte from a patient and a control,

Neutrophil granulocyte
Control person with no dental amalgam

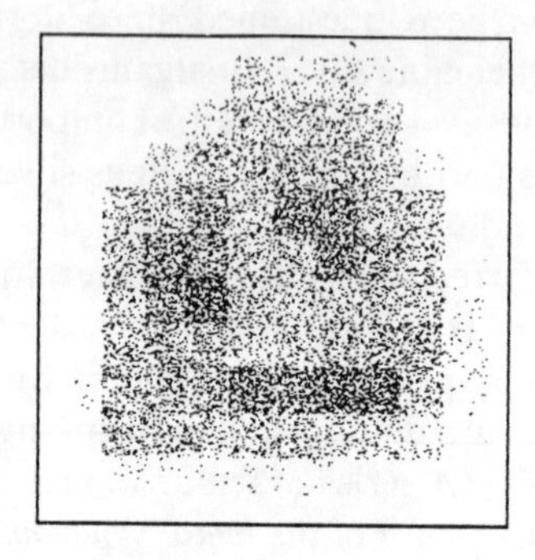

Zn map $15 \times 15\ \mu m^2$
$32.8 \pm 12.3\ \mu g/g$

Neutrophil granulocyte
CFS patient with dental amalgams

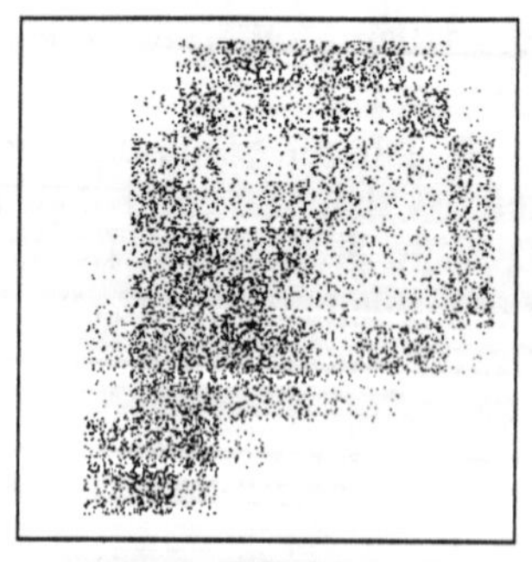

Zn map $15 \times 15\ \mu m^2$
$23.4 \pm 10.7\ \mu g/g$

Hg map $15 \times 15\ \mu m^2$
$1.2\ \mu g/g$

Fig. 7. Elemental maps of zinc and mercury in one neutrophil granulocyte from one control individual and one CFS patient.

respectively. There is no map for mercury in the cell from the control because the concentration was below the detection limit.

The levels of mercury found in one or a few cells from most of the patients range from just about detectable (≈ 0.5 µg/g dry weight) to sometimes quantitative concentrations (≈ 1.7 µg/g dry weight). Although these concentrations are low, they represent a high degree of accumulation in those cells with detectable amounts of mercury. The detection limit in the Slim-Up of 0.5 µg/g corresponds to roughly 6 µmol/1 of mercury. A mean concentration in whole blood in the patient group is something like 25 nmol/1. The detection limit thus corresponds to an accumulation factor of 240. Considering the low number of white cells in the blood, an accumulation of this kind, if it affects only a few of the white cells, would account for fractions of per cent. Thus, such an accumulation would remain completely undetected using a bulk analytical method.

The similarity of the symptoms of the patients suffering from the metal syndrome and those of patients with chronic mononucleosis, which is an Epstein–Barr virus associated disease [64], led us to make a study based on nuclear microscopy of isolated blood cells from the latter group. The Epstein–Barr virus (EBV) is very well characterized. It is a lymphotropic virus causing infectious mononucleosis. Chronic illness has been associated with serological evidence of persistent active EBV infection. The terms chronic mononucleosis and chronic fatigue syndrome (CFS) have been used. From a few years ago, there is a definition of CFS with both major and minor criteria [65].

The results of studies of isolated blood cells from CFS patients revealed a striking similarity to those of the metal syndrome. Apparently, both patient groups show the same alterations of the elemental profiles of blood cells. The alterations which we call an 'inflammatory picture' involve magnesium, calcium, manganese, iron, copper, zinc and strontium. In addition, there is the finding of detectable mercury concentrations in one or a few blood cells from about 80% of the patients. The common denominator shared between the patient groups is the effect on the immune system of immunomodulating heavy metals. In fact, there are reasons to believe that the two syndromes are the same.

10 NUCLEAR MICROSCOPY AS A MEANS TO STUDY PROTECTIVE EFFECTS

Considerable evidence has accrued to show that the excess of one essential trace element may greatly affect one or more of the other essential elements, particularly if they are present at a minimal level [66]. In adition, there is evidence of both beneficial and adverse effects of interactions between essential and toxic trace elements. An extensive review of trace element inter-relationships has been published [67].

Many processes involved in transport and utilization of the essential trace elements are susceptible to inhibition if local imbalances develop in the tissue concentrations of potentially competitive ions. Similar interactions influence the consequences of exposure to potentially toxic elements. The relevance of such interactions to the incidence and effects of trace element-related disorders in domesticated animals is clearly established [68]. In the study of the Sunde and Lindh [69], interactions between essential and 'toxic' trace elements were emphasized, as they were visualized in elemental maps at the tissue level. The main aim was to elucidate whether some essential trace elements exerted a protective effect against the well known adverse effects of heavy metals such as cadmium, lead and mercury. Several studies have, for example, shown that selenium has a protective effect against mercury toxicity from both organic and inorganic compounds [70–74].

The experimental model chosen was the Sprague–Dawley rat. The rats were randomly divided into nine groups, one of which served as control. The other groups were administered either salts of zinc, selenium, cadmium, mercury and lead or combinations of selenium, cadmium and mercury as well as zinc and lead. The administrations were made intraperitoneally and the doses chosen were scaled down from what is believed to be tolerable for man concerning the essential trace elements. For the non-essential trace elements, the doses were chosen to give toxic effects at the tissue level. The exposure was continued for 30 days and then the animals were sacrificed. Livers and kidneys were rapidly removed. Coronal sections of the kidney and a slice from the left lobe of the liver were immediately quench frozen in isopentane chilled with liquid nitrogen. The tissues were cryosectioned at 5–10 μm and freeze-dried. Adjacent sections were put onto microscope slides and stained with haematoxylin and eosin or toluidine blue to allow optical inspection and histopathological analysis.

The unsustained sections were analysed by nuclear microscopy to localize the essential and non-essential elements in those areas where toxicity was expected. The animals exposed solely to non-essential elements were clearly affected by the exposure, and tissue damage was detected as expected. Epithelial cells in the proximal tubules of the kidneys and parenchymal cells, periportal hepatocytes and Kupfer cells in the livers showed a high degree of cell damage. The same parts of the tissues from animals in the combination groups were inspected histopathologically and by nuclear microscopy. There were no signs at all of toxic effects in the cells, nor did the animals show signs of such effects. Nuclear microscopy showed the simultaneous presence of both essential and non-essential trace elements in those cells where toxicity was manifested in the animals exposed to non-essential elements only. Figure 8 shows examples of elemental maps of mercury and selenium in kidney from one animal in the combination group. The results clearly prove that selenium exerts protective effects against the toxicity of both mercury and cadmium. Also, zinc showed some protective effects against lead toxicity, but not as clearly as selenium [69].

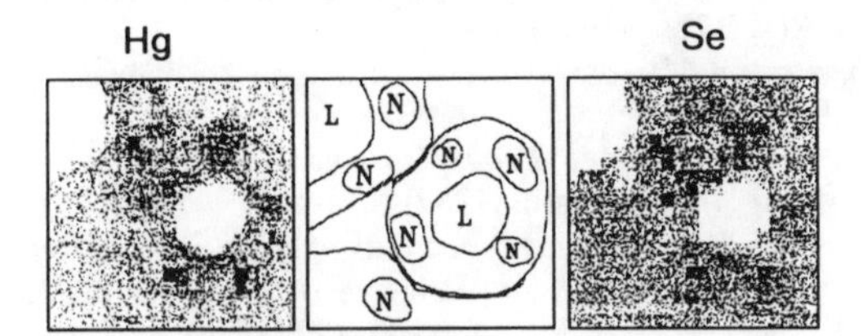

Fig. 8. Mercury (left part) and selenium (right part) elemental maps generated from a kidney section of one of the animals treated with selenium and mercury salts. The total area scanned was $50 \times 50\,\mu m^2$. The elemental maps cover (middle part) two parts of convoluted proximal tubules with their lumen (L) and epithelial cell nuclei (N).

11 CONCLUDING REMARKS

Although this presentation has pointed to only a few uses of nuclear microprobes, the versatility of nuclear microscopy is beyond doubt. When the resolution approaches 100 nm, use in cell biology will be drastically increased. Even with resolutions of some μm the nuclear microscope can be used for meaningful investigations in the realm of biomedicine. The problems with specimen preparation and specimen damage must be looked into thoroughly. Much can be learned from electron microsocpy, although the contamination of samples is not as serious a problem as in nuclear microscopy.

A serious disadvantage of nuclear microscopy is the restricted accessibility. Most of the probes are developed at nuclear structure and atomic physics laboratories which often lack the biomedical environment. When the accessibility increases, a development similar to that of electron microscopy can be foreseen.

REFERENCES

1. J. A. Cookson and F. D. Pilling, *UK At. Energy Res. Establ., Rep.* AERE-R 6300 (1970).
2. *Proceedings of the First International Conference on Nuclear Microprobe Technology and Applications*, Oxford, UK, 1–4 September 1987, *Nucl. Instrum. Methods Phys. Res.* **B30**, 227 (1988).
3. *Proceedings of the second International Conference on Nuclear Microprobe Technology*

and Applications, Melbourne, Australia, 5–9 February 1990, *Nucl. Instrum. Methods Phys. Res.* **B54**, 1 (1991).
4. *Proceedings of the Third International Conference on Nuclear Microprobe Technology and Applications*, Wik Castle, Uppsala, Sweden, 8–12 June 1992, *Nucl. Instrum. Methods Phys. Res.* **B77**, 1 (1993).
5. R. D. Vis, *The Proton Microprobe: Applications in the Biomedical Field*, CRC Press, Boca Raton (1985).
6. F. Watt and G. W. Grime, *Principles and Applications of High-Energy Ion Microbeams*, Adam Hilger, Bristol (1987).
7. J. A. Cookson, in *Principles and Applications of High-Energy Ion Microbeams*, edited by F. Watt and G. W. Grime, Adam Hilger, Bristol (1987), pp. 21–78.
8. T. A. Hall and B. L. Gupta, *J. Microsc.* **100**, 177 (1974).
9. Y. J. Uemura, Y. Kuno, H. Koyama, T. Yamazaki and P. Kienle, *Nucl. Instrum. Methods* **153**, 573 (1978).
10. S. B. Russel, R. S. Gibson, S. Faiq and J. L. Campbell, *Nucl. Instrum. Methods* **181**, 97 (1981).
11. A. J. J. Bos, C. C. A. H. van der Stap, W. J. M. Lenglet, R. D. Vis and V. Valković, *IEEE Trans. Nucl. Sci.* **30**, 1249 (1983).
12. A. J. Bos, R. D. Vis, F. van Langervelde, F. Ullings and H. Verheul, *Nucl. Instrum. Methods* **197**, 139 (1982).
13. N. Roos and A. J. Morgan, *Cryoprotection of Thin Biological Specimens for Electron Microscopy: Methods and Applications. Royal Microscopy Society, Microscopy Handbooks* No. 21, Oxford University Press, Oxford (1990).
14. A. W. Robards and U. B. Sleytr in *Low Temperature Methods in Biological Electron Microscopy. Practical Methods in Electron Microscopy*, Vol. 10, edited by A. M. Glauert, Elsevier, Amsterdam (1985), pp. 27–68.
15. W.-K. Chu, J. W. Mayer and M. -A. Nicolet, *Backscattering Spectrometry*, Academic Press, New York (1978).
16. J. C. Overley, R. C. Connolly, G. E. Seiger, J. D. MacDonald and H. W. Lefevre, *Nucl. Instrum. Methods* **218**, 43 (1983).
17. R. M. Sealock, A. P. Mazzolini and G. J. F. Legge, *Nucl. Instrum. Methods* **218**, 217 (1983).
18. K. M. Hambridge, C. E. Casey and N. F. Krebs, *Trace Elements in Human Nutrition*, Vol. 2, edited by W. Mertz, 5th Ed., Academic Press, New York, (1986), p. 1
19. G. V. Iyengar, W. M. Kollmer and H. J. M. Bowen, *The Elemental Composition of Human Tissues and Body Fluids*, Verlag Chemie, Weinheim (1978).
20. U. Lindh, E. Johansson and L. Gille, *Nucl. Instrum. Methods Phys. Res., Sect. B* **B3**, 631 (1984).
21. U. Lindh and E. Johansson, *Neurotoxicology* **4**(3) 177 (1983).
22. U. Lindh, E. Johansson, S. Jameson and R. Hällgren, *Nutr. Res., Suppl. 1* 156 (1985).
23. G. Anneren, U. Lindh and E. Johansson, *Acta Paediatr. Scand.* **74**, 259 (1985).
24. T. Foucard, M. Gebre-Medhin, K.-H. Gustavsson and U. Lindh, *Acta Paediatr. Scand*, **80**, 57 (1991).
25. J. R. J. Sorensen, *Inorg. Perspect. Biol. Med.* **2**, 1 (1978).
26. R. Hällgren, K. Svensson, E. Johansson and U. Lindh, *Arthritis Rheum.* **28**, 169 (1985).
27. R. Hällgren, K. Svensson, E. Johansson and U. Lindh, *J. Lab. Clin. Med.* **104**, 893 (1985).
28. R. Hällgren, K. Svensson, E. Johansson and U. Lindh, *J. Rheumatol.* **12**, 876 (1986).
29. K. Svensson, R. Hällgren, E. Johansson and U. Lindh, *Inflammation* **2**, 189 (1985).
30. N. Feltelius, R. Hällgren, and U. Lindh, *J. Rheumatol.* **15**, 308 (1985).
31. D. L. Massart, B. G. M. Vandeginste, S. N. Deming, Y. Michotte and L. Kaufman, *Chemometrics: A Textbook*, Elsevier, Amsterdam (1988).

32. G. L. Allan, J. Camarakis and G. J. F. Legge, *Nucl. Instrum. Methods Phys. Res., Sect. B* **B54**, 175 (1991).
33. U. Lindh and T. Sunde, *Nucl. Instrum. Methods Phys. Res.* **B10/ 11**, 703 (1985).
34. R. W. Jacobs, T. Duong, R. E. Jones, G. A. Trapp and A. B. Scheibel, *Can. J. Neurol. Sci.* **16**, 498 (1989).
35. A. H. Chafi, J. J. Hauw, G. Ranairel, J.-P. Berry and C. Galle, *Neurosci. Lett.* **123**, 61 (1991).
36. D. R. Crapper, S. S. Krishnan and A. J. Dalton, *Science* **180**, 511 (1973).
37. D. R. Crapper, S. S. Krishnan and S. Quittkat, *Brain* **99**, 67 (1976).
38. S. Ducket and P. Galle, *J. Neuropathol. Exp. Neurol.* **39**, 350 (1980).
39. J. M. Candy, A. E. Oakley, J. Klinowski, T. A. Carpenter, R. H. Perry, J. R. Atack, E. K. Perry, G. Blessed, A. Fairbairn and J. A. Edwardson, *Lancet* 1, 354 (1986).
40. J. P. Landsberg, B. McDonald, J. M. Roberts, G. W. Grime and F. Watt, *Nucl. Instrum. Methods Phys. Res.* **B54**, 180 (1991).
41. T. Pinheiro, U. A. S. Tapper, K. Sturesson and A. Brun, *Nucl. Instrum. Methods Phys. Res.* **B54**, 186 (1991).
42. W. R. Markesbury, W. D. Ehmann, T. I. Hassai, M. A. Canddin and D. T. Goodin, *Arch. Neurol.* **10**, 512 (1981).
43. R. D. Traub, T. C. Rains, R. C. Garruto, D. C. Gaydusek and C. J. Gibbs, *Neurology* **31**, 9086 (1981).
44. T. P. Flaten, *Environ. Geochem. Health* **12**, 152 (1990).
45. L. C. Neri and D. Hewitt, *Lancet* 338, 390 (1991).
46. D. R. Crapper McLachlan, A. J. Dalton, T. P. A. Kruck and M. Y. Bell, *Lancet* 337, 1304 (1991).
47. F. Watt and J. P. Landsberg, *Nucl. Instrum. Methods Phys. Res.* **B77**, 249 (1993).
48. B. McDonald, M. M. Esiri and R. A. J. McIlhinney, *J. Neurochem.* **57**, 1172 (1991).
49. P. Artaxo, F. Andrade and W. Maenhaut, *Nucl. Instrum. Methods Phys. Res., Sect. B* **B49**, 398 (1988).
50. F. Watt, J. P. Landsberg, G. W. Grime and B. McDonald, in *X-ray Microanalysis in Biology*, edited by D. C. Sigee, A. T. Summer and A. Warley, Cambridge University Press, Cambridge, pp. 62–79 (1992).
51. B. Forslind, T. G. Grundin, M. Lindberg, G. M. Roomans and Y. Werner, *Scanning Electron Microsc.* **2**, 687 (1985).
52. M. Lindberg, B. Forslind and G. M. Roomans, *Acta Derm.-Venereol., Suppl.* (135), 31 (1987).
53. Y. Werner-Linde, Thesis, Karolinska Institute, Stockholm (1989).
54. J. Pallon, J. Knox, B. Forslind, Y. Werner-Linde and T. Pinheiro, *Nucl. Instrum. Methods Phys. Res.* **B77**, 287 (1993).
55. R. D. Vis, A. J. J. Bos, F. Ullings, J. P. W. Houtman and H. Verheul, *Nucl. Instrum. Methods* **197**, 179 (1982).
56. M. Cholewa, G. J. F. Legge, H. Weigold, G. Holan and C. Birch, *Nucl. Instrum. Methods Phys. Res.* **B77** 282 (1993).
57. H. Weigold, G. Holan, S. M. Marcuccio, L. D. Gust and C. J. Birch, *Int. Patent Appl.* PCT/AU/00280 (28 June 1991).
58. G. L. Allan, J. Camarakis and G. J. F. Legge, *Nucl. Instrum. Methods Phys. Res., Sect. B* **B54**, 175 (1991).
59. B. Ahlrot Westerlund, B. Carlmark, S.-O. Grönquist, E. Johansson, U. Lindh, H. Theorell and K. deVahl, *Gustaf Werner Inst. Rep.* GWI-R 1/83 (1983), in Swedish.
60. B. Ahlrot Westerlund, B. Carlmark, S.-O. Grönquist, E. Johanssons, U. Lindh, H. Theorell and K. de Vahl, *Nutr. Res.* Suppl. I, 156 (1985).
61. U. Lindh and A.-B. Tveit, *J. Radioanal. Chem.* **59**, 167 (1980).
62. D. Brune, G. Brunell and U. Lindh, *Nucl. Instrum. Methods* **197**, 209 (1982).

63. U. Lindh, B. Carlmark, S.-O. Grönquist and A. Lindvall, *J. Trace Elem. Electrolytes Health Dis.*, in press.
64. A. Lindvall, S.-O. Grönquist, A. Linde, U. Lindh and G. Friman, submitted for publication.
65. G. P. Holmes, J. E. Kaplan, N. M. Gantz, A. L. Komaroff, L. B. Schonberger, S. E. Straus, J. F. Jones, R. E. Dubois, C. Cunningham-Rundles, S. Pahwa, G. Tosato, L. S. Zegans, D. T. Purtilo, N. Brown, R. T. Schooley and I. Brus, *Ann. Int. Med.* **108**, 387 (1988).
66. B. L. O'Dell in *Trace Elements in Nutrition of Children*, edited by R. J. Chandra, Raven Press, New York (1985), pp. 41–58.
67. C. H. Hill, in *Trace Elements in Human Health and Disease*, Vol. 2, edited by A. S. Prasad, Academic Press, New York (1976), pp. 281–312.
68. C. F. Mills, N. T. Davies, J. Quarterman and P. J. Agget, *Nutr. Res.*, *Suppl* 471 (1985).
69. T. Sunde and U. Lindh, submitted for publication.
70. R. F. Burk, H. E. Jordan, Jr. and K. W. Kiker, *Toxicol. Appl. Pharmacol.* **40**, 71 (1977).
71. J. Parizek and I. Ostadalova, *J. Nutr.* **104**, 638 (1974).
72. H. E. Ganther, C. Goudie, M. L. Sunde, M. J. Kopecky and P. Wagner, *Science* **175**, 1122 (1972).
73. M. M. El-Begearmi, M. L. Sunde and H. E. Ganther, *Poult. Sci.* **175**, 939 (1977).
74. S. L. Johnson and W. G. Pond, *Nutr. Rep. Int.* **9**, 135 (1974).

Part 4

Use of Radioactive (alpha, beta and gamma) Sources

Chapter 15

X-RAY FLUORESCENCE ANALYSIS WITH RADIOACTIVE SOURCES

T. Biran-Izak and M. Mantel
Soreq Nuclear Research Center, Yavne, Israel

1 INTRODUCTION

X-ray fluorescence as an analytical tool for elemental analysis may be considered to be as old as the recognition of the significance of X-ray spectra by Mosley in 1913. The techniques used to measure these spectra have developed along two paths: wavelength dispersive, which is the older, and energy dispersive (X-ray energy spectroscopy), or more specifically X-ray fluorescence (XRF), which is the subject of the present chapter.

A brief summary of the fundamental principles governing X-ray fluorescence will be given in order to provide the reader with a better understanding of the technique and of its possible applications.

2 GENERAL

Fluorescence, or the generation of secondary radiation, is accomplished by a two-step process. In the first step, a high energy particle such as a photon, a proton or an electron strikes an atom and knocks out an inner-shell electron (photoelectric effect). The second step is readjustment in the atom almost immediately (10^{-12}–10^{-14} s) by filling the inner-shell vacancy with one of the outer-shell electrons and simultaneous emission of an X-ray photon. The first step uses up the energy of the incident quantum, and in the second step energy is

Chemical Analysis by Nuclear Methods Edited by Z. B. Alfassi

emitted as the characteristic X-ray photon. The incident quantum may have any energy greater than the binding energy of the inner-shell electron; the excess energy is carried away as kinetic energy of the electron being removed. The energy released by the replacement of the inner-shell electron by one of the outer-shell electrons corresponds exactly to the difference in energy between the two levels.

The preference of emission of *K* or *L* X-rays will depend on the specific inner shell from which the electron is removed; *K* X-rays will be emitted if the vacancy is produced in the *K* shells and *L* X-rays (of lower energy) if the vacancy is in the *L* shells. As mentioned before, the energy of the emitted X-rays will be the difference between the binding energy of the inner- and outer-shell electron, which is characteristic of the atom. This fact forms the basis for XRF, as the emitted photon can be detected and measured according to its energy.

The probability for the photoelectric effect to occur is dependent on energy (approximately E^{-3}) and for a given energy is a function of the atomic number Z ($\approx Z^4$). This probability shows specific discontinuities called absorption edges, which occur at the critical energy for the shell in question. The maximum probability for the photoelectric effect occurs when the photon energy is just above this critical energy. This fact dictates one of the important considerations in XRF in order to obtain maximum analytical efficiency for a given element.

The actual X-ray yields are lower than those predicted (calculated from the photoelectric effect) owing to a competitive effect which takes place in the atom; i.e. the Auger effect. In this process the X-ray photon generated does not escape from the atom but instead knocks out an additional extranuclear electron from the *L* or *M* shell (Auger electron) with a kinetic energy equal to the characteristic X-ray energy minus its binding energy.

The fluorescence yield (ω) is the ratio of emitted X-rays to the number of primary vacancies created. The *K* fluorescence yield (ω_k) is the fractional probability for *K* X-ray emission and, likewise, the *L* fluorescence yield (ω_l) is the probability for *L* X-ray emission. The latter is several times smaller than the *K* yield for a given Z. The fluorescence yield increases with increasing atomic number. As a result, low values (a few percent or less) are obtained for low of Z elements, limiting the possibility of their determination by XRF.

There are three different modes in which photons of $E < 1.02$ MeV may interact with an atom: the photoelectric effect (see above), elastic scattering and inelastic scattering.

Elastic scattering (Rayleigh) takes place without change in energy from that of the primary beam, and is dependent approximately on Z^2. It results when a photon collides with a firmly bound electron in an atom; the energy of the scattered photon is the same as that of the primary photon. Elastic scattering is of importance in analytical X-ray fluorescence because of its contribution to the general background.

Inelastic (Compton) scattering results from the collision of photons with loosely held electrons of the atoms, with loss of photon energy to the recoiling electron. The absolute energy loss increases with the scattering angle, with the maximum energy shift for an 180° angle. The probability for inelastic scattering increases with the atomic number Z. The most important consequence of inelastic scattering is the appearance of scattered photons of lower energy than the incident photon beam, which may cause overlap and high background effects in the XRF spectra.

3 X-RAY FLUORESCENCE SYSTEM

The wide use and attractiveness of X-ray fluorescence is based on the two most important characeristics of X-rays:
(1) The small number of lines in the characteristic X-ray spectrum of each element, which makes the interpretation easier;
(2) The simple and direct correlation between wavelength λ and atomic number Z (Mosley's law), which permits identification of an element by merely measuring the energies of its characteristic X-rays.

The technique of X-ray fluorescence involves two principal components: the excitation system and the detection system.

Production of secondary characteristic X-rays may be achieved by excitation either with photons or with charged particles. Photons may be produced by radioisotopes or X-ray tubes; in this chapter, only radioisotope-excited X-ray fluorescence (REXRF) will be discussed.

3.1 Excitation

3.1.1 Sources

A few basic factors determine the possible use of a radioisotope for fluorescing purposes. The emission spectra should be as simple as possible, with either a simple strong gamma or a simple characteristic X-ray emission. A complex emission pattern may complicate in two ways the use of a radioisotope as a source for XRF:

- the presence of high energy gamma-ray emision may cause undesirable background in the X-ray region owing to the Compton escape spectrum;
- multiple emission lines in the X-ray region may limit the application range because of interference problems.

Another factor is the half-life of the radioisotope, which should be reasonably long to avoid the need for frequent replacement owing to decreasing intensity. However, the intensity of the radiation from a radioisotope is very stable and depends only on its half-life. The decrease in intensity may generally be neglected for the time of the experiment, or it may be taken into account very exactly by a simple correction.

Ideally each element should be analysed via emission lines with minimum interference problems, optimum fluorescence efficiency and fluorescence yield. Thus, effective coverage of a number of elements with different optimum fluorescing efficiencies is not to be sought by a source with emission lines spread over the range, but by several separate radioisotopes with individual emission lines which cover the same range.

Most radioactive isotopes used as sources in XRF decay by characteristic monoenergetic emission (X- or γ-rays); however, 'bremsstrahlung' sources, which decay via electron emission (beta-particle decay), have also some areas of application. In this case, the radioactive isotope which emits the bremsstrahlung is generally mixed with a target material. The latter emits X-rays which are used to induce fluorescence in the sample. It is obvious that the conversion efficiency in such sources is very low, needing the use of high activity radioisotopes in order to obtain fluorescence beams with an intensity high enough to be measured.

Table 1 shows the radioisotopes which may be used as excitation sources for XRF analysis. The isotopes are classified according to their mode of decay (X, γ or β emitters) and listed according to increasing energy. Most radioisotopes mentioned in the literature are given: commonly used, less frequently used and proposed.

In the choice of the most appropriate radioisotope for the determination of a given element, the following requirements have to be fulfilled:

- the energy of the emitted photons should be higher than (but as close as possible to) the K or L absorption edge of the element to be determined (see Section 2);
- the acvitity of the radioactive source should be in accordance with the analytical sensitivity required; the use of high activity sources should be avoided if not necessary, in order to minimize occupational health hazards and high shielding costs.

The range of application of the most commonly used radioisotopes to the determination of different elements (according to their atomic number) is shown in Fig. 1 [1]. The X-ray production yields obtained from 'infinitely thick' samples are shown, normalized to the yield of the element most efficiently caused to fluoresce by the emission energy (1.0) and adjusted to 100% detection efficiency. As an example, it is seen from the figure that Cd (22.1 keV) is the most efficient radioisotope for determining Mo ($Z = 42$, K absorption edge $= 20.0$ keV) by XRF. For elements with lower energy absorption edges the yields are lower, as expected; e.g. for Zn ($Z = 30$, K absorption edge 9.7 keV) the relative yield is approximately ≈ 0.35, as compared with 1.0 for Mo.

A radionuclide source is a very simple, small (tenth of mm) and light device

TABLE 1.
Radioactive sources emitting X, γ and β-rays

Radionuclide	Half-life $t_{1/2}$	Emission					Remarks
		X-rays		γ-rays		β_{max}	
		Energy (keV)	% emission	Energy (keV)	% emission	Energy (keV)	
^{55}Fe [1–5]	2.7 y	5.9; 6.5 Mn K X-rays	25				commonly used
^{57}Co [1–5]	271 d	6.4; 7.0 Fe K X-rays	48	121.9	85		commonly used
				136.3	11		
^{71}Ge [2, 3, 6]	11.8 d	9.2; 10.3 Ga K X-rays	≈ 45				
^{210}Pb [2, 3]	22.3 y	9.4; 15.7 Bi LX-rays	24	46.5	4		
^{238}Pu [1–5]	87.8 y	11.6; 20.7 U L X-rays	13				
^{241}Am [1–5]	458 y	11.9; 21.3 Np L X-rays	37	59.5	36		commonly used
^{244}Cm [3, 4]	18.1 y	12.1; 22.0 Pu L X-rays					possible
^{93m}Nb [2, 7]	13.6y	16.6; 18.6 Nb K X-rays	10	88.0	4		commonly used
^{109}Cd [1–5]	453 d	22.1; 24.9 Ag K X-rays	73	392	64		
^{113}Sn [2, 8]	115 d	24.2; 27.3 In K X-rays	≈ 28				
^{119m}Sn [2, 5]	245 d	25.3; 28.5 Sn K X-rays					
^{123m}Te [2–4]	119.7 d	27.5; 31.0 Te K X-rays	≈ 50	159	84		
^{125}I [2–4]	59.7 d	27.5; 31.0 Te K X-rays	141	35.4	7		
^{145}Sm [2, 5]	340 d	38.7; 43.8 Pm K X-rays	13	61.3			
^{153}Gd [1–4]	242 d	41.3 ; 47.0 Eu K X-rays	110	97.4	30		
				103.2	20		
^{170}Tm [2, 3]	129 d	52.4 ; 57.5 Yb K X-rays	5	84	3		
^{203}Hg [3, 4, 8]	47.6 d	72.8; 82.1 Tl K X-rays	13	279	82		
^{75}Se [3, 8]	120 d			121	16		
				136	56		
				265	59		
				280	25		
				401	11		
^{192}Ir [8, 9]	74.2 d			296	30		
				308	32		
				316	87		
				468	52		
^{133}Ba [2, 3, 8]	10.7 y			356	62		
				303	19		
^{3}H [1–3, 5]	12.3 y					18.6	usually Ti target
^{147}Pm [1, 3, 11]	2.6 y					244	
^{63}Ni [3, 11]	100 y					67	
^{32}P [3, 10]	14 d					1709	

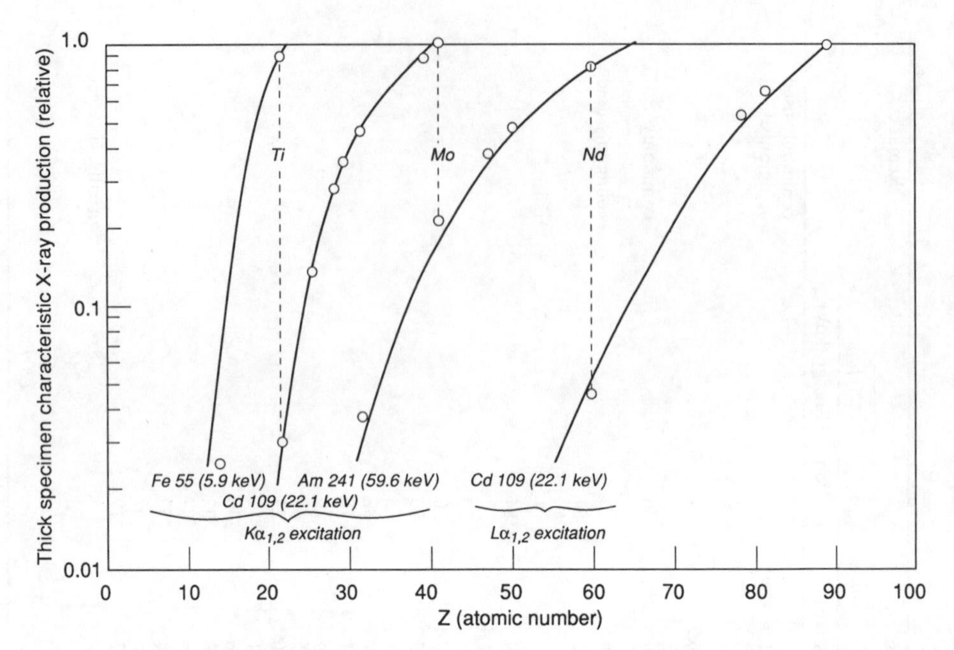

Fig. 1. X-ray production yields obtained by excitation of 'infinitely thick' samples with radioactive isotopes [1].

which is generally embedded in a special metal assembly. The encapsulation of the radioactive material may lead to absorption and to changes in the emitted spectrum which can interfere with the measurements. It follows that, in the choice of the metal to be used for construction of the 'housing' of the source, the energy of the photons emitted by the source and by the sample has to be taken into account. If the emitted photons are of low energy (< 30 keV) a thin window of a low-Z metal, generally Be [2, 4, 12, 13], is necessary in order to minimize the absorption. A Be [12, 13] or a Ti [4] window is used with a ^{125}I source and an Al [2] window with a ^{109}Cd or a ^{238}Pu source. Stainless steel is a suitable material for the encapsulation of ^{57}Co [4, 14], ^{133}Ba [8] and ^{241}Am [2] if the 60 keV radiation is required but is unsuitable as a window if the Np L X-rays have to be used. In some cases, in order to minimize even further the interference of the housing material, a source collimator is used [8, 14].

Radionuclides are generally not available with activities comparable with those produced by an X-ray tube, but in most cases an activity of ≈ 370 MBq is sufficient to obtain detection limits in the range of 10 µg/g. There are examples where such an activity is preferable considering possible radiation damage to the sample.

3.1.2 Configuration

In the design of an excitation system, the principal aim is to obtain the optimal ratio of fluorescence to scattering. For this purpose the angle between the excitation radiation and the sample should be so chosen as to obtain Compton scattering with a maximum energy below the energy of the fluorescent X-ray to be measured [8, 14]. The probability of Compton scattering is at a minimum around 90° scatter geometry. However, because of the various parameters which have to be taken into account it is difficult to obtain optimum conditions, and generally a compromise has to be accepted even though the intensity of the signal may sometimes be significantly reduced.

The optimum source–sample–detector geometry is generally determined experimentally [8]; however, a number of theoretical calculations have also been published [15, 16].

The most commonly applied configurations are the annular and the point source. Figure 2 [1] shows the geometry applied with annular sources. In this particular case, a 150° scattering angle was used to obtain optimum results. As may be seen, two different excitation modes are possible: direct irradiation and secondary target irradiation. In the direct irradiation mode the sample is directly irradiated by the X- or gamma-rays emitted from the source. The use of a secondary target reduces the background due to the primary source radiation, thereby increasing the signal-to-background ratio. The secondary target may also be used to 'shift' the excitation energy closer to the absorption edge of the element to be determined. Thus, a Eu_2O_3 target has been used for the determination of Sn in ores irradiated with an ^{241}Am source [17]. The energies of the X-rays emitted by the ^{241}Am source are too low (see Table 1) to cause fluorescence of the tin atom and the 60 keV gamma-ray is too far from the energy of the Sn K absorption edge (29.2 keV), whereas the Eu K_α X-rays (41.5 keV) have an energy

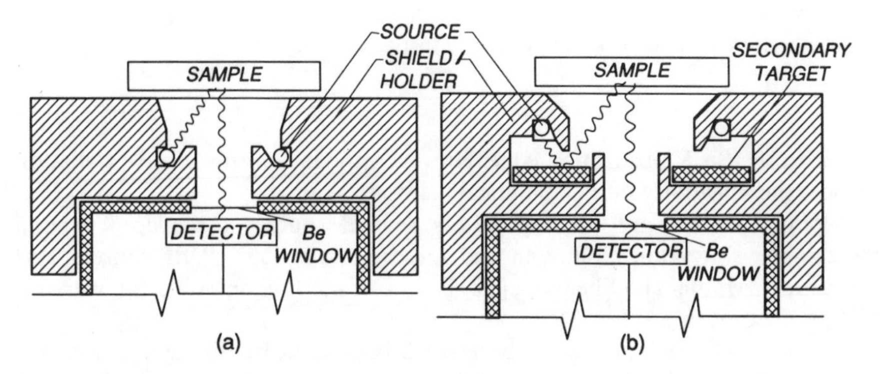

Fig. 2. Schematic representation of a radioisotope excitation system with an annular source configuration: (a) direct irradiation; (b) secondary target irradiation [1].

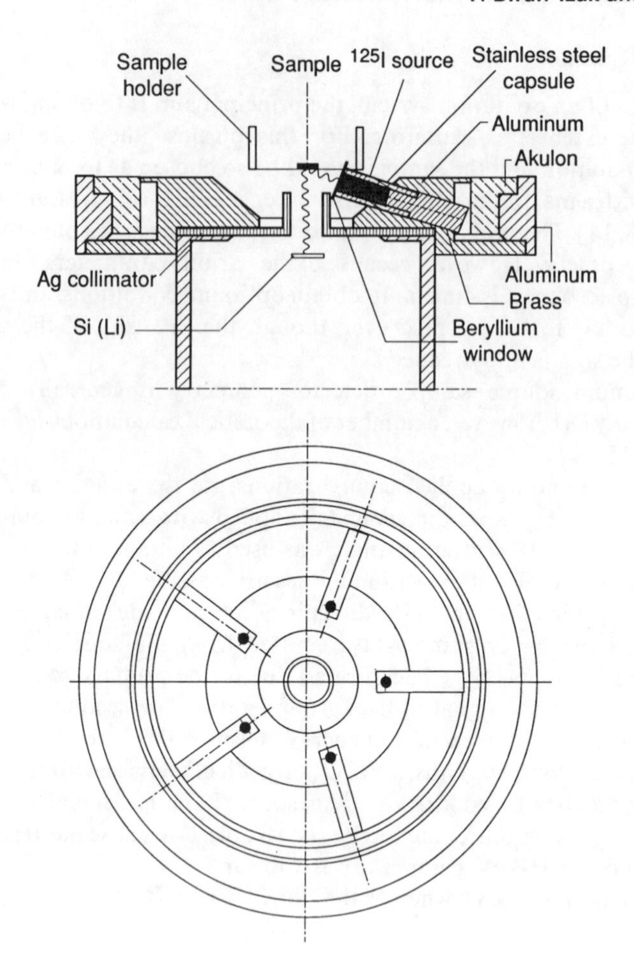

Fig. 3. Configuration of a ^{125}I source consisting of five individual point sources [12] (Reproduced by permission of the American Institute of Physics).

nearer to the Sn *K* absorption edge. However, the use of this type of radiation reduces the sensitivity of the analytical method.

If point sources are used, generally several sources are placed around the sample. Hofmann, Hoffmann and Lieser [8] use four ^{133}Ba sources, and Rappaport, Mantel and Shenberg [12] base their method on excitation with five ^{125}I point sources (see Fig. 3).

A number of special configurations have been used by different workers in order to solve specific analytical problems (see Section 6).

3.2 Detection

The detection assembly consists of the detector, the collimator, shielding and a data acquisition system.

3.2.1 Detectors

The three common types of detector used in X-ray fluorescence analysis are the solid-state counter, the proportional counter and the scintillation counter.

The choice of the most favorable detector is a question of resolution and required sensitivity. If only one element is to be determined, the resolution is not an important factor and the peak-to-background ratio becomes the decisive problem. In this case proportional counters are the most suitable detectors provided that the matrix in question does not contain neighboring elements in amounts which could interfere with the determination. The gases commonly used for filling this type of detector are argon for lower energies [18, 19], krypton [18, 20] and xenon [21] for higher energies. As an example, Kowalska and Urbanski [18] use an argon filled proportional counter to measure Pb *L* X-rays excited by a ^{109}Cd source, whereas a krypton-filled counter is used for the measurement of Sn *K* X-rays excited by an ^{241}Am source. A NaI(Tl) scintillation detector, which has higher efficiency than the proportional counter, may also be used when one element has to be determined in a matrix which does not contain interfering elements [17, 22].

If more than one element is to be determined, solid-state semiconductor detectors are the most suitable because of their high resolution. In this case the choice of the appropriate detector depends on the range of energies of the X-rays to be measured [23]: Si(Li) detectors for low energies (2–30 keV) [7, 24–26] and planar high purity Ge detectors for higher energies (30–125 keV) [8, 9, 27, 28]; Ge(Li) detectors may also be used for the measurement of X-rays with energies above 100 keV [29].

Experiments have been made with unconventional detectors. LaBrecque [30] compared the performance of a HgI_2 detector with that of a conventional Si(Li) one. The HgI_2 detector may be used more easily in the field because of its compact size and because it functions at room temperature and does not need liquid nitrogen cooling. On account of these advantages the author recommends its use for determination of Nb despite its much poorer energy resolution and efficiency as compared with Si(Li) detectors—0.59 keV vs 0.15 keV (FWHM) and 5/1 respectively.

3.2.2 Collimation and shielding

The detectors have to be shielded against the direct incidence of X- or gamma-rays from the source or from the environment. The dimension, shape and location of the shielding relative to the source, the sample, the collimator and the detector have to be chosen in such a way as to obtain maximum values for the peak-to-background ratio.

The shielding material and its thickness are chosen according to the energy of the source and the element to be determined. For excitation of *L* X-rays, sources of lower energy are sufficient and the shielding may consist of low-*Z* material, whereas higher energies and thicker shielding are needed for excitation of *K* lines.

Lead is the mostly used material for the construction of the shielding [8, 17, 24]. Obviously, if lead has to be determined, other materials have to be used for shielding. Thus, tungsten has been used [27, 31] for the determination of lead in vivo because it satisfies the dual criteria of maximum photon attenuation whilst having characteristic X-rays well separated from those of lead. Martinelli *et al.* [9] use depleted U/Mo or U/V alloys for shielding of very strong sources (> 740 GBq).

The detector itself may contribute to the background. The useful working region in this case is the region between the inelastic scattering peak from the sample and the Compton contribution from the detector. The background due to the detector may be reduced or eliminated by limiting the active volume of the detector, i.e. by screening out the edge region. This may be achieved by 'collimating' the detector by means of a beam collimator. The reduction in detector area may be compensated by using high intensity excitation sources. Furthermore, if resolution requirements permit, thicker detectors will allow equivalent noise reduction with smaller relative area loss.

As an example, the influence of a tin collimator of 40 mm thickness on the signal-to-background ratio in the determination of actinides excited by four ^{133}Ba sources was studied by Hofmann, Hoffmann and Lieser [8] by increasing the collimator width from 5 to 40 mm. The results showed that the intensity of the U K_{α_1} line increases in proportion to the square of the diameter of the collimator from 5 to 30 mm. For higher diameters the intensity approaches a saturation value.

The collimators may be made of different materials such as tin [8], graphite [24] and tungsten [27, 31]. As mentioned above, tungsten functions also as shielding in the case of lead measurements in vivo.

4 MATRIX EFFECT

One of the most important sources of error in X-ray spectrometry is the absorption of X-rays by the matrix. This phenomenon may represent a serious limitation owing to the low energy of X-rays.

The mass absorption coefficient μ_m of an element, expressed in cm^2/g units, defines the absorption process quantitatively. It depends on the photon energy, decreasing continuously with increasing photon energy. For a given energy, it increases with the atomic number of the absorbing element. The mass absorption coefficient of a compound is the weighted sum of the absorption coefficients of the constituent elements. It follows that, in a given matrix, the absorption of X-rays depends on their energy and on the compostion of the matrix, which means that, for the same element, the absorption in a matrix composed of light elements (for instance organic materials) will be much lower than in a heavy-element matrix (see Section 2).

In order to overcome the matrix effect, various correction methods based on theoretical calculations were developed [28, 32–36]. However, most workers try to overcome this source of error by appropriate preparation of the samples and choice of standards [26, 37].

As a result of the matrix effect, the intensity of the characteristic X-rays emitted from a sample varies with sample thickness. Three different regions may be distinguished. The linear region is the thickness range in which the intensities of the X-rays are proportional to the concentration of the elements and there is no effect of matrix compositions. This region varies from zero to $\approx 10\,\mu m$ for different elements and depends upon the mass absorption coefficients for the incident and excited X-rays and the angles made by the incident and emergent beams with the sample surface. As sample thickness increases, the intensity of the excited radiation will decrease owing to absorption and the linear correlation between X-ray intensity and concentration will not subsist. When the material is thick enough, the fluorescence intensity will arrive at a maximum. This thickness is called the 'critical depth' (or the 'infinite thickness') and is the depth at which maximum intensity is attained for the particular experimental conditions. At thicknesses exceeding the critical value, intensity is independent of thickness.

The two limiting thickness regions are separated by a region of transition in which the influence of the matrix effect has to be taken into account. The appropriate preparation of standards is imperative for carrying out XRF analysis in this region.

Another possible sources of error is the enhancement produced by secondary X-rays. This enhancement occurs when the characteristic radiation from one element has enough energy to excite characteristic radiation of one or more other elements in the sample. In general the intensities of the radioactive sources are too low to produce secondary X-rays in sufficient yield to interfere with the measurements.

Interference may also be produced by elastic and inelastic scatter (see Section 2). The latter can be minimized by using an optimum geometry (angle) and appropriate collimation.

5 SAMPLE PREPARATION

The adequate preparation of samples is of paramount importance for obtaining accurate analytical results by radioisotope-excited XRF.

One of the advantages of this method is the possibility of measuring samples in all forms: gas [14], liquid [12, 36], slurry [19] and solid [36]. Liquid samples have the advantage that in most cases no sample preparation is needed, standards are easy to prepare and samples and standards are generally perfectly homogeneous. Furthermore, matrix effects are reduced in solutions by the influence of the solvent. However, in most solutions the solvent consists of low-Z elements which produce very high scattering of the fluorescing radiation (see Section 2), so limiting the sensitivity of the analysis.

In the case of solid samples, the matrix effect and homogeneity are the two most important factors to be taken into account in the preparation of samples. Homogeneity is obtained by grinding or milling the samples and the standards until a powder with uniform grain size is obtained. The powder needs to be pressed into pellets in order to obtain tight packed samples.

As mentioned before (see Section 4), very thin and infinitely thick samples are not subject to matrix effects. The use of thin samples results in a severe reduction in element sensitivity and thus can generally not be applied in radioisotope-excited XRF. The use of infinitely thick samples is limited by the amount of sample necessary to obtain infinite thickness, which depends on the energy of the X-rays to be measured [25].

Most workers use samples of intermediate thickness. In this case the matrix effect is overcome by different methods, i.e. standard addition [26, 38], dilution [26, 39] and preparation of standards similar in composition to the samples [40].

The standard addition method, a widely used analytical technique, consists in the addition to the sample of increasing known amounts of the element to be determined.

In the dilution method, a low-Z substance is added to the sample in order to reduce the mean Z of the sample matrix. Cellulose, which is a good binder and light absorbing matrix, is often used as diluent for solid samples. As an example, it was added to U and Th ores to reduce the absorption effect by decreasing the average atomic number of the samples [26].

The preparation of standards similar in composition to the samples is based on the fact that the absorption coefficient of the matrix as a whole is only slightly influenced by small variations in the relative abundance of its major constituents. Thus, it will be possible to use as standards matrices with a composition similar to that of the sample without it being necessary for standard and sample to match perfectly; the same type of rock for geological samples or the same type of tissue, drug, or food for organic materials. As an example, Table 2 shows the similarity in absorption coefficients of organic matrices at 4.95 keV.

The presence of trace elements, even those of high atomic number, will

TABLE 2.
Absorption coefficients of different organic matrices at 4.95 keV [41]

	Kidney	Liver	Heart	Cellulose
Absorption coefficient, μ_m (cm^2/g)	36.8	34.6	36.4	31.2

influence only slightly the absorption of the matrix. However, great attention must be given to the presence of elements with absorption edges near the energy of the X-rays to be determined. The increase in the absorption with energy is sharp and the decrease is gradual. It follows that the highest absorption will be obtained from elements having absorption edges at energies just below those of the X-rays to be measured.

The choice of the method of sample and standard preparation depends on the degree of accuracy required from the analysis. If only semi-quantitative results are wanted, less elaborate methods may be applied.

The sensitivities obtained by the application of radioisotope-excited X-ray fluorescence are summarized in Table 3.

6 APPLICATIONS

As mentioned before, a radionuclide source is a very simple, small and light device compared with an X-ray tube with generator. The simplicity of this excitation mode allows the construction of portable XRF systems, which may be used for on-line determinations or for analysis 'on the spot'. Furthermore, a large number of elements can be detected simultaneously on the same spectrum. As an example, Fig. 4 shows the spectrum obtained by the analysis of a thorium ore by REXRF [26].

Another advantage of this system is its relatively low cost compared with other analytical methods.

Because of these features, REXRF has been applied to the solution of analytical problems in a wide range of disciplines. Examples are given; however, the present chapter is not intended to be an exhaustive bibliography of all the work published on the applications of this technique.

6.1 Industry

On-line determination of elements during industrial processes in the various process solutions and in the waste streams is of great importance for monitoring the quality of the products. Avoidance of the need for transportation of samples to a central laboratory permits immediate response to changes in density and

TABLE 3.
Sensitivities obtained by radioisotope-excited XRF analysis

Element (preceded by reference)	Matrix	Source: ^{241}Am Intensity (Bq)	^{241}Am Sensitivity	^{109}Cd Intensity (Bq)	^{109}Cd Sensitivity	^{55}Fe Intensity (Bq)	^{55}Fe Sensitivity	^{125}I Intensity (Bq)	^{125}I Sensitivity	$^{192}Ir/^{133}Ba$ Intensity (Bq)	$^{192}Ir/^{133}Ba$ Sensitivity	^{57}Co Intensity (Bq)	^{57}Co Sensitivity	^{170}Tm Intensity (Bq)	^{170}Tm Sensitivity	^{3}H Intensity (Bq)	^{3}H Sensitivity
[50] Al	thin film						1–2 mg/cm²										
[24] Mg	alloy															1.1T	0.5–3%
[74]	biolog.																0.1–5%
[25] P	biolog.					5G	0.2–1%										
[38] S	oil					370M	0.05%										
[25]	biolog.					5G	0.3–4%										
[38] Cl	oil					370M	0.05%										
[25]	biolog.					5G	0.05–1%										
[74]	biolog.																0.03–5%
[25] K	biolog.					5G	0.01–2%										
[39]	biolog.					200M	0.3%										
[74]	biolog.																0.5–2%
[25] Ca	biolog.					5G	0.01–2%										
[13] Cr	wire							400M	0.05–1%								
[39]	biolog.					200M	20 μg/g										
[13] Mn	wire							400M	0.05–1%								
[25]	biolog.							500M	1–100 μg/g								
[39]	biolog.							800M	40 μg/g								
[50] Fe	thin film				0.1–1 mg/cm²												
[56]	coal				6–50%												
[25]	biolog.							500M	40–4000 μg/g								
[39]	biolog.							800M	100 μg/g								
[39] Co	biolog.							800M	1 μg/g								
[39] Ni	biolog.							800M	1 μg/g								
[47] Cu	alloy		81–88%														
[50]	thin film				0.1–1 mg/cm²												
[25]	biolog.							500M	20–200 μg/g								
[39]	biolog.							800M	10 μg/g								
[19] Zn	slurry			70–185M	5–20%												
[47]	alloy		2–8%														
[25]	biolog.							500M	20–200 μg/g								
[39]	biolog.							800M	50 μg/g								
[12] Br	biolog.							180M	0.1 μg/g								
[25]	biolog.							500M	2–200 μg/g								

[25] Rb	biolog.					500 M	4–100 µg/g					
[39]	biolog.					800 M	5 µg/g					
[25] Sr	biolog.					5 G	1–50 µg/g					
[39]	biolog.					800 M	5 µg/g					
[48] Zr	alloy											5–25%
[39]	biolog.					800 M	1 µg/g					
[50] Mo	thin film				0.1–1 mg/cm^2							
[39]	biolog.					800 M	1 µg/g					
[67] Cd	biolog.[a]	11 G	30–150 µg/g									
[25]	biolog.					500 M	0.1–200 µg/g					
[18] Sn	thin film	75 M	0.1 mg/cm^2									
[20]	ore		0.03–0.5%									
[43]	slurry		0.1–0.5%									
[47]	alloy		4–8%									
[22] Sb	ore	18 G	0.15%									
[54] Au	ore	2–60 G	0.02 µg/g									
[18] Pb	thin film			185 M	0.1 mg/cm^2							
[19]	slurry			75–185 M	2–6%							
[21]	air			20 M	15 µg/m^3							
[27, 66]	biolog.[a]			3.7–7.4 G 7.4G	20 µg							
[29]	biolog.[a]									700 M	20 µg/g	
[31]	biolog.[a]									740 M	10–50 µg/g	
[8] U	solution							370 M	0.005–400 g/l			
[9]	solution							740 G	0.05 g/l			
[9]	ore							740G	50 µg/g			
[14]	gas									925 M	1–80 Torr	
[26]	ore				13 µg/g							
[28]	soil										500 µg/g	
[36]	solution	370 M	0.3–30%									
[36]	ore	370 M	0.2–0.4%									
[44]	solution		0.5–50 g/l									
[8] Th	solution							370 M	0.005–400 g/l			
[26]	ore				108 µg/g							
[28]	soil										500 µg/g	
[9] Pu	solution							740G	0.005 g/l			
[44]	solution		0.1–50 g/l									

[a] In vivo.

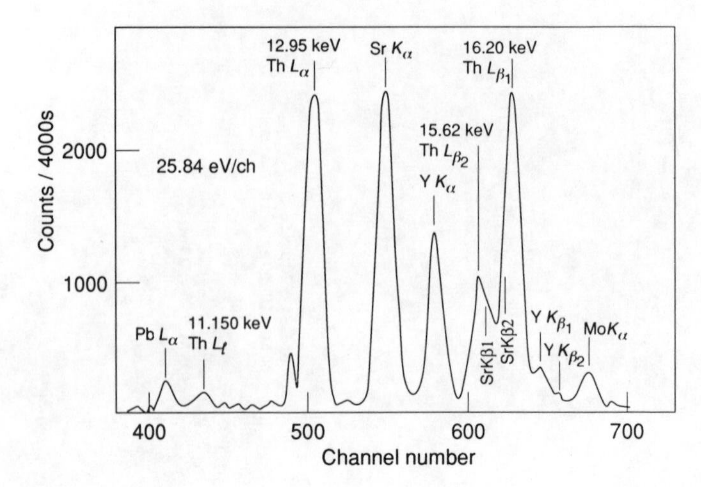

Fig. 4. X-ray spectrum of a thorium ore irradiated with a ^{109}Cd source and measured with a Si(Li) detector [26].

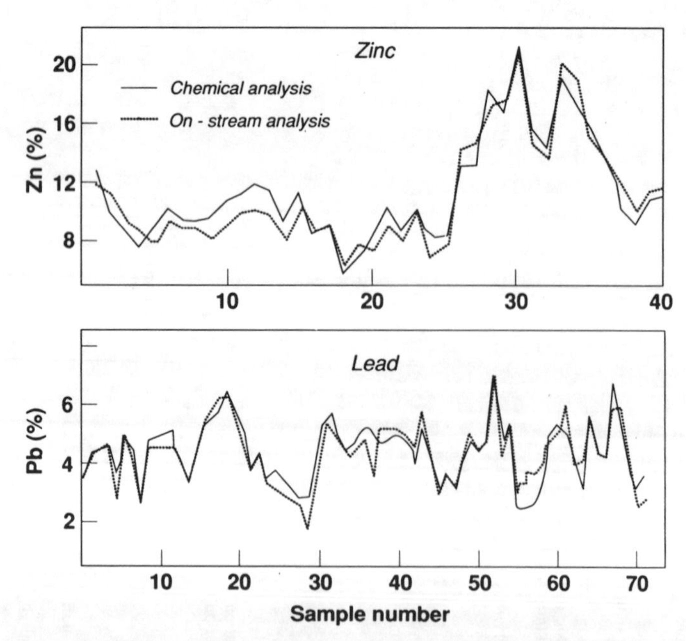

Fig. 5. Spectra of tailings from a Zn and Pb extraction process from zinc–lead ores [19]. (Reproduced by permission of the International Atomic Energy Agency.)

composition of the solution. Thus a better optimization of the process is achieved.

Watt [42] reviews the possibility of using on-stream analysis of metalliferous ore slurries based on REXRF techniques using three types of detector: scintillation, proportional and solid-state detectors are described and the economic benefits of the method are evaluated.

The percentage of iron, zinc and lead in zinc–lead ore slurries [19] has been determined on-line by using several XRF probes mounted in a flow cell which is fed through a by-pass line from the process streams. The probe consists of a 75–185 MBq ^{109}Cd source and an argon filled proportional counter. Figure 5 shows a comparison between the results obtained by on-stream XRF analysis and chemical determination of lead and zinc in tailings. The good agreement obtained emphasizes the possibility of using the REXRF technique for the monitoring of metals in an industrial process.

Tin in the range 0.1–0.5% has been determined on-line in the extraction solutions of a flotation plant [43]; ^{241}Am was used as primary source, barium salts as secondary targets and a NaI(Tl) crystal as detector. Slurries with up to 800 g/l of solids could be analyzed.

Uranium and plutonium have been determined on-line during their separation in nuclear fuel reprocessing plants [8, 9, 44, 45]. As highly radioactive solutions are involved in this case, sources of high activity have to be used. Martinelli *et al.* [9] used a 740 GBq ^{192}Ir source and a planar Ge detector for the excitation and measurement of *K* X-rays of uranium and plutonium with a sensitivity of 50 mg/l. In highly radioactive solutions containing up to 1.5 TBq of fission products, the two metals are determined by difference of two sectra: one obtained from an irradiated sample measured in the 'excitation position' and one obtained from a non-irradiated sample in the 'safety position' of the ^{192}Ir source. In this way the contribution of the self-excitation of U and Pu *K* X-rays due to the high gamma-ray activity of the fission products is overcome. Hofmann, Hoffmann and Lieser [8] developed an experimental set-up for the determination of uranium and thorium which may work as an on-line or off-line system. Two 370 MBq ^{133}Ba point sources or an annular ^{131}I source of 4–40 MBq activity were used and a sensitivity of 0.005–400g/l was obtained. Groll [44] used an ^{241}Am annular source with a Cd secondary target for the measurement of uranium and plutonium *L* X-rays with a sensitivity of 0.5–50 g/l.

A major problem in the uranium enrichment industry and in uranium fabrication plants is the determination of the amount of gaseous uranium inside cylinders [14]. X-ray fluorescence with a 925 MBq ^{57}Co fluorescing source was used for this purpose, based on the fact that the X-ray signal is directly proportional to the pressure, i.e. to the total amount of uranium present. If a highly collimated source, a highly collimated detector and a very rigid geometry are used, the measurement technique is independent of the uranium deposit that forms on the cyclinder wall through chemical reactions of the gas with impurities. The detection limit is <1 Torr gas pressure. The same principle was applied to the determination of total fission gas pressure in the rod plenum [46] via the measurement of xenon X-rays, based on the constant xenon/krypton ratio in fission gas.

The concentration of metals in alloys has been determined by various workers [13, 24, 47–49]. Magnesium and Si were determined in aluminum alloys in the range 0.1–0.7 wt% using a 1.1 TBq $^{3}H/Ti$ ring-shaped radioactive source and a Si(Li) semiconductor detector [23]. Chromium and Mn [13] were determined in aluminum wires and sheets in the range 0.05–1% using an 400 MBq ring-shaped ^{125}I source and a Si(Li) detector. Lehmann and Brendel [47] routinely determined Sn, Zn and Cu on the spot in the foundry during the melting process of bronze alloys. An ^{241}Am source with a Be window and different filter pairs—Ni/Co for Cu, Ag/Pd for Sn and Cu/Ni for Zn—were used. The range of concentrations analyzed was 81–88% of Cu, 2–8% of Zn and 4–8% of Sn.

Thickness measurements of thin films of pure elements [50] were carried out by irradiating the samples with an ^{55}Fe or a ^{109}Cd annular source and measuring the ratio between fluorescent X-rays and the coherent scatter of the primary radiation. Pure element films (0.1–2 mg/cm^{2}) for Al, Fe, Cu and Mo were prepared by vacuum evaporation on glass substrates. Very good agreement was obtain by comparing the results obtained by XRF and those obtained by weighing. Tin and lead thickness in thin films (>0.1 mg/cm^{2}) was also determined by irradiating with ^{241}Am or ^{109}Cd sources respectively (Kowalska and Urbanski [18]).

6.2 Geology

A number of papers have been published on the application of REXRF to the determination of trace elements in different geological mateials such as ores, soils, coal and crude oil.

In order to develop a method for the determination of barium and lanthanides in deposits of laterite material in Venezuela, LaBrecque, Rosales and Mejias [51] determined 14 rare earths in synthetic mixtures of their respective oxides. The samples were excited by the 88 keV gamma-rays of a 100 MBq ^{109}Cd source or by the β-particles (continuous to 225 keV) obtained from a 3.7 MBq ^{147}Pm source. In this way only the K X-rays of lanthanides were obtained, overcoming the interference of low energy K X-rays from elements with low atomic number, which is a limiting factor if L X-rays are excited. The measurements were carried out with a high purity planar germanium detector. Figure 6 shows a comparison between the spectra from a mixture of lanthanide oxides (La, Pr, Gd, Dy, Er and Yb) obtained by irradiation with the ^{109}Cd and ^{147}Pm sources respectively. As may be seen, up to Gd K_{α} X-rays at 43 keV, the net peak intensities obtained by excitation with ^{147}Pm are nearly twice those obtained by excitation with ^{109}Cd. Thus the ^{147}Pm source is recommended for the measurements of lanthanides

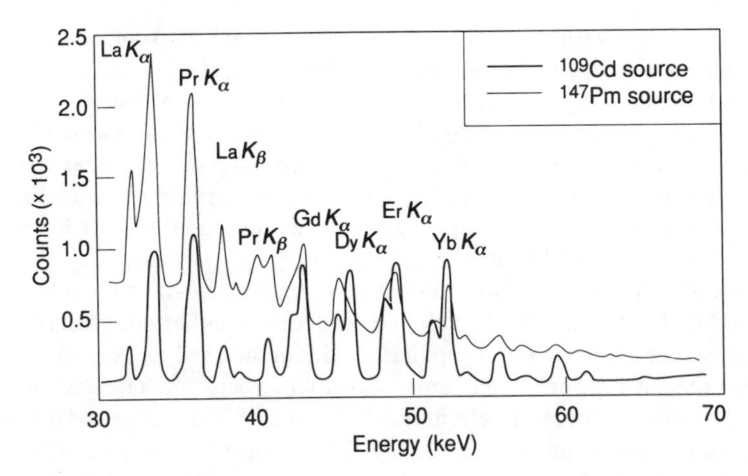

Fig. 6. Spectrum of a mixture of lanthanide oxides: La, Pr, Gd, Dy, Er and Yb (150 mg each), irradiated with two different sources: ^{109}Cd and ^{147}Pm; 10 min counting time [51]. (Reproduced by permission of Elsevier Science Publishers BV.)

up to gadolinium ($Z = 64$), where the Cd source is more appropriate for higher-Z lanthanides.

LaBrecque [30] also used REXRF for the determination of other trace elements in the same laterite material from Venezuela. A ^{109}Cd source and a Si(Li) detector were used for the measurements of thorium, niobium, lead and zinc. Thorium and the lanthanides were also determined in britholite ores from the Quebec region of Canada [52] by using a ^{57}Co radioactive source.

Rolle *et al.* [53] used a portable XRF analyzer for the in situ determination of gold and lead in Witwatersrand gold deposits in South Africa. The 88 keV gamma-ray of ^{109}Cd was used for the excitation of gold and lead *K* X-rays. Gold was also determined in geological materials by Ivanova *et al.* [54] with a limit of detection of 2.5×10^{-2} μg/g by using ^{238}Pu and ^{241}Am with activities from 1.8 to 60 GBq as radioactive sources.

Kowalska and Urbanski [20] used an ^{241}Am point source and a krypton-filled proportional counter for the determination of tin in geological samples in the range 0.03–0.5%. Kruger *et al.* [17] also applied an ^{241}Am source for the determination of tin (0.1–5%) in ores using an Eu_2O_3 secondary target (see Section 3.1.2).

Arikan and Alkan [26] studied the optimum conditions for the determination of uranium and thorium by REXRF in radioactive ores from Turkey. The results of the experiments showed that the best analysis system is a ^{109}Cd source and a Si(Li) detector and the best sensitivity (13 μg/g of U and 108 μg/g of Th) is attained by measuring *L* X-rays and using the method of internal standard addition (see Fig. 4). Lazo, Roessler and Berven [28] determined uranium and thorium in unprocessed soil samples using a ^{57}Co radioactive source and measuring the induced K_α X-rays with a planar Ge detector. An attenuation correction method was applied which overcomes the need for soil standards.

Ayala-Jimenez [38] determined sulfur and chlorine in crude oils with a sensitivity of 0.05% using a 370 MBq ^{55}Fe radioactive source and a Si(Li) detector. A standard addition method was recommended in order to obtain quantitative results, and CS_2 and CCl_4 were used as internal standards. The great advantage of the method is its rapidity: 20 min as compared to $\approx$16 h necessary for the conventional ASTM method.

Total sulfur in coal [55] was determined using a two-source configuration (^{55}Fe and ^{238}Pu) housed in two collimators placed on either side of the detector. The two-source configuration simplifies the design and reduces the sulfur determination time. A proportional counter was used as detector. The ash content in coal in the range 6–50% was determined by using a ^{109}Cd source and measuring the Compton scattering and the X-ray fluorescence intensity of iron in the samples [56].

Radioisotope-excited X-ray fluorescence was applied to borehole logging techniques [22, 57, 58]. Zhang *et al.* [57] designed a borehole logger which can be used to determine the content of tin and antimony in ores and to measure their thickness in the borehole. The logging of tin and antimony was also carried out by using an 18 GBq ^{241}Am source with a holmium secondary target and a NaI(Tl) scintillation detector [22].

6.3 Environment

Another field of application of REXRF is the determination of the concentration of metals in air by analyzing aerosols trapped on air filters [21, 59–61].

Kostalas [21] developed a compact portable unit for the detection of lead in air samples. A 22 MBq ^{109}Cd source and a xenon-filled proportional counter were used. Figure 7 shows the fluorescence spectrum of an air sample collected on a glass fiber filter in the stack of a lead factory. The dotted line spectrum is produced by scattering of the source photons by the air and the supporting materials. The peak at the high energy end of the spectrum is due to the silver K X-rays emitted by the source. The low energy peak was attributed to the fluorescence radiation emitted by the stainless steel case of the source, and its energy corresponds to nickel K X-rays. The solid line spectrum is the spectrum of an air sample. As shown in Fig. 7, the lead L_α of the air sample cannot be properly resolved from the fluorescence peak of the source case, and an energy range was therefore selected over which the L_α peak was integrated. A sensitivity of 15 μg/g of lead per m^3 of air was obtained for a 2 h sampling time, a flow rate of 20 ft^3 min and a counting time of 1 h.

The elements Cu, Zn, Pb, As, Br, Sr and Zr were identified in the aerosol spectrum obtained by irradiation with a 165 MBq ^{109}Cd source and a Si(Li)

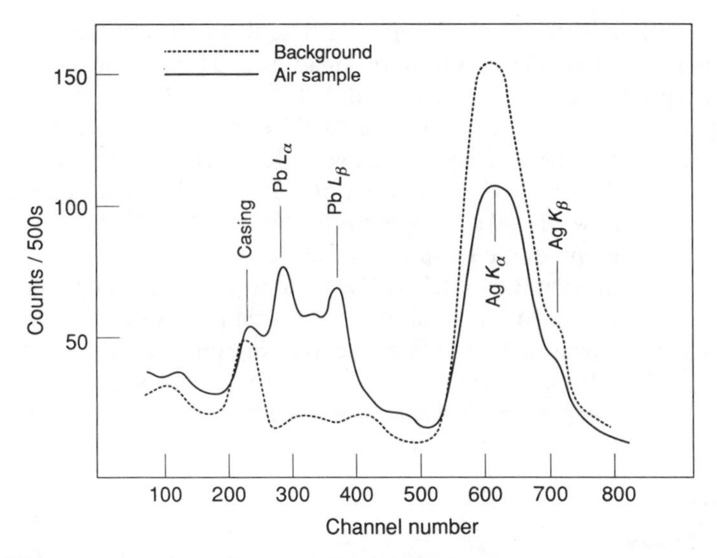

Fig. 7. XRF spectrum of an air sample [21]. (Reprinted from Kostalas, *Int. J. Appl. Radiat. Isot.*, **33**, 1475, copyright 1982, with permission from Pergamon Press Ltd, Headington Hill Hall, Oxford OX3 OBW, UK.)

detector [59]. Trace amounts of selenium were determined without any prior chemical separation [60].

The chemical quality of the air breathed by a worker welding chromium—nickel steels [61] is assessed by measuring the amount of chromium released during the welding process. A ring-shaped ^{241}Am source with an activity of 18 GBq and a Si(Li) detector were used to measure the Cr K X-rays with a sensitivity of 0.05 mg for a period of measurement of 1000 s.

6.4 Archaeology

Works of art and archaeological samples were analyzed by REXRF for the evaluation of their composition and determination of the period of their manufacture.

Yap [62] studied Nonya ware (Straits Chinese or Ch'ing porcelains) in order to detect modern fake reproductions of the original works. A 1.1 GBq ^{241}Am annular source and a Si(Li) detector were used for the analysis of the porcelain samples, which were exposed on the unpainted regions. As the variation of the major components is small, the relative fluorescent intensity of the trace elements analyzed versus the intensity of the coherently scattered gamma-radiation at 26.4 keV (I_γ) were calculated. Traces, up to a few hundreds of μg/g, were found for Sn, Sb, Fe, Cs, Ba, La, Ce, Pr, Nd and Pm. However, plots of the ratio of fluorescent intensity to I_γ against time of manufacture show that a significant time correlation exists only for Cs and Ba. The Ba content of the Nonya wares was found to be 100–135 μg/g from the analysis of 34 pieces, whereas the three modern reproductions pieces contained 250–430 μg/g. The Ba/I_γ ratio for the modern pieces was 2.4–3.8, as compared with ≈ 1 for Ch'ing pieces.

Hoffmann [3] applied the same principle to the analysis of ancient Roman tiles and clays. The Zr/Nb, Cs/Ce and Ba/La ratios were calculated and correlation diagrams were drawn in order to corroborate the hypothesis of the archaeologists concerning the origin of the samples studied.

As one of the advantages of REXEF is the possibility of analysis on site, and because of the narrow localization of the analyzed area, valuable objects of art could be analyzed without fear of destruction. An example is the analysis of the treasure of St. Vitus cathedral for gold [63]. The results showed that gold of different purity was used in different parts of the relic. These results were of great interest to the study of the crown cross relic and its jewels.

6.5 Biology

A unique application of REXRF is the determination of trace elements, especially toxic metals such as lead and cadmium, in the human body *in vivo*.

In *in vivo* measurements, the characteristic photon generated must penetrate relatively thick layers of soft tissue in order to reach the detector. This means that there is a lower limit for the energy of the emitted photons, i.e. of the atomic number of the elements which may be determined *in vivo* by REXRF. The lower limit of atomic number is generally $Z = 40$.

It was found [29] that in most situations the contribution of coherent and incoherent scattered primary photons is a major limiting factor in *in vivo* applications owing to the large scattering volumes and the low concentrations of the trace elements. It follows that a standard which simulates the composition and properties of human soft tissue has to be prepared. This standard, known as a 'phantom', has been constructed from different materials [29, 31, 40, 64] to which the element to be determined is added. The shape of the phantom differs according to the specific part of the body to be analyzed (tibia, finger, kidney etc.).

Ahlgren and Mattsson [40] compared the properties of different phantom materials with those of water, which is a good simulant for soft tissue. The authors recommend the use of silica paraffin wax, which is easy to handle and has acceptable radiophysical properties. For the determination of lead in finger bone, they propose a finger phantom which consists of two parts, one made only of silica paraffin wax, simulating the surrounding soft tissue, and the central bone-simulating part, silica paraffin wax mixed with 10–30% of ground bone ash.

The phantom and the experimental set-up proposed by Ahlgren and Mattsson [40] were applied to the *in vivo* measurement of lead in the forefinger of workers from the metal industry. The mean concentration found was 50 μg/g. Thomas *et al.* [31] determined Pb in the phalanx of the index finger of 15 patients suffering from lead nephropathy as compared with that of patients suffering from renal failure due to other causes (controls). Two ^{57}Co sources, 740 MBq each, a W collimator and a Ge(Li) detector were used. Mean elevated values of 47 μg/g were found in patients exposed to lead, as compared with 16 μg/g for controls.

Lead in tibia bone was determined using a 2–8 GBq ^{109}Cd source [27, 65, 66]. This source was chosen as the optimum one because the energy of its photons is just above the lead K absorption edge and, in an appropriate geometry, the Compton scattering peak from the source is well below the lead K X-ray peaks. To relate the *in vivo* measurement of lead X-rays to an absolute bone lead concentration, the authors used the coherently scattered gamma-rays. Calibration was accomplished with plaster of Paris phantoms doped with known concentrations of lead. Irradiation for 20 min of 15 occupationally exposed patients and 22 non-exposed persons resulted in mean lead concentrations of 30 and 10 μg/g, respectively.

Wielopolski *et al.* [67, 68] studied the possibility of using L X-rays for the determination of lead *in vivo* in tibia bone by measuring lead post-mortem in human legs. A ^{125}I or a ^{109}Cd source was used. In practice, the use of L X-rays is restricted to superficial bones such as the tibia, for which the overlying tissue is 2–3 mm thick. The advantage of reducing the energy of the incident radiation is that the dose to the bone marrow is significantly lower. A detection limit of 20 μg/g was obtained.

Ahlgren and Mattsson [69] determined cadmium *in vivo* in the right kidney by using a collimated beam from an 11 GBq ^{241}Am source and a Ge(Li) detector, which were directed towards the center of the kidney as indicated by a previous ultrasonic measurement. The Cd concentrations in the kidneys of five occupationally exposed persons were between 30–150 μg/g, as compared with 10–20 μg/g for controls. Further improvements of the determination of cadmium *in vivo* by XRF were studied by Nilsson *et al.* [70].

Margaliot *et al.* [64] proposed a method to determine stable iodine *in vivo* in the thyroid gland using an ^{241}Am source and a Si(Li) detector. The errors stemming from variations in gland depth and shape are overcome by two different methods: the K_α/K_β intensity is measured in order to calculate the effective depth of the gland, taking into account the difference of attenuation between the iodine K_α and K_β X-rays in tissue; the K_α intensity is measured at two different distances from the detector. The second method was found to be superior; however, it requires twice the exposure. A phantom representing the throat region and two types of gland, normal and hyperthyroidal, was used for calibration.

Morgan *et al.* [71] developed an *in vivo* analyzer for measuring the uptake of platinum in the kidneys and in superficial tumours receiving *cis*-Pt chemotherapy. A 740 MBq ^{57}Co source is directed at the posterior lumbar region and a pure

Ge detector records the fluorescent Pt *K* X-rays emitted at 90° to the incident beam. A detection limit of 20 μg/g was obtained.

The *in vivo* trace element measurements involve exposure to external radiation that depends on the intensity and energy of the source and the time of exposure. It is obvious that the dose absorbed should be kept at a minimum. However, the low levels to be measured in the different tissues and the measurement requirements imply a minimum dose which cannot be further reduced. As an example, in the determination of Pb in a forefinger, the doses (to the finger) range from 0.1 [27] to 2.5 mGy [29], and in the determination of Cd in kidney the dose absorbed by the kidney during one measurement is 0.6 mGy [69].

REXRF methods were also used to determine trace elements in different biological materials *in vitro*.

Uzonyi [25] determined quantitatively 13 elements of $Z > 15$ in different IAEA and NBS biological standards. Different annular radioactive sources were used: ^{55}Fe, 5 GBq, for elements of $Z < 23$ and ^{125}I, 0.5 GBq, for elements of $Z > 23$. Pellets of 0.25 g/cm^2 were prepared. This value assures 'infinite thickness' for X-ray energies below 10 keV. For higher energies the inter-element effects are corrected for mathematically. In this way biological samples of any thickness may be analyzed.

Hegedus *et al.* [39] also used ^{55}Fe and ^{125}I annular sources (0.2 and 0.8 GBq respectively) for the determination of 15 elements in wheat. The standard addition method and boric acid dilution were used to overcome the matrix effect.

Five 370 MBq ^{125}I point sources arranged in a special radial configuration were used for the determination of bromine in very small blood samples (< 50 μl) with a sensitivity of 0.05 μg of Br [12]. Havranek, Bumbalova and Harangozo [72] determined Fe, Zn, Br and Rb in the blood of donors from the Bratislava region using a ^{109}Cd source and a semiconductor Si(Li) detector. The same source was used by Perez-Navara [73] for the determination of heavy metals in fish.

Mumba, Csikai and Peto [74] determined light elements (Mg, K and Cl) in medicinal plants using a ^{3}H source on Ti backing under vacuum.

REFERENCES

1. R. Woldseth, *X-ray Energy Spectrometry,* Kevex Corp., Burlingame, CA (1973).
2. J. Leonowich, S. Pandian and I. L. Preiss, *J. Radioanal. Chem.* 40, 175 (1977).
3. P. Hoffmann, *Fresenius' Z. Anal. Chem.* **323**, 801 (1986).
4. Amersham Corp. Tech Bull. Amersham, UK (1992).
5. S. L. Faddeev, V. V. Fedorov and G. E. Shchukin, *Appl. Radiat. Isot.* **41**, 1153 (1990).
6. P. Z. Hien, N. T. Anh, D. D. Thao and T. Truong, *J. Radioanal. Nucl. Chem.* **118**, 217 (1987).
7. Y. G. Sevastyanov, A. A. Razbach, G. A. Molin and A. I. Leonov, *Appl. Radiat. Isot.* **41**, 1149 (1990).

8. T. Hofmann, P. Hoffmann and K. H. Lieser, *J. Radioanal. Nucl. Chem.* **109**, 419 (1987).
9. P. Martinelli, J. L. Boutaine, G. Gousseau, J. C. Tanguy and C. Tellechea, *Nucl. Instrum. Methods. Phys. Res., Sect. A* **A242**, 569 (1986).
10. J. J. LaBrecque and P. A. Rosales, *X-Ray Spectrom.* **18**, 133 (1989).
11. J. J. LaBrecque, W. C. Parker and P. A. Rosales, *Radiochem. Radioanal. Lett.* **49**, 261 (1981).
12. M. S. Rappaport, M. Mantel and C. Shenberg, *Med. Phys.* **9**, 194 (1982).
13. I. Szaloki, J. Patko and L. Papp, *J. Radioanal. Nucl. Chem.* **141**, 279 (1990).
14. D. A. Close, J. C. Pratt, J. J. Malanify and H. F. Atwater, *Nucl. Instrum. Methods Phys. Res., Sect. A* **A234**, 556 (1985).
15. A. B. Hollak and N. S. Saleh, *J. Phys. E. Sci. Instrum.* **18**, 9 (1985).
16. A. Zararsiz and E. Aygun, *J. Radioanal. Nucl. Chem.* **129**, 367 (1989).
17. G. Kruger, W. Bohme, M. Thomae and G. Loos , *Isotopenpraxis* **19**, 198 (1983).
18. E. Kowalska and P. Urbanski, *Nukleonika* **32**, 131 (1987).
19. B. Holynska, M. Lankosz, J. Ostachowicz, T. Wesolowski and J. Zalewski, *Tech. Rep. Ser.–I.A.E.A.* (TECDOC-520), 43 (1989).
20. E. Kowalska and P. Urbanski, INR-1954/15-A/C/B (1982); INIS **18**: 018003.
21. H. Kostalas, *Int. J. Appl. Radiat. Isot.* **33**, 1475 (1982).
22. V. M. Aboleshin, *Geofiz. Appar.* **72**, 177 (1981).
23. *Nuclear Instruments and Systems.* Radiation detection catalogue, Ortec Company, USA (1992).
24. I. Szaloki, *X-Ray Spectrom.* **17**, 75 (1988).
25. I. Uzonyi, *Isotopenpraxis.* **24**, 79 (1988).
26. P. Arikan and I. Alkan, *J. Radioanal. Chem.* **67**, 403 (1981).
27. L. J. Somervaille, D. R. Chettle and M. C. Scott, *Phys. Med. Biol.* **30**, 929 (1985).
28. E. N. Lazo, G. S. Roessler and B. A. Berven, *Health Phys.* **61**, 231 (1991).
29. S. Mattsson, *Tech. Rep. Ser.–I.A.E.A.*, (IAEA/RL/131), 41 (1986).
30. J. J. LaBrecque, *Proc. Second Symposium on Anal. Chem. in Exploration and Mining processing of materials*, Pretoria, S. Africa (1985), pp. 308–309.
31. B. J. Thomas, B. W. Thomas, J. F. Davey, H. Baddeley, V. Summers and P. Craswell , *Tech. Rep. Ser.–I.A.E.A* (IAEA/RL/131), (1986).
32. J. Lucas-Tooth and C. Pyne, *Adv. X-Ray Anal.* **7**, 523 (1964).
33. A. Lubecki, *J. Radioanal. Chem.* **2**, 3 (1969).
34. P. Frigieri, F. Rossi and R. Trucco, *Spectrochim. Acta, Part B* **35B**, 351 (1980).
35. H. A. Ashry and F. A. S. Soliman, *Isotopenpraxis* **25**, 485 (1989).
36. C. Shenberg, H. Feldstein and S. Amiel, *J. Radioanal. Chem.* **55**, 183 (1980).
37. A. B. Hallak and N. S. Saleh, *X-Ray Spectrom.* **12**, 148 (1983).
38. R. E. Ayala-Jimenez, *INIS Atomindex INIS-mf-10151* (1986); INIS **17**: 042793.
39. D. Hegedus, J. Solymosi, N. Vajda *et al., Period. Polytech., Chem. Eng.* **28**, 193 (1984).
40. L. Ahlgren and S. Mattsson, *Phys. Med. Biol.* **24**, 136 (1979).
41. M. Mantel, in *Activation Analysis*, Vol. 1, edited by Z. B. Alfassi, CRC Press, Baton Rouge (1990) pp. 112–128.
42. J. S. Watt, *Int. J. Appl. Radiat. Isot.* **34**, 309 (1983).
43. G. Kruger and J. Schwarzlose, *Isotopenpraxis* **19**, 24 (1983).
44. P. Groll, *Atomkernenerg. Kerntech.* **46**, 94 (1985).
45. J. W. Leonhardt, H. Bruchertseifer, B. Eckert, P. Morgenstern and W. Riedel, *Tech. Rep. Ser.–I.A.E.A* (TECDOC-520), 49 (1989).
46. R. Wuertz, DE Patent 3326737/A/ (1985); INIS: **18**, 034853.
47. E. Lehmann and W. Brendel, *Giessereitechnik* **32**, 281 (1986).
48. A. A. Shumakher, V. N. Muzkin and T. I. Palnikova, *Zavod. Lab.* **51**, 28 (1986).
49. S. E. A. Morales, *INIS Atomindex* INIS-mf-11092 (1986); INIS **19**: 016413.
50. N. S. Saleh and A. B. Hallak, *X-Ray Spectrom.* **12**, 170 (1983).

51. J. J. LaBrecque, P. A. Rosales and G. Mejias, *Anal. Chim. Acta* **188**, 9 (1986).
52. Ankara Nuclear Research and Training Center. Quzi University, Ankara, Turkey, N. Efe, Analysis of thorium and lanthanides in britholite ore by radioisotope excited X-ray fluorescence technique, M.Sc. Thesis (1985); INIS **18**: 012269.
53. R. Rolle, V. C. Hammond, M. Nami, A. W. Lamos, D. A. Roberts and J. M. Stewart, *Proc. Second Symposium on Anal. Chem. in Exploration and Mining of Materials*, Pretoria, S. Africa (1985), pp. 324–315.
54. E. F. Ivanova, A. V. Kuznichenko, V. N. Lebedev, V. N. Majsterenko and Z. O. Bour-Moskalenko, *Ukr. Khim. Zh.* **52**, 1186 (1986).
55. J. Hally, L. Simon, T. Cechak and L. Koc, CS Patent 262733/B1 (1986); INIS **22**: 038540.
56. L. Jimin, L. Guizhen, L. Guodong and L. Lijuan. *Nucl. Technol.* **12**, 703 (1989); INIS **22**: 018438.
57. Z. Ye, X. Tingzhou, L. Zhirong, H. Gouqiang, and Z. Sichun, *Nucl. Technol.* **9**, 9 (1985), INIS **18**: 013015.
58. J. Majer, V. Skaba, A. Janata and M. Vidra CS Patent 226834/B (1985); INIS **17**: 082003.
59. V. Balgava, M. Kern and M. Melich, *Radioaktiv. Zivotn. Prostr.* **9**, 213 (1986); INIS **19**: 004514.
60. B. Holynska, A. Markowicz and B. Ostachowicz, INT-181/I (1983); INIS 19: 098463.
61. L. Olah, *Jad. Energ.* **31**, 47 (1986); INIS **17**, 030915.
62. C. T. Yap, *Archaeometry* **28**, 197 (1986).
63. B. Stverak, D. Tluchor and O. Vavrikova, *Jad. Energ.* **35**, 196 (1989); INIS **20**: 067711.
64. M. Margaliot, T. Schlesinger, Y. Eisen and E. Lubin, *Trans. Isr. Nucl. Soc.* **10**, 201 (1982).
65. D. R. Chettle, M. C. Scott and L. J. Somervaille, *Phys. Med. Biol.* **34**, 1295 (1989).
66. K. W. Jones, G. Schidlovsky, R. P. Wedeen and V. A. Batuman, *Brookhaven Natl. Lab.*, [*Rep.*] BNL-39428 (1986), p. 25.
67. L. Wielopolski, D. N. Slatkin, D. Vartsky, K. J. Ellis and S. H. Cohn, *IEEE Trans. Nucl. Sci.* **28**, 114 (1981).
68. L. Wielopolski, J. F. Rosen, D. N. Slatkin *et al.*, *Med. Phys.* **16**, 521 (1989).
69. L. Ahlgren and S. Mattsson, *Phys. Med. Biol.* **26**, 19 (1981).
70. U. Nilsson, L. Ahlgren, J. O. Christoffersson and S. Mattsson, *Basic Life Sci.* **55**, 297 (1990).
71. W. D. Morgan, A. M. EL-Sharakawi, M. Yaib *et al.*, *Brookhaven Natl. Lab* [*Rep.*] BNL-49428 (1986), D3.
72. E. Havranek, A. Bumbalova and M. Harangozo, *Radioakt. Zivotn. Prostr.* **12**, 241 (1989); INIS **21**: 092657.
73. A. M. Perez-Navara, ININ-GSTN-RX-1-86 (1986); INIS **20**: 014754.
74. N. K. Mumba, J. Csikai and G. Peto, *Radiochem. Radioanal. Lett.* **52**, 373 (1982).

Chapter 16

SCATTERING OF ALPHA-, BETA- AND GAMMA-RADIATION FOR CHEMICAL ANALYSIS

Esam M. A. Hussein

Department of Mechanical Engineering, University of New Brunswick, Fredericton, New Brunswick, Canada E3B 5A3

1 INTRODUCTION

A chemical reaction is inherently an electronic interaction. Determination of chemical properties requires, therefore, the employment of electron-sensitive probes. Ionizing radiation, such as alpha-particles, beta-particles and gamma-rays, is particularly suited for chemical analysis, since these types of radiation interact directly with atomic electrons.

Radiation methods for chemical probing can depend on one of the following radiation interaction modes: (a) absorption, (b) scattering and (c) activation (production of secondary radiation). Radiation emission can also be utilized in tracer and imaging techniques to determine the distribution of a radiotracer as it is transported within a medium.

Absorption is usually monitored by measuring the reduction in the amount of radiation transmitted through an object. A well collimated beam of radiation then needs to be employed to eliminate the effect of scattering. Direct utilization of the scattering signal is particularly attractive because of the rich information contained therein. Moreover, in scattering (unlike transmission), the radiating source and detector need not be on opposite sides of the object. This provides flexibility in arranging the system set-up and enables the monitoring of thick and extended structures which are difficult to interrogate using transmission techniques. Activation techniques are not very widely used with ionizing radiation, as

Chemical Analysis by Nuclear Methods Edited by Z. B. Alfassi

the probability of producing secondary particles is quite low and most such particles are usually absorbed within the inspected object and do not reach the detector at high rates. Tracer techniques and emission techniques have also been successful in many applications, but they require the injection, or presence, of a radioactive material in the interrogated medium.

This chapter examines the use of scattering techniques for chemical analysis. The three well known forms of ionizing radiation, alpha-, beta- and gamma-rays, are examined. A section is devoted to each type of radiation. The physical nature of the scattering process is first described and the exploitation of the characteristics of scattering for chemical analysis is explained. Some typical applications are then given.

2 ALPHA-PARTICLES

Alpha-particles are heavy charged particles, with a positive charge equal to twice the electronic charge and a mass four times that of a hydrogen nucleus. An alpha-particle is essentially a positive ion of a helium-4 nucleus ($^4_2$He). For practical utilization of alpha-particles in chemical analysis, the particle must have sufficient energy to be able to enter the interrogated object, interact with it, and then exit for subsequent detection. The energy of alpha-particles emitted from most radionuclides is between 4 and 11 MeV. The distance an alpha-particle travels in a substance until it delivers all its energy and is absorbed is called the range, R. The value of the range, in mm, is given approximately by

$$R = \sqrt{A}\rho^{-1} E^{1.5} \tag{1}$$

where A is the mass number of the material, ρ is its density in kg/m^3 and E is the energy of the alpha-particle in MeV. Therefore the range of a 4 MeV alpha-particle in aluminum ($A = 27, \rho = 2700$ kg/m^3) is $\approx$0.015 mm, while that in air (at standard temperature and pressure, $A \approx 3.8$) is $\approx$24 mm. This shows that, within the energy range of alpha-particles emitted from radionuclides, alpha-particles will not deeply penetrate the surface of most materials. The information deduced from the interaction of alpha-particles pertains only to the surface of the material, i.e. within the range of the alpha-particle.

2.1 Scattering modalities

The overwhelmingly predominant mode of scattering of an alpha-particle is that of inelastic collision with bound atomic electrons. As a result of such collision, an electron experiences a transition to an excited state or to an unbound state, causing in the latter case ionization of the atom. The interacting alpha-particles lose some energy in this process but, because of their large mass compared with the affected electron, they hardly change their straight path through the material. For this reason, alpha-particles involed in inelastic scattering with the atoms can be essentially considered as undergoing a transmission process. The change in the intensity and energy of the scattered particle, can however, provide useful information on the nature of the traversed medium.

The energy loss per unit length due to the excitation/ionization process can be approximated by [1]

$$\frac{dE}{ds} = (2\pi e^4 z^2 m/m_0 E) N Z \ln(m_0 E/mI) \tag{2}$$

where e is the electronic charge, m_0 is the electron's rest mass, m is the mass of the alpha-particle, z is its atomic number ($z = 2$), Z is the atomic number of the target nucleus, N is the number of material atoms per unit volume, E is the energy of the alpha-particle and I is the ionization potential of the material. The value of I for most elements is linearly proportional to the atomic number. The above relationship shows the dependence of the energy loss on the chemical composition of the material, through Z. The density of the material, through N, also affects the amount of energy loss. Therefore, by measuring the energy of alpha-particles after passing through a material, one can obtain some relevant properties of the medium.

An alpha-particle may also be elastically deflected in the electric field of the atomic electrons. Energy and momentum are then conserved, but the energy transferred is generally much less than the excitation energy of the electrons. Such collisions are therefore irrelevant and are usually neglected.

In an elastic collision of an alpha-particle with a nucleus, the charged particle is deflected but it does not radiate, nor excite the nucleus. The incident particle loses only the kinetic energy needed for the conservation of momentum between the two particles. Some nucleus-specific information may then be deduced.

Two possibilities exist for the inelastic collision with a nucleus. The first possibility is a close, non-capture, encounter. The incident charged particle then invariably experiences a deflection in accordance with Rutherford's law of scattering. In some of these deflections a quantum of radiation is emitted (bremsstrahlung), and a corresponding amount of kinetic energy is lost by the colliding particle. The collision is considered to be inelastic owing to this radiative emission. The amount of bremsstrahlung produced by an alpha-particle is very small because of its large mass. Therefore the effect of radiative collision by alpha-particles is usually neglected. The other possibility is a nuclear excitation in non-capture collisions. The probability of the latter interaction is very low in most materials. Nuclear scattering of alpha-particles is therefore mainly an elastic scattering process.

Because of their heavy weight and large charge, alpha-particles are easily stopped within a small distance in any material. The only practical scattering signal that can be obtained from alpha-particles in a large system is backscattering. Forward scattering can be detected only if the object is very thin, and this requires the preparation of a special sample for proper examination. This is however difficult for liquids and gases, since they require a container which will not absorb the alpha-particles before they can reach the sample. Nevertheless, the scattering mechanisms are examined in view of the utilization of backscattering for thick objects and at any angle of scattering for specially prepared samples.

2.2 Angular distribution

The elastic scattering of alpha-particles by a nucleus can be described by the classical Rutherford scattering formula. The probability of scattering of an alpha-particle into a solid angle $d\Omega$ around the direction of a scattering angle ϕ, the differential cross-section, is given by [1]

$$\frac{d\sigma}{d\Omega} = \frac{(Zze^2)^2}{16E_0^2} \sin^{-4}\phi/2 \tag{3}$$

where E_0 is the initial energy of the incident particles. Here, the nucleus of the scatterer is so heavy that its motion during the interaction (considered as a collision) is neglected. The above scattering formula is therefore applicable only to materials of high atomic number. This formula indicates that, by monitoring the scattering of a beam of alpha-particles at some angle of scattering ϕ, one can deduce the atomic number Z, i.e., the composition of the target. However, the probability of small-angle scattering is clearly much greater than that of scattering at larger angles. Unfortunately, however, small-angle scattering can be detected only if the object is very thin; otherwise the scattered particle will be absorbed within the object.

For the scattering of alpha-particles by light elements, the thermal motion of the target nucleus can no longer be ignored. This results in some dependence of the probability of scattering on the mass of the target nucleus. The probability of scattering of an alpha-particle by the nuclei of light elements can be approximated by the differential cross-section [2]

$$\frac{d\sigma}{d\Omega} = \frac{(Zze^2)^2}{16E^2 \sin^4\phi} \frac{\left[\cos\phi \pm \sqrt{1-(m/M)^2 \sin^2\phi}\right]^2}{\sqrt{1-(m/M)^2 \sin^2\phi}} \tag{4}$$

The positive sign applies for $m < M$ and the negative sign is taken when $m > M$. Direct deduction of the Z, or composition, of the target materials is not possible owing to the effect of the target nucleus. However, the probability of scattering is still uniquely defined by the target. Scattering information can therefore be utilized for identifying the elemental composition of the target material.

The number of alpha-particles scattered to a particular angle depends, not only on the value of $d\sigma/d\Omega$, but also on the number of nuclei N per unit volume of the target material, the thickness of the object, and the distance between the detector and the object. The thickness of the scatterer needs however, to be sufficiently small so that the scattered particles are not absorbed within the scatterer before reaching the detector. Scattering measurements may, however, be exploited for measuring the thickness of a coating material on a more dense substrate. Most of the backscattering would then arise from the coating material, while alpha-particles reaching the substrate would most likely be absorbed. It is also possible to utilize scattering measurements to determine the density, or more precisely the electron density NZ, of a material.

2.3 Kinematics

Another interesting physical observation that can be exploited for providing chemical composition information is the energy of the scattered alpha-particles. Conservation of momentum and energy, for a target of mass M, leads to [1]

$$E_s = E[1 - 2(1 - \cos\phi)m/M] \tag{5}$$

where E_s is the energy of an alpha-particle scattered by an angle ϕ. The relationship for targets of smaller mass number is more complex but carries essentially the same physical information. That is, the energy of scattering is uniquely related to the angle of scattering for a given incident energy. Moreover, the energy of the scattered particle along a particular direction is dependent on the mass of the target nucleus.

The above energy–angle relationship can be exploited to obtain density maps if a well collimated alpha beam is made incident on a target and the spectrum of scattered particles is recorded. The direction from which the scattered particle originated can be determined from the measured energy of the particle. The distribution of material density along the direction of the incident beam can be obtained without having to collimate the scattered beam to focus at a particular location. Alternatively, one can fix the angle of scattering at a given direction by collimating the detector. The energy spectrum can then provide information on the mass number, and consequently on composition, since according to equation (5) elements of different mass number will result in alpha-particles of different energy for the same angle of scattering. A number of detectors are available for measuring the spectrum of alpha-particles [3], including proportional counters, scintillation detectors, Frisch-grid ionization chambers and silicon–gold surface barrier detectors.

For light elements the contribution from inelastic scattering becomes significant, especially as the energy of the alpha-particle increases. This makes interpretation based on the above relationships, which are applicable to elastic scattering, less relevant. Techniques for chemical, or compositional, analysis relying on inelastic scattering need therefore to be designed carefully, perhaps with prior calibration for a known set of elements.

2.4 Applications

In spite of the rich information the scattering of alpha-particles can provide in chemical analysis, only a limited number of applications is reported in the literature. This is perhaps because of the weak penetration power of alpha-particles emanating from isotopes and the material damage and ionization introduced by high energy particles generated by accelerators. Moreover, the Rutherford scattering of alpha-particles, as indicated above, has low probability and is primarily in the forward direction. This, combined with the small range of alpha-particles, makes the detection of forward scattered particles possible only in gases and very thin foils, such as paper sheets. On the other hand, the probability of backscattering is quite low. A source of considerable strength is therefore needed if one is to take advantage of the backscattering mode of alpha-particles for determining the property of the skin depth (range equivalent) of an object. One should also point out that the absorption of alpha-particles occurs mostly through successive collisions in which the alpha-particle gradually loses energy. Techniques involving absorption or transmission of alpha-particles can be considered as applications of the scattering process, and are accordingly considered here.

Although the art of alpha-particle spectrometry and the associated detectors have considerably advanced in the past few years, its application has been limited to the identification of alpha emitters, rather than to scattering techniques. The development in this technology, together with the availability of commercial alpha emitters with considerable strength and reasonable half-life such as americium-241 (458 year half-life), should encourage the exploitation of the scattering of alpha-particles for material identification and analysis.

Recently reported applications of alpha-particle scattering using isotopic sources include the measurement of the area of apertures of diaphragms [4] and the determination of the atomic number of thick metal backings [5]. Alpha particles were also used for obtaining complete and detailed in situ chemical analysis of surface and thin atmospheres of extraterrestrial bodies [6, 7]. More applications of alpha-particles are found, however, when helium-4 ions are produced in an accelerator. Rutherford backscattering spectrometry of alpha-particles is used for studying the structure and composition of mechanically polished and chemically etched surfaces of CdTe single crystals [8]. Alpha-particle backscattering is used for near-surface analysis of oxygen-containing high-Z compounds; thus, ^{4}He backscattering resonance with oxygen-16 was used for determining the concentration of thin films on high-Z substrates and high-Z

bulk oxides [9]. Surface smoothness on rough and polished silicon surfaces was also monitored using the backscattering of energetic alpha-particles [10]. Rutherford scattering of helium was used for determining the concentration of trace and minor elements in cigarette tobacco [11]. The backscattering of ions has also been used for elemental trace analysis of blood, wheat and well water samples [12]. More information on the design criteria and applications of Rutherford backscattering spectrometry can be found Ref. 13 and in chapter 11 of this book.

3 BETA-PARTICLES

Beta-particles are electrons emitted as a result of the decay of a radionuclide. They are in practice distinguished from electrons produced by some other means by their continuous spectrum of energy. The energy spectrum for a particular nuclide has a Maxwellian-type distribution curve, which starts from zero energy, rises to a peak and then decreases to a well defined maximum energy. This upper limit of energy is uniquely defined by the beta emitter. Most beta emitters have a maximum energy limited to ≈ 3 MeV.

The continuous distribution of the energy of beta-particles results in varying stopping distances of particles in matter. The range of a beta-particle is usually taken as that corresponding to the maximum beta energy, as this results in the largest penetration distance. This range R_m can be approximated by

$$R_m = 4 \times 10^3 E_{max}^{1.4} \rho^{-1} \tag{6}$$

where ρ is the material density in kg/m^3, E_{max} is in MeV and R_m is in mm.

The wide energy range of the emitted beta-particles allows an empirical expression of the attenuation coefficient of the intensity of beta-particles in an exponential fashion, permitting the definition of an attenuation coefficient. That is, the intensity of a beta beam, with initial intensity I_0, after traversing a distance x in matter can be expressed as

$$I = I_0 e^{-\mu x} \tag{7}$$

where μ is the attenuation coefficient and can be approximated by

$$\mu = 2.2 \times 10^{-3} \rho E_{max}^{-4/3} \tag{8}$$

where μ is in units of mm^{-1}, ρ is the material density in kg/m^3 and E_{max} is the maximum energy of the beta source in MeV. Taking a beta source with E_{max} 4 MeV, for the sake of comparison with alpha-particles, the attenuation coefficients for aluminum and air, respectively, are 0.94 and 10^{-3} mm^{-1}. Assuming that for all practical purposes a beta beam is fully attenuated as its intensity drops to e^{-5} of its original value, the required distance in aluminum is ≈ 5.3 mm whereas that in air is 5000 mm. Comparing these values with the corresponding values for alpha-particles (which are lower by two orders of magnitude), it is clear that beta-particles have a much larger penetration power than alpha-particles. This makes beta-particles useful for application in techniques for determining physical properties.

The utilization of the scattering of beta-particles for inferring material property information is more complicated than that of alpha-particles. This is because beta-particles change their direction significantly after scattering and are more likely to be scattered again before leaving the object. This straggling effect also makes it difficult to define a precise range of a beta-particle over a staight line. Moreover, because of the complex energy spectrum, relating the energy to the angle of scattering for singly scattered beta-particles is a difficult task. Adding to the complexity is the relativistic effect of the change in mass of the beta-particle as its velocity changes. One therefore has to rely on gross (integral) scattering information in material evaluation. The three main basic scattering mechanisms of beta-particles are first reviewed before discussing different integral scattering techniques. For simplification, relativistic effects for high speed beta-particles are not considered in the following discussion.

3.1 Scattering modalities

The law of Rutherford for scattering with a nucleus, already discussed for alpha-particles, applies also to beta-particles [1]. Therefore equation (3), with $z = 1$, becomes

$$\frac{d\sigma}{d\Omega} = \frac{(Ze^2)^2}{16E^2} \sin^{-4} \phi/2 \tag{9}$$

Inelastic collision of beta-rays with atomic electrons is the chief mode of interaction. The energy transferred to the atomic electron in such a process causes excitation or ionization of the atom. However, the energy transferred to the atomic electron is much larger that its binding energy and the collision can be considered as a collision between two particles of equal mass. Equation (4), with $m = M = m_0$ and $z = Z = 1$, then becomes

$$\frac{d\sigma}{d\Omega} = \frac{e^4}{4E^2} \cos \phi \sin^{-4} \phi \tag{10}$$

This relationship shows that backscattering ($\phi > \pi/2$) is not possible, since the two colliding particles have equal mass. Conservation of momentum and energy of the collision between two particles of equal mass requires that the angle between the directions of the two particles after collision be always equal to $\pi/2$. Experimentally the two emerging electrons are indistinguishable. Therefore the probability of scattering or differential angular cross-section, per atom, of electrons into a solid angle $d\Omega$ around the direction Ω at an angle ϕ becomes

$$\frac{d\sigma}{d\Omega} = \frac{Ze^4}{4E^2}(\sin^{-4}\phi + \cos\phi^{-4})\cos\phi \tag{11}$$

The Z number of the target material appears in the above equation in order to provide a probability per atom, rather than per electron.

Comparison of the probability of nuclear scattering of beta-particles, equation (9), to that with the atomic electrons, equation (11), shows that nuclear scattering increases with Z^2 whereas electronic scattering increases only with Z. Therefore, in general, nuclear scattering predominates over the scattering by electrons by a factor of Z. The two mechanisms of scattering have the same probability of occurrence only with hydrogen, $Z = 1$.

A third possible mode of scattering of beta-rays is that of inelastic collision with the nucleus, which results in the production of a continuous X-ray spectrum or bremsstrahlung. The Coulomb field of the nucleus accelerates the incident electron, which consequently emits a pulse of radiation (photon) as it slows down. In such a radiative collision the photon carries a very small momentum and can be emitted in any direction. The problem can be considered as a collision of an electron with the nucleus, but with some energy loss to the emitted photon. The angular distribution of the electron is then expected to be similar to that described by equation (9). The probability of occurrence of this radiative collision process is much lower, however, than that of the other forms of scattering.

3.2 Applications

The above forms of scattering indicate some interesting deflection characteristics which may be exploited for material diagnosis. Unfortunately, the wide distribution of the energy of beta-particles makes it difficult to relate the energy of the scattered beta-particle to a particular direction, or to determine the probability of emission of beta-particles at a given direction and energy. One can, however, take advantage of the overall scattering signal to obtain some relevant information.

Backscattering of beta-rays is used in a number of applications where access to only one side of the object is possible. Such scattering is due mainly to the elastic scattering by the nucleus, equation (9). This equation shows that the probability of scattering is proportional to Z^2, which indicates good sensitivity of the scattering signal to variations in the composition of the scattering object. On a macroscopic level, the thickness of the object and its atomic density (number of atoms per unit volume) also affect the scattering signal. Nevertheless, these are parameters that can be determined by beta-particle scattering for materials of known Z.

The probability of forward scattering of beta-rays is much larger than that of backscattering, equations (9) and (11). This is not only because of the forward-peaked nature of the angular distribution, but also because of the contribution of scattering by the atomic electrons, which is possible only in forward scattering.

Absorption of beta-rays is usually the result of successive collisions in which the particle loses its energy. Absorption techniques can therefore be considered in effect to be scattering techniques and are thus included here.

Many applications of isotopic beta-particle sources have been recently reported. Beta-particles emitted from a strontium-90 source were used for the measurement of frost density distribution on the surface of a vertical pipe [14]. A strontium-90/yttrium-90 source in backscatter geometry was used with a cadmium telluride solid-state detector for precise determination of the effective atomic number of materials with $Z = 9–30$ [15]. Automatic monitoring of aerosol samples was achieved by beta-particle attenuation [16]. Low energy beta sources, tritium, nickel-63 and carbon-14, were utilized for the radiography of forensic documents [17]. A beta-ray gauge was employed as an evaporimeter when a thallium-204 source was used to measure the evaporation rate through turgid tobacco leaf [18]. A beta-ray gauge with a thallium-204 source was applied as a leaf surface wetness detector [19]. An extended beta source was used for radiographic examination of surface layers by capturing diffused-back beta-particles on a film [20]; for example, a sandstone rock sample containing uraninite was photographed to identify uranium-bearing phases. Compositional analysis of tin/lead plating on printed circuit boards by the beta backscatter method, using an nickel-63 source, has also been reported [21]. Non-destructive detection of faults in conventional explosive fuses was performed using a strontium-90/yttrium-90 beta source [22].

Thickness gauging with scattered beta-rays is very common. For example, a beta gauge was used for measuring the thickness of high quality polymer sheets for the predictability and control of layflatness in calendering [23]. On-line measurement of thin organic films on metallic sheets was reported [24], in which organic coatings on aluminum and steel substrates were measured by beta backscattering. The thickness of glass and aluminum containers was measured by beta backscattering with strontium-90 and yttrium-90 sources [25]. It was also shown that analog squaring of the spectrum transmitted through an absorber, followed by pulse-height analysis, can yield a thickness gauge for aluminum using phosphorus-32/yttrium-90 sources [26].

4 GAMMA-RAYS

Gamma-rays are very short electromagnetic waves. They are distinguished from X-rays by the fact that they originate from nuclear, rather than atomic electron, transitions. Gamma-rays are also emitted with well defined discrete energies, whereas X-rays have a wide energy spectrum. The scattering of gamma-rays can occur either coherently, without a change of wavelength, or incoherently with change of wavelength. The latter is known as Compton scattering and the former is called Rayleigh scattering. Owing to their extremely short wavelengths, their energy (= frequency times Planck's constant) is usually used in describing the kinematics of gamma-ray interactions.

4.1 Compton scattering

Compton scattering is the dominant mode of interaction of photons with most materials in the energy range 300 keV to 2 MeV. In Compton scattering, photons interact with the relatively free electrons of the matter. As a result of this scattering, the photon energy changes from the original energy E to an energy E_s and to a corresponding angle of scattering ϕ according to the well known Compton relationship

$$E_s = \frac{Em_0c^2}{m_0c^2 + E(1 - \cos\phi)} \tag{12}$$

where m_0c^2 is the rest mass energy of the electron. This relationship shows a unique relationship between the energy and the angle of scattering; that is, knowing one leads to knowing the other.

The differential cross-section, the probability of scattering from E to E_s or the equivalent probability of scattering by an angle ϕ, is given by the well known Klein–Nishina formula

$$\frac{d\sigma}{d\Omega} = r_0 \frac{1 + \cos^2\phi}{2}\left[\frac{1}{1 + \alpha(1 - \cos\phi)}\right]^3\left[1 + \frac{\alpha^2(1 - \cos\theta)^2}{[1 + \alpha(1 - \cos\theta)](1 + \cos^2\theta)}\right] \tag{13}$$

where r_0 is the classical electron radius (2.818×10^{-15} m) and $\alpha = E/m_0c^2$. The probability of occurrence of Compton scattering increases with increasing Z of the target material, owing to the increase in the number of electrons available for interactions. Therefore measurement of the intensity of Compton scattered photons can be utilized for determining the electron density of a material. The unique relationship between the angle and energy of scattering can be further exploited for determining the distribution of the electron density, as each photon energy identifies a particular direction of the first scatter of the photon.

4.2 Rayleigh scattering

Another useful mode of photon interactions is the coherent, or Rayleigh, scattering process. In this process the photon behaves strictly as an electromagnetic wave and does not change its wavelength, and consequently maintains its energy upon colliding with an atom. The recoil momentum imparted to the atom by the incident radiation does not, therefore, produce excitation or ionization. The permissible angles of scattering are very small. At least 75% of Rayleigh scattering is confined to scattering angles which are smaller than the angle θ_c given by [27]

$$\theta_c = 2\tan^{-1}(0.0133Z^{1/3}E^{-1}) \tag{14}$$

with E being the incident energy of the photon in MeV. This angle, for example, is 4.5° for a 1 MeV beam incident on iron. This small change in angle, combined with the fact that the interacting photon does not change its energy, makes the measurement of coherent scattering somewhat difficult. In fact, in many instances, coherent scattering is entirely neglected on the assumption that it does not cause a change in the photon energy and hardly causes a change in the beam direction.

The probability of occurrence of Rayleigh scattering has, however, a strong dependence on Z. The probability of Rayleigh scattering increases with Z^3 but decreases with E^3. This makes the use of Rayleigh scattering of low energy photons, particularly X-rays, attractive for analyzing the composition of high-Z substances. A few examples of the utilization of this phenomena are discussed in the following section.

4.3 Applications

Some recently reported applications of Compton scattering include systems for lung densitometry [28, 29]. Dual-energy Compton scatter densitometry was examined for use in determination of the density of a substance consisting of two components [30]. The scattering of high energy photons extracted from a reactor was proposed for dynamic density measurement of boiling water [31], and in situ density measurement in aqueous solutions was performed by the gamma backscattering method [32]. A gamma scattering technique for detecting pipe-thinning problems was recently reported [33]. Some density and imaging techniques using Compton scattering have been developed; see for example Refs. 34 and 35.

Measurement of the multiple scattering of gamma-rays was exploited for in situ determination of bulk density of surface formations and lithology (average Z number) [36]. A similar technique was used for measurement of the density and water content of soil [37, 38]. As indicated in the review article of Clayton and Wormald [39], Compton scattering was used in coal mining to provide information on the correspondence of strata in the regions surrounding adjacent boreholes and to indicate the thickness of particular coal strata and the presence of 'dirt bands' within the strata. In addition, gamma backscattering was used to determine the grade of iron ore in samples and to detect shale on conveyer belts [40]. The energy spectrum of backscattered photons was also used for the determination of rock properties, such as heavy element content, density, borehole diameter and grain size [41]. Borehole analysis for copper and nickel using gamma-ray resonance scattering was also reported [42]. In this last application, gaseous zinc-65 and cobalt-60 gamma sources were used to produce photons with energies that coincide with the critical (resonant) absorption energies of the nuclei concerned, copper and nickel, respectively. For boreholes in rock of constant type and density, the number of resonantly scattered gamma-rays is proportional to the concentration of the wanted element. More information on the use of gamma scattering in assaying of mine boreholes can be found in Ref. 43.

Compton scattering was utilized for surface inspection of aluminum blocks with imperfections up to 1.6 mm in size [44]. Backscattering was used for measuring the density of a layer, or thin-lift, of known and constant thickness on a planar substrate, such as in the repaving of roadways and bridges. Two measurements are required, one before the layer was applied and one after [45]. A similar technique was applied to locate groups of termite-damaged sleepers during continuous scans of a railway track [46]. Gamma backscattering was used for measuring the thickness of glass and aluminum containers [25], as well as concrete, carbon blocks and wood [47]. A backscattering gauge was also developed for monitoring and control of material conditions in a steel sintering plant [48] and for locating steel re-bars in reinforced concrete blocks [49, 50].

Applications of Rayleigh scattering include a system for bone mineral analysis by utilizing the Rayleigh/Compton scattering ratio [51]. The same scattering ratio was used for measuring changes in titanium-rich alloys based on aluminum and zirconium, as well as for measuring the surface composition of heavy elements [52, 53] and for the analysis of bronze or silver–copper alloys [54]. The same method was applied to the determination of the effective Z of compounds with effective $Z < 20$ [55]. The ratio technique was also utilized for measuring the concentration proportions in substances comprising major compounds of low Z, such as fat and water content in milk or meat and reaction products in organic chemistry [56]. In principle, the effective atomic number of materials of $Z = 6–83$ can be determined by the Rayleigh-to-Compton ratio. Medical applications of the ratio technique include the measurement of fat in liver [57], trabecular bone mineral density [58], bone density [59] and stable iodine content in tissue [60].

Small-angle Rayleigh scattering has been recently proposed as a method to complement conventional transmission tomography [61]. Although this method relies on the use of X-rays, it can be equally applicable to gamma-rays with appropriate energy. This technique enables the determination of composition characteristics of some biological tissues and plastics that cannot be resolved by transmission tomography alone [62]. A similar technique was used for measuring changes in the composition of a bone-equivalent material, potassium orthophosphate solution [63]. This technique has the potential of wide use in material characterization owing to its superior sensitivity to compositional changes [64].

The reader may refer to the two review articles written by the author of this chapter for more detail on the theory and use of gamma scattering techniques in medical [65] and industrial [66] applications.

REFERENCES

1. R. B. Evans, *The Atomic Nucleus*, McGraw-Hill, New York (1955).
2. J. B. Rajam, *Atomic Physics*, S. Chand & Co., Delhi, 1968.
3. K. M. Glover, *Int. J. Appl. Radiat. Isot.* **35**, 239 (1984).
4. A. Rytz, *Int. J. Appl. Radiat. Isot.* **35**, 311 (1984).
5. M. J. R. Hutchinson, C. R. Nass, D. H. Walker and W. B. Mann, *Int. J. Appl. Radiat, Isot.* **19**, 517 (1968).
6. T. E. Economou and J. Iwanczyk, *Nucl. Instrum. Methods Phys. Res., Sect. A* **A283**, 352 (1989).
7. T. E. Economou and A. L. Turhevich, *Nucl. Instrum. Methods* **134**, 391 (1976).
8. E. Perillo, G. Spadaccini, M. Vigilante *et al.*, *Vacuum* **39**, 125 (1989).
9. B. Blanpain, P. Revesz, L. R. Doolittle *et al., Nucl. Instrum. Methods Phys. Res., Sect. B* **B34**, 459 (1988).
10. K. Schmid and H. Ryssel, *Nucl. Instrum. Methods* **119**, 287 (1974).
11. K. A. Al-Saleh and N. S. Saleh, *Nucl. Instrum. Methods Phys. Res., Sect. B* **B18**, 77 (1986).
12. A. Anttila, A. L. Kairento, M. Piiparinen *et al., Int. J. Appl. Radiat. Isot.* **23**, 315 (1972).
13. M. D. Strathman, *Nucl. Instrum. Methods Phys. Res., Sect.*, **B10/11**, 600 (1985).
14. T. Y. Bong, N. E. Wijeysundera, E. L. Saw and K. O. Lau, *Exp. Heat Transfer* **4**, 567 (1991).
15. M. S. Soller, L. Cirignano, P. Lieberman and M. R. Squillant, *IEEE Trans. Nucl. Sci.* **37**, 230 (1990).
16. G. S. Spagnolo, *J. Aerosol Sci.* **18**, 899 (1987).
17. F. M. Kerr, *Nucl. Instrum. Methods Phys. Res., Sect. A* **A257**, 26 (1986).
18. N. N. Barthakur, *Int. J. Appl. Radiat. Isot.* **36**, 162 (1985).
19. N. N. Barthakur, *Int. J. Appl. Radiat. Isot.* **34**, 1549 (1983).
20. J. Pant, G. Pregl, P. Leskovar and F. Zitnik, *Int. J. Appl. Radiat. Isot.* **33**, 207 (1982).
21. A. Damkjaer, *Int. J. Appl. Radiat. Isot.* **27**, 631 (1976).
22. A. Sy Ong and G. Duravaldo, *Int. J. Appl. Radiat. Isot.* **26**, 313 (1975).
23. A. A. Tseng, *Proc. Polymer and Polymeric Composites*, Winter Annual Meeting, American Society of Mechanical Engineers Dallas, TX (1990).
24. S. P. Strum, *Light Met. Age* **47**, 19 (1989).
25. H. Mohammadi, *Int. J. Appl. Radiat. Isot.* **32**, 524 (1981).
26. A. N. Brown and T. J. Kennett, *Int. J. Appl. Radiat. Isot.* **27**, 126 (1976).
27. J. Hubbell, Photon cross section attenuation coefficients, *U.S., Natl. Bur. Stand.*, [Tech. Rep.] SNRDS-NBS 29 (1969).
28. B. W. Loo, F. S. Goulding, N. W. Madden and D. S. Simon, *IEEE Trans. Nucl. Sci.* **36**, 1144 (1989).
29. B. W. Loo, F. S. Goulding and D. S. Simon, *IEEE Trans. Nucl. Sci.* **33**, 531 (1986).
30. A. L. Huddleston and J. B. Weaver, *Int. J. Appl. Radiat. Isot.* **34**, 997 (1983).
31. T. J. Kennett, W. V. Prestwich and A. Robertson, *Int. J. Appl. Radiat. Isot.* **27**, 529 (1976).

32. A. Gayer, S. Bukshpan and D. Kedem, *Nucl. Instrum. Methods* **192**, 619 (1982).
33. H. Lee and E. S. Kenney, *Nucl. Technol.* **100**, 70 (1992).
34. T. H. Prettyman R. P. Gardner, J. C. Russ, and K. Verghese, *Trans. Am. Nucl. Soc.* **65** (1), 55 (1992).
35. N. V. Arendtsz and E. M. A. Hussein, Compton scatter imaging with energy sensitive high resolution detectors, in *Proc. Gamma-Ray Detector Conf., SPIE*, International Society for Optical Engineering, San Diego, CA (July 1992), pp. 249–260.
36. D. V. Ellis, Gamma ray scattering measurement for density and lithology determination, *IEEE Trans. Nucl. Sci.* **35**, 806 (1988).
37. C. Ertek and N. Haselberger, *Nucl. Instrum. Methods* **227**, 182 (1984).
38. L. Daddi, *Int. J. Appl. Radiat. Isot.* **24**, 295 (1973).
39. C. G. Clayton and M. R. Wormald, *Int. J. Appl. Radiat. Isot.* **34**, 3 (1983).
40. M. Borsaru, R. J. Holmes and J. Mathew, *Int. J. Appl. Radiat. Isot.* **34**, 397 (1983).
41. J. Charbucinski, *Int. J. Appl. Radiat. Isot.* **34**, 353 (1983).
42. B. D. Sowerby and W. K. Ellis, *Nucl. Instrum. Methods* **115**, 511 (1974).
43. A. W. Wylie, *Nuclear Assaying of Mining Boreholes*, Elsevier, Amsterdam (1984), Chapter 13.
44. M. J. Anjos, R. T. Lopes and J. C. Borges *Nucl. Instrum. Methods Phys. Res., Sect. A* **A280**, 535 (1989).
45. W. L. Dunn and J. E. Hutchinson, *Int. J. Appl. Radiat. Isot.* **33**, 563 (1982).
46. R. A. Fookes, J. S. Watt, B. W. Seatonberry *et al.*, *Int. J. Appl. Radiat. Isot.* **29**, 721 (1978).
47. M. Kato, O. Sato and H. Saito, A study of the application of backscattered gamma-rays to gauging, *Proc. ERDA X- and Gamma-Ray Symposium*, Ann Arbor, MI (May 1976), pp. 230–233.
48. K. Yui, Y. Shirakawa, Y. Matsuo *et al.*, *Trans. Am. Nucl. Soc.* **56**, (3), 21 (1988).
49. S. Tuzi and O. Sato, *Int. J. Appl. Radiat. Isot.* **41**, 1013 (1990).
50. E. M. A. Hussein and T. M. Whynot, *Nucl. Instrum. Methods Phys. Res., Sect. A* **283A**, 100 (1989).
51. I. K. MacKenzie, *Nucl. Instrum. Methods Phys. Res., Sect. A* **A299**, 377 (1990).
52. M. J. Cooper, A. J. Rollason and R. W. Tuxworth, *J. Phys. E: Sci. Instrum.* **15**, 568 (1982).
53. M. Cooper, R. S. Holt and G. Harding, *J. Phys. E: Sci. Instrum.* **18**, 354 (1985).
54. L. Confalonieri, R. Crippa and M. Milazzo, *et al.*, *Int. J. Appl. Radiat. Isot.* **38**, 139 (1987).
55. S. Manninen, T. Tikanen, S. Koikkalainen and T. Paakkari, *Int. J. Appl. Radiat. Isot.* **35**, 965 (1984).
56. H. P. Schatzler, *Int. J. Appl. Radiat. Isot.* **30**, 115 (1979).
57. P. Puumalainen, H. Olkkonen and P. Sikanen, *Int. J. Appl. Radiat. Isot.* **28**, 785 (1977).
58. S. S. Ling, S. Rustgi, A. Karelles *et al.*, *Med. Phys.* **9**, 208 (1977).
59. J. J. Stalp and R. B. Mazess, *Med. Phys.* **7**, 723 (1980).
60. P. Puumalainen, P. Sikanen and H. Olkkonen, *Nucl. Instrum. Methods*, **163**, 261 (1979).
61. G. Harding, J. Kosanetzky and U. Neitzel, *Med. Phys.* **14**, 515 (1987).
62. J. Kosanetzky, B. Knoerr, G. Harding and U. Neitzel, *Med. Phys.* **14**, 526 (1987).
63. J. R. Mossop, S. A. Kerr, D. A. Bradley *et al.*, *Nucl. Instrum. Methods Phys. Res., Sect. A* **A255**, 419 (1987).
64. G. Harding and J. Kosanetzky, *J. Opt. Soc. Am. A* **4**, 933 (1987).
65. E. M. A. Hussein, Compton scatter imaging systems, in *Bioinstrumentation: Development and Applications*, edited by D. Wise, Butterworth Boston (1990), Chapter 35, pp. 1053–1086.
66. E. M. A. Hussein, *Int. Adv. Nondestr. Test.* **14**, 301 (1989).

Chapter 17

MÖSSBAUER SPECTROSCOPY IN CHEMICAL ANALYSIS

Ernö Kuzmann, Sándor Nagy and Attila Vértes
Department of Nuclear Chemistry, Eötvös University, Budapest, Hungary

1 PRINCIPLES OF MÖSSBAUER SPECTROSCOPY

Mössbauer spectroscopy [1–15], a method based on the recoil-free resonance fluorescence of γ-photons observed with certain atomic nuclei, is a tool making possible the measurement of the energy of nuclear levels to an extremely high accuracy (to 13–15 decimals). This accuracy is required to measure the slight variation of nuclear energy caused by electric monopole and electric dipole as well as magnetic dipole interactions between electrons and the nucleus. Such interactions reflect changes in electronic, magnetic, geometric or defect structure and also in the lattice vibrations, serving as a basis for a variety of analytical applications.

In a typical transmission Mössbauer experiment (Fig. 1), we use the Doppler effect in order to modify the energy of an emitted γ-ray by moving the source which contains the excited-state nuclei of a Mössbauer-active nuclide 'frozen' in a solid matrix, thus providing the conditions for recoilless γ-emission. In this way we can use this modified energy to scan the absorption characteristics of the studied sample (i.e., the absorber). (Naturally, the ground-state nuclei of the Mössbauer-active nuclide contained by the absorber must be also 'frozen' in a solid matrix so that the conditions for recoilless absorption are satisfied.) The counts registered by a detector placed behind the absorber are then recorded as a function of Doppler velocity, and the line pattern thus obtained is called a Mössbauer spectrum.

Chemical Analysis by Nuclear Methods Edited by Z. B. Alfassi

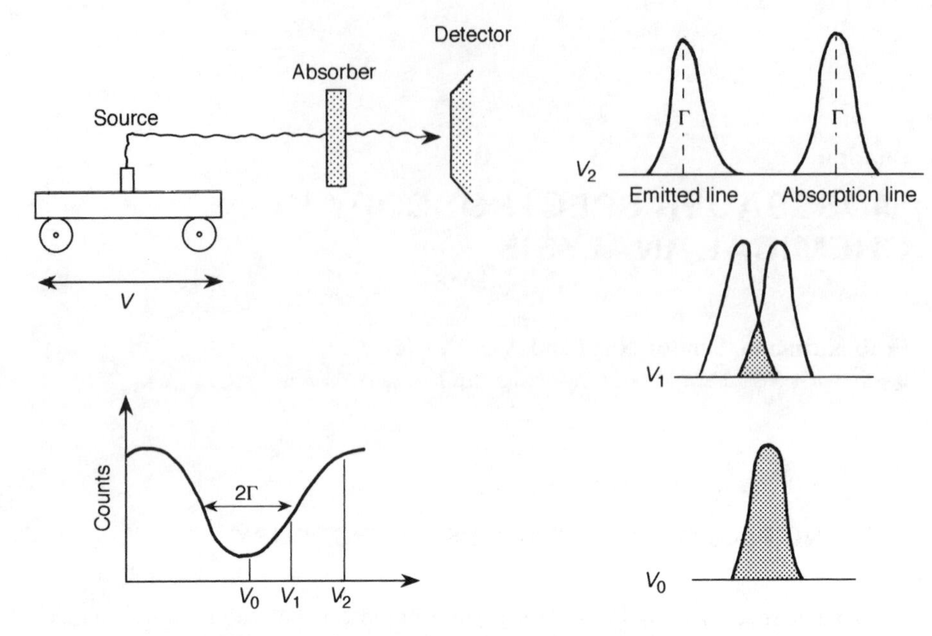

Fig. 1. Typical transmission Mössbauer experiment.

1.1 Mössbauer parameters

From the analytical point of view, the main parameters (P) which can be derived from the Mössbauer spectrum are as follows: Mössbauer–Lamb factor (f), line shift (δ), quadrupole splitting (ΔE_Q), magnetic splitting (ΔE_m), line width (Γ), and intensity (I).

Such a Mössbauer parameter can be represented as a function of a number of externally adjustable physical quantities such as temperature (T), pressure (p), external magnetic field (H), polar angles of sample relative to the direction of the γ-ray (θ, ϕ), frequency of h.f. field (ν), etc.:

$$P = (P(T, p, H, \theta, \phi, \nu, \ldots)$$

Some of the externally adjustable physical parameters have standard values. For instance $T^\circ = 293\,\mathrm{K}$, $p^\circ = 1\,\mathrm{bar}$, $H^\circ = 0$.

The main Mössbauer parameters are summarized in Fig. 2.

1.2 Dependence of the Mössbauer parameters on external physical parameters

The *temperature dependence* of Mössbauer parameters is very important for the analysis of samples, owing to the relatively easy measurement of temperature dependence.

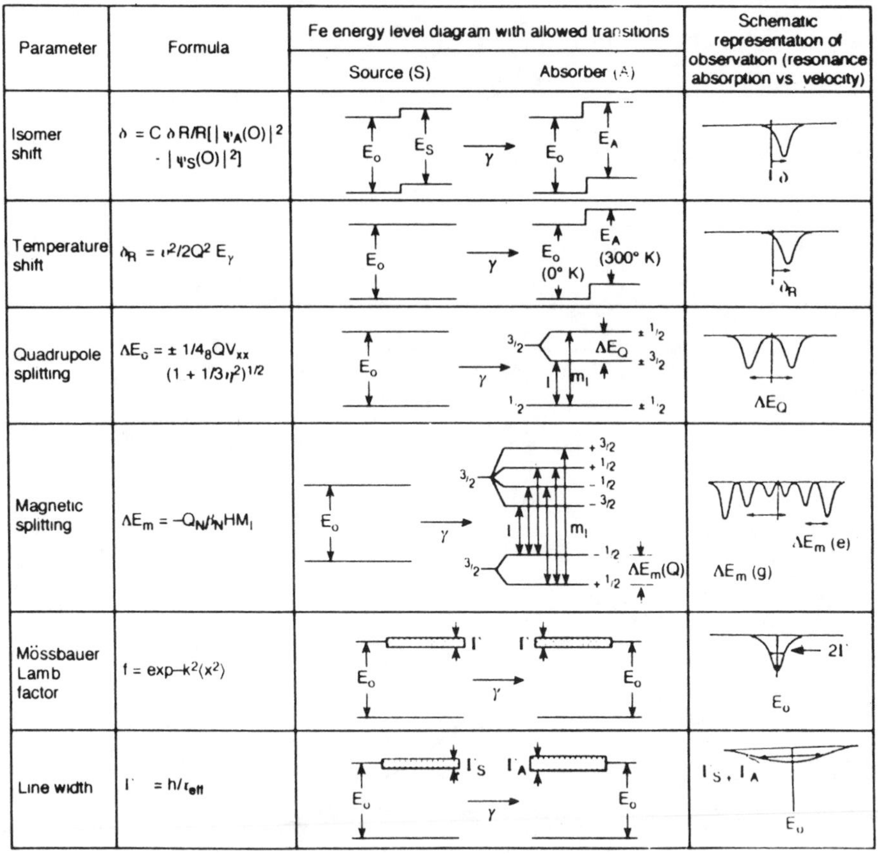

Parameter	Formula	Fe energy level diagram with allowed transitions: Source (S)	Fe energy level diagram with allowed transitions: Absorber (A)	Schematic representation of observation (resonance absorption vs velocity)
Isomer shift	$\delta = C \, \delta R/R[\lvert \psi_A(0) \rvert^2 - \lvert \psi_S(0) \rvert^2]$	E_0, E_S → γ	E_0, E_A	δ
Temperature shift	$\delta_R = v^2/2c^2 \, E_\gamma$	E_0 → γ	E_0 (0° K), E_A (300° K)	δ_R
Quadrupole splitting	$\Delta E_Q = \pm 1/4 e^2 Q V_{zz} (1 + 1/3 \eta^2)^{1/2}$	E_0 → γ	$3/2$, $\pm 1/2$, ΔE_Q, $\pm 3/2$, m_I, $1/2$, $\pm 1/2$	ΔE_Q
Magnetic splitting	$\Delta E_m = -g_N \mu_N H M_I$	E_0 → γ	$3/2$: $+3/2$, $+1/2$, $-1/2$, $-3/2$; m_I; $1/2$: $-1/2$, $+1/2$; $\Delta E_m(Q)$	ΔE_m (e), ΔE_m (g)
Mössbauer Lamb factor	$f = \exp -k^2 \langle x^2 \rangle$	Γ, E_0 → γ	Γ, E_0	2Γ, E_0
Line width	$\Gamma = h/\tau_{eff}$	Γ_S, E_0 → γ	Γ_A, E_0	$\Gamma_S + \Gamma_A$, E_0

Fig. 2. The main Mössbauer parameters.

The temperature dependence of the *line shift* (see Fig. 3) is characterized by that of the second order Doppler shift. The temperature dependence of the *quadrupole splitting* arises from the temperature dependent population of the different valence levels if the valence contribution is not negligible in comparison with the electric field gradient. The temperature dependence of quadrupole splitting is illustrated in Fig. 4 for $Cd_3\,[Fe(CN)_6]_2$.

The temperature dependence of the *Mössbauer–Lamb factor* [16, 17] (see Fig. 5) can be calculated from the Debye model [19] of solids:

$$f(T) = \exp\left\{ -\frac{3E_R}{2k\Theta_D}\left[1 + 4\left(\frac{T}{\Theta_D}\right)^2 \int_0^{\Theta_D/T} \frac{x\,dx}{e^x - 1} \right]\right\} \tag{1}$$

where $E_R = E_0^2/mc^2$ is the recoil energy of the nucleus.

The temperature dependence of the *magnetic splitting* depends on the nature of

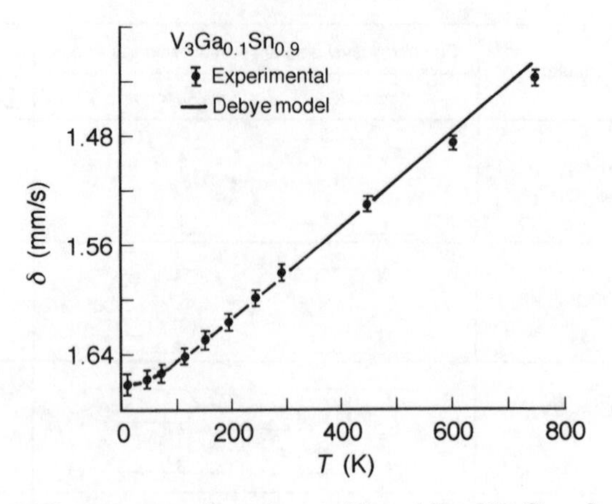

Fig. 3. Temperature dependence of line shift of $V_3Ga_{0.1}Sn_{0.9}$.

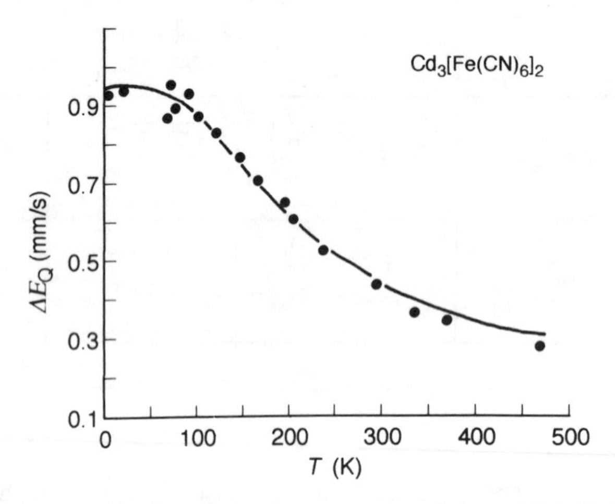

Fig. 4. Temperature dependence of quadrupole splitting of $Gd_3(Fe(CN)_6)_2$.

the magnetic interaction of the investigated materials. In the case of a simple ferromagnetic interaction, for example, when the Bloch law [19] is valid for the field, we get

$$\Delta E_m(T)/\Delta E_m(0) = 1 - CT^{3/2} \tag{2}$$

where C is a factor including a number of physical quantities.

In the case of some paramagnetic materials exhibiting magnetically split Mössbauer spectra, the magnetic splitting is due to the high paramagnetic spin relaxation time τ_{SR} relative to the lifetime (τ_M) of the excited state of the

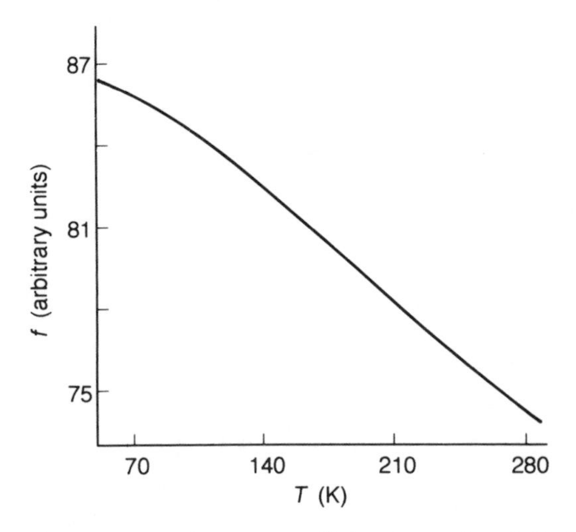

Fig. 5. Temperature dependence of Mössbauer–Lamb factor.

Mössbauer nucleus. In this case, temperature dependence is governed by the spin–lattice relaxation temperature of the material [20].

The temperature dependence of the *line width* caused by the diffusive motion of resonant nuclei can be expressed either by

$$\Delta\Gamma = \exp(-E_{\text{act}}/kT)$$

or by

$$\Delta\Gamma = \frac{2\hbar}{\tau_0}\left[1 - \int e^{i\mathbf{kr}}\, h(\mathbf{r})\, dv\right] \tag{3}$$

where E_{act} is the activation energy, $\mathbf{k}$ is the wave vector of the photon, $h(\mathbf{r})$ is the probability density of finding an atom after one jump at point $\mathbf{r}$ relative to the origin of the jump, and τ_0^{-1} is the jump frequency.

The pressure dependence of the Mössbauer parameters can be obtained from the pressure dependence of physical quantities determining these parameters.

The pressure dependence of the *line shift* [15] is illustrated in Fig. 6.

The *external magnetic field dependence* of the Mössbauer parameters is very important because, by playing with the external magnetic field, one often obtains better resolution of the spectral lines.

By applying an external magnetic field one can produce sources of polarized ^{57}Fe γ-rays. This can be done, for instance, by embedding the ^{57}Co parent nuclide in a ferromagnetic host (e.g. α-Fe) and by magnetically saturating the host material with an appropriate external field. This source will emit six lines (instead of just one) which, when overlapping with a simple six-line absorber hyperfine pattern, will lead to 36 possible absorption lines (instead of six).

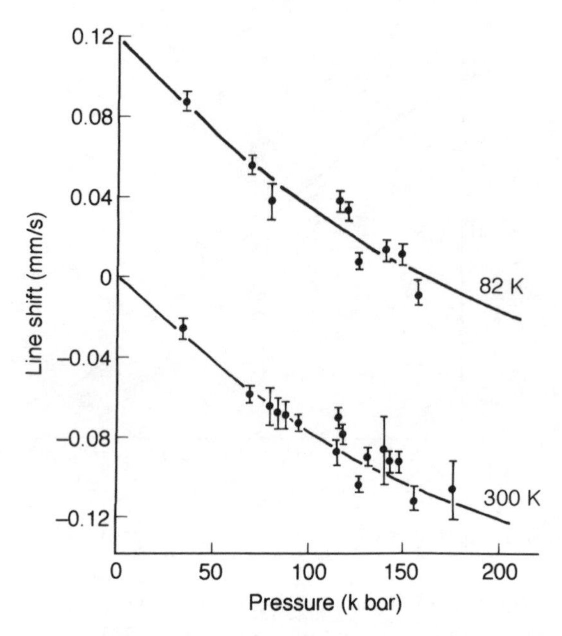

Fig. 6. Pressure dependence of second order Doppler shift.

The *effective magnetic field strength* H_{eff} measured at the nucleus consists of two terms:

$$H_{eff} = H + H_{int} \tag{4}$$

where H is the external magnetic field and H_{int} is the internal magnetic field which, again, consists of a number of components of different origin [13, 21].

The use of an external magnetic field makes it possible to determine the sign of the quadrupole splitting in the case of ^{57}Fe and ^{119}Sn spectroscopy of non-magnetic materials from the induced magnetic pattern (see Fig. 7).

The effect of external magnetic field on the magnetic splitting of magnetic materials depends on the nature of magnetic coupling (i.e. whether the material is ferromagnetic, ferrimagnetic or antiferromagnetic) [8].

Radio-frequency electromagnetic radiation can have two kinds of effect on the Mössbauer spectrum [22]. It can collapse the Mössbauer pattern, but it can also produce side bands in the spectrum.

The angular dependence of *line intensity* is connected with the angular dependence of the transition probability of the different nuclear transitions.

The angular dependence of the intensity of magnetically split ^{57}Fe or ^{119}Sn spectra is given by the following expressions

$$I_{1,6}(\theta,\varphi) = \alpha\,(1 + \cos^2\theta) \tag{5}$$

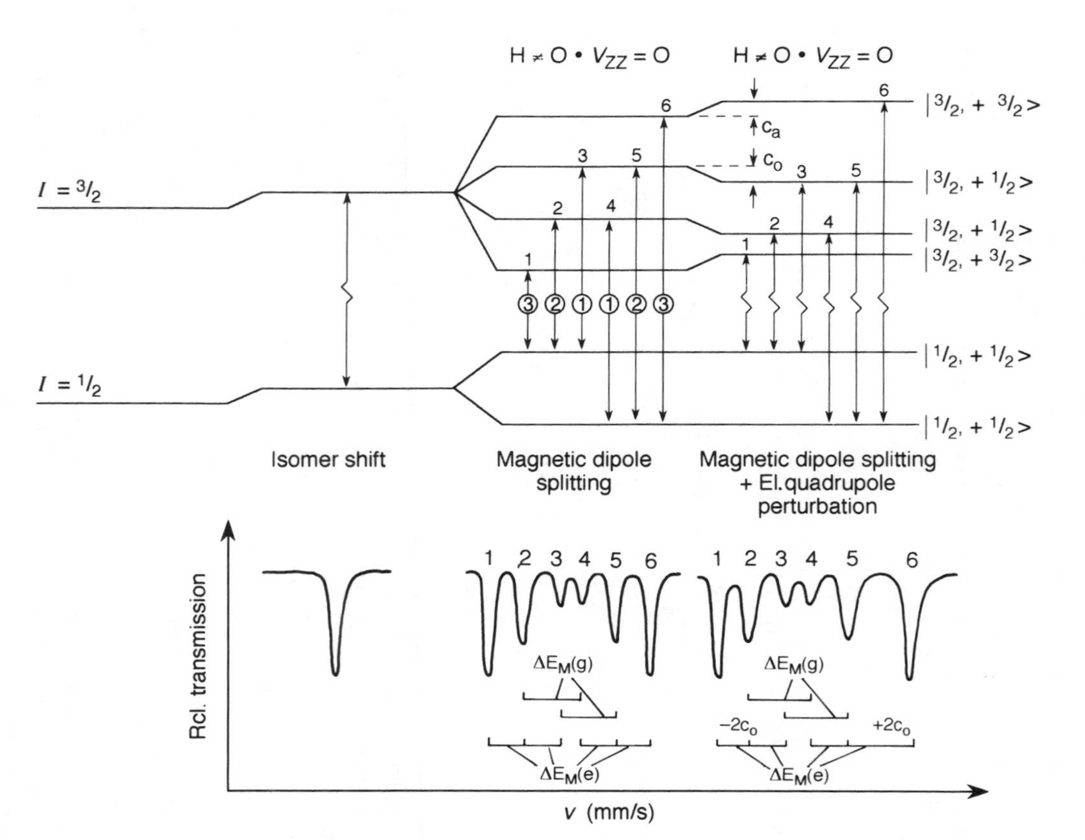

Fig. 7. Magnetic splitting without and with quadrupole interaction.

$$I_{2,5}(\theta,\varphi)=\alpha' \sin^2\theta \tag{6}$$

$$I_{3,4}(\theta,\varphi)=\alpha'' (1+\cos^2\theta) \tag{7}$$

where the polar angles θ and φ define the orientation of the wave vector **k** of the γ-ray when the z-axis is taken parallel to the direction of the magnetic field; α, α' and α'' are constants involving also the Clebsch–Gordon coefficients of the magnetic dipole nuclear transition for $\Delta m = -1, 0, 1$ [23]; the intensity subscripts denote the line number of the sextet.

The relative line intensities within the resultant sextet of a sample with a completely random distribution of magnetic field average out in the following way:

$$I_1:I_2:I_3:I_4:I_5:I_6=3:2:1:1:2:3 \tag{8}$$

In the case of magnetic anisotropy, the relative intensity of the second or the fifth line compared with that of the first or the sixth line is [8]

$$I_{2,5}/I_{1,6}=\frac{4\ \sin^2\theta}{3\ (1+\cos^2\theta)} \tag{9}$$

In the case of a pure quadrupole spectrum the intensity ratio of the doublet is [11]

$$I_2/I_1=\frac{\int_0^\pi 3(1+\cos^2\theta)\ h(\theta)\ f(\theta)\ \sin\theta\ \mathrm{d}\theta}{\int_0^\pi (5-3\cos^2\theta)\ h(\theta)\ f(\theta)\ \sin\theta\ \mathrm{d}\theta} \tag{10}$$

where θ is the angle between the direction of the electric field gradient and the direction of the γ-ray propagation, $h(\theta)$ is the probability density of the angle θ, and $f(\theta)$ gives the angular dependence of the Mössbauer–Lamb factor.

In a well mixed powder sample, where there is a random distribution of electric field gradient, we expect a doublet with equal line intensities:

$$I_2/I_1=1$$

In anisotropic crystals the amplitudes of atomic vibrations are essentially the function of vibrational direction. This can result in asymmetric quadrupole doublets.

Another possible reason for the asymmetry of quadrupole doublets in ^{57}Fe or ^{119}Sn spectra is a non-uniform $h(\theta)$ density function in the case of single crystals or textured materials [8].

The effect of absorber thickness on the *line intensity* is given by equations (19) and (20) through the effective thickness $\tau = v_a f_a \sigma_0 d$, where σ_0 is the resonant cross-section.

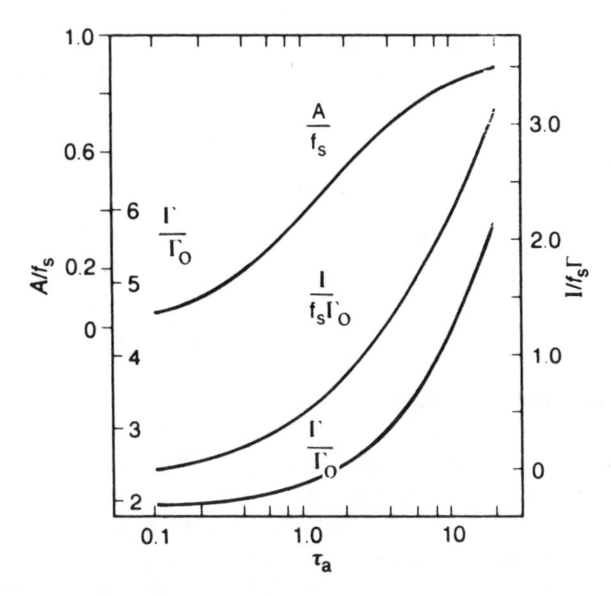

Fig. 8. The dependence of intensity, amplitude and line width on the effective thichness

The thickness dependence of *line width* (Γ) for Lorentzian lines is given [13] by the following equation involving the modified Bessel functions I_0 and I_1:

$$\Gamma = \Gamma_0 \frac{\tau \exp(-\tau/2)\ [I_0(-\tau/2) + I_1(\tau/2)}{1 - \exp\ (-\tau/2)\ I_0(\tau/2)} \tag{11}$$

The line width can be approximated by the following formulae:

$$\Gamma = \Gamma_0\,(2 + \tau/4 + \tau^2/96 - \tau^3/64) \qquad \text{if } \tau < 2$$

$$\Gamma = 2\Gamma_0\tau/(\sqrt{\pi\tau} - 1) \qquad \text{if } \tau > 2$$

Figure 8 shows the intensity, amplitude and line width as functions of effective thickness.

The dependence of Mössbauer parameters on geometric arrangements is mainly due to *cosine smearing* of the velocity (Fig. 9). Since the Doppler energy shift is given by

$$\Delta E = E_0(v_0/c)\cos\theta \tag{12}$$

(where θ is the angle between the γ-ray propagation and the direction of the relative velocity), smearing can always be observed with the line shape, line width and line shift [24] even under 'ideal' conditions when we use an anisotropic point source and a circular absorber placed coaxially to the source displacement as illustrated by Fig. 9.

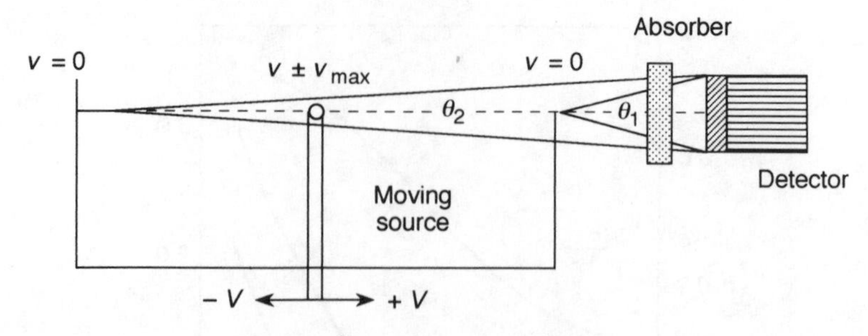

Fig. 9. The origin of 'cos' effect.

1.3 Measurement of Mössbauer spectra

Mössbauer spectra can be recorded in different ways (Fig. 10). In the most frequently used transmission technique, the γ-rays are counted after passing through an absorber; in the scattering technique the re-emitted radiation produced by an earlier Mössbauer resonance is detected.

With the scattering technique it is possible to detect γ-rays (backscattering γ-ray Mössbauer spectroscopy), X-rays (X-ray Mössbauer spectroscopy) or conversion electrons (conversion electron Mössbauer spectroscopy, CEMS), which radiations have characteristic penetration depths. Consequently, the

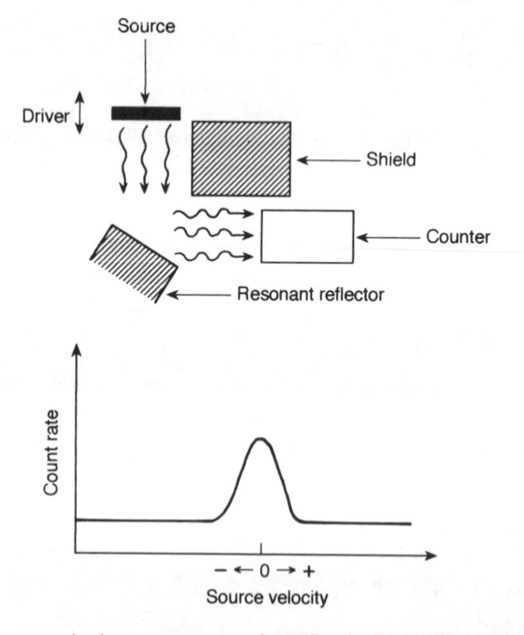

Fig. 10. Geometrical arrangement of a reflection Mössbauer measurement.

surface of the samples can be investigated at different depths. Another important advantage of the scattering Mössbauer technique is that this method provides a non-destructive way of material testing.

In the ^{57}Fe scattering experiment the 14.4 keV γ-rays penetrate to $\approx 20-30\ \mu m$, whereas the conversion electrons come from a depth of $\approx 0.1\ \mu m$. Depth selective conversion electron Mössbauer spectroscopy (DCEMS) makes it possible to get analytical information from a layer that is ≈ 0.2 nm thick situated at a readily measurable depth between 0 and 100 nm. This can be achieved by a CEM spectrometer placed in an electron spectrometer which is used for the separation of the electrons coming from different depths (i.e. having different energies).

A number of techniques also use Mössbauer spectroscopy combined with other methods [25].

Mössbauer photometry is a useful method with a principle similar to that of optical polarimetry. In this case the polarization and the analysis can be performed by using appropriately oriented external magnetic fields. Using ^{57}Fe polarization spectroscopy, instead of six lines, 36 lines are observed with magnetic materials [25].

In the case of the *Mössbauer thermal scan method*, only the sepectrum point belonging to zero velocity is measured, at different temperatures (without Doppler moving). By this simple technique transition temperatures can be detected.

A typical Mössbauer spectrometer is shown schematically in Fig. 11.

The transmitted or scattered radiation is detected mainly by scintillation or proportional counters. The signals of the detector are amplified and gated by a single channel analyzer set to transmit the Mössbauer transition energy range. These signals are fed to a multichannel analyzer working in multiscaler regime. The opening of channels is synchronized with the drive system. The drive is

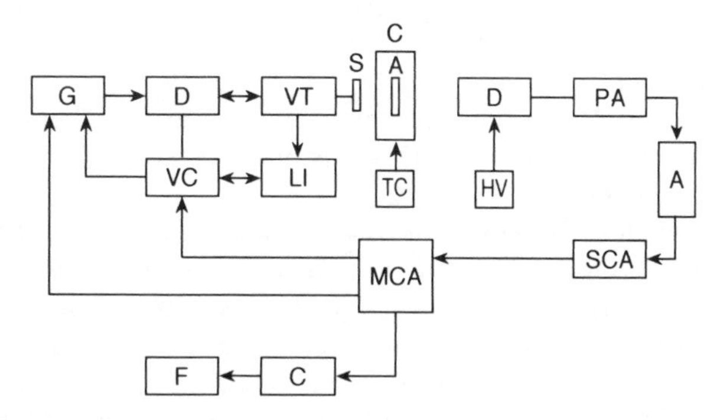

Fig. 11. Block schematic arrangement of a Mössbauer spectrometer. S, source; VT, vibrator; D (left), drive; A (centre), absorber; D (right), detector; PA, preamplifier; A, amplifier; SCA, single channel analyser; MCA, multi-channel analyser; C (bottom), computer.

usually a loudspeaker-type electromechanical vibrator which moves the source relative to the absorber periodically (with frequency 10–100 Hz) according to a given waveform. In the most typical case a triangular waveform is used, providing constant acceleration for the movement in the evaluation, the triangular waveform of velocity (with a maximum of the order of several mm/s) is approximated by a stepwise function of time, at each step of which (50–100 μs) the velocity is considered to be constant. An appropriate synchronization ensures a one-to-one correspondence between the velocity and the channels of the analyzer. The Mössbauer spectra are usually recorded in 512 channels in each of which 10^5–10^7 counts are collected in a period of several hours.

The Mössbauer spectrum is conventionally given by the transmitted or reflected intensity (in counts) as a function of the Doppler velocity (in mm/s). However, the Mössbauer spectrum obtained from the analyzer is in the form of counts as a function of the channel number. The correspondence between channel number and velocity is determined by velocity calibration.

The *velocity calibration* of the Mössbauer spectrometer is performed either by measuring the spectrum of standard materials or by using calibrator instruments which measure the absolute velocity of the source relative to the absorber.

In present day Mössbauer equipment, a personal computer including an analyzer card (PCA card) is used for the pulse height and multichannel analysis as well as for directing the measurement and evaluating the recorded spectra.

The Mössbauer source is a solid material containing (or rather continuously generating) the excited-state nuclei of the studied nuclide and thus emitting the γ-radiation characteric of the nuclear transition in question. Although Mössbauer transition has been observed with more than 40 nuclides (Fig. 12), only a few of them (e.g. ^{57}Fe: 2%, ^{119}Sn: 6%, and ^{151}Eu: 3%) are widely used—those having a relatively long lived parent nuclide making it possible to produce a source lasting for at least a couple of months. The commercial source for the 14.4 keV transition of ^{57}Fe is ^{57}Co diffused into a thin metallic matrix (Pd, Pt or Rh) in which the Mössbauer line is sufficiently narrow (FWHM = 0.21 mm/s). The usual activity of such sources is 200 MBq to 10 GBq.

Normally the sample to be studied is used either as an absorber or a scatterer. The absorber has to contain the ground-state nuclei of the nuclide responsible for the Mössbauer radiation of the source.

The Mössbauer absorbers used for transmission experiments can be sheets or powders of solids as well as frozen solutions. The ideal sample has uniform thickness, circular shape, texture-free homogeneous distribution of material and optimum surface density of the Mössbauer nuclide as well as an optimum thickness [26].

For example, in the case of a natural α-iron foil 7 mg/cm^2 of natural iron content is the optimum (i.e. a linear thickness of 9 μm); however, 0.2 mg/cm^2 samples in which the iron content is merely 5% of the optimum can also be measured, with some difficulty. From the point of view of Mössbauer resonant

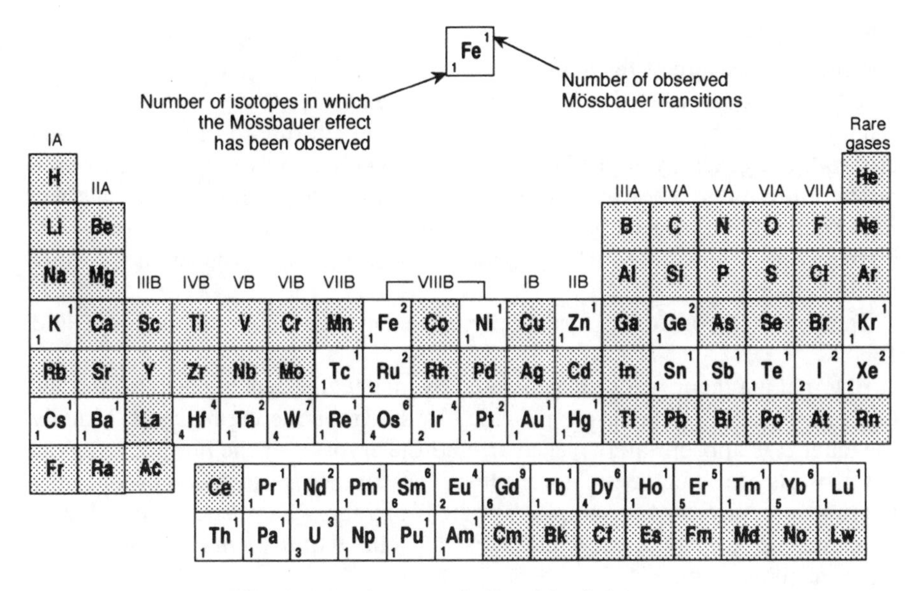

Fig. 12. Mössbauer periodic table of elements.

transition, only the ^{57}Fe content is important (which has a 2.19% abundance in natural iron), therefore a significant lowering of minimum iron content (down to a factor of 1/45) can be achieved by enrichment with the ^{57}Fe isotope. Thus, e.g., the spectrum of an Al sample containing 20 ppm of ^{57}Fe can be easily measured by a modern commercial spectrometer.

Powders to be measured are usually spread out with uniform thickness between thin foils of Mylar or Al. Sometimes, in order to avoid vibrational effects, the powder is embedded in a low atomic absorption material (e.g. paraffin wax) containing no resonant nuclide.

In the case of absorbers made from frozen solutions, the most important requirement of sample preparation is to cool the solution at a rate higher than 8 K/s in order to preserve the original structure.

For samples used in scattering experiments, the only requirement is that the available surface density of the Mössbauer nuclide be high enough.

Materials can also be analysed by *source experiments* provided that a radiating Mössbauer nuclide is present in the sample to be investigated. In this case the sample can contain one or two orders of magnitude less of the Mössbauer nuclide than the corresponding sample used as an absorber. This can be a very important point when the Mössbauer nuclide—an impurity—is introduced only in order to probe the material.

Different kinds of cryostats and furnaces are used to determine the temperature dependence of Mössbauer spectra. A very simple cold finger cryostat can be designed to measure at the temperature of solid CO_2 or liquid nitrogen. In this case the previously cooled sample can be immediately put into the cryostat. This is imporant, e.g., with frozen solution samples. Temperature controlled evacuated cryostats are generally applied between the millikelvin and a few hundred kelvin temperature range. They have to be supplied with special cooling materials (e.g. liquid helium or nitrogen). Nowadays closed circuit *refrigerator cryostats* operating between 12 and 400 K are more frequently used because they do not require liquid helium.

Temperature controlled furnaces operate between room temperature and 1500 K.

External magnetic fields are generated by magnets producing field from a few Oe to 150 kOe. High fields are provided by superconducting magnets. Usually the magnet is appropriately placed around the cryostat or the furnace.

Cryostats are sometimes combined with *pressure cells* which can produce the required pressure on the sample. The high-pressure chambers are capable of pressures up to 200 kbar while satisfying the requirements of Mössbauer measurements.

As to instrumentation designed for varying a parameter other than those mentioned above, reference should be made to the literature [7].

2 ANALYTICAL INFORMATION FROM MÖSSBAUER SPECTRA

In this part we shall give an insight into the topic of how to use Mössbauer spectroscopy for analytical purposes.

2.1 The fingerprint method

As with other analytical techniques, the fingerprint method is one of the most powerful methods of qualitative analysis in Mössbauer spectroscopy. The fingerprint method operates with patterns characteristic of individual species. Typical patterns ('fingerprints') are associated with different single speices (e.g., crystallographic sites, phases, etc.) present in a given sample. The correspondence between individual patterns and species is determined by the nature of interactions serving as a basis for the analytical method.

In order to understand better this correspondence in the case of Mössbauer analysis, let us first discuss the question of what physical or chemical entities are reflected by a simple Mössbauer pattern.

As mentioned in the previous section, the Mössbauer spectrum is affected by the hyperfine interactions existing between the nucleus and the electrons. These interactions are monitored at the site of the Mössbauer-active nucleus. Since the hyperfine interactions for a given nucleus are mainly determined by the electronic density and the inhomogeneous electric field, as well as by the effective magnetic field at the site of nucleus, the Mössbauer spectrum patterns differ from site to site where these physical quantities are different. Thus, e.g., a single-phase material can be characterized by a complex Mössbauer spectrum superimposed from a number of simple Mössbauer spectra belonging to resonant atoms experiencing different hyperfine interactions.

We will refer to a family of Mössbauer nuclei experiencing the same hyperfine interaction as a *micro-environment*. By this definition, a simple Mössbauer pattern is always associated with such a micro-environment of the Mössbauer nuclei in the studied sample.

2.2 Pattern analysis

The basic task of the analysis is to identify the individual physical or chemical species from the corresponding patterns present in the spectrum. Ideally, this can be done if we know the exact correspondence between *patterns* and *species*.

However, such a one-to-one correspondence between species of atoms and individual patterns can be non-existent at the given set of externally adjusted physical parameters at which the Mössbauer spectrum is recorded. However, when the whole range of these parameters is considered, we may find points in the space of parameters at which only one pattern is associated with one species and vice versa, and thus we can get round the problem of ambiguity.

From the analytical point of view we can classify the Mössbauer patterns in the following way:

- spontaneous pattern
    - elementary pattern
    - superimposed pattern
- induced pattern
- transformed pattern

The *spontaneous pattern* is the Mössbauer spectrum obtained at a given set of externally adjusted parameters (usually under standard conditions). The spontaneous pattern can be a simple spectrum (elementary pattern) reflecting only a unique hyperfine interaction or a complex spectrum (superimposed pattern) which consists of a number of subspectra.

The *elementary pattern* is the spectrum of one type of micro-environment. The superimposed pattern is the superposition of elementary patterns.

The *induced pattern* is the Mössbauer spectrum obtained under conditions other than standard (or the ones selected for measuring the spontaneous pattern).

In this case, the differences between the induced and the spontaneous pattern can make an important contribution to the analysis. We call the pattern induced because, on the one hand, we 'induce' it by changing the external conditions and, on the other hand, because a split pattern can be 'induced' in the spectrum by applying external fields or by lowering the temperature (see Section 2.1).

The *transformed pattern* is obtained from the measured Mössbauer spectrum by mathematical (e.g., Fourier) transformation. The magnetic hyperfine field distribution and the quadrupole splitting distribution are transformed patterns. The transformed pattern can provide a better basis for the analysis.

Figure 13 illustrates schematically the relationship between the different patterns and the object of the analysis. As shown in Fig. 13, we can classify these objects as species and subspecies including phases with different crystallographic sites. Note that the same crystallographic site may represent more than one micro-environment owing to variations in the hyperfine interactions (e.g., different electronic and spin density, defect arrangement, etc.). Each micro-environment has an elementary pattern. The pattern of an individual phase is a superimposed pattern consisting of elementary patterns. Generally, the Mössbauer spectrum of a multi-phase material is a superposition of superimposed patterns. The spectrum of a material taken under standard conditions can be considered a *spontaneous pattern.*

As an example, let us consider a steel sample in which ferrite and austenite phases are present. The spontaneous pattern (^{57}Fe Mössbauer spectrum) of the sample is a complex spectrum that is the sum of subspectra (Fig. 14). The

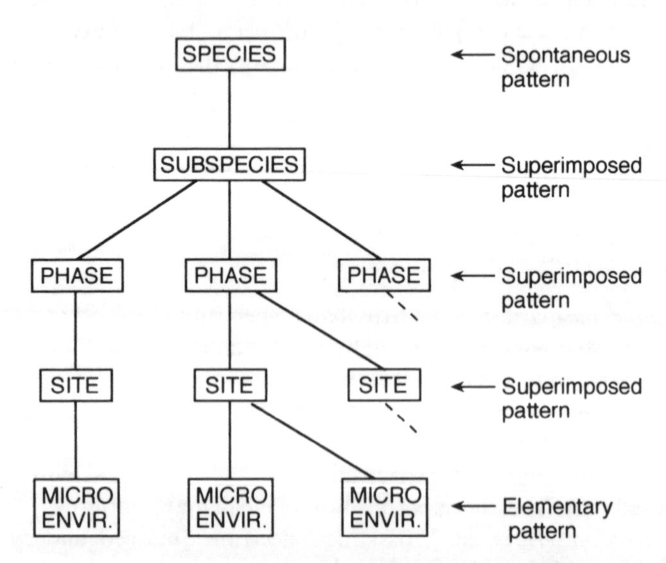

Fig. 13. Relation between the different patterns and the objects of analysis.

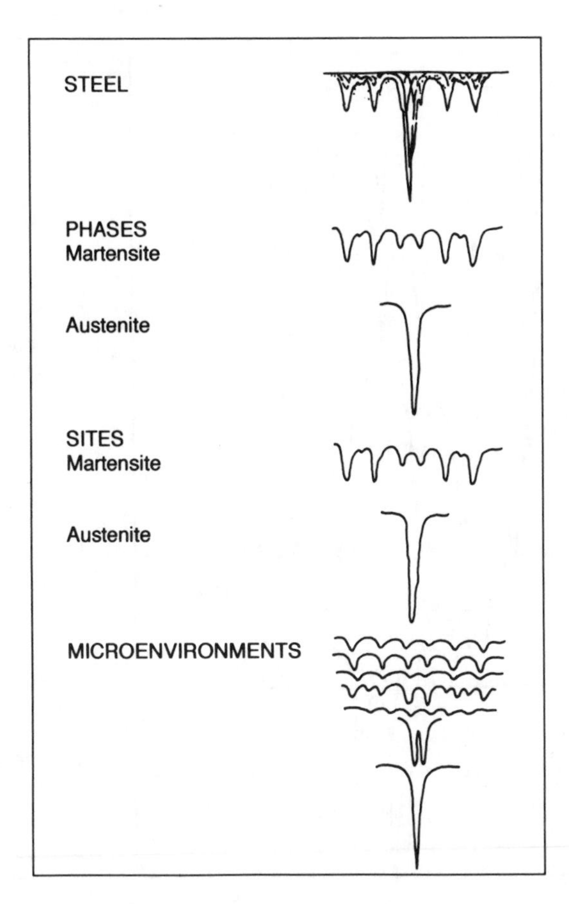

Fig. 14. The spontaneous pattern.

subspectra of the ferromagnetic ferrite and the paramagnetic austenite are also superimposed from elementary patterns because of the effect of alloying elements (a more detailed explanation will be given later). In some cases, when the characteristic subspectra of components are clearly distinguishable in the spontaneous pattern, the qualitative analysis can be performed by evaluating the spontaneous pattern.

In general, however, the subspectra overlap each other in the spontaneous pattern. When the exact decomposition of the spectrum becomes ambiguous, it is still possible to get further analytical information from an *induced pattern*.

For example, the room temperature spontaneous pattern of the akaganeite is a paramagnetic spectrum (Fig. 15). However, the induced pattern taken at different temperatures (which temperatures are below the magnetic transition

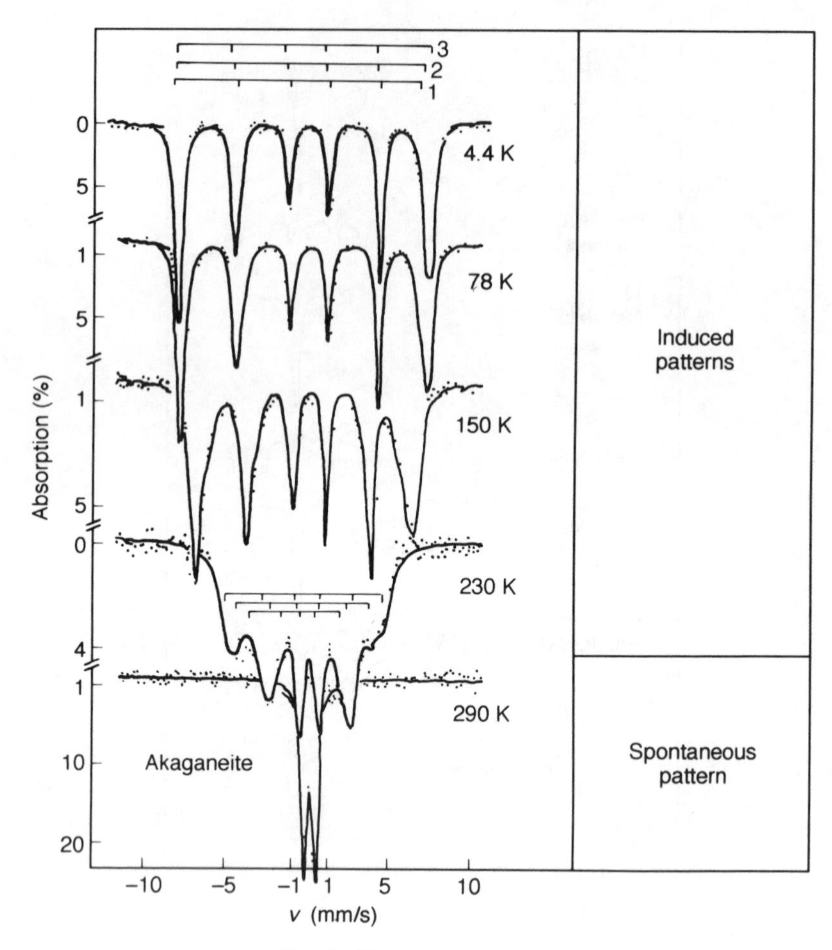

Fig. 15. The induced pattern.

point of akaganeite) exhibit magnetically split subspectra. Now, based upon the evaluation of the induced pattern, the analytical problem can be solved more easily.

In some cases the spontaneous pattern consists of hundreds of elementary patterns. This is the typical situation with microcrystalline and amorphous systems. The *transformed pattern* can help get more information about the short-range ordering (by characterizing some of the most probable arrangements); thus it can enhance analytical applicability.

For example, the spontaneous pattern of the $Fe_{83}B_{17}$ amorphous alloy (Fig. 16(c)) exhibits a very broad magnetically split spectrum which can be considered as the superposition of a large number of subspectra belonging to iron

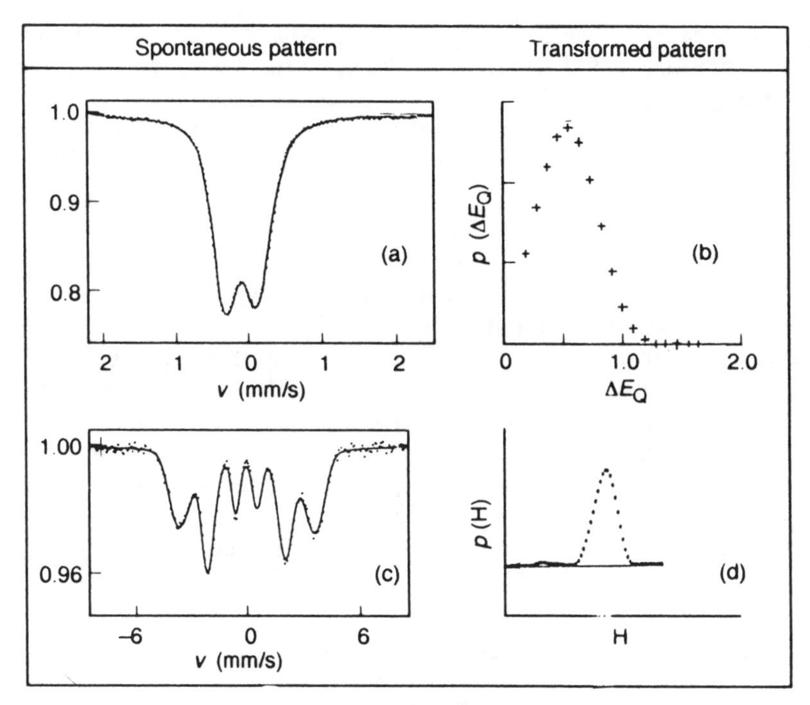

Fig. 16. The transformed pattern.

atoms having different surroundings in this alloy. The hyperfine field distribution (transformed pattern) shows three main peaks representing the main short-range orders. Figure 16(d) illustrates that a better analysis of structural changes can be performed using the hyperfine field distributions than on the basis of the corresponding unprocessed spectra.

2.3 Spectrum decomposition

The Mössbauer spectra encountered in practice are usually fairly complex. In order to get any analytical information from such a spectrum it has to be decomposed into elementary patterns first.

In the practice of Mössbauer spectroscopy, the Mössbauer parameters are determined by the computer evaluation of spectra. The Mössbauer parameters are derived from the line parameters (line position, line width, line intensity) during the fitting process.

For the spectral evaluation an *a priori* knowledge of both the line shape function and the connection between the Mössbauer parameters and the line parameters is necessary.

The Mössbauer *line shape* used in the fitting process depends on the sample and on the computing facilities that are available to the spectroscopist. The most precise way of fitting is based on the *transmission integral* [equation (9)]. Relatively fast and large computers are needed to carry out such a calculation.

Very often, however, *Lorentzian line shape* is considered:

$$L(v) = \frac{\Gamma^2}{4} \frac{A}{(v-\delta)^2 + \Gamma^2/4} \tag{13}$$

where δ is the position, Γ is the width and A is the amplitude of the peak. This can be a good approximation if the absorber is sufficiently thin (the effective thickness τ should be less than 1) and if the number of overlapping lines is not very high (fewer than 10). It is a great technical advantage of this method that personal computers can be applied for such calculations.

For the analysis of materials (e.g. amorphous metals) giving spectra that consist of a large number of overlapping lines, the product of Lorentzian and Gaussian curves is used to fit the lines [27].

The spectrum fitting based on a model function (e.g., Lorentzian) is done by the *least squares method.* In this procedure the sum

$$\chi^2 = \sum_{i=1}^{M} \frac{1}{N_i} [N_i - f(v_i; b, p_1, p_2, \ldots p_n)]^2 \tag{14}$$

is minimized, where f is the model function determined by the vector of the baseline (b) and peak parameters ($p_1, \ldots p_n$); n is the number of peaks, N_i and v_i are the number of counts and the velocity belonging to the ith channel and M is the total number of channels.

In the thin absorber approximation, an inhomogeneous linear transformation is used to make connection between the physical parameters a_j (isomer shift, quadrupole splitting, magnetic splitting, etc.) and the individual peak parameters p_k (position, width and intensity of each peak)

$$a_j = \sum_{k=1}^{n} t_{jk} \, p_k + c_j \tag{15}$$

where the constraints are determined by the elements t_{jk} and c_j of the matrix T and the constant vector **c**, respectively.

The *a priori* knowledge of the matrix T and the vector **c** is an essential condition of spectrum evaluation.

From the point of view of analytical applications, the *a priori* determination of the *T* matrix is of fundamental importance. Namely, this transformation enables the experimenter to specify a set of peaks as a pattern representing one particular subspecies in the sample.

In other words, in order to be able to evaluate a spectrum, we must have some kind of a hypothesis about the relationships among the spectral lines. Then the validity of this hypothesis can be verified by the fitting itself. The hypothesis must be based upon our previous knowledge about the material to be studied.

Looking at the spectrum, we have to determine first the total number of spectral lines. This is not so easy in the case of overlapping spectra. The incorrectly chosen number of lines makes the analysis itself incorrect, in contrast to some other methods where such a problem is of less importance.

If the available information is insufficient, a systematic successive trial can lead to a possible set of lines composed from a minimum number of lines still giving a satisfactory fit. However, in such a case, the correct grouping of the lines can rarely be obtained from the spontaneous pattern alone. It is necessary to compare the results with the evaluations of induced patterns and with spectra of standards as well as with theoretical considerations.

Any preliminary information about the sample (history, results obtained with other methods) can help identify prospective subspecies whose pattern can be considered when setting up the T matrix. Such subpatterns can sometimes be recognized by visual inspection. If this is the case the *stripping technique* can help further explore the components present. This technique *involves* the recognized and appropriately scaled subspectrum (or a reference spectrum) is subtracted from the recorded spectrum. A new component can then be more easily discovered in the residual spectrum than in the original one. The stripping of the spectrum is continued by the subtraction of the newly found component, and the iteration procedure is continued until the residuals become satisfactorily small. Finally, every spectral component can be assigned.

The assignment of the subspectrum to one particular subspecies of the material is based on the comparison of the decomposed subpattern with the reference spectrum. The reference spectrum can be obtained by measuring a standard material or from data published in the literature. In some cases the reference spectrum can be constructed theoretically (computer simulation).

To obtain the reference spectrum from the literature is one of the most convenient solutions. Although the parameters of a few thousand materials can be found in the Mössbauer Reference and Data Index [7], the majority of these represent spontaneous patterns and relatively few induced pattern parameters are published.

A special way of spectrum evaluation involves obtaining and analyzing transformed patterns. For a thin absorber, the spectrum $s(v)$ can be considered as the convolution of a distribution function $p(v)$ with a Lorentzian function $L(v)$:

$$s(v) = \int_{-\infty}^{\infty} p(v')\, L(v - v')\, \mathrm{d}v' \tag{16}$$

One of the methods [28–30] for obtaining the hyperfine field or quadrupole splitting distribution is the Window method [28], based on a Fourier technique. The *hyperfine field distribution*

$$p(H) = \sum_{i}^{N} a_i \left[\cos \frac{H_i - H_{\mathrm{min}}}{H_{\mathrm{max}} - H_{\mathrm{min}}} - (-1)^{-i} \right] \tag{17}$$

can be obtained by fitting the spectrum (H_{min} and H_{max} are the minimum and maximum hyperfine fields considered). Although it is very time consuming to obtain the correct distribution function, this method has a very important advantage, namely, no *a priori* knowledge is required to get the transformed pattern.

2.4 Quantitative analysis

In Mössbauer spectroscopy, similarly to other spectroscopic methods, quantitative analysis is based upon the determination of peak areas.

In the case of an ideally thin source and a not so ideally thin absorber (with Mössbauer–Lamb factors f_s and f_a, respectively), the line shape can differ from the simple Lorentzian. The probability of detecting a photon of the resonant transition behind the absorber can be given as

$$T(v) = e^{-\mu_a d}\left\{(1 - f_s) + \frac{f_s \Gamma}{2\pi}\int \exp\left[f_a v_a \sigma_a d \frac{(\Gamma/2)^2}{(E - E_0)^2 + (\Gamma/2)^2}\right] \cdot \frac{1}{(E - E_0(1 + v/c)]^2 + (\Gamma/2)^2} dE\right\} \tag{18}$$

which is known as the transmission integral, where μ_a is the electronic absorption coefficient of the absorber, v_a is the number of atoms of the resonant nuclide in unit volume of the absorber, v is the velocity of the source relative to the absorber, and d is the thickness of the absorber.

The intensity (I) of the transmission peak can be expressed as

$$I_T = \int_{-\infty}^{+\infty} \frac{T(\infty) - T(v)}{T(\infty)} dv = \frac{\pi \Gamma f_s}{2} - \tau_a \exp\left(-\frac{\tau_a}{2}\right)\left[I_0\left(\frac{\tau_a}{2}\right) + I_1\left(\frac{\tau_a}{2}\right)\right] \tag{19}$$

where τ_a is the effective thickness and I_0 and I_1 are modified Bessel functions of an imaginary argument.

The normalized Mössbauer spectrum can be expressed as

$$S(v) = (1 - b)\frac{T(\infty) - T(v)}{T(\infty)} \tag{20}$$

where b denotes the background fraction of photons (i.e. the fraction of photons that do not originate from the Mössbauer transition) detected behind the absorber. The area of the normalized spectrum is

$$I_s = (1 - b)\, I_T \tag{21}$$

Implicitly, the above equation defines the line area as an increasing function of the effective thickness (the latter being proportional to the concentration of the resonant atoms), thus making it the single most important parameter of quantitative analysis. The calculation giving the number of resonant atoms in the unit volume of the absorber (v_a) from the area is done by computer.

In the thin absorber approximation ($\tau \leq 0.1$) it can be shown that the line area is given by

$$I \approx k f_a \tau_a = k f_a \sigma_0 \, d v_a = k f_a \sigma_0 d a n \tag{22}$$

where σ_0 is the resonant cross-section, d is the absorber thickness, f_a is the Mössbauer–Lamb factor of the absorber, a is the fractional abundance of the isotope capable of resonance and n is the number of atoms of the studied element per unit volume. The constant k includes other constants and parameters involving the background and matrix effects. This equation indicates that the area of Mössbauer lines is proportional to the number of Mössbauer atoms in the thin sample.

Although the above approximation is widely used whenever Mössbauer spectroscopy serves as a tool of quantitative analysis, the relative error involved is $\approx 5\%$ even at $\tau = 0.1$ [12].

For the determination of the concentration n, all the parameters in equation (22) must be known. However, in most cases the exact values of f_a and k are unknown. In such a case, calibration graphs are made by measuring the spectra of a series of analogous samples having different known concentrations. For the same purpose internal standards can also be used.

The relative concentrations belonging to the subspecies of the sample can be more accurately determined by the Mössbauer method than the absolute values. Consequently, most applications are directed to the determination of relative concentrations. There are, however, cases when the determination of the relative concentrations of the subspecies of the resonant nuclide yields, as a spin-off, the absolute concentration of another (non-resonant) element.

3 EXAMPLES FOR ANALYTICAL APPLICATIONS

In this section a few selected analytical applications of Mössbauer spectroscopy will be given in order to show that, besides phase analysis (and in a few cases elemental analysis), this technique can provide important information on the valence state, the crystallographic, magnetic or defect site occupancy of the Mössbauer element, and the purity of the compound. It can also settle the question of whether a compound is present in another structure at the same composition, or whether the Mössbauer nuclei in polynuclear compounds are in equivalent environments. We can also get information about the location (surface, bulk, interlayers) of a given compound within the sample, or about its amorphous or crystalline character, as well as about the grain size of the sample.

However, in this section we will restrict ourselves to examples from corrosion studies, phase analysis and elemental analysis of alloys.

3.1 Corrosion

The study of corrosion in metals (mainly iron) represents a very large area of the analytical application of Mössbauer spectroscopy. This topic involves electrochemical, environmental, archaeological and fine-arts studies as well as industrial applications in the field of surface and bulk analysis [31]. In most of these applications the fingerprint technique is used to identify the compounds or species in a complex material.

Figure 17 and Table 1 show the Mössbauer spectra and parameters of metallic iron as well as some iron oxide and hydroxide compounds that are the most common components of corrosion products. It can be seen that these compounds have characteristic spontaneous patterns. The spectra of some of the species (e.g. α-iron and hematite) are elementary patterns, indicating that there is only one micro-environment of iron in any of these phases. The difference in the isomer shift shows that α-iron contains metallic iron whereas hematite has Fe^{3+}. The cubic and non-cubic symmetry of the arrangement of ligand charges results in zero and non-zero quadrupole splittings, respectively. Both materials have magnetically split spectra but they have different effective fields, owing to a difference in the polarization interactions responsible for the hyperfine fields. The difference between the parameters of α- and γ-Fe_2O_3 reflects differences in their structure.

The Mössbauer spectrum of magnetite consists of two elementary patterns belonging to Fe^{2+} and Fe^{3+} ions situated at different crystallographic positions (octahedral and tetrahedral).

The difference in the Mössbauer parameters of these phases gives good possibilities for their qualitative and quantitative analysis. To perform a complete analysis the spontaneous patterns will usually do, and in most cases no induced pattern is needed.

For example, the thermal oxidation of α-ion in an O_2 atmosphere at different temperatures can be followed by conversion electron Mössbauer spectroscopy (Fig. 18) [32]. By performing careful qualitative and quantitative analysis, the kinetics of the growth of oxide layers could be understood under different conditions [33].

Figure 19 shows transmission Mössbauer spectra belonging to cases of atmospheric corrosion that had occurred on mild steel and on the door and muffler of a car after six years. Whereas the corroded mild steel contained α- and γ-FeOOH as the main phases and a small amount of γ-Fe_2O_3 and Fe_3O_4, the door rust had γ-Fe_2O_3 as main component plus α- and γ-FeOOH [34].

As an example of electrochemical corrosion, the influence of sulfite (modeling

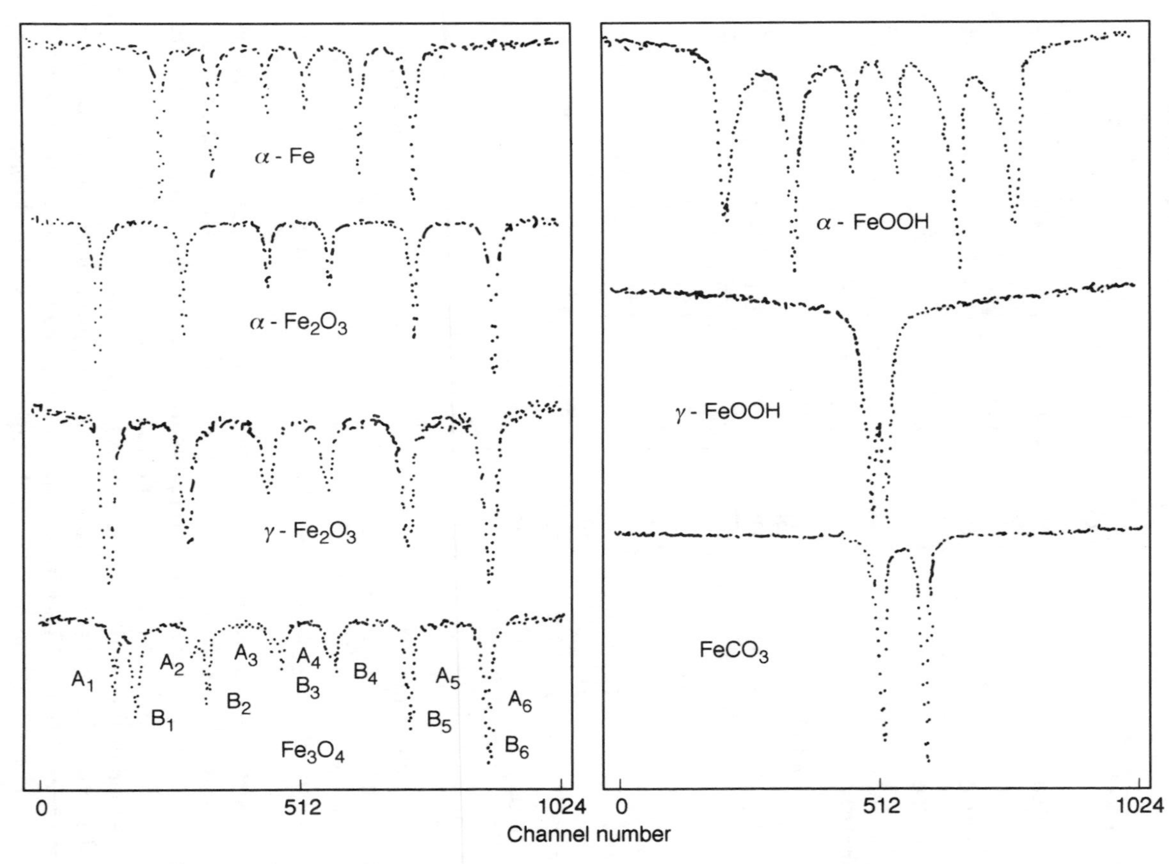

Fig. 17. Mössbauer spectra of α-iron, iron oxides, hydroxides and oxihydroxides at room temperature.

TABLE 1.
Mössbauer parameters of iron oxides, hydroxides and oxyhydroxides

Compound	Position	T (K)	δ ($mm\,s^{-1}$)	ΔE (mm/s)	H (kOe)	Magnetic transition (K)
α-FeOOH	–	0	–	–	510	393.3
	–	80	–	–	500	–
	–	300	+0.40	–	367	–
β-FeOOH	–	80	+0.47	0.11	476	295
		300	+0.38	0.53	–	–
			+0.39	0.88	–	
δ-FeOOH	–	80	–	–	525	–
		300	+0.35	0.60	–	–
γ-FeOOH	–	77	+0.52	–	–	–
		295	+0.38	0.60	–	
$Fe(OH)_2$	–	95	+1.36	3.13	–	–
$Fe(OH)_3$		300	+0.37	0.51	–	–
			+0.37	+0.85	–	–
$Fe(OH)_3 \cdot nH_2O$	–	300	+0.37	0.64	–	–
FeO	–	295	+0.93	0.8	–	198
Fe_3O_4	Fe^{3-}(Tet)	298	+0.27	0.01	493	
		77	+0.44	0	514	
		4.2	+0.042	−0.06	516	
	$Fe^{2+/3-}$ (Oct)	298	+0.67	0.04	460	
		77	+0.94	+0.76	499	
		4.2	+0.99	−0.89	510	
$Fe_{2-x}O_4$	Fe^{3-}(Tet)	300	+0.39	0.11	503	
	Fe^{2+3-}(Oct)	300	+0.78	0.28	465	
α-Fe_2O_3	–	296	+0.39	+0.24	518	956
		83	–	−0.06	542	
γ-Fe_2O_3	–	300	+0.43	0.06	506	

the strong corrosion effect of acidic rain containing also H_2SO_3) on the passivation of iron was studied by CEMS [35]. The main phase of the passive layer was found to be γ-FeOOH, but $FeSO_4 \cdot H_2O$ on the surface and $FeSO_3 \cdot 3H_2O$ (Fig. 20) inside the layer were identified as minor phases in the case of polarization at pH 3.7.

The *particle size distribution* of iron oxides (and other compounds) can also be determined, which can be important in some of the other (geological, archaeological, fine-arts, etc.) applications. Figure 21 shows the room temperature spectra of α-Fe_2O_3 samples of different grain size. The spectral differences can be understood by taking into consideration *superparamagnetism*, at which the small particle size influences the relaxation time of the paramagnetic spin fluctuation by $\tau = \tau_0 \exp[2KV/kT]$ (where K is the anisotropy constant and V is the volume) in such a way that the hyperfine splitting disappears. This happens because hyperfine splitting appears in the spectrum only for $\tau > 1/\omega_L$, where ω_L is the Larmor frequency. Since τ is also a function of the temperature T, the size distribution can be determined from the temperature dependence of the spectra

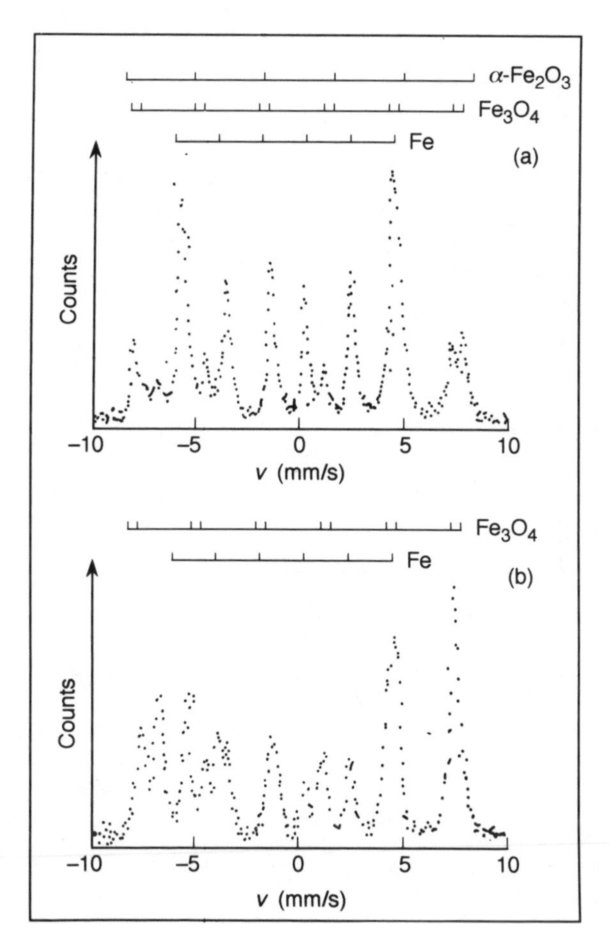

Fig. 18. Conversion electron Mössbauer spectra of thermally oxidized α-iron at different temperatures.

(Fig. 22). With many analytical applications the temperature dependence of the spectrum helps understand whether a component is in the superparamagnetic state or not.

3.2 Phase analysis in alloys

Most problems studied by Mössbauer spectroscopy in the field of metallurgy involve phase analytical applications. An interesting example for the unique application of Mössbauer phase analysis is the work [36] in which the compound

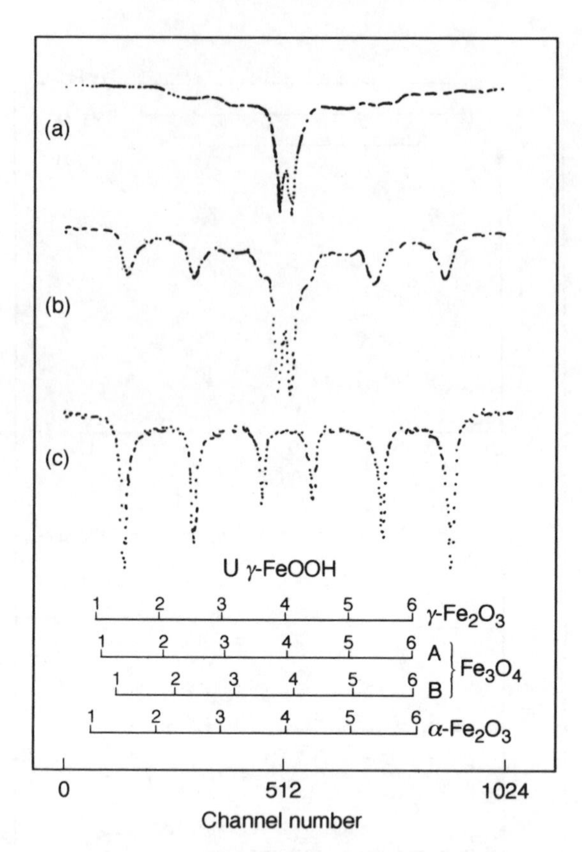

Fig. 19. Mössbauer spectra of atmospheric corroded mild steel (a), car door (b), and muffler (c).

in the layer between diffusion aged iron and aluminum foils was identified with the help of a Fe_2Al_5 calibration spectrum.

The quantitative determination of retained austenite in steels by Mössbauer spectroscopy is of great importance in industrial applications due to the unique precision of the Mössbauer method. In plain carbon steel, extra sensitivity (0.1% of austenite) can be achieved because the spontaneous patterns of martensite and austenite are very well separated in the spectrum (Fig. 23). Whereas the part of the spectrum corresponding to martensite is a ferromagnetically split pattern, austenite exhibits paramagnetic spectral lines. The quantity of retained austenite is determined from the formula

$$M_{\text{aust}} = \frac{I_{\text{aust}}}{I_{\text{aust}} + dI_{\text{mart}}}$$

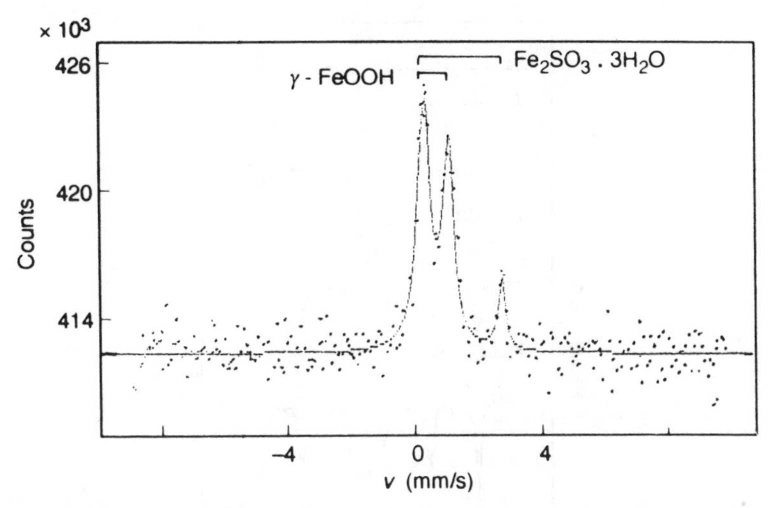

Fig. 20. Conversion electron Mössbauer spectrum of a passive layer formed on iron due to electrochemical corrosion.

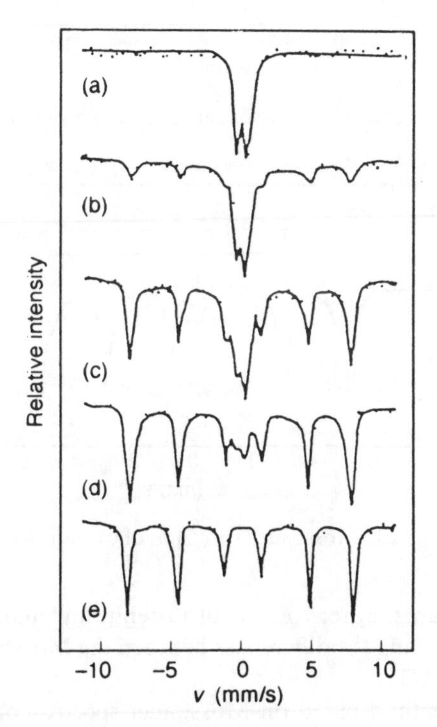

Fig. 21. Room temperature Mössbauer spectra of Fe_2O_3 having different grain sizes.

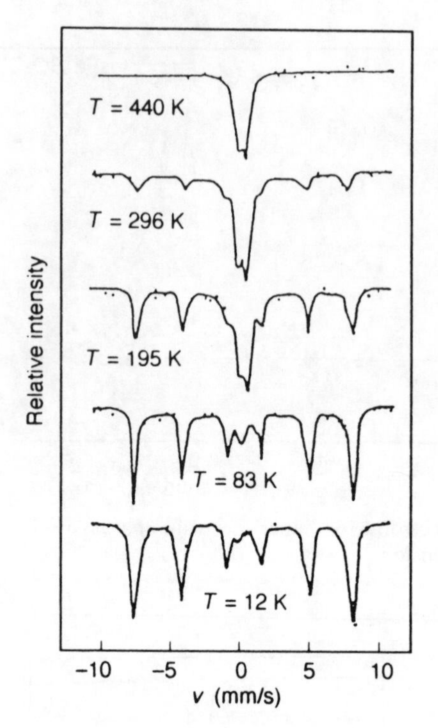

Fig. 22. Temperature dependent Mössbauer spectra of a superparamagnetic Fe_2O_3.

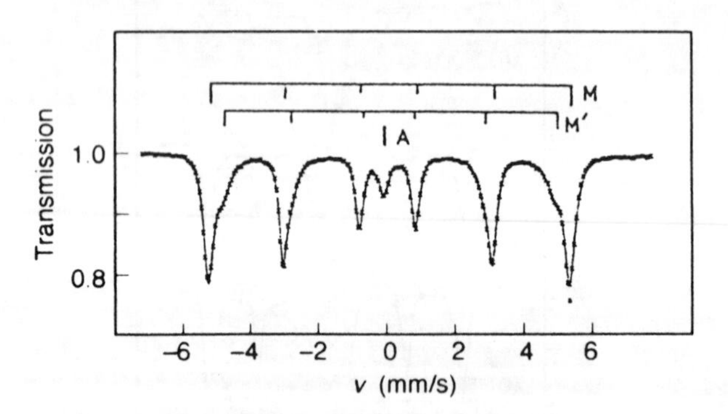

Fig. 23. Mössbauer spectrum of carbon steel.

where I_{aust} and I_{mart} are the areas of lines of austenite and martensite, respectively, and d is a constant taking the differences between the Mössbauer–Lamb factors into consideration [37].

Using a special method based on Mössbauer spectroscopy, the quantitative determination of retained austenite can also be performed in steels alloyed with carbide-forming elements. The disturbing effect of the paramagnetic iron-bearing carbides can be overcome by making use of some additional processes (e.g. heat treatment) [38]. In this case the sensitivity and accuracy of the analysis are still the best among the available methods, as illustrated in Table 2.

The determination of retained austenite can be used directly for the quality control of steels. For this purpose portable Mössbauer spectrometers, so-called *austenitometers*, have been developed [39].

Another procedure which can also be applied for the quality control of steels is the qualitative and quantitative Mössbauer analysis of iron-containing carbides occurring in steels [40]. An analytical method was developed by using the spontaneous and induced patterns of standard M_3C, M_6C, $M_{23}C_6$ and M_7C_3 types of carbides [40]. Figures 24 and 25 show the spectra of some typical carbides and the quantitative applicability of the method when M_3C and M_6C are present simultaneously, which is a difficult situation to handle by X-ray diffractometry. Another important advantage is that the Mössbauer carbide analysis of the steel matrix can be performed in a non-destructive way without chemical isolation of the carbides.

Mössbauer spectroscopy is often used to monitor metallurgical processes (e.g. phase transformation, precipitation, ordering) due to the aging of iron-containing alloys and steels [41]. Figure 26 illustrates the transformation of martensite, the decomposition of austenite and the formation of carbides in a plain steel aged at different temperatures.

The study of *magnetic transformations* is also a favored application of Mössbauer spectroscopy. In many cases the temperature of transformation can be determined (without moving the source) by the thermal scan method; an abrupt change in the detected intensity indicates the transformation taking place at the adjusted temperature.

Although the accuracy and sensitivity of Mössbauer analysis cannot reach those provided by chemical analysis, there are cases in which phase analysis can be performed with extremely high precision. For example, in aluminum containing a few ppm of ^{57}Fe, the ratio of the solid solution and $Al_{13}Fe_4$ phases can be determined very accurately (Fig. 27(a)). The Al corner of the equilibrium phase diagram of Al–Fe alloys has been improved by using Mössbauer data (Fig. 27(b)) [42].

TABLE 2.
The quantity of retained austenite determined by different methods in steel alloyed with carbide-forming elements (wt %)

Mössbauer spectroscopy	X-ray diffraction	Magnetic method	Light microscopy
1.8 ± 0.3	2.3 ± 2	1.5 ± 8	$0. \pm 10$

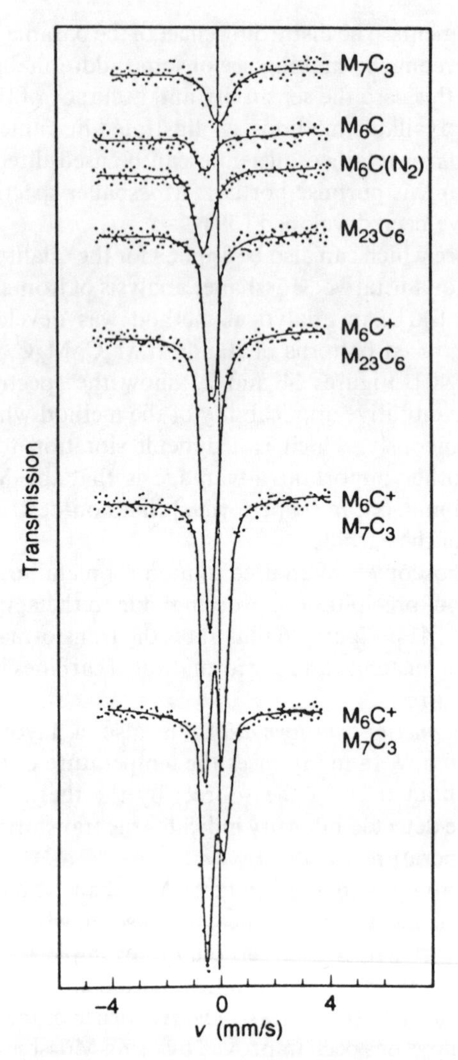

Fig. 24. Mössbauer spectra of standard paramagnetic carbides.

Besides phases, the *defects* in the lattice of alloys can also be analyzed by the Mössbauer method [43]. Figure 28 shows a few defect arrangements and phases in aluminum with the corresponding Mössbauer parameters, giving the possibility of obtaining valuable information about the processes of radiation damage and aging.

The study of *amorphous alloys* is a unique application area of Mössbauer

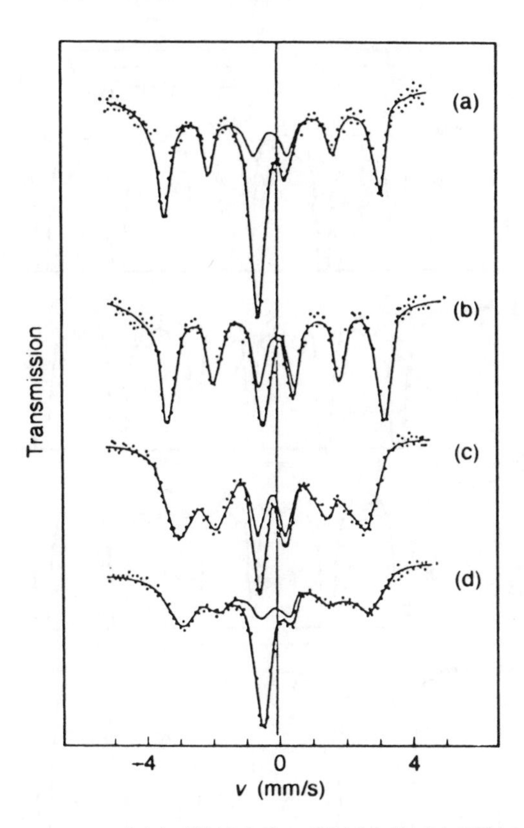

Fig. 25. Mössbauer spectra of (a) 60% M_6C + 40% M_3C, (b) 20% M_6C + 80% Fe_3C, (c) 20% M_6C + 80% $Fe_{2.9}Cr_{0.1}C$ and (d) 60% MC + 40% $Fe_{2.9}Cr_{0.1}C$.

spectroscopy because diffraction techniques cannot be applied effectively in this case owing to the absence of long-range ordering. Amorphous alloys (mainly ferromagnetic) have typical patterns as illustrated in Fig. 29. These spectra consist of a large number of elementary patterns belonging to different types of short-range order. The transformed patterns (Fig. 30) are more useful in distinguishing between the main types of short-range ordering.

The *crystallization* process of amorphous alloys, especially at a very early stage, can be studied with success. Figure 31 shows the Mössbauer spectra of an Fe–B amorphous alloy taken simultaneously of the surface and the bulk of the same material, exhibiting the initial stage of crystallization on the surface while the bulk is still amorphous [44]. The spectra of the $Fe_{90}Zr_{10}$ amorphous alloy in

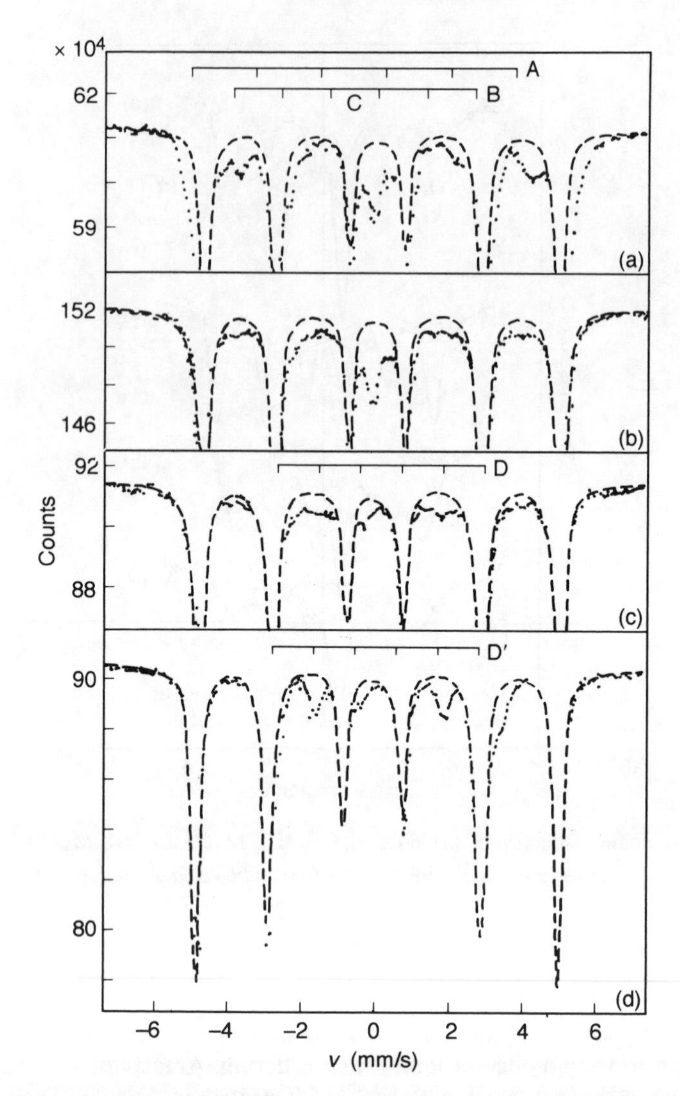

Fig. 26. Mössbauer spectra of 5 at% C containing steel as (a) quenched, (b) aged at 140 °C for 1 h, (c) at 220 °C for 1 h and (d) at 340 °C for 1 h. A and B patterns are due to martensite, C is austenite, D is Fe_5C_2 and D′ is Fe_3C.

Fig. 32 illustrate the effect of introducing different amounts of hydrogen into this material. On the basis of this observation, a method had been elaborated to analyze the quantity and the location of hydrogen [45].

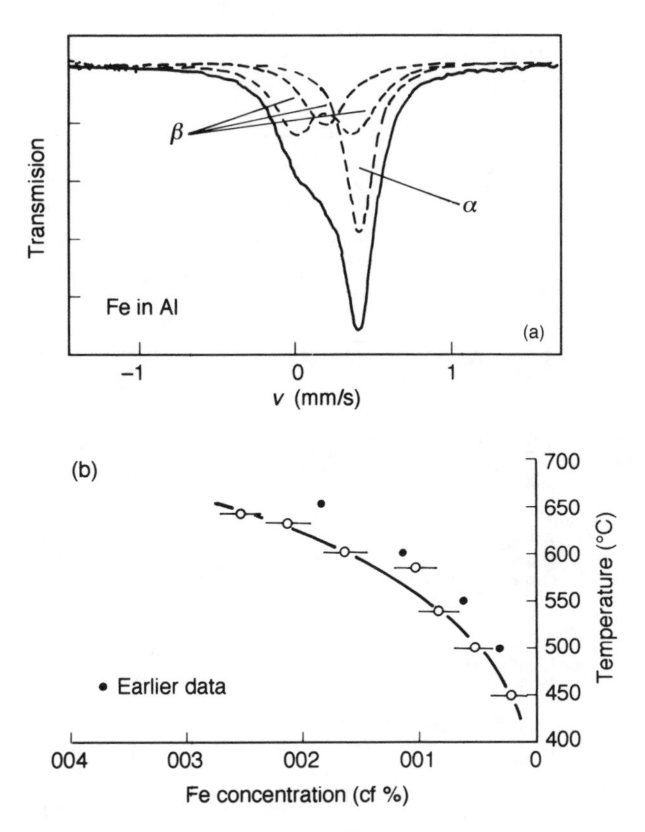

Fig. 27. (a) Mössbauer spectrum of 100 ppm ^{57}Fe containing Al. (b) Phase diagram of Al–Fe.

3.3 Elemental analysis in alloys

Mössbauer spectroscopy is usually not a very good tool for performing elemental analysis. However, in some special cases it is still possible to use this method to obtain elemental concentrations which cannot be obtained by other conventional methods.

One such example is the determination of the concentration of M (Mn, Cr or Fe) in $(Fe_{1-x}M_x)_3C$ carbides from the dependence of the Curie temperature on the concentration of the alloying elements. From the Mössbauer determination of the Curie temperature the concentration of the alloying elements can be obtained. This method is useful if the carbide cannot be separated from other phases in the material.

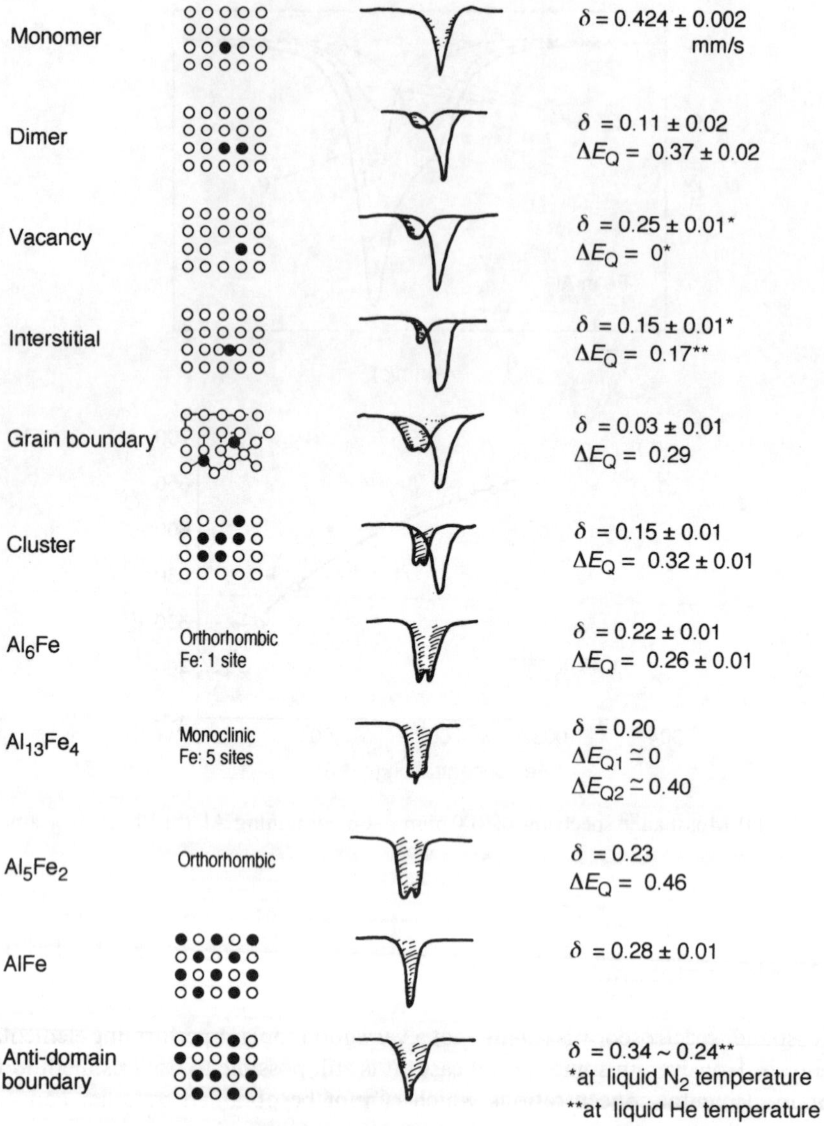

Monomer		$\delta = 0.424 \pm 0.002$ mm/s
Dimer		$\delta = 0.11 \pm 0.02$ $\Delta E_Q = 0.37 \pm 0.02$
Vacancy		$\delta = 0.25 \pm 0.01^*$ $\Delta E_Q = 0^*$
Interstitial		$\delta = 0.15 \pm 0.01^*$ $\Delta E_Q = 0.17^{**}$
Grain boundary		$\delta = 0.03 \pm 0.01$ $\Delta E_Q = 0.29$
Cluster		$\delta = 0.15 \pm 0.01$ $\Delta E_Q = 0.32 \pm 0.01$
Al_6Fe	Orthorhombic Fe: 1 site	$\delta = 0.22 \pm 0.01$ $\Delta E_Q = 0.26 \pm 0.01$
$Al_{13}Fe_4$	Monoclinic Fe: 5 sites	$\delta = 0.20$ $\Delta E_{Q1} \simeq 0$ $\Delta E_{Q2} \simeq 0.40$
Al_5Fe_2	Orthorhombic	$\delta = 0.23$ $\Delta E_Q = 0.46$
AlFe		$\delta = 0.28 \pm 0.01$
Anti-domain boundary		$\delta = 0.34 \sim 0.24^{**}$

*at liquid N_2 temperature
**at liquid He temperature

Fig. 28. Mössbauer patterns of defects and phases in $Al^{57}Fe$ alloy

Another example is the elemental analysis of one of the phases in the two-phase alloy system, namely the determination of chromium in Fe–Ni–Cr ferritic alloy. This method is based on the observation that, in iron-based alloys, the hyperfine magnetic field at the nucleus of a host iron atom is an approximately linear

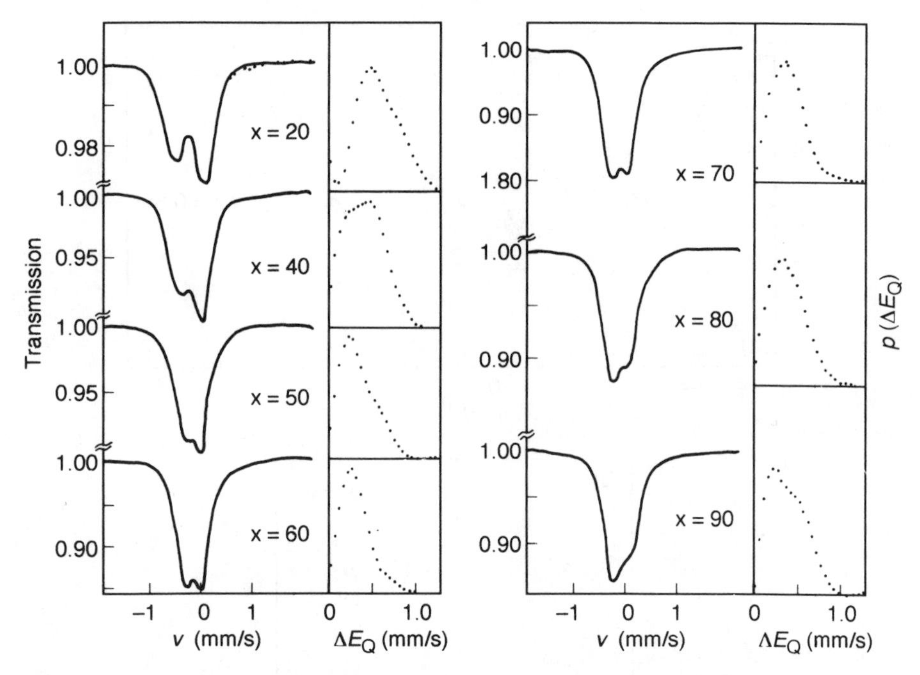

Fig. 29. Mössbauer spectrum and quadrupole splitting distributions of amorphous alloys Fe_xZr.

function of the number of sites occupied by the alloying elements in the first and the second coordination sphere. The hyperfine magnetic field can be expressed as

$$H = H_0 + \Sigma(n_{1k}H_{1k} + n_{2k}H_{2k})$$

where H_0 is the internal magnetic field experienced by iron atoms the first two neighborhoods of which contain only iron atoms, n_{ik} is the number of atoms of the kth alloying element in the ith coordination sphere and H_{ik} is the increment of magnetic field caused by the kth element situated in the ith coordination sphere [46].

The Mössbauer spectrum of a random iron-based alloy can be considered as a set of a finite number of discrete components (elementary patterns) belonging to different alloying element configurations. By assuming a random binomial distribution of elements with known concentration in the alloy, the numbers of

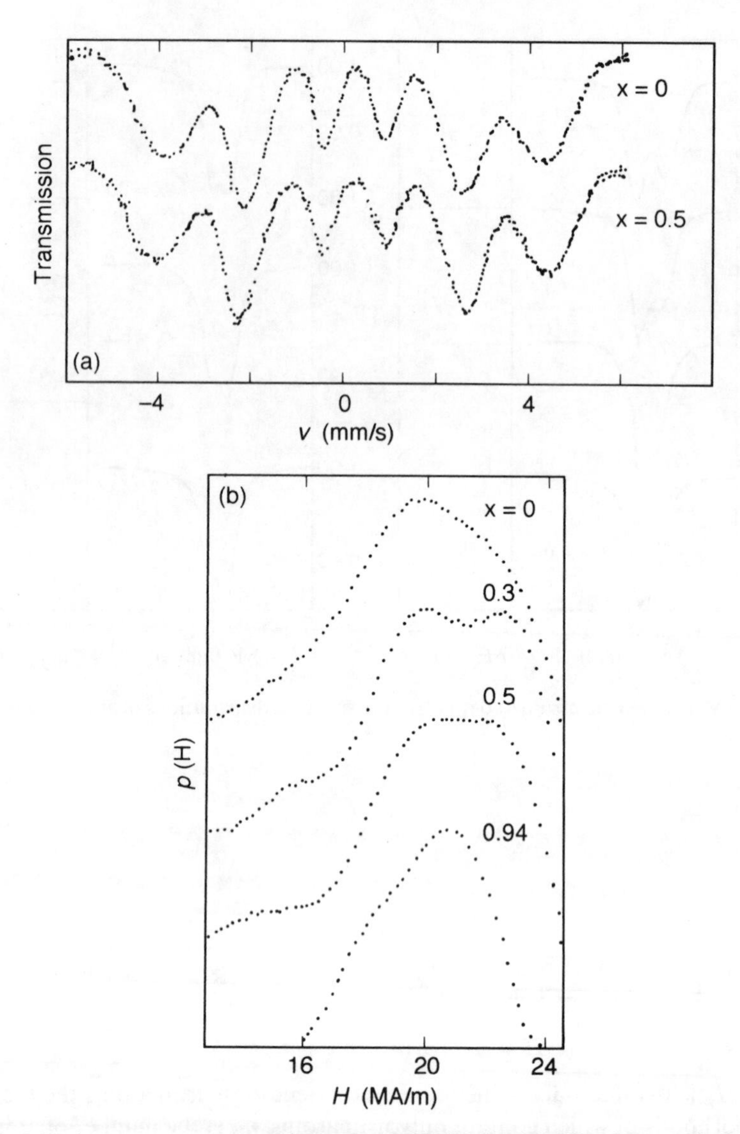

Fig. 30. Mössbauer spectra (a) and hyperfine field distributions (b) of $Fe_{1-x}Co_xB_{10}Si_{10}$ amorphous alloys.

alloying atoms situated in different surroundings can be calculated. Consequently, by measuring the spectral areas of the components (which are proportional to the number of iron atoms having different surroundings), the concentrations of the alloying elements can be determined [47].

It has been shown [48, 49] that a more accurate method can be proposed for the determination of the concentration of alloying elements (Cr, Mn). The

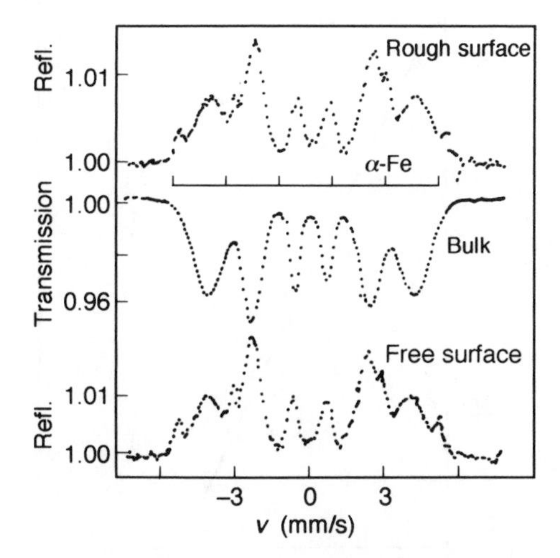

Fig. 31. Mössbauer spectra of amorphous alloy $Fe_{80}B_{20}$ taken from the surface and from the bulk.

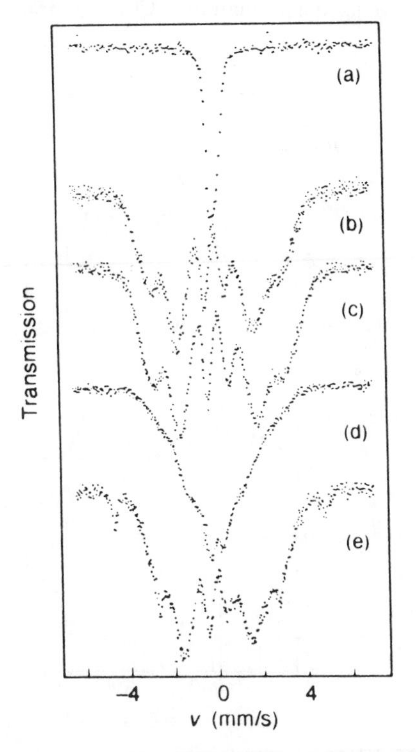

Fig. 32. Mössbauer spectra of amorphous $Fe_{90}Zr_{10}$ alloys hydrogenated at different cathodic potentials up to different hydrogen content. (a), 0 mV; (b), − 500 mV; (c), − 750 mV; (d), − 1000 mV; (e), − 1250 mV.

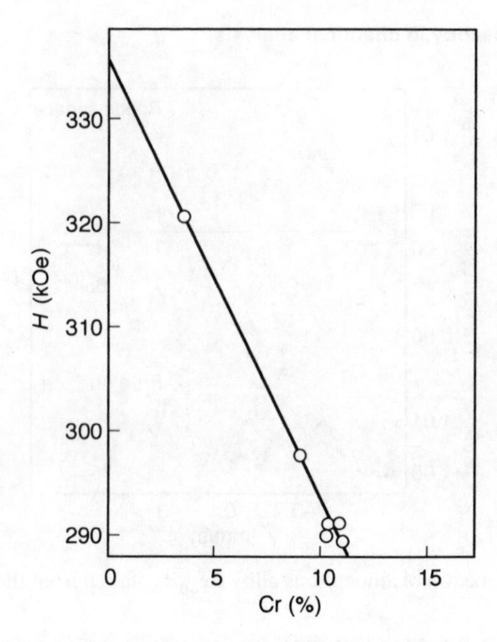

Fig. 33. Calibration curve for determination of Cr concentration in Fe–Ni–Cr alloys

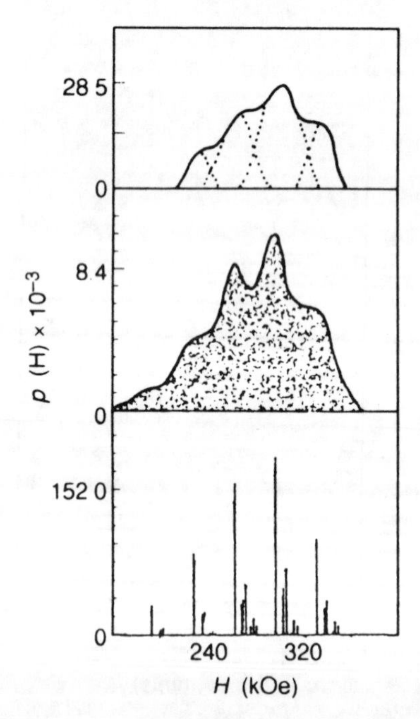

Fig. 34. Hyperfine field distribution of Fe–4Ni–12Cr alloy obtained from the Mössbauer spectra (a) and simulated by taking into theoretical considerations (b) and (c).

TABLE 3.
Results of chromium analysis in *bcc* FeCrNi alloys

Sample	Atomic percentage of Cr by	
	Chemical analysis	Mössbauer method
1	10.36	8.8
2	11.65	10.4
3	12.34	10.6
4	12.44	11.1
5	12.34	10.8
6	3.0	3.6

method is based upon the evaluation of transformed patterns, namely by applying the hyperfine field distributions derived from the spectra.

Figure 33 shows the calibration curve for the determination of Cr concentration from the average hyperfine field obtained from the hyperfine field distributions (Fig. 34) [49]. Although the determination of concentration carried out in this way is relatively inaccurate if compared with chemical analysis (see Table 3), the method can perform elemental analysis at any heat-treated stage of a ferritic phase coexisting with an austenitic phase, whereas chemical analysis cannot determine separately the concentration of the ferritic phase.

This and similar methods have been successfully applied to the determination of the precipitated phase in the age-hardening of Fe–Ni–Mn alloys [48] and to the study of the kinetic behavior of aged Fe–Ni–Cr steels [50].

REFERENCES

1. R. L. Mössbauer, *Z. Phys.* **151**, 124 (1958).
2. L. May, *An Introduction to Mössbauer Spectroscopy*, Plenum Press, New York, (1971).
3. G. K. Wertheim, *Mössbauer Effect*, Academic Press, New York (1964).
4. N. N. Greenwood and T. C. Tibb, *Mössbauer Spectroscopy*, Chapman and Hall, London (1971).
5. V. I. Goldanskii and R. H. Herber *Chemical Application of Mössbauer Spectroscopy*, Academic Press, New York (1968).
6. I. J. Gruverman (Ed.), *Mössbauer Eff. Methodol.* Plenum Press, New York (1965–).
7. J. G. Stevens and V. E. Stevens, *Mössbauer Effect Data Index and Reference Journal*, Adam Hilger, London, North Calfornia (1969–).
8. U. Gonser (Ed.), *Top. Appl. Phys.*, Vol. 5, *Mössbauer Spectrosc.* Springer, Berlin, New York (1975).
9. T. C. Tibb, *Principles of Mössbauer Spectroscopy*, Chapman and Hall, London (1976).
10. G. M. Bancroft, *Mössbauer Spectroscopy*, McGraw-Hill, London (1973).

11. P. Gütlich, R. Link and A. Trautwein, *Mössbauer Spectroscopy and Transition Metal Chemisry*, Springer, Berlin, Heidelberg, New York (1978).
12. A. Vértes, L. Korecz and K. Burger, *Mössbauer Spectroscopy*, Elsevier, Amsterdam, Akadémiai Kiadó Budapest (1979).
13. A. Vértes, and D. L. Nagy (Eds.), *Mössbauer Spectroscopy of Frozen Solutions*, Akadémiai Kiadó, Budapest (1990).
14. B. V. Thosar, J. K. Srivasatva, P. K. Iyengar and S. C. Bhargava (Eds.), *Advances in Mössbauer Spectroscopy*, Elsevier, Amsterdam (1983).
15. G. K. Shenoy and F. E. Wagner (Eds.), *Mössbauer Isomer Shifts*, North-Holland, Amsterdam (1978).
16. H. J. Lipkin, *Quantum Mechanics*, North-Holland, Amsterdam (1973).
17. K. S. Singwi and A. Sjölander, *Phys. Rev.* **120**, 1093 (1960).
18. S. Margulies and J. R. Ehrman, *Nucl. Instrum. Methods* **12**, 131 (1961).
19. C. Kittel, *Introduction to Solid State Physics*, Wiley, New York, (1968).
20. M. Blume and J. A. Tjon, *Phys. Rev.* **165**, 446 (1968).
21. A. J. Freeman and R. E. Watson, *Hyperfine Interaction in Magnetic Materials, Magnetism*, Academic Press, New York (1965).
22. L. Pfeiffer, *Mössbauer Effect Methodology*, Vol. 4, Plenum Press, New York (1968).
23. E. U. Condon and G. H. Shortley, *Theory of Atomic Spectra*, University Press, Cambridge (1935).
24. D. Crespo and J. Parellada, *Hyperfine Interact* **29**, 1539 (1986).
25. U. Gonser (Ed.), *Mössbauer Spectroscopy*, Vol. II, *The Exotic Side of the Method*, Springer, Berlin, Heidelberg, New York, (1981).
26. S. Nagy, B. Lévay and A. Vértes, *Acta Chim. Acad. Sci. Hung.* **85**, 273 (1975).
27. N. Seagusa and A. H. Morrish, *Phys. Rev.* B**26**, 10 (1992).
28. B. Window, *J. Phys. E: Sci. Instrum.* **4**, 401 (1971).
29. J. Hesse and A. Rübartsch, *J. Phys. E: Sci. Instrum.* **7**, 526 (1974).
30. R. A. Brand and G. Le Caer, *Nucl. Instrum. Methods Phys. Res., Sect. B* **B34**, 272 (1988).
31. A. Vértes and I. Czakó-Nagy, *Electrochim. Acta* **34**, 721 (1989).
32. G. W. Simmons, E. Kellerman and H. Leidheiser, *Corrosion* **29**, 227 (1973).
33. A. Sette Camera and W. Keune, *Corrosion Sci.* **15**, 441 (1975).
34. M. J. Graham and M. Cohen, *Corrosion* **32**, 432 (1976).
35. E. Kuzmann, M. L. Varsányi, A. Vértes and W. Meisel, *Electrochim. Acta* **36**, 911 (1991).
36. R. H. Bush, C. A. Stickels and L. W. Hobbs, *Scr. Metau*, **1**, 75 (1967).
37. L. H. Schwartz, *J. Nondest. Test.* **1**, 353 (1970).
38. E. Kuzmann, L. Domonkos, M. Kocsis, S. Nagy, A. Vértes and H. Mehner, *J. Phys. (Paris)* **40**, C2–627 (1979).
39. T. Zemcik and S. Halicek, *Proc. Int. Conf. Appl. Mössbauer Eff.* Bucharest (1977), p. 405.
40. E. Kuzmann, E. Bene, L. Domonkos, Z. Hegedüs, S. Nagy and A. Vértes, *J. Phys. (Paris)* **37**, C6–409 (1976).
41. H. Ino, T. Moriya, F. E. Fujita, Y. Ono and Y. Inokuti, *J. Phys. Soc. Jpn.* **24**, 83 (1968).
42. S. Nasu, U. Gonser, A. Blasius and F. E. Fujita, *J. Phys. (Paris)* **40**, C2–619 (1979).
43. U. Gonser, *Proc. Int. Conf. Appl. Mössbauer Eff.*, Cracow, Vol. 2 (1975) p. 113.
44. M. Ackermann, H. G. Wagner and U. Gonser, *Hyperfine Interact.* **27**, 397 (1986).
45. E. Kuzmann, A. Vértes, Y. Ujihira, P. Kovács, H. Kimura and T. Masumoto, *Struct. Chem.* **2**, 295 (1991).
46. I. Vincze and I. A. Campbell, *J. Phys. F*3, 647 (1964).

47. S. Nagy, K. Romhányi, A. Vértes, E. Kuzmann and Z. Hegedüs, *Acta Metall.* **31**, 529 (1983).
48. E. Kuzmann, R. Oshima and F. E. Fujita, *Proc. Int. Conf. Appl. Mössbauer Eff.*, Jaipur (1981), p. 553.
49. S. Nagy, E. Kuzmann, A Vértes, G. Szabó and G. Konczos, *Nucl. Instrum. Methods Phys. Res., Sect. B* **B34**, 217 (1988).
50. E. Kuzmann, J. Jaen, A. Vértes, L. Csöme, B. Tibiássy and M. Káldor, *Hyperfine Interact.* **58**, 2593 (1990).

Chapter 18

CHEMICAL ANALYSIS BY POSITRON ANNIHILATION

Y. C. Jean

Department of Chemistry, University of Missouri–Kansas City, Kansas City, MO 64110, USA

1 POSITRON ANNIHILATION AND POSITRONIUM CHEMISTRY

1.1 Positrons

Positrons were predicted by the Dirac theory of the electron in 1930 [1], and were discovered by Anderson from cosmic-ray tracks in a cloud chamber in 1932 [2]. Positrons are the anti-particles of electrons; they have precisely the same mass and spin as electrons but the oppossite charge (positive) and magnetic moment. The physical properties of the positron and the electron have been measured very carefully and are found to be identical except for the charge [3]. Positrons do not exist naturally in our world except in relatively small numbers for short times. They can be generated by various nuclear reactions, such as radioisotope decays, and by pair production by gamma-rays with energy greater than $2mc^2$.

When a positron encounters electrons, annihilation occurs, and this process is called positron–electron annihilation, or often just positron annihilation. The positron annihilation event and process are governed by quantum electrodynamics [4]. Conservation of momentum and energy must be satisfied in the annihilation process. Two-photon and three-photon annihilations are the most common. According to the Einstein $E = mc^2$ equation, the energy of a photon is between 0 and 511 keV. Positron annihilation can be easily detected by monitoring the annihilation gamma-rays. Therefore positron annihilation spectroscopy (PAS) is a branch of gamma-ray spectroscopy in monitoring nuclear events as a function of time and of energy. In the late 1940s, it was discovered that the

Chemical Analysis by Nuclear Methods Edited by Z. B. Alfassi

characteristics of positron annihilation depend on the physical and chemical properties of materials [5]. An annihilation rate is a measure of the electron density where positron annihilation takes place. Because energy loss cross-sections are high, positrons slow down to thermal energies before annihilation from a few hundred eV energies in beta decay. Typically the time of slowing down in solids or liquids is only a few picoseconds (10^{-12} s). The positron lives in the range of 100 ps to 10 ns in condensed matter. The lifetime and momentum distribution of the annihilation give useful information about the electronic environment in which the positron has thermalized. Therefore the technique of positron annihilation has become an interesting tool in the study of condensed matter.

In the 1960s, positron annihilation got a boost from the discovery that the positron was found preferentially localized in defects, such as an atomic vacancy [5]. In a real solid, a thermalized positron finds its favorable site, a defect, and then annihilates there. Therefore an annihilation signal will arise mainly from a defect. In this aspect, PAS has become a special tool in determining the vacancy formation enthalpy and concentration (between 10^{-4} and 10^{-6} atomic fraction) in most metals and semiconductors.

In the 1970s, the study of the interactions of positrons with matter revealed a very important property of the positron: the work functions of the positron in many solids are negative [5]. The strongly repulsive interaction between the core electrons and the positron means that the positron may become energetically more stable in vacuum than in the bulk. This discovery leads to a new technique in generating slow positrons at a well defined energy. Following these developments, positrons have been successfully used in many experimental investigations of surfaces. Almost every electron spectroscopy, such as diffraction energy loss, Auger electrons, microscopy, etc., has been developed in replacing electrons by positrons.

1.2 Positronium

A positron may be bound to an electron in forming an atom called positronium (Ps). The existence of the positronium atom was detected in 1950 by Deutch [6] from the collisions between positrons and gases. The Ps has been observed in molecular solids and liquids. The quantum mechanics describing Ps are identical with those of the H atom. There is a one-to-one correspondence between the states of hydrogen and those of Ps, but the energy level of the ground-state binding energy for Ps is 6.803 eV, which is half the energy of 1 s H. Because of the much smaller binding energy of Ps compared to H, the chemistry of Ps is very different from that of H. For example, in a chemical reaction, Ps often reacts as an electron donor whereas H behaves as an electron acceptor. As Ps has ultra-light mass, the chemical kinetics and dynamics of Ps are dominated by the quantum effect, such as tunneling and Ps molecule formation. Therefore it is important to realize that Ps behaves very differently from H in chemical reactions. It is often inappropriate to consider Ps as an isotope of H. Ps chemistry stands for itself and needs special consideration in many chemical and physical respects [7].

Ps exhibits two spin states, which are called '*ortho*' and '*para*' for the triplet and the singlet state, respectively. The large magnetic moment of the positron compared with that of the proton (657 times as great) makes the spin–spin interaction in Ps comparable with the spin–orbital interaction. The hyperfine splitting of Ps has the same order of magnitude as its fine structure splitting. This phenomenon leads to many anomalous effects of Ps annihilation under magnetic fields.

The selection rules and annihilation cross-section of ground-state Ps follow the quantum electrodynamic theory: (1) *para*-positronium (*p*-Ps) and *ortho*-positronium (*o*-Ps) annihilate to yield two and three photons respectively; (2) the average annihilation rates of *p*-Ps and *o*-Ps equal $7.9852 \times 10^9\,s^{-1}$ and $7.0386 \times 10^7\,s^{-1}$ (including radiative corrections), respectively [8]. Therefore the intrinsic lifetimes (defined as the reciprocal of the annihilation rate) of free *p*-Ps and *o*-Ps are 0.1252 ns and 142.1 ns respectively. In dilute gases and highly porous materials, the observed Ps lifetimes are close to those given above. As the lifetime of *o*-Ps is relatively long, measurements of its lifetime become an important tool for analytical chemistry. However, in most condensed phases *o*-Ps lifetimes are shortened to a few nanoseconds owing to the interactions of *o*-Ps with electrons in the medium. The positron in Ps can annihilate with one such electron in a two-photon process, a process called 'pick-off' annihilation. In organic materials, the pick-off lifetimes of *o*-Ps are between 1 and 10 ns. On the other hand, Ps cannot form in materials with very high electron densities, such as metals or semiconductors, in the absence of defects. Table 1 lists the important properties of Ps in condensed media.

1.3 Quenching of Ps

As described above, Ps lifetimes become shorter because of pick-off annihilation with electrons in the medium. Since Ps is a neutral atom, it can react chemically

TABLE 1.
Fundamental physical properties of positronium

State	Annihilation event	Lifetime	Size	Ionization potential
p-Ps (singlet state)	two gammas	0.1252 ns	1.06 Å	6.804 eV
o-Ps (triplet state)	three gammas or two gammas (pick-off)	142.1–1 ns (pick-off)	1.06 Å	6.803 eV

with certain types of molecules and its lifetime becomes even shorter. A process for reaction of Ps with molecules or functional groups, resulting in a significant reduction of Ps lifetime, is called 'chemical quenching'. From the analytical point of view, pick-off lifetimes give direct information about physical properties, whereas the chemical quenching lifetimes are related to the chemical properties of molecules. The current theory that accounts for a chemical quenching postulates a positronium molecular complex formation mechanism: Ps forms a bound (transition) state with certain molecules and as a result has a short lifetime (of the order of 0.4 ns) [9]. This transition state theory is schematically shown as [10]:

$$2\gamma \xleftarrow{\lambda_p} \mathrm{Ps} + \mathrm{M} \underset{k_b}{\overset{k_a}{\rightleftharpoons}} \mathrm{PsM} \xrightarrow{\lambda_c} 2\gamma \tag{1}$$

If λ_p is the rate constant for Ps annihilation with molecules in the medium (pick-off annihilation), λ_c is the rate constant for Ps annihilation with a PsM complex, and k_a and k_b are the rate constants for formation and dissociation of the PsM complex, the overall reaction rate constant between Ps and molecules (M) can be expressed with pseudo first order kinetics as:

$$k_{Ps} = \frac{k_a \lambda_c}{k_b + \lambda_c} \tag{2}$$

This quantity appears in the equation for the annihilation decay rate which is measured in positron lifetime experiments

$$\lambda = \lambda_p + k_{Ps}[\mathrm{M}] \tag{3}$$

Therefore k_{Ps} represents a measure of chemical reactivity of Ps with a molecule. It is a characteristic property of a molecule and depends on molecular structure and functional groups. In principle, each molecule possesses a specific value of k_{Ps}. However, only a selected family of molecules shows significant chemical quenching. As the analytical sensitivity depends on the observable difference in the Ps annihilation rate, we may classify all chemicals into two groups in terms of chemical quenching, i.e. strong ($k_{Ps} \geq 10^8\ \mathrm{M^{-1}\,s^{-1}}$), and weak chemical reacting groups ($k_{Ps} < 10^8\ \mathrm{M^{-1}\,s^{-1}}$). These are shown in Table 2 [10].

TABLE 2.
Ps chemical reactivities with molecules

Strong reactivity $k_{Ps} \geq 10^8\ \mathrm{M^{-1}\,s^{-1}}$	Weak reactivity $k_{Ps} < 10^8\ \mathrm{M^{-1}\,s^{-1}}$
Conjugated nitroaromatics	Aliphatic hydrocarbons
Halogens	Simple aromatics
Quinones	Alcohols
Maleic anhydrides	Haloalkanes
Strong electron acceptors	double/triple bonds
Inorganic ions with redox potentials < -0.9 V	Inorganic ions with redox potentials > -0.9 V

It is also important to realize that there are two other types of quenching besides a chemical process. In a molecule with an unpaired spin, such as a free radical, or a paramagnetic state, such as the ground state of O_2, the Ps can be quenched via a spin conversion process. The k_{Ps} for spin conversion quenching is of the order of 10^7 or $10^8 M^{-1} s^{-1}$. Although its effect is noticeable in most experiments, in a system containing a chemical quencher a spin conversion process becomes insignificant. Another type of quenching results from applying an external magnetic field, to the system of interest. The magnetic–spin interaction results in mixing of the $m = 0$ states between *p*-Ps and *o*-Ps. Under an external magnetic field, the overall *o*-Ps lifetime significantly decreases and the *p*-Ps lifetime slightly increases as a function of field strength. Both external and internal magnetic quenching processes can be used to analyze the magnetic states of molecules.

1.4 Inhibition

It has been realized that the process of the positron picking up an electron in forming Ps also depends on the chemical properties of molecules in the media. When positrons slow down from a few hundred keV of kinetic energy, the radiation effect of the positrons can be studied by observing the fraction of Ps formation. The fraction of Ps formation in most organic systems is in the range ≈ 10–70%. This large fraction of Ps formation has been an advantage in the application of PAS for chemical analysis.

The formation mechanism for Ps in condensed media is a complicated process. It depends not only on the radiation effect of the positron but also on the energy levels of the molecules of interest. In a similar way to the radiation chemistry of the electron, one often expresses the *o*-Ps formation fraction as a function of the concentrations of additives, such as an electron or a radical scavenger. A process that reduces the probability of Ps formation is called 'Ps inhibition', and its counterpart of increasing Ps formation is called 'anti-inhibition.' Most inhibitions are expressed in an empirical equation as a function of concentration $[M]$ as:

$$\frac{I_{o-Ps}}{I^o_{o-Ps}} = \frac{1}{(1 + K[M])} \tag{4}$$

The K is often called the 'total inhibition constant.' By analogy with the chemical reactivity of Ps with a molecule, inhibitors can be calssified as strong or weak according to the value of K. This classification is shown in Table 3 [11].

The inhibition of Ps formation depends not only on the properties of additives but also on the solvent. The observable effect of inhibition at a certain concentration is relatively small compared with that of chemical quenching. Noticeable change of Ps formation is observed at a concentration of the order of 0.1 M or

TABLE 3.
Ps inhibition constants in common liquids

Strong inhibitors $K \geq 3\,M^{-1}$	Weak inhibitors $K < \pm 3\,M^{-1}$	Strong anti-inhibitors $K \geq -3\,M^{-1}$
Electron scavengers such as $CHCl_3$, CCl_4 or NO_3^-	Simple hydrocarbons, alcohols, aromatics, acids	Fluoro compounds such as SF_6 or C_6H_5F

higher, whereas chemical quenching can be observed in solutions in the order of mM. Although Ps inhibition is not a sensitive method in terms of chemical analysis, it can be used to study the radiation chemistry of molecules.

2 EXPERIMENTAL TECHNIQUES FOR POSITRON ANNIHILATION

Positron annihilation spectroscopy (PAS) is a family of three nuclear techniques: positron annihilation lifetime (PAL) spectroscopy, Doppler broadening spectroscopy (DBS), and angular correlation of annihilation radiation (ACAR). These techniques detect photons with energy in the order of hundreds of keV. The source of the positrons affects the choice of instrumentation for PAS.

2.1 Positron sources

Positrons can be obtained in two ways: the decay of neutron-deficient radioisotopes, and pair production by high energy γ-rays. Radioisotope decay is the more convenient source of positrons, but pair production is important when high positron fluxes are required. Positrons obtained by these two means have energy of the order of hundreds of keV to MeV and a stopping range in the approximate range 0.1 g/cm^2 (200 keV) to 1 g/cm^2 (1 MeV). For a thin sample or for surface characterization, one requires a slow positron with a well defined energy. A monoenergetic slow positron beam can be produced for this application.

2.1.1 Fast positrons

Various radioisotopes for positron sources are listed in Table 4. Most of these radioisotopes are generated by bombarding protons, deuteron or neutron sources via nuclear reactions.

Sodium-22 is the most common choice of PAS source because it has a long half-life, a large fraction of positron emission, and yields prompt γ-rays for lifetime counting purposes. The decay scheme of ^{22}Na is shown in Fig. 1. The mass absorption coefficient (in units of cm^{-1}) is approximated to be 1.65 dZ_{av}

TABLE 4.
Characteristics of some common positron sources

Radioisotope	Fraction of e^+	$t_{1/2}$	$E_{max}(e^+)$	Prompt γ
^{11}C	99%	20 m	0.97 MeV	
^{22}Na	90%	2.6 y	0.54 MeV	1.28 MeV
^{44}Ti (as ^{44}Sc)	88%	47 y	1.47 MeV	1.16 MeV
^{57}Ni	46%	36 h	0.40 MeV	1.4 MeV
^{58}Co	15%	71 d	0.48 MeV	0.81 MeV
^{64}Cu	19%	12.8 h	0.66 MeV	
^{65}Zn	1.7%	245 d	0.33 MeV	
^{68}Ga (as ^{68}Ge)	88%	275 d	0.98 MeV	

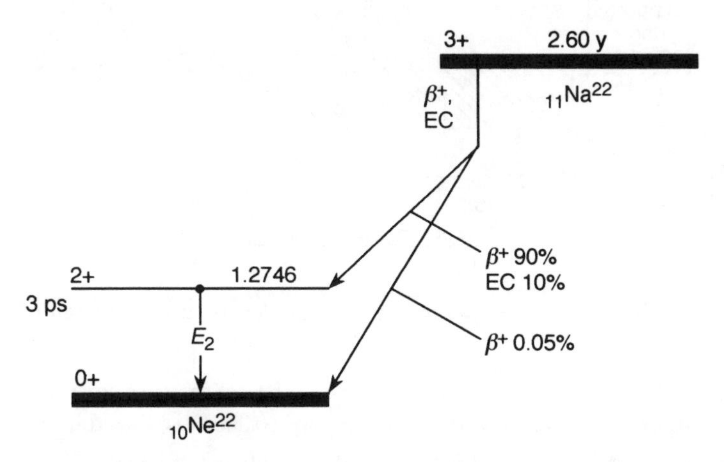

Fig. 1. The nuclear decay scheme of a positron source, ^{22}Na.

(d = density in g/cm^3, Z_{av} = average atomic number of the medium). In a typical sample with thickness 1 mm and density = 1.0 g/cm^3, all positrons (> 99.9%) are essentially absorbed and annihilated in the sample.

2.1.2 Slow positrons

When fast positrons are caused to bombard a solid surface, they slow down, and a fraction of them may diffuse back to the surface. In some solids, such as W, Ni, Ne, Cu, etc., the positron work function is negative, i.e. the positron is more stable in vacuum than in the bulk of these solids [5]. A large fraction of positrons is emitted from these solids with an energy between thermal energy and a few eV, and can be accelerated to a monoenergetic positron beam. Solids with a negative positron work funtion have been used as moderators in generating slow positrons. Figure 2 shows a schematic diagram for a reflection mode of a slow positron

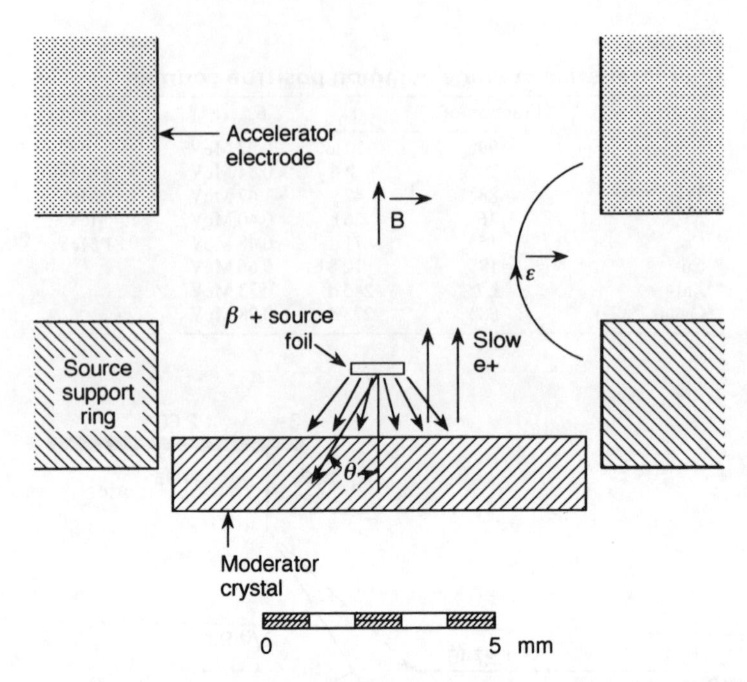

Fig. 2. Schematic diagram for a reflection mode slow positron moderator [12].

moderator [12]. Another moderator design is that of the transmitting mode, in which single crystal W (100) with a thickness of 10 000 Å is placed in front of the fast positron source [13]. With good design, one can achieve a conversion ratio of 0.1% from a fast positron (radioisotope) to a slow positron. The re-emitted slow positrons can be focussed magnetically and/or electrostatically to a monoenergetic slow positron beam. In shaping the beam structure of positrons, such as an enhancement of brightness, positron optics is essentially the same as that for electrons. A small beam cross-section (such as a diameter of 1 mm), at a flux of 10^6 e^+/s and with an energy resolution of 0.1 eV, can be achieved in most university laboratories.

Another method of generating slow positrons is by pair production of bremsstrahlung radiation generated by high energy particles. For example, using an electron beam with an energy > 10 MeV, one can generate a slow positron beam with a flux of 10^9 e^+/s [14].

With a well defined slow positron beam, many thin film samples and surfaces can be characterized by PAS. Today, almost every technique in electron spectroscopy has been developed into positron spectroscopy. There are many advantages in using positron spectroscopy: (1) positrons are surface sensitive, (2) additional annihilation gamma-rays can be used for signal detection, and (3) positrons are particularly defect-sensitive. On the other hand, the advantage of the positron lifetime technique associated with fast positrons does not exist in the slow positron technique, although a sophisticated timing technique has been developed to overcome this difficulty [15].

2.2 Positron annihilation lifetime spectroscopy (PAL)

Positron lifetime measurements are performed using a fast–fast coincidence method which entails monitoring the signal (1.28 MeV gamma-ray) from positron decay in the isotope ^{22}Na (see Fig. 1) as the start time and the signal (0.51 MeV gamma-ray) from positron annihilation as the end time. The schematic diagram of a positron lifetime apparatus is shown in Fig. 3. A sample (typically 0.5 g) surrounds a specially designed positron source ($\approx 10\ \mu$Ci ^{22}Na). The thickness of the sample, ≈ 1 mm, is adequate to absorb more than 99.9% of the positrons emitted. A typical positron lifetime spectrum in water is shown in Fig. 4. The positron lifetime spectrum reveals a function containing a multi-exponential:

$$N(t) = \sum_{i=1,n} I_i e^{-\lambda_i t} \tag{5}$$

where n is the number of exponential terms, and I_i and λ_i represent the number of

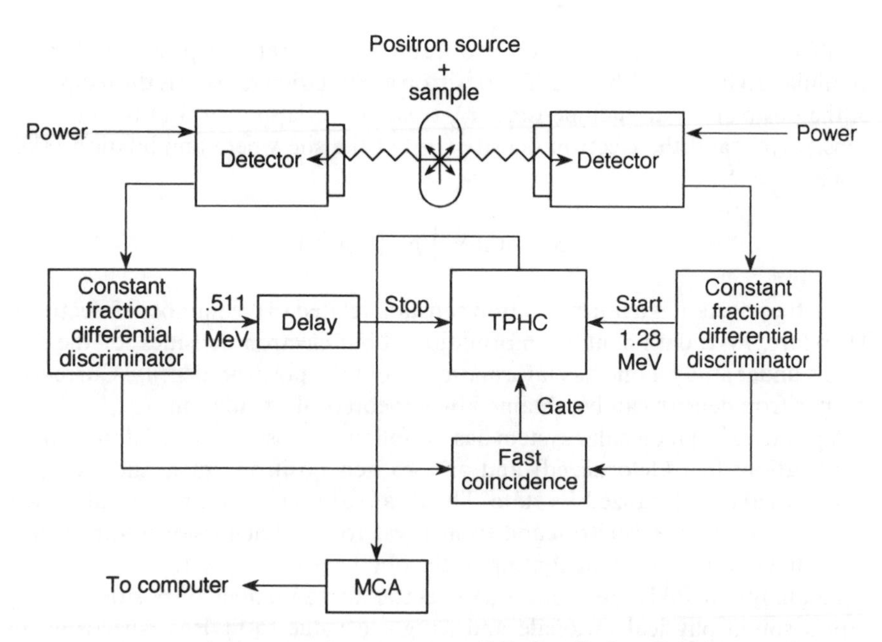

Fig. 3. Schematic diagram of a positron annihilation lifetime (PAL) spectrometer.

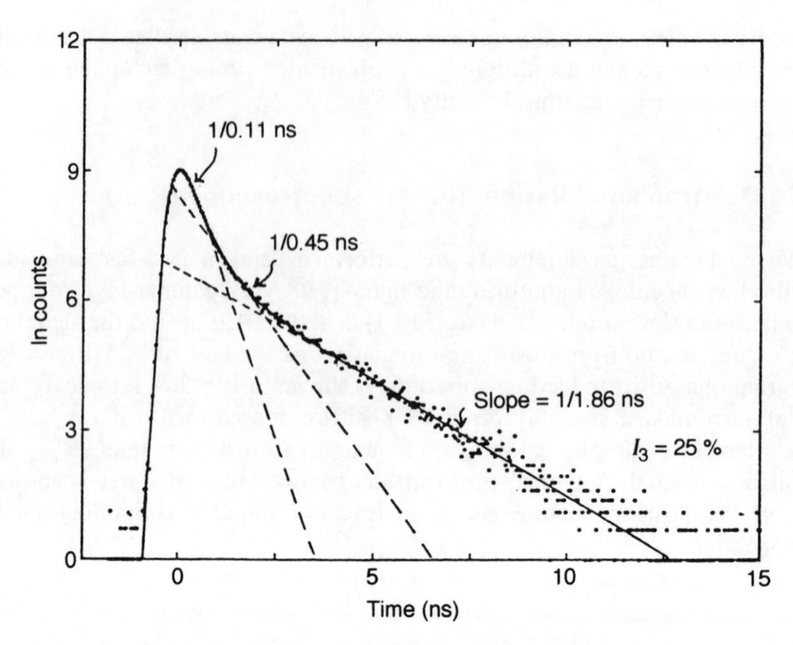

Fig. 4. A typical positron lifetime spectrum in water.

positrons (intensity) and the positron annihilation rate, respectively, for the annihilation from the ith state. The positron annihilation rate (λ_i) is the reciprocal of the positron mean lifetime time (τ_i). It is an overlap integral of the positron density (ρ_+) and the electron density (ρ_-) at the site where annihilation takes place:

$$\lambda = \text{constant} \times \int \rho_-(r)\,\rho_+(r)\,\mathrm{d}r \tag{6}$$

where the constant is a normalization constant related to the number of electrons. Therefore, PAL data contain information about electron densities for the materials under study if one has information about the positron density. Currently, the positron density can be obtained by a theoretical calculation.

A positron in a molecular system has the following possible states at the time of annihilation: free (delocalized) and/or localized positron state, and free (delocalized) and/or localized Ps state. The localized sites are more favorable sites than the bulk for the positron and Ps in a real system. Each Ps or positron state has a different lifetime contributing to the obtained PAL spectra.

To analyze a PAL spectrum, one selects a finite number of positron states with a sound physical rationale and assigns a value to fit PAL spectrum. In molecular systems, it is customary to fit PAL spectra into three lifetimes, i.e.,

$n = 3$ in equation (5), by using a computer program (PATFIT) [16]. The shortest lived component. $\tau_1 = 0.125 \pm 0.010$ ns with an intensity $I_1 = 3-20\%$, is attributed to the annihilation of $p-$Ps. The intermediate component, lifetime $\tau_2 =$ 0.3–0.5 ns and intensity $I_2 = 10-60\%$, is attributed to the direct annihilation of positrons and positron–molecular species. The long lived component, with a lifetime $\tau_3 = 0.5-10$ ns and an intensity $I_3 = 10-70\%$, is due to *o*-Ps annihilation.

For a more accurate data analysis, a PAL spectrum can be expressed in a continuous distribution of λ_i vs. I_i. In this case we can replace the discrete summation in equation (5) by an integration:

$$N(t) = R(t)^* \int_0^{\infty} \alpha\lambda e^{-\lambda_i t}\, dt \quad (7)$$

where $R(t)$ is the instrument resolution ($*$ denotes a convolution) and $\alpha\lambda$ is the probability density function for the positron annihilation with rate λ. To obtain $\alpha\lambda$ vs. λ involves a Laplace inversion of a PAL spectrum $N(t)$ in solving equation (7). To solve equation (7) directly is a very difficult problem because the exact $R(t)$ is unknown in PAL. One approach is to measure PAL in a standard sample with a known positron lifetime and then utilize it to deconvolute the unknown spectra. A computer program CONTIN developed in fluorescence spectroscopy [17] has been adopted for deconvoluting a PAL spectrum into a continuous lifetime distribution [18]. Figure 5 shows the positron density function $\alpha(\lambda)\lambda^2$ vs $\tau\ (= 1/\lambda)$ deconvoluted by using a standard single crystal Cu spectrum ($\tau_r = 122$ ps) [19]. Other useful standards are Ni, Al and ^{207}Bi radioisotopes.

2.3 Angular correlation of positron annihilation radiation (ACAR)

An angular correlation (ACAR) experiment measures the coincidence count as a function of the angle θ, the deviation from collinearity of the two annihilation radiations (0.511 MeV energy). An ACAR spectrum maps out the momentum of the annihilating positron–electron pair. Since the positron is thermalized at the time of annihilation, the angular deviation reflects the momentum of the electron. In a one-dimensional angular correlation measurement, the momentum is measured at a direction perpendicular to the direction of the two annihilation photons, i.e., p_z. The relationship between the angle θ and p_z is given by

$$\theta = \tan^{-1}(p_z/mc) \approx p_z/mc \quad (8)$$

A long-slit one-dimensional ACAR apparatus is schematically shown in Fig. 6. A typical distance between the detectors and the sample is 5–10 m and the width of the long slits is 0.5 cm. Thus a typical instrument resolution is 0.5–1 mrad.

A one-dimensional ACAR spectrum can be expressed as a two-dimensional

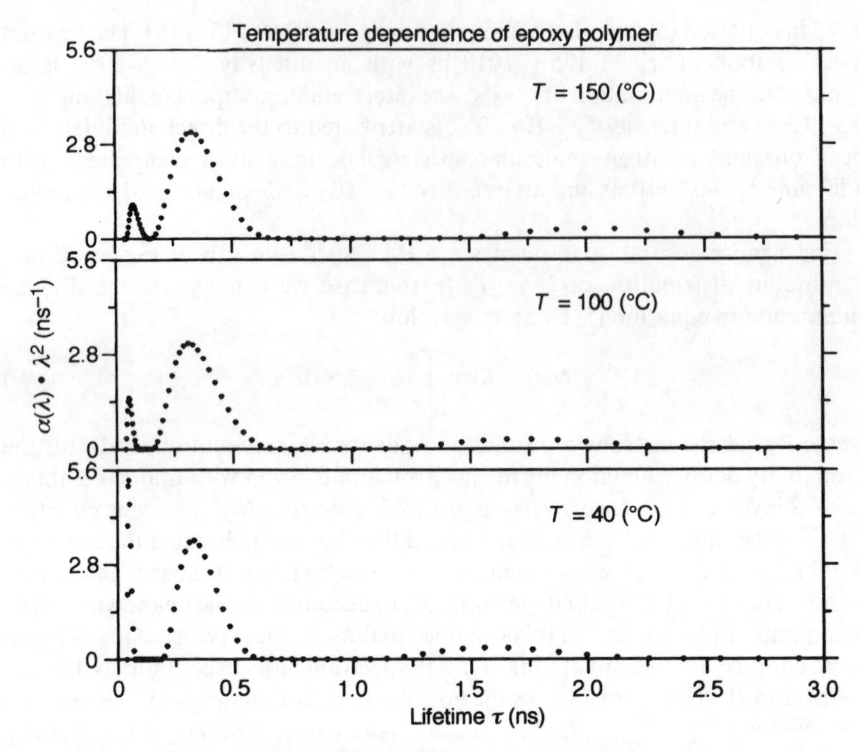

Fig. 5. Positron lifetime density function vs. positron lifetime in an epoxy polymer at different temperatures. The spectrum was deconvoluted by using the CONTIN program [19].

integral of positron and electron densities in momentum space:

$$N(p_z) = \text{constant} \int_x \int_y |\Psi_+(\mathbf{P})\Psi_-(\mathbf{P})|^2 \, dp_x \, dp_y \tag{9}$$

where the constant is the normalization constant related to the number of electrons, and $\Psi_+(\mathbf{P})$ and $\Psi_-(\mathbf{P})$ are positron and electron wave functions, respectively, in momentum space. Note that the positron and electron wave functions in momentum space shown in equation (9) are simple Fourier transforms of those in real space.

In data analysis, an ACAR spectrum from an organic material is expressed in a function containing multi-Gaussians:

$$N(\theta) = \sum_{i=1,n} I_i e^{-a_i \theta^2} \tag{10}$$

where I_i and a_i are the intensities and parameters related to the positron–electron

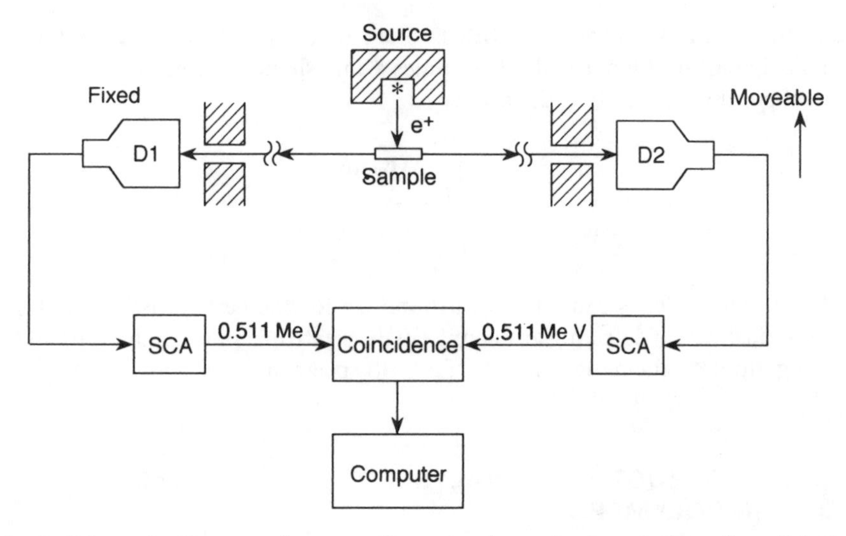

Fig. 6. Schematic diagram of an one-dimensional angular correlation of annihilation radiation (ACAR) spectrometer [21].

momentum distributions of the material (e.g. polymer). The FWHM (full width at half maximum) of each component, for a Gaussian function, is given by $2\sqrt{(\ln 2/a)}$. A typical ACAR spectrum is fitted to three Gaussians ($n = 3$) with FWHM $\approx$ 2–5, 7–10 and 15–20 mrad respectively. As with the PAL analysis, we use the program package PATFIT [16] which gives results of three intensities and FWHM values. The narrow component is contributed from the annihilation of *p*-Ps with an intensity of typically 3–20% (note that the ratio of *p*-Ps to *o*-Ps formation probability is equal to $\frac{1}{3}$). The intermediate component arises from the annihilation of positrons and positron–molecular species, and the broad component comes from the annihilation of core electrons. The resolved FWHM due to the electronic momentum of a typical valence electron in a chemical bonding is equivalent to 5–10 mrad. When Ps is localized in a void, the smallest FWHM, called $\theta_{1/2}$, arises solely from the uncertainty of the momentum due to *p*-Ps annihilation. The $\theta_{1/2}$ is used to determine the size of voids in the direction perpendicular to the annihilation radiation.

2.4 Doppler broadening spectroscopy (DBS)

Doppler broadening spectroscopy (DBS) measures the longitudinal momentum density $N(p_x)$ along the annihilation radiation. The resulting Doppler shift due to the electronic momentum from valence electrons in the energy spectra at 0.511 MeV has the same order of magnitude as the energy resolution of solid-state detectors ($\approx 1\,\text{keV}$). Therefore DBS is often used as a qualitative method in chemical analysis. One usually determines a line shape parameter (S), near the central point of 0.511 MeV, defined as:

$$S = \frac{\int_{E_a}^{E_b} N(E)\,\mathrm{d}E}{\int_{-\infty}^{\infty} N(E)\,\mathrm{d}E} \tag{11}$$

The S parameter is chosen to be around 0.5 for the best statistical analysis. DBS is a fast way of detecting any change in the electronic state of materials under investigation. A schematic diagram of a DBS spectrometer is shown in Fig. 7.

3 MICROSTRUCTURAL ANALYSIS OF FREE VOLUME HOLES IN POLYMERS

One of the key problems in the utilization of polymers for industrial applications is the existence of free volumes with a hole size of the order of a few Å [20]. Characterizations of such atomic scale free volume holes have been successfully performed by PAL and ACAR in recent years [21]. The unique property of Ps localization in free volume holes enables us to use PAS to determine the free volume hole size, distribution, shape, and fraction directly.

3.1 Hole size, distribution, and fraction

When Ps is localized in a free volume hole, its lifetime changes as a function of hole size: a larger hole results in a longer lifetime. A simple quantum mechanical model has been developed to obtain the relationship between the observed o-Ps

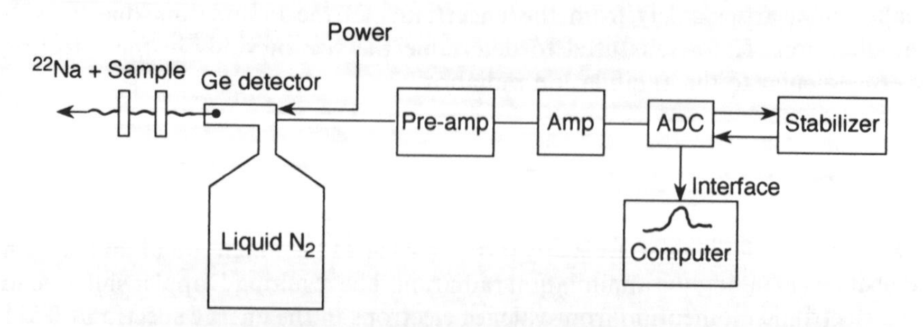

Fig. 7. Schematic diagram of a Doppler broadening spectrometer (DBS).

lifetime (λ) and the radius of the free volume hole (R), and is expressed as [22]:

$$\lambda = 2\left[1 - \frac{R}{R_0} + \frac{1}{2\pi}\sin\left(2\pi\frac{R}{R_0}\right)\right] \tag{12}$$

In equation (12), $R_0 = R + \Delta R$, where ΔR is an empirical parameter which is determined by fitting the observed lifetime $\tau(=1/\lambda)$ with the known hole cavity sizes in molecular substrates. The best fitted value of ΔR for all known data is found to be 1.656 Å. The correlation between $\tau_3(=\lambda^{-1})$ and free volume (spherical) is shown in Fig. 8. Equation (12) and Fig. 8 become the foundation for the determination of free volume hole size by using PAL. Further development, made by including a finite potential depth instead of infinite potential [22], gave results which were similar to the model presented here.

Based on the same quantum model, we obtained the relationship between the experimental $\theta_{1/2}$, the FWHM of the narrow component in an ACAR spectrum, and the hole radius (R) as [22]:

$$R = 16.60/\theta_{1/2} - 1.656 \tag{13}$$

where R and $\theta_{1/2}$ are expressed in units of Å and mrad, respectively. Using the above equation, we can determine the free volume radius in polymers by measuring $\theta_{1/2}$ of the narrow component from ACAR experiments along a crystallographic orientation. The advantage of ACAR over PAL is that ACAR

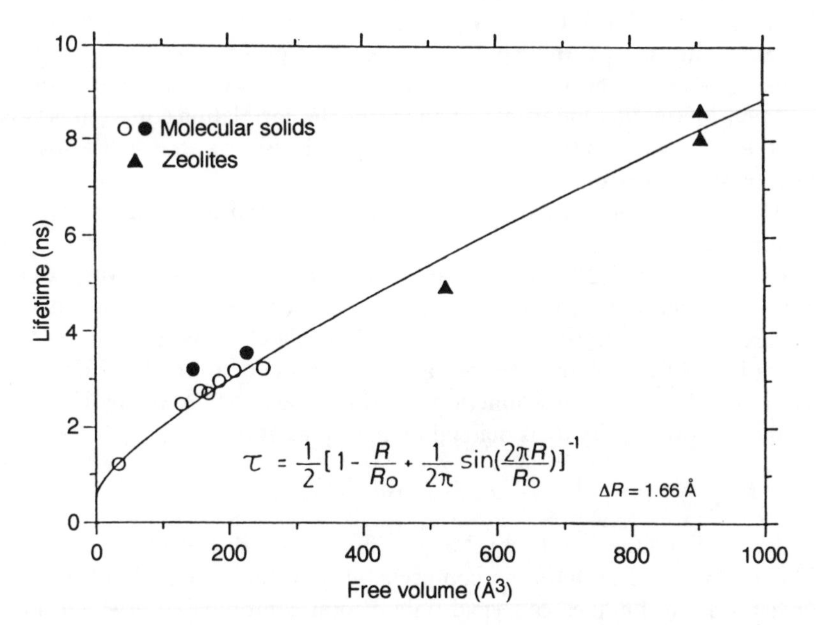

Fig. 8. A universal correlation between the observed *o*-Ps lifetimes and hole volumes [22].

gives an orientational dependence of R (along the axis of measured momentum) whereas PAL gives a mean value of R. On the other hand, an ACAR spectrum takes a few days of data acquisition, whereas a PAL experiment takes only an hour.

Since free volumes exist only in the amorphous material, we describe the fractional free volume $f(\%)$, as a product of free volume V_f and I_3, as [23]:

$$f = V_f(AI_3 + B) \tag{14}$$

where V_f is determined by using the value of τ_3, and A and B are the constants to be determined by including certain physical parameters. Note that I_3 from PAL is not equal to f, the fractional free volume. I_3 is usually a few times f, owing to the Ps being preferentially annihilated in free volume holes. In order to determine the absolute value of f by using equation (14), one needs two physical parameters as well as the positron technique. One way of calibrating A and B [equation (14)] is to use the thermal expansion coefficients for A and to use a reference free volume fraction for B. A simplified equation for epoxy polymers [23] gives $A = 0.0018 \pm 0.0002$ and $B = 0$ (where V_f and I_3 are in units of Å and % respectively).

In a series of epoxy samples of composition DGEBA/DDH/DAB, the lifetime spectra were analyzed in three components as described earlier. The *o*-Ps lifetime τ_3 varied significantly as a function of temperature. The variation of τ_3 with temperature in these polymers is shown in Fig. 9. In Fig. 9 we also have calculated the hole volume according to equation (12) and plotted it on the right ordinate axis. It is seen in Fig. 9 that τ_3 varies greatly with respect to temperature. The steepest change in τ_3 with respect to temperature occurs near T_g.

Free volume distributions of polymers can be obtained by deconvoluting PAL into a continuous lifetime spectrum and using the correlation equation between τ and R. For example, in the same epoxy sample, the free volume distributions as a function of temperature are shown in Fig. 10.

PAL studies of pressure dependence have been reported in an epoxy sample containing a DGEBA: DAB:DDH equivalent ratio of 5:2.2:2.8 [24]. The T_g of the epoxy sample was 62 °C as measured by PAS. The pressure was externally applied uniaxially from 0 to 15 kbar. A rapid decrease of τ_3 as a function of pressure is due to a decrease of free volume hole under stress. The corresponding hole radius of free volumes can be calculated from equation (12). The resulting free volume fractions f as a function of pressure are shown in Fig. 11. The free volume compressibility β_f as defined by the equation:

$$\beta_f = -(\mathrm{d}f/\mathrm{d}p)_T \tag{15}$$

is calculated and the results are shown in Fig. 12. These β_f values are for the free volume alone and exclude the compressibility of the occupied volume. Many conventional techniques can yield only a total compressibility which includes both volumes. The results for β_f are larger than the beta obtained by the volume

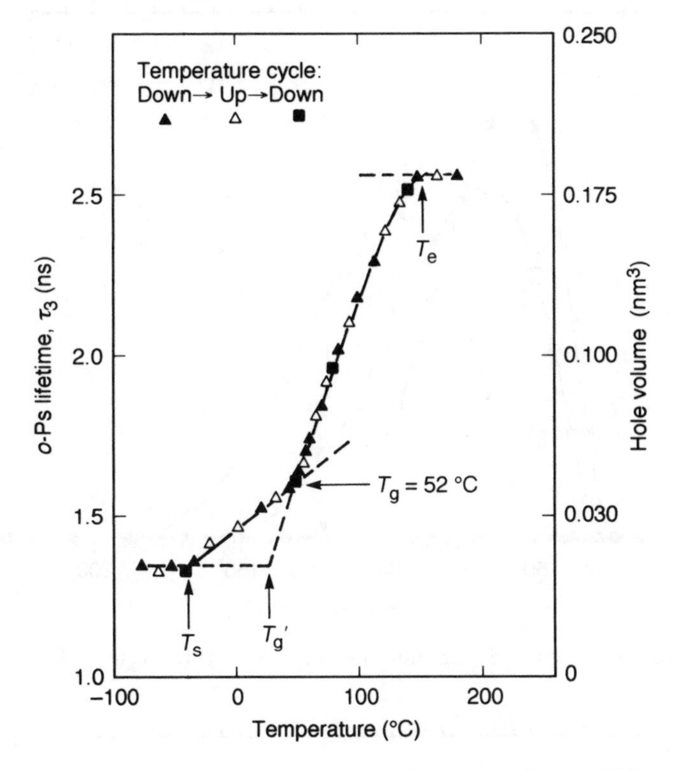

Fig. 9. *o*-Ps lifetimes vs. temperature in epoxy polymers [21].

dilation measurements [20]. Furthermore, the distributions of free volume holes as a function of pressure have been reported [25] and are plotted in Fig. 13.

3.2 Anisotropic structure of free volume holes

The results from PAL measurements give a mean value of hole size and a fractional content of free volumes at a certain condition of pressure and temperature. The free volume hole has been assumed to be spherical. It is known that in reality a free volume hole contains a three-dimensional structure. In an ACAR experiment, it is possible to map out the three-dimensional structure of free volumes in oriented polymers by measuring ACAR as a function of sample orientation.

Samples of PEEK [poly(ether ether ketone)] and PMMA (polymethyl methacrylate) were stretched uniaxially and biaxially, respectively, and ACAR measurements were performed at room temperature [26]. The ratio of the long axis to the narrowed axis of PEEK samples was measured as being 2.60 and the

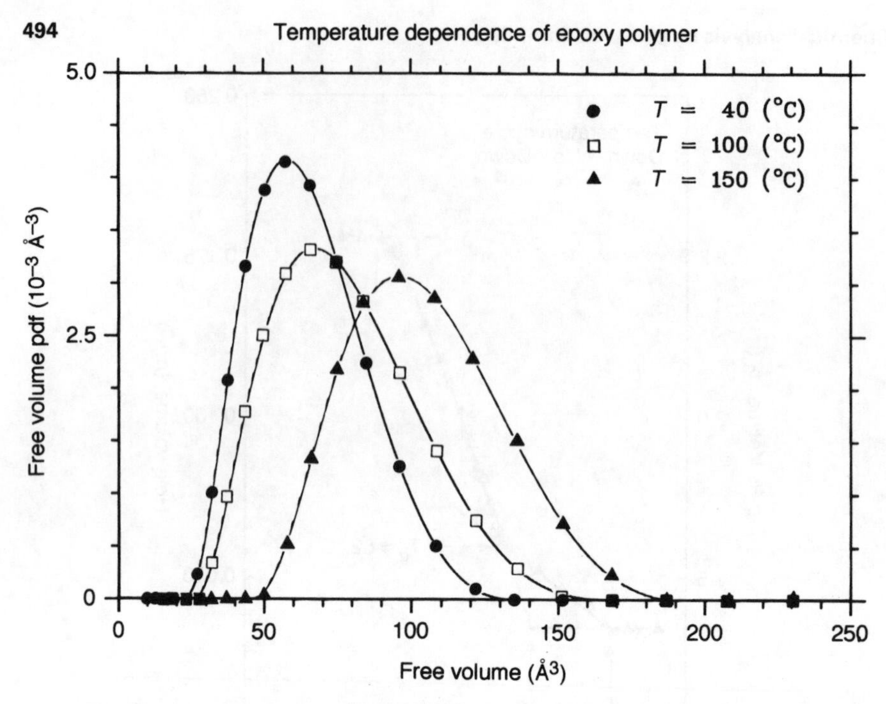

Fig. 10. Hole volume distributions in an epoxy polymer ($T_g = 62\,°C$) [19].

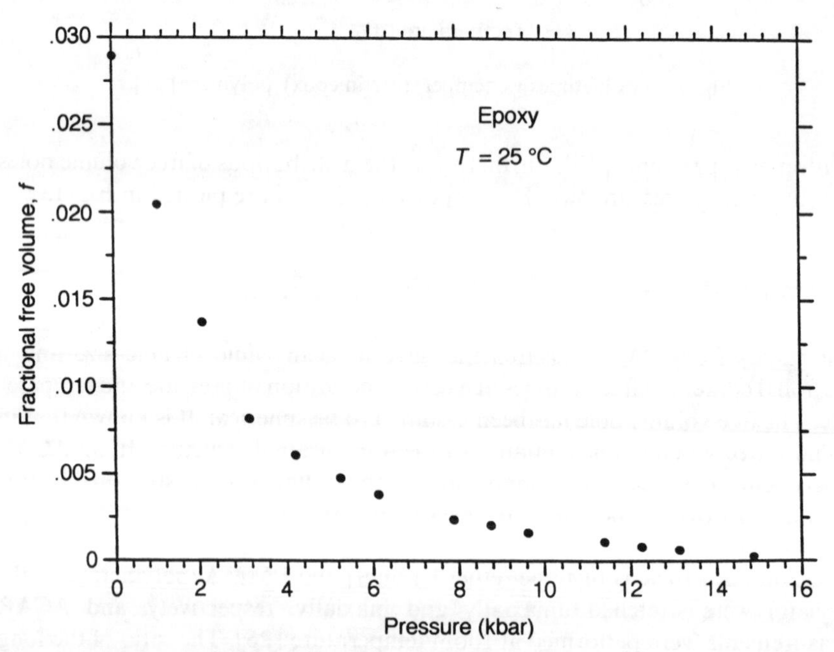

Fig. 11. Free volume hole fraction vs. pressure in an epoxy polymer ($T_g = 62\,°C$) [24].

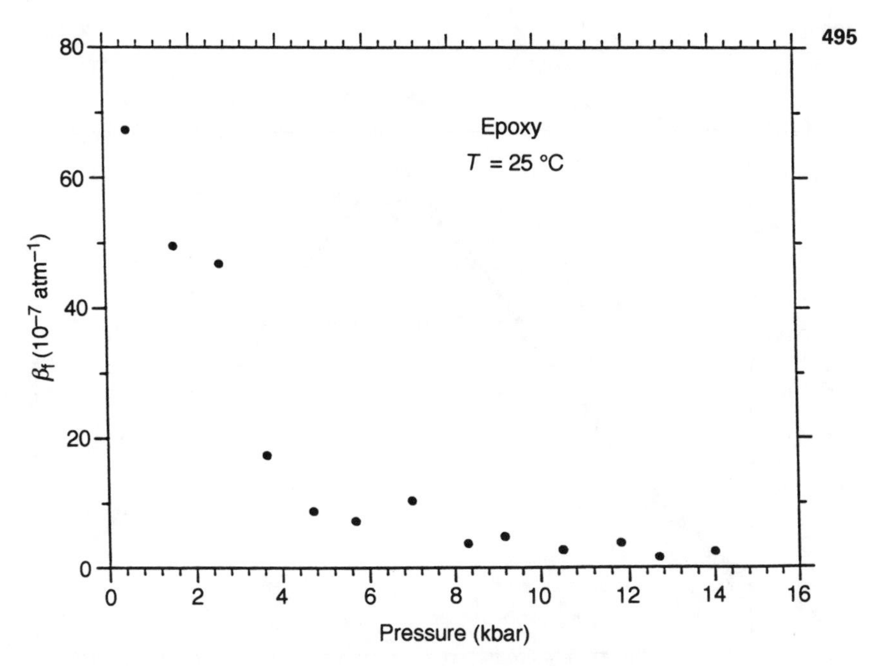

Fig. 12. Free volume hole compressibility vs. pressure in an epoxy polymer ($T_g = 62\,°C$) [24].

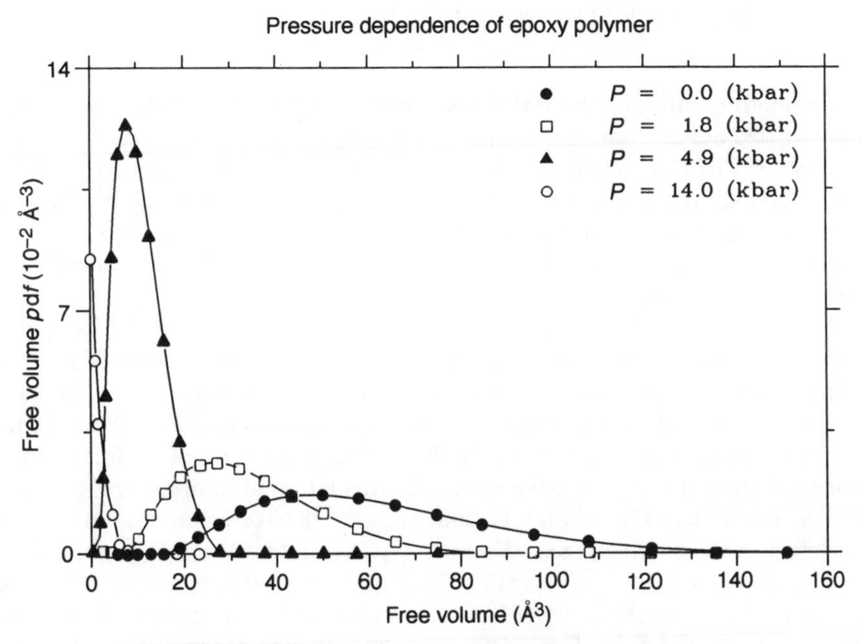

Fig. 13. Free volume hole distributions in an epoxy polymer as a function of pressure [25].

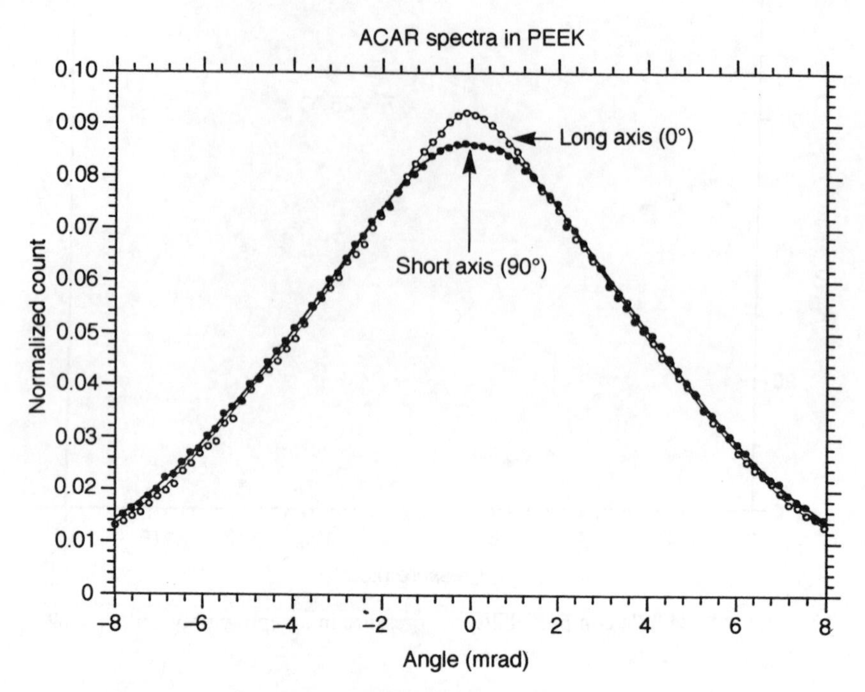

Fig. 14. ACAR spectra of a PEEK polymer at two orientations [26].

corresponding ratio for the PMMA sample was 1.70. The ACAR experiments were carried out as a function of sample orientation. Figure 14 shows the ACAR spectra for PEEK at two orientations: 0° and 90° refer to the long axis and the narrowed axis, respectively. As seen in Fig. 14, the long axis orientation shows a narrower ACAR spectrum than the short axis orientation. This difference is due to the uncertainty of the momentum distribution when *p*-Ps is localized to a different spatial event.

The obtained ACAR spectra were fitted into three Gaussians by using the computer program PATFIT [16]. The results for $\theta_{1/2}$ (the FWHM of narrow components) as a function of orientation are shown in Fig. 15. As discussed earlier, the FWHM of the narrow component, $\theta_{1/2}$, is related to the free volume dimension according to equation (13). We calculate the radius of the free volume using equation (13) and plot the radius vs. orientation of samples in Fig. 15 for PEEK. It is seen in Fig. 15 that the dimensions of the free volume vary from the stretched to the narrowed axis. From the ACAR data, we calculate the anisotrophy, the ratio of the dimension at the long axis to that at the short axis, as being 2.6 and 1.7 for PEEK and PMMA, respectively. The anisotropies found are consistent with the ratio of stretched dimensions in macroscopic measurements. The shape of free volume in stretched PEEK is thus established as being

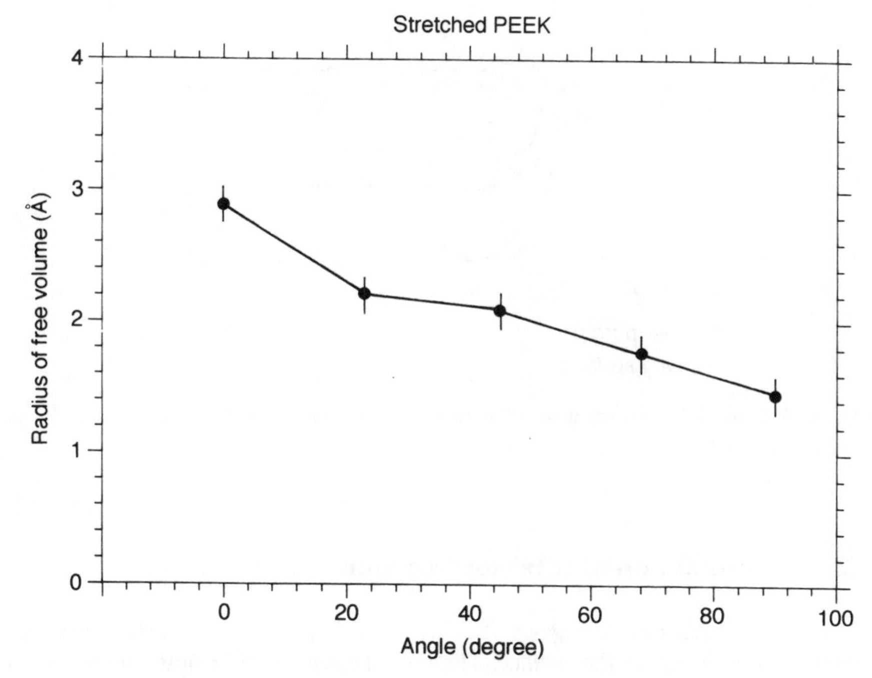

Fig. 15. Variations of free volume radius vs. orientations of a PEEK polymer [26].

cylindrical with an anisotropy of 2.6, and the shape in PMMA is spherical in two dimensions with an anisotropy of 1.7.

4 SURFACE-STATE ANALYSIS OF POROUS MEDIA

4.1 In situ characteristics for Ps

One of the most intriguing problems in surface research is that many active surfaces available for catalytic reactions are not accessible by the conventional surface probes, such as electrons, neutrons, photons, and ions [27]. These conventional probes interact with surfaces via an external approach, categorized as an ex situ analytical method. On the other hand, PAS is classified as an in situ technique, where the positron and Ps diffuse into surfaces internally after they are thermalized. The basic principle of in situ characteristics is the strong repulsive force between the positron and the core electrons of the media. The trapping property of the positron and Ps for cavities, holes and surfaces has made PAS a unique technique for surface analysis in porous materials. Figure 16 illustrates this in situ ability to probe the surfaces in a porous medium.

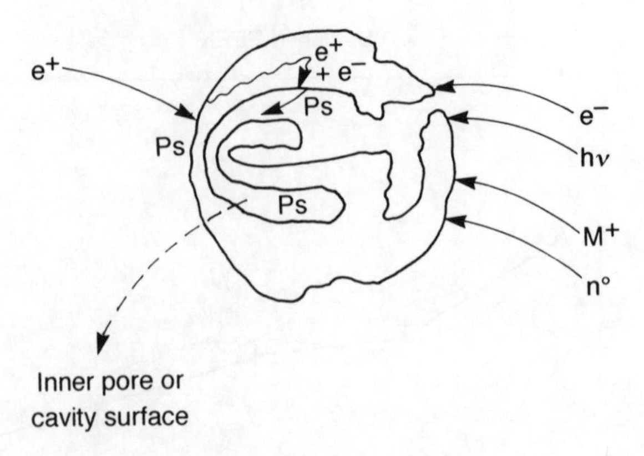

Fig. 16. Schematic demonstration of in situ characteristics of the positron and Ps in porous media [9].

4.2 Determination of total surface area

For catalytic reactions, a large surface area is always desirable so that chemical reactions will occur on the surface. The size of cavity is distributed from 1 Å to several thousand Å. The conventional technique for surface area analysis is the BET (Brunauer, Emmett and Teller) method. The BET method monitors the absorbed volumes as a function of pressure by nitrogen gas. The smaller cavities and hidden surfaces escape BET measurements even though they are still important for many catalytic reactions. Ps, on the other hand, has a size of 1 Å and is able to determine areas of all surfaces from 1 Å to several thousand Å in size. A linear correlation between the observed long lived *o*-Ps intensity, I_3, and known surface area is observed, and is plotted in Fig. 17. In order to use this universal slope for the surface area determination, empirically fitted equations are available:

$$I_3 = 3.00 + 0.33\delta \left(> 70 \frac{m^2}{g} \right) \tag{16}$$

$$I_3 = 0.080\delta \left(< 70 \frac{m^2}{g} \right) \tag{17}$$

where δ is the surface area in m^2/g and I_3 is the observed longest *o*-Ps intensity in the media. The esimated accuracy of surface area determination is $\approx 7\%$. More data for different systems will help improve the accuracy of this application. The pore size and its distribution in porous media can be determined by using the same method as described in Section 3 for free volume holes in polymers.

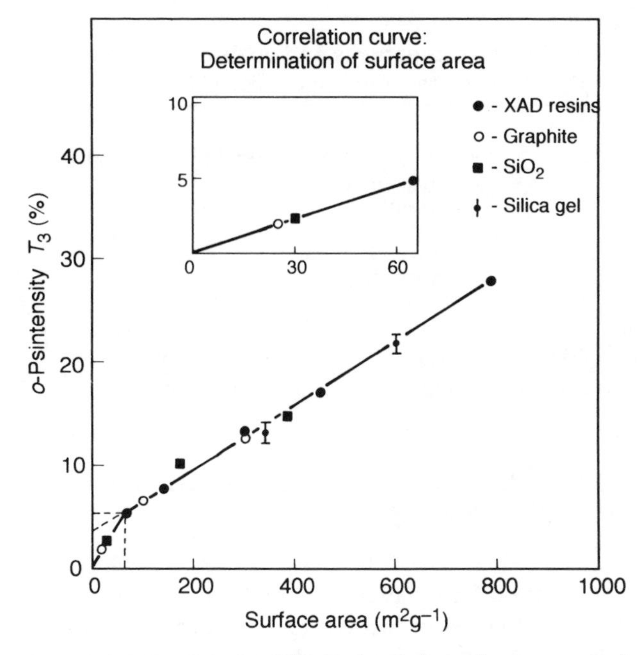

Fig. 17. Universal correlation between observed o-Ps intensity vs. reported surface areas [28].

4.3 Determination of chemical state in catalysts

In real catalytic reactions on surfaces, cations impregnate the surfaces. The chemical states of these impregnating ions are very important for understanding catalytic reactions. PAS has been used to determine the oxidation state and the acidity of impregnated surfaces. Inhibition consants K for Ps formation are found to correlate with the parameters that correlated with the oxidation states and the acidity in porous media. For example, in a zeolite Y system, one observes an increase of o-Ps annihilation rate and an increase of surface Ps formation probability as a function of impregnating cationic concentrations. Results for k_{Ps} and K from these plots for different impregnating ions are listed in Table 5.

The formation of Ps on surfaces is related to the electronic properties of local electrons on the surface. An increase of Ps formation is a result of increased electron and positron mobility. If we consider the positron on the surface as a Brønsted acid, then we can correlate the Brønsted acidity of these ions with K values [29]. Furthermore, it is also better to correlate log(k) with the charge-to-radius ratio [29].

TABLE 5.
Ps reaction rate constants and anti-inhibition constant in metallic ions absorbed on zeolite [29]

Sample	k_{Ps} (mol/kg^{-1}s^{-1} ($\times 10^9$))	K (10^3 kg zeolite/moles of ions)
Cs^+–NaY	0.086–0.36	0.0033
Li^+–NaY	0.050–0.20	0.0027
Ni^{2+}–NaY	1.05	0.0654
Cu^{2+}–NaY	5.58	0.387
Fe^{3+}–NaY	0.065–0.44	0.0704
Cr^{3+}–NaY	1.80	0.323
Th^{4+}–NaY	0.125	0.0190
Zr^{4+}–NaY	6.25	0.913

Note: The units are mol of impregnating ions per kg of substrate.

5 POSITRON ANNIHILATION-INDUCED AUGER ELECTRON SPECTROSCOPY

Recently, a new method for the excitation of Auger electron emission has been developed by using slow positrons to create the necessary core ionizations by positron–electron annihilation [31]. In this application, one establishes a well defined slow positron beam connecting to the conventional Auger electron spectrometer. Positron annihilation-induced Auger electron spectroscopy (PAES) has advantages over the conventional Auger electron method: (1) the elimination of the very large secondary electrons increases the sensitivity of the surface and the defects; (2) PAES can be used for the elemental analysis of the topmost atomic layer of surfaces [32].

The PAES mechanism is different from that of the conventional Auger method in that the hole of the core electron is generated by positron annihilation instead of by incident electrons, as in the conventional method. Figure 18 contrasts the PAES excitation mechanism with that of conventional electron-induced Auger spectroscopy. The PAES mechanism is: (1) positrons are implanted at low energy (below the energy of Auger electrons) and then diffuse and get trapped in the system; (2) a few percent of the positrons annihilates with core electrons and leaves atoms in excited states, and (3) the atom relaxes via emission of an Auger electron. The additional annihilation photon signals are used to gate the coincident signals between the Auger electron signal and the electrons of positron annihilation. This has the effect of significantly reducing background. Figure 19 shows a comparison of spectra between PAES and the electron-induced Auger method. As shown in Fig. 19, the background of secondary electrons is significantly decreased in PAES. Surface sensitivity of PAES to the outermost layer has also been achieved. Applications of PAES in chemical analysis appear very promising for the near future.

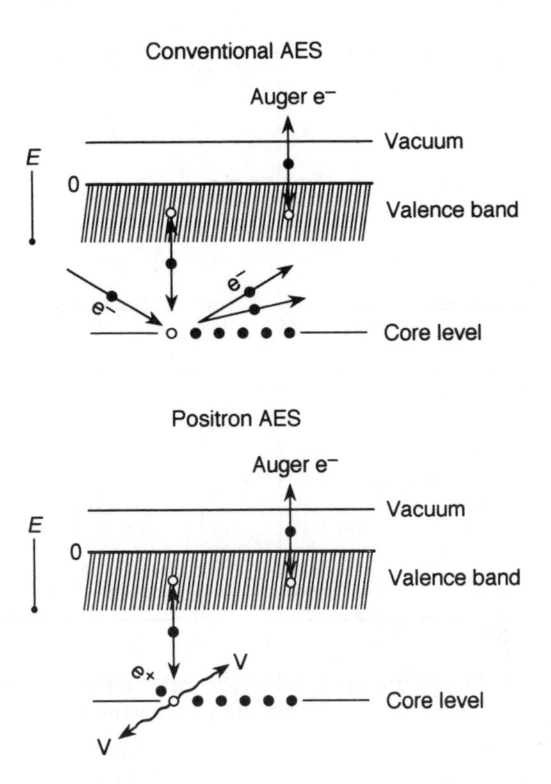

Fig. 18. Comparison of the core hole creation mechanism in electron Auger electron and positron Auger electron. In PAES the core hole is created by an annihilation process, whereas in electron Auger electron the core electrons are removed by collisions with an electron [31].

6 POSITRON IONIZATION MASS SPECTROMETRY

The positron has been used as an ionization agent for mass spectrometric application to polyatomic molecules [33]. This application also requires a monoenergetic slow positron beam of energy ranging from a few tenth of an eV to several keV. This study was carried out at Oak Ridge National Laboratory (ORNL), TN, USA. The design of a positron ionization mass spectrometer requires a Penning trap which allows multiple collisions between the positron and molecules. Figure 20 shows the design of such a positron mass spectrometer used at ORNL.

The mass spectra from positrons having keV energy resemble those from electrons. However, spectra at low energy ($\approx 10\,\mathrm{eV}$) show a special feature. For example, for decane, the mass spectra of ion fragments produced by positrons having energy above and below the Ps formation threshold are quite different. As

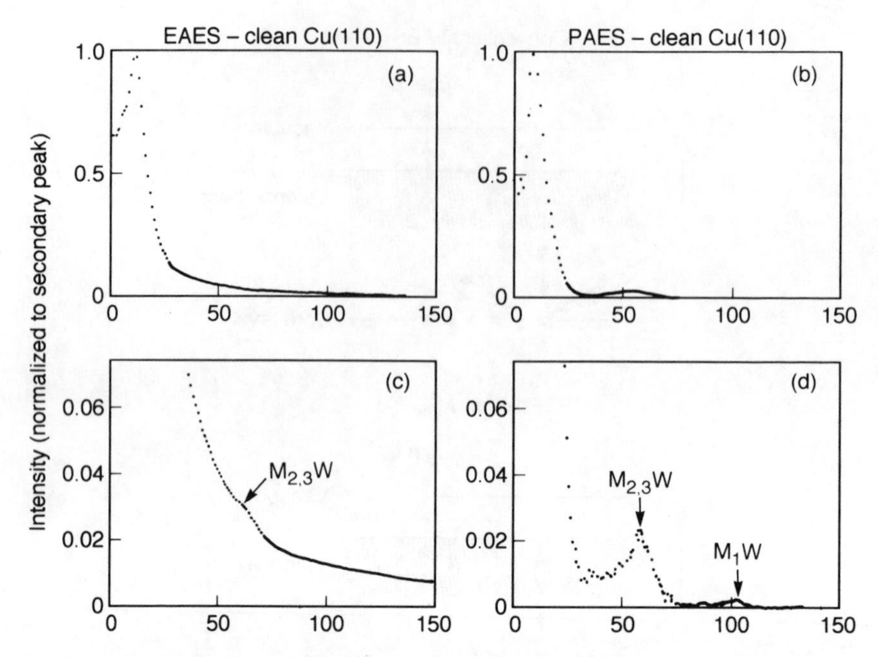

Fig. 19. Comparison of (a) electron-induced vs. (b) positron annihilation-induced Auger spectra [32] on the surface of Cu. Parts (c) and (d) are the magnified data of parts (a) and (b) respectively.

shown in Fig. 21 [34], when a molecule is ionized by Ps formation (bottom figure), the electron is extracted from the molecule and the Ps atom formed escapes from the site of the ionization process. When the Ps finally annihilates, it does so at a location completely isolated from the molecule for which the ionization process took place. In this case, the spectra have a better resolution and the parent ion (C_{10}) is highest in population. On the other hand, when the ionization process does not involve Ps formation, the mechanism of ionization is not clearly understood. The parent ion is undetectable and small ion fragments predominate. These results show that positrons are valuable sources for generating ions from organic molecules in mass spectrometry.

7 OTHER POSITRON SPECTROSCOPIES SIMULATING ELECTRON SPECTROSCOPY

7.1 Positron microscopy

Positrons have been employed to construct various types of microscopes similar to electron microscopes, such as the transmitting positron microscope [35], the

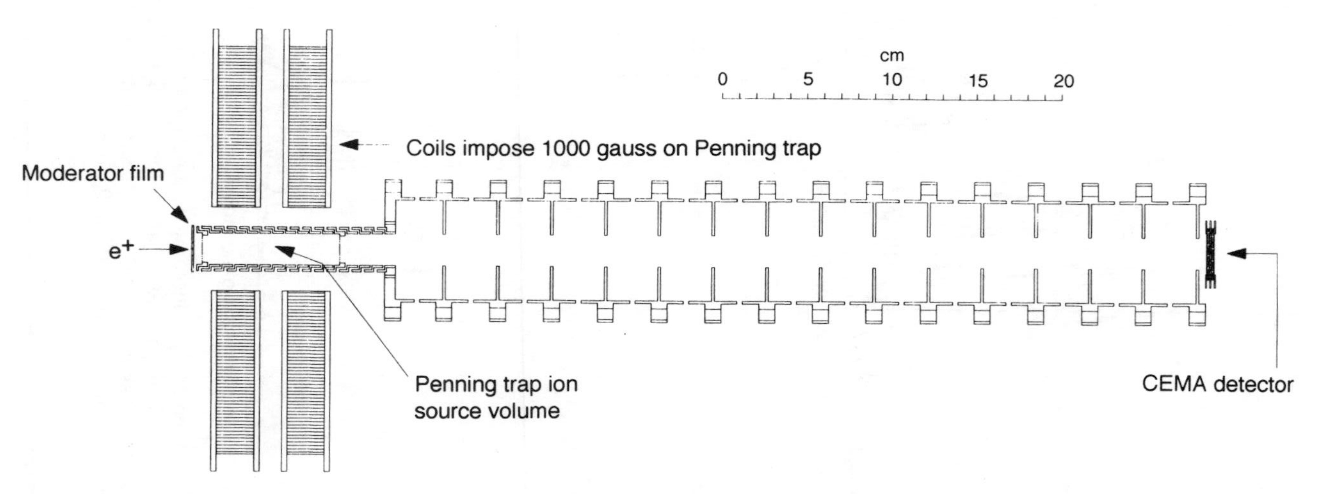

Fig. 20. Diagram of the time-of-fight mass spectrometer [33]. The spectrometer allows the pulses of positrons from the LINAC to be moderated at low energies and captured in a Penning trap. Positrons are contained in the Penning trap for periods >0.1 s.

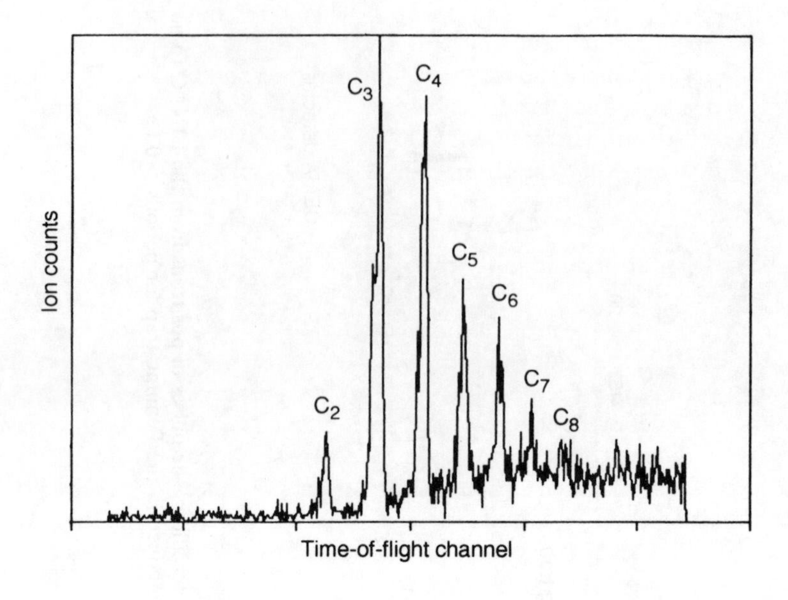

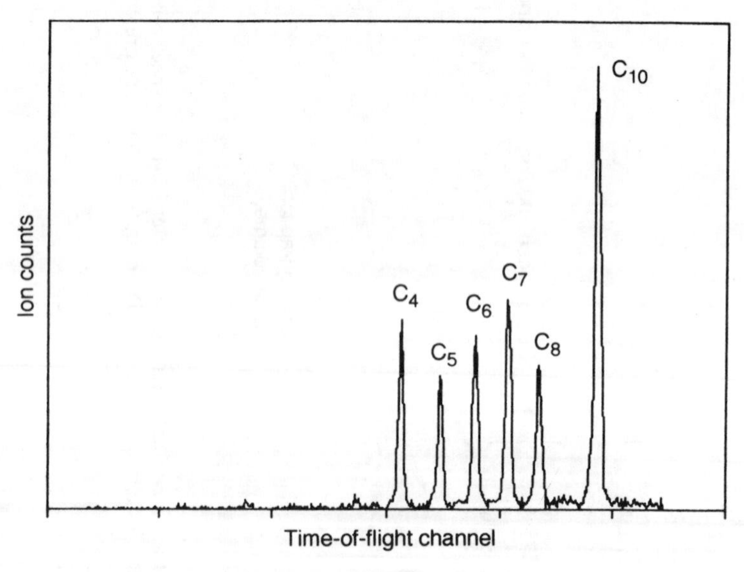

Fig. 21. Mass spectra of ion fragments produced by the ionization of decane by positrons having energy 0.7 eV below (top figure) and 5.0 eV above (bottom figure) the threshold for Ps formation [34]. The distinct difference in the mass spectra is due to the difference in mechanism involved without (top) and with (bottom) Ps formation [34].

re-emission positron microscope [36], and the tunneling positron microscope [37]. In order to achieve the same level of spatial resolution as electron microscopes, a positron microbeam with high brightness must be achieved. The advantage of positron microscopes is that the positron is more sensitive to surfaces and defects than the electrons.

7.2 Positron energy loss spectroscopy

Electron energy loss spectroscopy is a known powerful analytical method for studying the electronic states and chemical bonding of surfaces. Positron energy loss spectroscopy has been developed as a method similar to that involving electrons [38, 39], but the energy loss spectra obtained with the positron and with electrons are different. In general, positron spectra are more complicated. Vibrational levels of chemisorbed CO on Ni(100) surfaces have been observed [39], but the electronic levels in W, Si and Ni could not be detected by using the positron [38, 40]. The practical application of positron energy loss spectroscopy needs further exploration.

7.3 Low energy positron diffraction

Low energy electron diffraction (LEED) is known to be a very powerful tool for the determination of atomic structures in solids. Low-energy positron diffraction (LEPD) has also been recently developed [41, 42]. Because of the opposite charge of the positron, LEPD spectra are different from LEED spectra. In general, the features of LEPD spectra are simpler than those of LEED. The structural information deduced from LEPD often complements that from LEED. On the other hand, theoretical interpretation for LEPD spectra can be achieved more accurately than for LEED because of the absence of electron–positron exchange terms in the calculations.

7.4 Positron polarimetry

The positron emitted from radioisotope decays in highly polarized. A positron polarimeter has been constructed and used to study the magnetic property of Ni and the origin of life related to chiral molecules [43]. In principle, polarized positrons can be used as a spin probe for the study of magnetic materials.

8 CONCLUSIONS

Positron annihilation has been used as an analytical probe in a variety of applications, from the determination of atomic scale free volume holes to the medical imaging of brain tumors [44]. The main advantages of using positron annihilation are two fold: (1) sensitivity: additional experimental information from annihilation photons, and (2) specificity: the localization of the positron and Ps in open spaces, defects, holes and surfaces. New scientific and technological information has been contributed by the study of positron annihilation. Many fundamental properties of the positron and of Ps in the condensed phase are still not well understood. It is our hope that more research will be pursued in positron annihilation so that its applications to chemical analysis can progress in the near future.

9 ACKNOWLEDGMENT

This research was supported by the National Science Foundation (grant DMR-9004083). I am grateful also for fruitful input from Drs L. D. Hulett, Jr, A. P. Mills, K. F. Canter, A. Weiss, and D. M. Schrader.

REFERENCES

1. P. A. M. Dirac, *Proc. Cambridge Philos. Soc.* **26**, 361 (1930).
2. C. D. Anderson, *Science* **76**, 238 (1932); *Phys. Rev.* **43**, 491 (1933).
3. See, for example, *Encyclopedia of Physics*, edited by R. G. Lerner and G. L. Trigg, Addison-Wesley, Reading, MA (1981).
4. P. A. M. Dirac, *The Principle of Quantum Mechanics*, Oxford University Press, Oxford (1935).
5. See, for example, *Positron Solid-State Physics*, edited by A. Dupasquier and W. Brandt, North-Holland, Amsterdam (1984).
6. M. Deutch, *Phys. Rev.* **83**, 207 (1951).
7. See, for example, *Positron and Positronium Chemistry*, edited by D. M. Schrader and Y. C. Jean, Elsevier, Amsterdam (1988).
8. See, for example, A. Rich, *Rev. Med. Phys.* **53**, 127 (1981).
9. D. M. Schrader, in *Positron and Positronium Chemistry*, edited by D. M. Schrader and Y. C. Jean, Elsevier, Amsterdam (1988), p. 27.
10. H. J. Ache, in *Positronium and Muonium Chemistry*, *Adv. Chem. Ser.* (175), (1979).
11. Y. Ito, in *Positron and Positronium Chemistry*, edited by D. M. Schrader and Y. C. Jean, Elsevier, Amsterdam (1988), p. 120.
12. A. P. Mills, Jr., *Appl. Phys. Lett.* **35**, 427 (1979).
13. See, for example, Positron beams for solids and surfaces, *AIP Conf. Proc.* No. 218, edited by P. S. Schultz, G. R. Massouni and P. J. Simpson, American Institute of Physics, New York (1991).

14. See, for example, R. H. Howell, M. J. Fluss, I. J. Rosenberg and P. Meyer, *Nucl. Instrum. Methods Phys. Res., Sect. B* **B10**, 373 (1985).
15. See, for example, R. Suzuki, Y. Kobayashi, T. Mikado, H. Ohgaki, M. Chiwaki, T. Yamazaki and T. Tomimasu, *Jpn. J. Appl. Phys.* **30**, 1532 (1991).
16. A PATFIT package can be purchased from Risø National Laboratory, Roskilde Denmark.
17. S. W. Provencher, *Comput. Phys. Commun.* **27**, 229 (1982).
18. R. B. Gregory and Y. Zhu, *Nucl. Instrum. Methods Phys. Res., Sect. A* **A290**, 172 (1990).
19. Q. Deng and Y. C. Jean, *Macromolecules* **25**, 1090 (1992).
20. See, for example, J. D. Ferry, Viscoelastic Properties of Physics, 3rd Ed., Wiley, New York (1989).
21. Y. C. Jean, *Microchem. J.* **41**, 27 (1990).
22. H. Nakanishi and Y. C. Jean, in *Positron and Positronium Chemistry*, edited by D. M. Schrader and Y. C. Jean, Elsevier, Amsterdam (1988), p. 159.
23. Y.Y Wang, H. Nakanishi, E. Smith and T. C. Sandreczki, *J. Polym. Sci. B*, **28**, 1431 (1990).
24. Q. Deng, C. S. Sundar and Y. C. Jean, *J. Phys. Chem.* **96**, 492 (1992).
25. Q. Deng and Y. C. Jean, *Macromolecules* **26**, 30 (1993).
26. Y. C. Jean, H. Nakanishi, L. Y. Hao and T. C. Sandreczki, *Phys. Rev. B* **42**, 9705 (1990).
27. See, for example, G. A. Somorjai, *Chemistry in Two-Dimensional Surfaces*, Cornell University Press, Ithaca (1982).
28. K. Venkateswaran, K. L. Cheng and Y. C. Jean, *J. Phys. Chem.* **86**, 4446 (1984); **89**, 3001 (1985); *Chem. Phys. Lett.* **126**, 33 (1986).
29. K. Venkateswaran, Ph.D. Dissertation in Chemistry, University of Missouri–Kansas City (1985).
30. A. Weiss, R. Meyer, M. Jibaly, C. Lei, D. Mehl and K. G. Lynn, *Phys. Rev. Lett.* **61**, 2245 (1988).
31. A. Weiss, D. Mehl, A. R. Koymen, K. H. Lee and C. Lei, *J. Vac. Sci. Technol., A* 2517 (1990).
32. D. Mehl, A. R. Koymen, K. O. Jensen, F. Gotwald and A. Weiss, *Phys. Rev. B* **41**, 799 (1990).
33. D. L. Donohue, L. D. Hulett, Jr., S. A. McLuckey, G. L. Glish and H. S. McKown, *Int. J. Mass. Spectrom. Ion Processes* **97**, 227, 237 (1990).
34. L. D. Hulett, Jr., D. L. Donohue, G. L. Glish, S. A. McLuckey and T. A. Lewis, Positron Annihilation, edited by Zs. Kajcsos, Cs. Szeles, Sz. Vass, B. Levay and A. Vertes, Proc. 9th Int. Conf. Positron Annihilation, Traus. Tech. Pub., Switzerland (1992).
35. J. Van House and A. Rich, *Phys. Rev. Lett.* **60**, 169 (1988).
36. G. R. Brandes, K. F. Canter and A. P. Mills, Jr., *Phys. Rev. Lett.* **61**, 492 (1988).
37. W. E. Frieze, G. W. Gidley and B. D. Wissman, *Solid State Commun.* **74**, 1079 (1990).
38. J. M. Dale, L. D. Hulett and S. Pendyala, *Appl. Surf. Sci.* **8**, 472 (1981).
39. D. A. Fischer, K. G. Lynn and W. E. Frieze, *Phys. Rev. Lett.* **50**, 1149 (1983).
40. E. M. Gullikson, A. P. Mills, Jr., W. S. Crane and B. L. Brown, *Phys. Rev. B* **32**, 5484 (1985).
41. A. H. Weiss, I. J. Rosenberg, K. F. Canter, C. B. Duke and A. Paton, *Phys. Rev. B* **27**, 867 (1983).
42. D. L. Lessor, C. B. Duke, P. H. Lippel, G. R. Brandes, K. F. Canter and T. N. Horsky, *J. Vac. Sci. Technol., A* **A9**, 867 (1983).
43. M. Skalsey, T. A. Girard, D. Newman and A. Rich, *Phys. Rev. Lett.* **49**, 708 (1982).
44. See, for example, I. Lemahieu, *Physicalia Mag.* **12**, 225 (1990).

Part 5

Use of Radio Tracers

Chapter 19

ISOTOPE DILUTION ANALYSIS

K. Masumoto

1 INTRODUCTION

The use of a radioisotope as a tracer is based on the fact that all isotopes of an element behave chemically in the same manner. Various tracer techniques have been developed for chemical analysis. Within these techniques, isotope dilution analysis (IDA) is one of the most important and useful analytical methods, and a number of methods and their combinations have been developed by using the isotope dilution principle.

Originally, IDA was based on the measurement of the specific activity of an element of interest in a spike (or sample) and in the spiked sample. In this case, the 'specific activity' of the element is defined as the radioactivity per unit amount, and 'isotope dilution' is performed by mixing of the radioisotope and the corresponding non-radioactive element.

On the other hand, the direct measurement of the isotope ratio of an element by using the mass difference or a distinctive nuclear property of each isotope has been applied in IDA. In this case, 'isotope dilution' means a change in the isotope ratio by mixing with an enriched isotope.

In this chapter, IDA is divided mainly into two subsections, (1) IDA based on the measurement of specific activity, and (2) IDA based on the measurement of isotope ratio, and several important selected methods are introduced.

Chemical Analysis by Nuclear Methods Edited by Z. B. Alfassi

2 IDA BASED ON SPECIFIC ACTIVITY MEASUREMENT

2.1 General types of IDA

Isotope dilution can be divided into two main types, direct IDA (DIDA) and reverse IDA (RIDA). The sample is diluted with radioisotopes in DIDA and a radioactive sample is diluted with the non-radioactive element in RIDA.

2.1.1 DIDA

When a known amount (W_0) of radioisotope with known specific activity [$S_0 = A_0/W_0$] is added to a sample containing W_x of the element to be determined, the new specific activity (S_1) becomes [$S_1 = A_0/(W_x + W_0)$]. From these two relationships, the ratio of the specific activities (S_0/S_1) is

$$S_0/S_1 = (W_x + W_0)/W_0 \tag{1}$$

The unknown amount (W_x) of the element can be calculated from

$$W_x = W_0[(S_0/S_1) - 1] \tag{2}$$

2.1.2 RIDA

When W_1 of the stable isotope (or element) is added to the radioisotope which has specific activity $S_0 = A_0/W_x$, the specific activity changes to $S_1 = A_0/(W_1 + W_x)$. The unknown amount of the radioisotope can be determined from the equation

$$W_x = W_1/[(S_0/S_1) - 1] \tag{3}$$

RIDA has been used mainly for the determination of the amount of radioactive tracer. If the unknown sample has no radioactivity, radioisotopes can be introduced by (1) activation or (2) synthesis of a labeled compound. The latter is also known as 'derivative IDA' and various modifications have been proposed to date.

2.1.3 Conditions for IDA

In IDA, (1) the added amount of spike must be precisely known, (2) isotopic equilibrium must be achieved after spiking, and (3) the isolated fraction must be chemically pure. A complete recovery of the element is unnecessary and the chemical yield of separation of the element need not be measured in IDA because the specific activity is always constant in any portion of spiked solution. Therefore IDA is especially useful in instances when quantitative separation is difficult or very time consuming.

The sensitivity of IDA is dependent on radioactivity measurement and determination of mass. The sensitivity of the former is dependent on (1) the specific activity of the tracer, (2) the emission probability of radiation from the isotope and (3) the counting efficiency of the detector. For mass determination, as the sensitivity of a chemical balance is of submilligram order, reagents of high molecular weight are preferred for the isolation of elements. For more sensitive determination, other methods such as spectrophotometry should be adopted. In spectrophotometry, reagents with large absorption coefficients are preferred.

2.1.4 Application of IDA

In modern IDA, the analysis of chemical species is one of the most important aspects. The combined use of chromatography and a composite sensor of mass and radioactivity should be promising for routine analysis, because separation and specific activity determination can be done simultaneously. Peters *et al.* used such a method for the determination of fecapentaenes-12 and -14 (FPs) in human feces [1]. In order to study the formation and excretion of FPs by human subjects and the influence of diet, FPs labeled with ^{3}H were separated by liquid chromatography, and the effluent was passed successively through an UV detector and the flow cell of a liquid scintillation detector. The specific activities of the FPs were defined as the count rate (cpm) of the flow-through counter per absorbance unit at 340 nm from the UV detector, and the detection limits were ≈ 1 μg/g.

2.2 Substoichiometric IDA

In trace analysis by IDA, mass determination becomes very difficult. If an equal amount of element can be separated from the spike and the spiked sample, mass determination can be omitted in the assessment of specific activity. This approach was independently proposed by Suzuki [2] and by Ruzicka and Benes [3]. Ruzicka and Stary [4, 5] developed the theory and called this method substoichiometric IDA (Sub-IDA). Suzuki and co-workers have proposed useful variations of this method, especially in solvent extraction systems, activation analysis and chemical speciation. Sub-IDA has brought improvements in simplicity, selectivity and accuracy to IDA.

2.2.1 Sub-IDA

If the radioactivities of substoichiometrically isolated fractions from the spike and the spiked sample are denoted as a_0 and a_1, the specific activities S_0 and S_1 in equation (2) are replaced by a_0 and a_1, as follows:

$$W_x = W_0[(a_0/a_1) - 1] \tag{4}$$

From this equation, it is clear that this method is one of DIDA, and the amount of element can be directly determined from the activity ratio without weighing the separated mass.

2.2.2 Substoichiometry after activation

(a) Sub-RIDA after activation

In this method, activation is used for the production of an appropriate radioisotope of the element to be determined. The sample is divided into two portions after the activation and one portion is diluted with the inactive element. If the fractions isolated substoichiometrically from the diluted portion and the undiluted portion have radioactivity a_1 and a_0 respectively, the unknown amount of the radioisotope can be determined from the equation

$$W_x = W_1/[(a_0/a_1) - 1] \tag{5}$$

In the case of sub-RIDA, it is not necessary to use comparative standards or flux monitors because of induced radioisotope is used only as the tracer. Therefore the same radioisotope produced from other elements, which causes a severe interference problem in activation analysis, does not interfere with the determination but improves the detection sensitivity. Of course, matrix effects, such as the self-shielding effect in neutron activation and the particle range in charged particle activation, do not cause systematic errors. The first example was reported for the determination of Cu impurity in Zn metal by Suzuki and Kudo [6]. The content of Cu was determined by using ^{64}Cu produced through the $^{63}Cu(n, \gamma)\,^{64}Cu$ reaction from Cu and the $^{64}Zn(n, p)\,^{64}Cu$ reaction from Zn.

(b) Sub-AA

This method was proposed by Ruzicka and Stary [7] and by Suzuki and Kudo [6]. The procedure for this method consists in (1) activation of sample and standard, (2) addition of carrier to the sample and standard and (3) substoichiometric separation from sample and standard. A large excess amount (W) of carrier compared with the amount of element already present in the sample (W_x) and in the standard (W_s) is added. The specific activities are expressed as $A_x/(W_x + W) \approx A_x/W$ and $A_s/(W_s + W) \approx A_s/W$. After the same substoichiometric amount of element has been isolated from the sample and the standard, the radioactivities of the separated fractions (a_x and a_s) are measured. The unknown amount W_x is calculated from

$$W_x = W_s(a_x/a_s) \tag{6}$$

The beam fluence of the activating particles and the extent of interfering reaction must be monitored in the same way as in ordinary activation analysis. The most significant feature of this method is that the substoichiometric separation can be done with a macro amount of the element to be determined. This substoichiometric condition leads to a reduction in interference and an improvement in selectivity.

In order to clarify the merit of this method, consider the determination of a

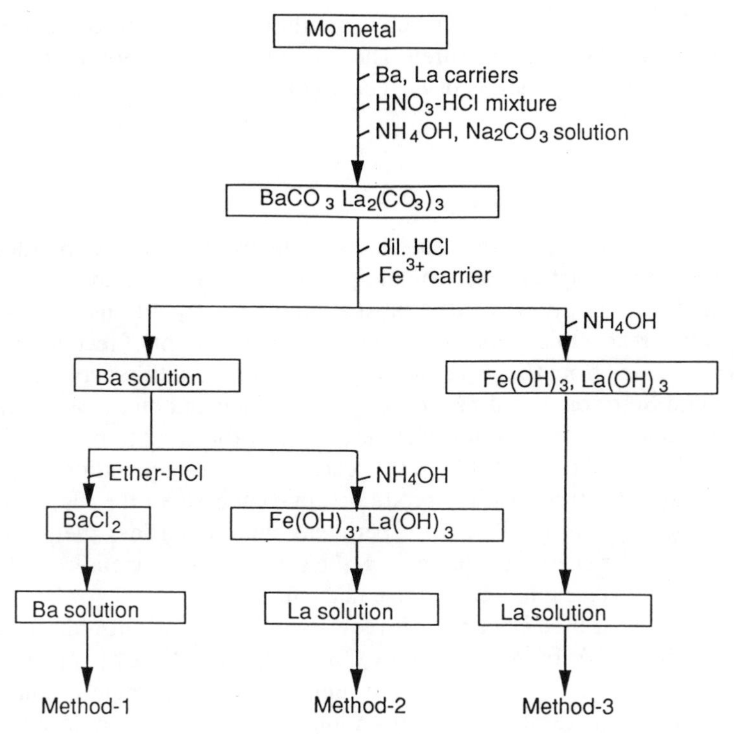

Fig. 1. Analytical procedure for the determination of uranium using substoichiometric separation of barium and lanthanum. (From [9])

trace amount of Yb in a manganese nodule by sub-AA [8]. After neutron activation of the sample (0.1–0.2 g), 25 mg of Yb carrier was added to the sample and the standard. The substoichiometric separation of Yb(III) was effected by extraction with a substoichiometric amount of thenoyltrifluoroacetone and an excess amount of phenanthroline in benzene after dissolving the sample and adjusting the pH to 6–8. Although this separation method is not specific for Yb(III), the interference by other rare earth ions becomes negligibly small, because their amounts in the sample are in the μg range and much smaller than that of the Yb carrier.

In sub-AA, the radioisotope used for the determination is not always the same as the element to be determined. As a typical example, the amount of uranium can be determined by the separation of a fission product, such as ^{140}Ba, and its daughter nuclide ^{140}La after neutron activation [9]. As shown in Fig. 1, three types of separation procedure were tried: (1) substoichiometric separation of ^{140}Ba, (2) quantitative separation of ^{140}Ba and substoichiometric separation of ^{140}La after reaching radioactive equilibrium, and (3) substoichiometric separation of ^{140}La after reaching radioactive equilibrium. In all three methods the radioactivity of ^{140}La was measured. The first method was simple and rapid and applicable down to ppb levels of U. The second and third methods were even more sensitive than the first.

2.2.3 Conditions for sub-IDA

Various separation techniques, such as solvent extraction, precipitation, ion exchange, redox reaction, electroplating, etc. have been used for the substoichiometric separation. Under substoichiometric conditions, reagent and element must react stoichiometrically and an equal amount of reaction product must be separated from the spike and the spiked sample without contamination. Ruzicka and Stary discussed the conditions of substoichiometric separation by solvent extraction [4] and ion exchange [10] and the conditions for the trace determination of 24 metals [11]. In practice, the best conditions of substoichiometric separation should be studied in each case. In the case of solvent extraction, the typical preliminary experiment should consider the pH dependence, the time for chemical equilibrium, the stability of the metal complex, the interference of diverse ions and suitable masking agents.

In order to determine elements with low extraction constants, 'displacement substoichiometry' has been proposed by Obrusnik and Adamek [12] and Braun *et al.* [13]. In this case, an unknown amount (W_x) of the element is completely extracted into the organic phase after adding its radioactive standard solution (W_0). After removing the excess of reagent in the organic phase, the metal complex is partially replaced by a substoichiometric amount of another element with a much higher extraction constant. Then the radioactivity (a_x) of the element back-extracted into the aqueous phase is measured. The standard solution is treated in the same manner as above and the radioactivity (a_0) in the aqueous solution is measured. Finally, the unknown amount (W_x) of the element can be calculated by using equation (4).

2.2.4 Application of sub-IDA

More than 60 elements have been determined by sub-IDA so far. The application by Suzuki [14] of sub-IDA to the analysis of chemical states, which is known as 'selective substoichiometry' or 'substoichiometric speciation', is very attractive and significant. For example, the selective determination of tin species by sub-IDA has been applied to a commercial organotin chemical and a PVC sample [15]. The separation scheme for the speciation of tin compounds is shown in Fig. 2. Inorganic Sn, butyltin, and dibutyl- and tributyl-tin compounds were extracted into benzene from sulfuric acid as the iodide, bromide and chloride, respectively. After complex formation of inorganic, butyl- and dibutyl-tin with a substoichiometric amount of salicylideneamino-2-thiophenol (SATP) in the

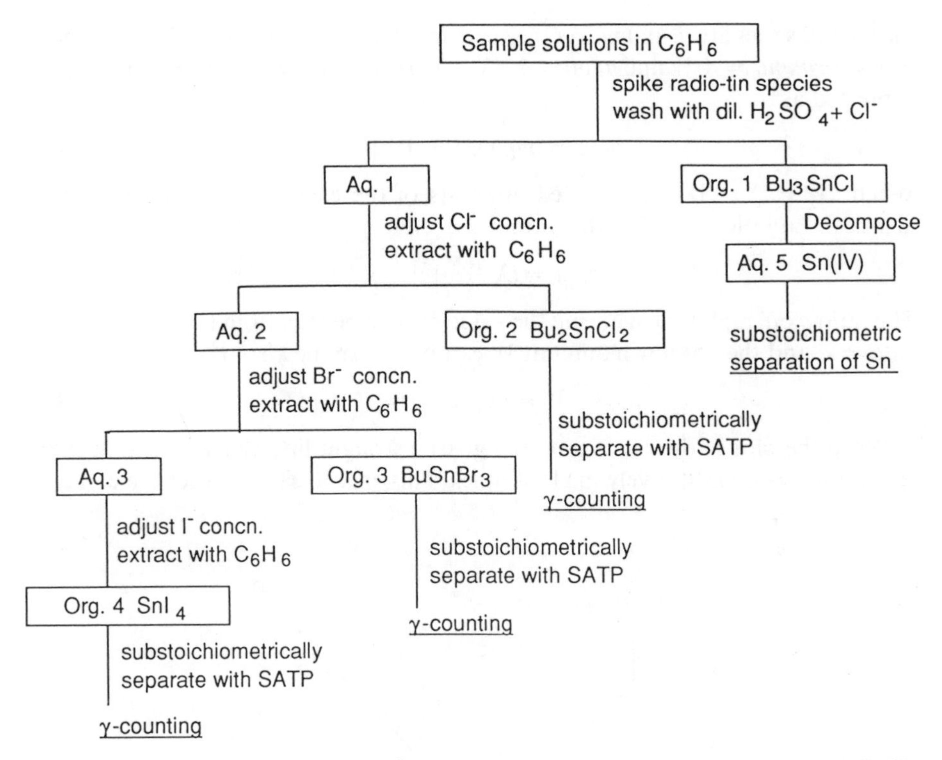

Fig. 2. Scheme of substoichiometric speciation of a series of tin compounds. (From [15])

organic phase, unreacted compounds were removed into the aqueous phase. Only tributyltin was decomposed to inorganic tin by bromine and separated substoichiometrically.

2.3 Sub- and super-equivalence (SSE) method

2.3.1 Principle of SSE

In order to solve the difficult problem of attaining substoichiometric conditions, the sub- and super-equivalence method (SSE) was proposed by Klas, Tolgyessy and Klehr in 1974 [16]. In SSE, two series of solutions are prepared. One is a series composed of one solution containing radioactivity kA and weight kW of an element, and the other is a series of solutions containing radioactivity A and weight $iX + W$, where W is a unknown weight of the element to be determined, k is a constant greater than 1 and i is a series of integers. When the element is isolated from the above two series of solution by using a constant amount of reagent, the activity and the amount in the first series are a_0 and m_0 and those in the second series are a_i and m_i. As the specific activities of the first and the second series are $a_0/m_0 = A/W$ and $a_i/m_i = A/(iX + W)$ respectively, the ratio a_0/a_i can be expressed as

$$a_0/a_i = (m_0/m_i)(iX/W + 1) \quad (7)$$

When $kW = iX + W$, the isolated amounts of the element in both series are identical. Equation (7) takes the form:

$$a_0/a_i = iX/W + 1 = k \quad (8)$$

The ratio a_0/a_i is plotted against i; the value of i can be read from the point where $a_0/a_i = k$, and the unknown amount W can be determined from

$$W = iX/(k-1) \quad (9)$$

When the plot of the ratio a_0/a_i vs. i gives a straight line, this means that the reagent reacts quantitatively in all samples. This special case satisfies the condi-

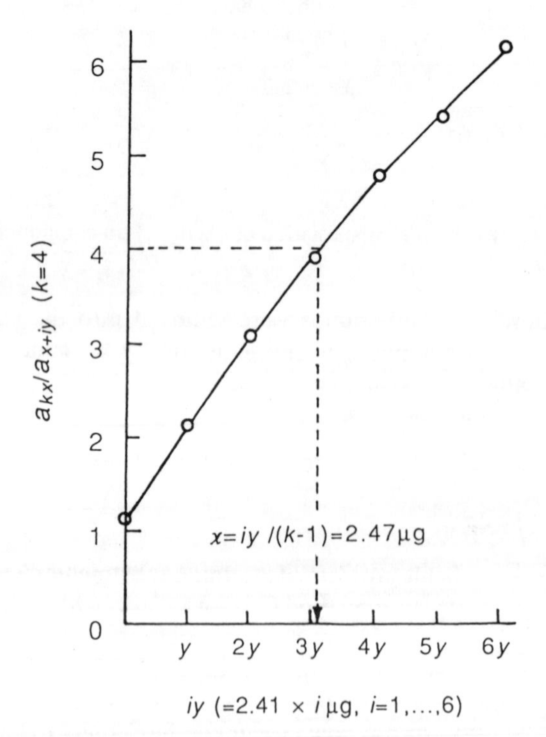

Fig. 3. An example of the determination of Sb(III) in the presence of an interfering element [As(III)] by redox SSE-IDA: 0.2 ml of [Sb(III) + ^{125}Sb(III) 12.04 µg (total) + As(III) 7.49 µg]/ml solution in 6 M HCl was taken. $K_2Cr_2O_7$, 6×10^{-2} µequiv; $[HCl]_{ox} = 3.3$ M; $Vol_{ox} = 3.3$ ml; $[HCl]_{ex} = 0.9$ M; $Vol_{ex} = 11.3$ ml. (From [18])

tion of sub-IDA. Kyrs and Prikrylova discussed the analytical error in SSE, and 2–4 is recommended as in the k value [17]. They concluded that SSE allows determination at a concentration level two or three orders of magnitude lower than that in sub-IDA.

2.3.2 Application of SSE

Several applications to the determination of elements and organic compounds have been reported. Yoshioka and Kambara compared the characteristics of sub-IDA and SSE by using the example of Sb determination by a redox method [18]. Known amounts of Sb(III) were oxidized by a substoichiometric amount of potassium dichromate in hydrochloric acid media, and the unreacted Sb was extracted by BPA in chloroform. Then the radioactivity of Sb(V) in the aqueous phase was measured. An example of the graphical method for calculating the result is shown in Fig. 3. The result was in good agreement with the added amount of 2.41 μg. In contrast, sub-IDA resulted in large errors.

3 IDA BASED ON ISOTOPE RATIO MEASUREMENT

3.1 Isotope dilution mass spectrometry

Mass spectrometry is the most popular method for the measurement of isotope ratios. Isotope dilution in mass spectrometry has been used for highly accurate and very sensitive analyses of elements.

When isotope j is used as a spike, the unknown mass is given by

$$W_x = W_s(AW_x/AW_s)(\theta_{js}/\theta_{jx})(R_s - R_m)/(R_m - R_x) \qquad (10)$$

where the subscripts x, s and m mean sample, spike and their mixture, respectively; W = weight of element; AW = atomic weight; R = mass peak ratio (isotope i/isotope j); θ_{js}, θ_{jx} = isotope abundance of isotope j in spike and sample.

Various important organic compounds in biological or ecological systems consist mainly of light elements such as H, C, N and O. Analysis for these compounds is of significant concern in metabolic research. In such cases, the isotope dilution method is preferred because quantitative recovery from living organisms is very difficult. Labeling with stable isotopes of $^{2}H(D)$, ^{13}C, ^{15}N or ^{18}O is adopted in order to avoid radioactive contamination.

For the monoisotopic element, a long lived radioisotope can be used as spike. For example, Fasset and Murphy reported the determination of iodine by IDMS [19]. As natural iodine is a monoisotopic element (^{127}I), the long lived radioisotope ^{129}I (half-life = 1.59×10^{7} years) was used as spike.

Recently, mass spectrometry has been used for radioisotope detection, instead of radioactivity measurement. For example, an ultra-trace level of ^{99}Tc has been

TABLE 1.
Alpha-emitting isotopes used as spikes for isotope dilution alpha spectrometry

Element	Isotope	Half-life	Principal α-ray (MeV)	Reference
Th	^{229}Th	7.3×10^3 y	4.845	21
U	^{233}U	1.592×10^5 y	4.824	22
	^{236}U	2.342×10^7 y	4.494	21
Pu	^{236}Pu	2.85 y	5.768	28
	^{238}Pu	87.74 y	5.499	23, 24
	^{239}Pu	2.41×10^4 y	5.135	24, 25
	^{242}Pu	3.76×10^5 y	4.901	27, 28
Am	^{243}Am	7.37×10^3 y	5.276	27

determined by using ^{97}Tc as isotope spike [20]. Negative thermal ionization has high ionization efficiency for Tc as TcO_4^- and can eliminate the isobaric interference of molybdenum, because Mo forms MoO_3^- and not MoO_4^-. The detection limit was 5×10^6 atoms of ^{99}Tc.

3.2 Isotope dilution alpha spectrometry

The determination of actinide elements is important for the evaluation of the burn-up of nuclear fuel, the study of environmental pollution caused by nuclear accidents and the geological study of U and Th. Alpha-particle spectrometry is a popular method for the detection of these elements and is often coupled with the isotope dilution method.

When isotope j is used as a spike, the appropriate equation is,

$$W_x = W_s(AW_x/AW_s)(1 - R_m/R_s)/(R_m - R_x)(\theta_{is}\lambda_i)/(\theta_{jx}\lambda_j) \tag{11}$$

where the subscripts x, s and m mean sample, spike and their mixture, respectively; W = weight of element; AW = atomic weight; R = α-particle emission ratio (isotope i/isotope j); $\theta_{i,j}$ = isotopic abundance of isotope j or i; λ = decay constant of isotope. This isotope dilution method can also be applied in γ-ray spectrometry by using γ-emitting nuclides.

The experimental procedure comprises (1) isotope dilution, (2) chemical separation, (3) electroplating and (4) α-particle spectrometry.

Alpha-emitting nuclides suitable as spikes are listed in Table 1. For the determination of plutonium, ^{238}Pu or ^{239}Pu is often used as spike. Additionally, enriched ^{236}Pu and ^{242}Pu are also available.

3.3 Stable isotope dilution activation analysis

In stable isotope dilution activation analysis (SIDAA), it is necessary that (1) the analyzed element consists of at least two stable isotopes, and (2) these isotopes can be converted to radioisotopes that are separately detectable by conventional γ-ray spectrometry.

When isotope j is used as spike, the following equation applies:

$$W_x = W_s(AW_x/AW_s)(\theta_{js}/\theta_{jx})[(R_m/R_x) - (\theta_i/\theta_j)_s/(\theta_i/\theta_j)_x]/[1 - (R_m/R_x)] \quad (12)$$

where the subscripts x and m indicate the sample and the mixture, respectively; W = weight of element; AW = atomic weight; R = activity ratio (isotope i/isotope j); θ_{js}, θ_{jx} = isotopic abundance of isotope j in spike and sample. This equation was introduced by Masumoto and Yagi in 1983 [29]. Activity ratios R_x and R_m can be substituted for the γ-ray peak area ratio in the natural and the spiked sample, respectively.

If the enriched isotope is irradiated simultaneously, equation (10) for IDMS is also available. Yonezawa and Komori applied this equation to the neutron activation analysis of Hf in Zr [30]. Tsukada *et al.* proposed the same method as the 'NAA using activable tracer method' and applied it to the analysis of Cd [31] and rare earth elements [32].

As is seen clearly from equation (12), it is not necessary to monitor the activating fluxes. When the element to be determined is separated from the sample matrix before or after activation, it is not necessary to determine the chemical yield of the separation. It is also not necessary to consider the matrix effect. Radioactivity measurement using a Ge detector is easier than mass spectrometry. However, the nuclear interference problem which is caused by the same radioisotope being produced from different isotopes can not be overcome.

TABLE 2.
Elements determined by stable isotope dilution activation analysis and their isotope pairs used for analysis

Element	Isotope	Reaction	Product	Half-life	Principal γ-ray (keV)	Reference
Ca	$^{48}Ca^a$	(γ, n)	^{47}Ca	4.536 d	1297	29, 35
	^{44}Ca	(γ, n)	^{43}K	22.3 h	373, 618	
Zn	^{66}Zn	(γ, n)	^{65}Zn	244.1 d	1116	29
	$^{68}Zn^a$	(γ, n)	^{67}Cu	2.580 d	185	
Rb	^{85}Rb	(γ, n)	^{84}Rb	32.87 d	882	35
	$^{87}Rb^a$	(γ, n)	^{86}Rb	18.66 d	1077	
Sr	$^{86}Sr^a$	(γ, n)	^{85m}Sr	1.126 h	232	33, 35
	^{88}Sr	(γ, n)	^{87m}Sr	2.795 h	388	
Ce	^{140}Ce	(γ, n)	^{139}Ce	137.7 d	166	29, 35
	$^{142}Ce^a$	(γ, n)	^{141}Ce	32.50 d	145	
Sr	$^{86}Sr^a$	(p, n)	^{86}Y	14.74 h	1077	34
	^{87}Sr	(p, n)	^{87m}Y	12.9 h	381	
Cd	^{114}Cd	(n, γ)	^{115}Cd	2.228 d	336 (^{115m}In)	31
	$^{116}Cd^a$	(n, γ)	^{117}Cd	2.49 h	273	
Dy	$^{156}Dy^a$	(n, γ)	^{157}Dy	8.1 h	326	32
	^{164}Dy	(n, γ)	^{165}Dy	2.33 h	362	
Hf	$^{174}Hf^a$	(n, γ)	^{175}Hf	70 d	343	30
	^{180}Hf	(n, γ)	^{181}Hf	42.4 d	482	

[a] Enriched isotope used for determination.

The sensitivity of this method depends on the isotope abundance of the target nuclide, the activation cross-section, the nuclear characteristics of the product nuclide, such as half-life, branching ratio and energy of γ-ray, detection efficiency and so on.

Examples of SIDAA are listed in Table 2. SIDAA has been applied to cases with chemical separation after irradiation [30], before activation [31–34] and without chemical separation [35]. SIDAA has been applied to the simultaneous determination of four elements [35].

4 SPECIAL FEATURES OF ISOTOPE DILUTION

4.1 Isotope effect

IDA is based on the fact that all isotopes of an element behave chemically in the same manner. Usually the isotope effect caused by mass difference is negligible except for light elements, which are the major elements in organic substances. This is especially significant in the study of metabolism in biological or ecological systems. These compounds pass through many processes from the initial isotope dilution step to the detection step in IDA. Therefore attention should be paid to possible fractionation by mass difference.

In atomic or molecular emission or absorption spectra, the characteristic line of each atom or molecule shows a slight difference depending on the isotope involved. This isotope effect is called an 'isotope shift', and can be utilized in IDA for the selective detection of isotopes.

4.2 Isotopic equilibrium

The chemical and physical state of the spike is often different from that of the element in the sample. In IDA, complete isotopic exchange must be attained before separation. If an element exists in several oxidation states, the redox cycle is indispensable for attaining a uniform oxidation state after isotope dilution. The achievement of complete dissolution and isotopic equilibrium is the most important factor in IDA.

On the other hand, if great care is taken not to cause the isotopic exchange reaction, a given chemical species can be determined selectively by IDA, as shown in the application of sub-IDA.

REFERENCES

1. J. H. Peters, H. W. Nolen III, G. R. Gordon, W. W. Bradford III, J. E. Bupp and E. J. Reist, *J. Chromatogr.* **488**, 301 (1989).

2. N. Suzuki, *Nippon Kagaku Zasshi* **80**, 370 (1959).
3. J. Ruzicka and P. Benes, *Collect. Czech. Chem. Commun.* **26**, 1784 (1961).
4. J. Ruzicka and J. Stary, *Talanta* **8**, 228 (1961).
5. J. Ruzicka and J. Stary, *Substoichiometry in Radiochemical Analysis*, Pergamon Press, Oxford (1968).
6. N. Suzuki and K. Kudo, *Anal. Chim. Acta* **32**, 456 (1965).
7. J. Ruzicka and J. Stary, *Talanta* **10**, 287 (1963).
8. N. Suzuki, K. Miura and S. Nakamura, *J. Radioanal. Nucl. Chem.* **112**, 555, (1987).
9. M. Katoh and K. Kudo, *J. Radioanal. Nucl. Chem.* **84**, 277 (1984).
10. J. Ruzicka and J. Stary, *Talanta* **8**, 775 (1961).
11. J. Stary and J. Ruzicka, *Talanta* **11**, 697 (1964).
12. I. Obrusnik and A. Adamek, *Talanta* **15**, 433 (1968).
13. T. Braun, L. Ladanyi, M. Marothy and I. Osgyani, *J. Radioanal. Chem.* **2**, 235 (1969).
14. N. Suzuki, *J. Radioanal. Nucl. Chem.* **124**, 197 (1988).
15. H. Imura and N. Suzuki, *Anal. Chem.* **55**, 1107 (1983).
16. J. Klas, J. Tolgyessy and E. H Klehr, *Radiochem. Radioanal. Lett.* **18**, 83 (1974).
17. M. Kyrs and K. Prikrylova, *J. Radioanal. Chem.* **79**, 103 (1983).
18. H. Yoshioka and T. Kambara, *Talanta* **31**, 509 (1984).
19. J. D. Fasset and T. J. Murphy, *Anal. Chem.* **62**, 386 (1990).
20. D. J. Rokop, N. C. Schroeder and K. Wolfsberg, *Anal. Chem.* **62**, 1271 (1990).
21. J. N. Rosholt, *Nucl. Instrum. Methods Phys. Res.* **223**, 572 (1984).
22. H. M. Shihomatsu and S. S. Iyer, *J. Radioanal. Nucl. Chem. Lett.* **128**, 393 (1988).
23. F. Sus, J. Krtil and J. Moravec, *J. Radioanal. Nucl. Chem.* **111**, 105 (1987).
24. J. Parus, W. Raab, H. Swietly, J. Cappis and S. Deron, *Nucl. Instrum. Methods Phys. Res., Sect. A* **A312**, 278 (1991).
25. S. K. Aggearwal, G. Chourasiya, R. K. Duggal, R. Rao, P. A. Ramasubramanian and H. C. Jain, *J. Radioanal. Nucl. Chem. Lett.* **119**, 1 (1987).
26. J. G. Jia, C. Testa, D. Desideri and M. Assunta, *Anal. Chim. Acta* **220**, 103 (1989).
27. J. Krtil, F. Sus and J. Moravec, *J. Radioanal. Nucl. Chem. Lett.* **127**, 379 (1988).
28. B. R. Harvey and M. B. Lovett, *Nucl. Instrum. Methods Phys. Res.* **223**, 224 (1984).
29. K. Masumoto and M. Yagi, *J. Radioanal. Nucl. Chem.* **79**, 57 (1983).
30. C. Yonezawa and T. Komori, *Anal. Chem.* **55**, 2059 (1983).
31. M. Tsukada, D. Yamamoto, H. Yoshikawa, K. Endo and H. Nakahara, *J. Radioanal. Nucl. Chem.* **84**, 223 (1984).
32. M. Tsukada, H. Yoshikawa, M. Yanagawa, K. Endo and H. Nakahara, *J. Radioanal. Nucl. Chem.* **125**, 351 (1988).
33. M. Yagi and K. Masumoto, *J. Radioanal. Nucl. Chem.* **91**, 91 (1985).
34. M. Yagi and K. Masumoto, *J. Radioanal. Nucl. Chem.* **91**, 369 (1985).
35. M. Yagi and K. Masumoto, *J. Radioanal. Nucl. Chem.* **99**, 287 (1986).

Chapter 20

RADIOIMMUNOASSAYS AND RELATED RADIOACTIVE METHODS FOR THE QUANTITATION OF HORMONES AND OTHER SUBSTANCES IN BIOLOGICAL FLUIDS

Joseph Levy, Symour Glick and Yoav Sharoni
Clinical Biochemistry Department Faculty of Health Sciences,
Ben-Gurion University of the Negev, Soroka Medical Center of Kupat Holim,
Beer-Sheva 84105, Israel

1 RADIOIMMUNOASSAY

Historically, the development of radioimmunoassay (RIA) some 35 years ago was made possible by advances in our understanding of the chemistry of peptide hormones, the development of methods for labeling purified hormones with radioactive isotopes, progress in immunization methodology, and the application of these advances by scientists who appreciated their enormous potential. Today, radioimmunoassay is probably the most extensively applied technique for the quantitation of hormones and a variety of other substances of biological interest. Radioimmunoassay in its classic sense is a measurement technique which uses the radiolabeled form of the antigen of interest and an antibody which recognizes this moiety specifically.

Berson and Yalow pioneered the description of this methodology for the measurement of insulin [1]. In their original study they used an immune globulin circulating in the peripheral blood of insulin-treated diabetic patients as a binding protein for ^{131}I-labeled insulin. This observation and their subsequent work led to the award of the Nobel Prize in medicine. After their breakthrough, many other hormone assays were developed in their laboratories and those of

Chemical Analysis by Nuclear Methods Edited by Z. B. Alfassi

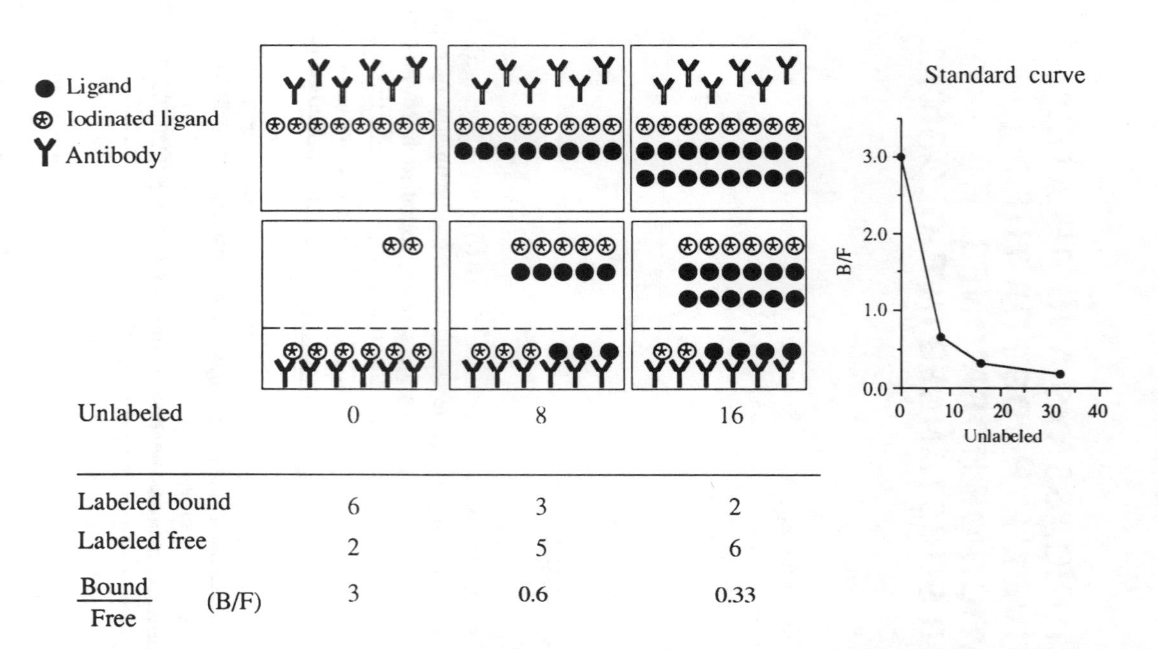

Unlabeled	0	8	16
Labeled bound	6	3	2
Labeled free	2	5	6
Bound/Free (B/F)	3	0.6	0.33

Fig. 1. Standard curve for radio-immunoassay. The model is shown with a small number of antibody and ligand molecules for simplicity. The upper boxes show the situation in the solution at the beginning of incubation. The lower boxes show the situation after separation of the antibodies and the antibody-bound ligands, which appear below the dashed line. The table below shows the calculation of the bound and free labeled ligand. The standard curve is constructed from these results with an additional value of unlabeled ligand which is not shown in the model.

other researchers. These assays used either antibodies developed in animals or a variety of relatively specific binding proteins.

1.1 Principles

Kinetically, the interaction of antigen and antibody may be considered a pseudo-first-order reaction. The antigen–antibody reaction is reversible and exothermic. The rate of association of the antigen–antibody complex is considerably faster than its rate of dissociation. The observed amount of hormone bound to the antibody is the resultant of the rates of association and dissociation.

The basis of radioimmunoassay is competitive inhibition of the binding of the labeled ligand to antibody by unlabeled ligand present in standards or in unknown samples. A typical radioimmunoassay is performed by the simultaneous preparation of standard and unknown mixtures in test tubes to which are added fixed amounts of antibody and radiolabeled antigen (Fig. 1). After an appropriate incubation time, antibody-bound (B) and unbound (F = free) labeled antigens are separated by any of a variety of techniques. The number of counts bound to the antibody varies inversely with the concentration of unlabeled hormone in the standard or unknown sample; the higher the concentration of unlabeled hormone in the standard or unknown sample; the higher the concentration of unlabeled hormone, the fewer the number of counts bound to the fixed concentration of antibody. There is no requirement in radiommunoassay for biological or chemical identity between standards and unknowns. All that is required for a properly validated radioimmunoassay is that the unknown be immunochemically identical with the standard. This dependence of the RIA on immunochemical determinants may sometimes result in a divergence between levels of a substance as measured by RIA and those as measured by a bioassay, receptor assay or chemical assay based on other determinants on the molecule. Validation of a RIA requires that the dilution curve of the unknown sample be superimposable on that of standards, but this superimposition alone can never prove identity. The prerequisites for sensitive radioimmunoassays include the following.

(1) Ligand: highly purified and radioactively labeled to sufficiently high specific radioactivity so that only trace amounts of the labeled hormone need to be added to each assay tube. The immunoreactivity of the labeled hormone should not be markedly altered by the radiolabeling procedure but the immunoreactivity often is not identical with that of the standard or unknown.
(2) Antibody: of sufficiently high sensitivity (affinity) and specificity (minimal cross reactivity) to measure the desired antigen at the level of relevance to the investigator.

(3) A reference preparation to be used for the calibration curve against which concentrations of unknown samples may be dose-interpolated.
(4) A method for separation between antibody-bound and free antigen.

1.2 Reagents and antibody

Antibodies used for radioimmunoassays belong to the class of immunoglobulins which are predominantly gamma globulins (IgG). The IgG molecules consist of four polypeptide chains, two heavy and two light chains symmetrically arranged and covalently bound by disulfide bonds. The antigen-antibody interaction is not covalent but is a combination of electrostatic, hydrogen-bonding, and van der Waals interactions [2]. As the shape of the antibody combining site is complementary to that of the antigen, the interaction of an antigen or ligand with an antibody may be considered a 'lock and key' arrangement.

An 'immunogen' is a substance capable of inducing the formation of antibody. Most peptide hormones are satisfactorily immunogenic in a variety of experimental animals when the hormone is administered as an emulsion in Freund adjuvant. It is a historical irony that the pioneer reports of Yalow and Berson were initially greeted with skepticism because the immunological community at that time doubted that peptides as small as insulin could be 'immunogens'. Commercial or low purity hormone preparations can be used as immunogens, and conceivably may be advantageous because of the possible slight denaturation of such preparations, rendering the hormones more 'foreign' and thereby enhancing their antigenicity. There appears to be little or no advantage to immunizing with highly purified and more expressive antigens, as contaminants are not likely to lead to immunological reactions that interfere with the assay when the labeled antigen is highly purified. The purity of the labeled antigen is important in order to avoid interaction of labeled contaminants with non-specific antibody.

Peptides of low molecular weight (less than 1000 daltons) or non-peptide substances that are not themselves readily antigenic may be rendered so by coupling them to a large protein. A variety of methods may be used to bind the small molecules to immunogenic carriers [3, 4]. These molecules, or haptens, must be conjugated to substances such as albumin or thyroglobulin to induce antibody formation. It is preferred to use a non-mammalian protein, such as keyhole limpet hemocyanin, to increase the probability of producing antibodies [5]. In general, the immunological response is better for those carrier proteins which are intrinsically highly immunogenic. As a rule, steroids, as well as many drugs and vitamins, need to be conjugated to a carrier protein. Care should be taken to conjugate the steroid or other hapten through that part of the molecule which is least likely to alter its unique stereospecificity. For example, for the female sex hormone 17β-estradiol, antibody is best generated by conjugating its

6-keto analogue covalently to albumin. In this way the characteristic 3- and 17-hydroxyl groups, as well as the aromatic A ring of that steroid, are undisturbed.

The production of a satisfactory antiserum remains as much an art as a science. Although numerous papers have described specialized procedures for immunization, there is no general agreement concerning the most suitable animal species or the optimal technique for producing the best antiserum for each of the diverse substances for which radioimmunoassay has been described. Nor have there been carefully controlled studies comparing various immunization procedures for RIA purposes. Antibody concentration usually increases with repeated immunization, generally reaching a plateau after three to five doses of antigen. The use of large animals for immunization offers the advantage of yielding large volumes of antiserum, guaranteeing uniformity of results in the laboratory over a long period of time. However, the degree of foreignness may be a more important factor in the choice of a suitable animal for the preparation of antisera for peptide hormones. Because guinea-pig hormones such as insulin, glucagon, vasoactive intestinal peptide and gastrin differ from the corresponding peptides of other mammals, the guinea-pig has proved to be an extremely useful animal for antibody production. In the case of antibody production for drugs the selection of animal is less crucial.

Several techniques have been developed over the years from the original technique described by Freund [6]. The emulsion containing the hormone or immunogen, appropriate oils, and heat-killed tubercle bacilli is injected either intradermally or subcutaneously, the former method requiring smaller doses of antigen in general. There is good experimental evidence supporting the development of specific high affinity antibodies with small doses of immunogen. Antibody-forming cells with the highest affinity for the immunogen are more likely to be induced to form specific antibody when low doses of immunogen are used. Conversely, with higher doses of immunogen, there is a greater likelihood of developing lower affinity antibody. With very high doses of immunogen a phenomenon of tolerance may be induced in the immunized animal, resulting in no detectable circulating antibody or very low titers of that antibody [7]. With initial immunization, a minimum of 6 weeks is usually needed for significant antibody titers to appear. The affinity of the antibody continues to increase and usually peaks between 8 and 12 weeks after the initial immunization. Empirically, re-challenge or boosting results in the best anamnestic responses when the animal is re-challenged at a time when its circulating antibody titer is falling. When the animal is re-challenged with one-half to one-quarter of the initial immunizing dose of hormone, a significant rise in circulating antibody titer is observed within a few days of re-immunization. The animal should be re-immunized over the same anatomic areas used for the initial immunization.

During recent years there has been considerable interest in the use of monoclonal antibodies for radioimmunoassay. This method ensures a virtually everlasting supply of monospecific antibodies with uniform and predictable qualities of specificity and sensitivity. When an antiserum is generated with an in vivo immunization procedure, a population of immunoglobulin molecules with differing affinities and specificities is generated. In contrast, monoclonal antibodies represent a single clone of immunoglobulin molecules with a single affinity and specificity for one antigenic site of the immunogen. Typically, the monoclonal antibody is isolated in vitro after an animal, usually a mouse, has been immunized and antibody-producing cells from that animal's spleen or lymph nodes have been fused with a non-immunoglobulin secreting myeloma cell line. Like the parent myeloma cells, the hybridoma cells proliferate endlessly. Clones are selected from the hybridomas on the basis of their specificity and affinity for the desired antigen [8]. However, the sensitivity obtained with the use of such antisera is often less than that attainable by proper selection among the polyclonal antisera produced by traditional immunization. When the usual heterogeneous antiserum of high titer is used, it is diluted sufficiently so that only antibody binding sites with the highest equilibrium constants, and hence the highest sensitivities, are able to bind the antigen. If optimal sensitivity is not required, the use of monoclonal antibodies may be advantageous. For routine clinical assays, a monoclonal antibody specific for intact hormone may fail to detect altered forms of circulating hormone if the antigenic site recognized by that monoclonal antibody has been altered during secretion or metabolism. However, that shortcoming can be theoretically overcome by mixing several monoclonal antibodies with different specificities, thus creating a 'defined' polyclonal antibody. For the best results one should match the specific assay to the specific purpose and conditions for which the assay will be applied. At times, high specificity may be critical; in other situations such specificity may be a disadvantage.

1.3 Reagents and antigen labeling

The high specific radioactivity (cpm/mol) needed for the assay of compounds present in body fluids in minute concentration is illustrated by the following example. The use of a tracer of 10 pmol/l (10^{-11} M) to measure a hormone concentration of 1 pmol/l (10^{-12} M) would mean that a random 10% error in the tracer (1.0 pmol) would produce a 100% error in the calculated hormone concentration. With the same random error, if the concentration of the tracer had been only 1 pmol/l, the error in the calculated hormone concentration would have been only 10%. From this example it is obvious that ideally one should use the labeled tracer antigen at a concentration as far below the concentration of the unknown unlabeled antigen as possible.

The compound $Na^{131}I$ was the isotopic substance used for the labeling of insulin in the first RIA. Currently $Na^{125}I$ rather than $Na^{131}I$ is preferred because of its significantly longer half-life (60 vs. 8 days). However, care must be taken with assays using ^{125}I, since differential quenching of ^{125}I may be observed with glass reaction tubes of variable thickness or lead content. Plastic reaction tubes are therefore preferred, but one must always be aware of the possible effects of changes in manufactures' quality control and specification in tube supply.

Radioiodine can readily be substituted onto a tyrosyl or histidyl residue using a variety of procedures. The chloramine-T technique for oxidation of the radioiodide [9] is the most commonly used. For peptide hormones that do not contain a suitable tyrosyl or histidyl residue for iodination, the Bolton–Hunter reagent is preferred [10]. This pre-iodinated acylating reagent is readily condensed with the amino groups of peptides such as the ε-amino side chains of lysine or the NH_2-terminal amino group. This reagent is also useful if the tyrosine in the peptide is sulfated and hence not readily iodinated, or if the tyrosyl residue is in the antigenic site and the presence of an added iodine atom would diminish the reaction of antigen with antibody, thus decreasing the sensitivity of the assay. Use of the Bolton–Hunter reagent, which is pre-iodinated before coupling, may also be desirable if the peptide does not tolerate even the gentle oxidation associated with the chloramine-T reaction.

An additional technique in use is enzymatic oxidation with lactoperoxidase [11]. This enzyme, which uses hydrogen peroxide added directly to the reaction mixture, oxidizes $Na^{125}I$ to molecular iodine for selective incorporation of those atoms into tyrosyl residues. As hydrogen peroxide is a strong oxidizing reagent, exposure of the hormone to high concentrations of that reagent may lead to chemical damage. For this reason, the hydrogen peroxide is usually added in two or three small increments and is quickly consumed within the reaction. Lactoperoxidase may also use the hydrogen peroxide that is formed gradually and slowly by the reaction by glucose and glucose oxidase. This procedure is more gentle and may prevent oxidation damage. With any technique of iodination there may be significant variations depending on the nature and quality of the iodine label used, and there are occasional variations from one iodination to the next which are not readily explainable.

The specific activity of a ^{125}I-labeled antigen may be increased by increasing the number of radioiodine substitutions. However, the more highly iodinated preparations generally show diminished immunoreactivity as well as increased susceptibility to radiation damage [12]. For maximal stability and immunoreactivity, it is therefore preferable for the labeled molecule to contain only one radioiodine atom.

It must be appreciated that iodination at an average of one radioiodine atom per molecule does not mean that all or even the major fraction of the radioactivity is incorporated into molecules containing one radioactive atom. Thus, purification of monoiodinated molecules is generally required. For large molecules, methods that separate on the basis of charge, such as starch gel electrophoresis or ion-exchange chromatography, are used. For small peptides (≈ 1000 daltons) such as vasopressin, the monoiodinated peptide is easily separated from the non-iodinated peptide and other iodinated products by gel filtration [13].

For non-peptide hormones and for drugs, which are generally present at much higher concentrations in plasma than the peptide hormones, the experimental requirements for labeled with high specific radioactivity are much less stringent than those for the peptide hormones. For these assays, tritium-labeled tracers prepared and purified in commercial laboratories have frequently been used; ^{3}H-labeled antigens have longer shelf life than the ^{125}I tracers. However, most commercial kits for assay of these substances use ^{125}I-coupled tracers to obviate the need to use liquid scintillation counters, which are required for the detection of ^{3}H.

1.4 Separation methods

The classic immunological method for the separation of antibody-bound from free antigen was based on spontaneous precipitation of antigen–antibody complexes. However, at the low concentration of antigen and antibody necessary for RIA, spontaneous precipitation does not occur and the antigen–antibody complexes remain soluble. The wide variety of methods used for the separation of antibody-bound and free antigen does not affect the principle of the RIA.

(1) precipitation of antigen–antibody complexes with a second antibody directed against the first specific antibody (double antibody);
(2) the use of organic solvents, large concentrations of salts, or high molecular weight polymers such as polyethylene glycol to precipitate the complexes;
(3) adsorption or complexing of antibody to a solid-phase matrix or to magnetic particles (see Figs. 3 and 4).
(4) adsorption of free antigen to solid-phase material such as cellulose, charcoal, or ion-exchange resin.

The double antibody method is considered the classic method of separation for radioimmunoassay procedures. However, the cost of the second antibody may make this method quite expensive when thousands of samples are to be analyzed. In this technique, a second antibody raised in response to the IgG of the animal species in which the first antibody was generated is added after equilibrium has been attained between the hormone and the first antibody. The second antibody binds to the first antigen–antibody complex, creating an insoluble molecular species. Bound and free hormone can now be separated by simple centrifugation, with the bound hormone being precipitated. For example, if an insulin antibody is generated in guinea-pig, a second antibody to guinea-pig IgG is generated in another animal species such as horse, sheep, or goat. The horse antiguinea-pig globulin is added to the test tube after the reaction mixture containing ^{125}I-labeled insulin and guinea-pig anti-insulin serum has attained equilibrium. The time required for the second antibody reaction to attain equilibrium is a function of the concentration of carrier guinea-pig serum in the reaction mixture. The carrier serum or globulin is added to shorten the time necessary for the incubation and to increase the volume and completeness of the precipitate. The time required to reach equilibrium is initially determined empirically, and the minimal optimum volume of second antibody needed is ascertained for each batch of second antibody harvested or purchased from a commercial source. Lower concentrations of carrier serum may be added to the assay tubes, and consequently a smaller volume of second antibody will be needed, but the time required for the second antibody reaction to attain equilibrium becomes greater.

For separation of steroids, dextran-coated charcoal will adsorb the free ligand. Consequently, antibody-bound hormone will be present in the supernatant in the reaction tube.

Immobilization of the antibody may be achieved by coating the inner surface of the test tubes so that bound and free hormone are separated by simple decantation of the supernatant, which contains the free hormone. However, most solid-phase techniques have the disadvantage of decreased sensitivity that results from possible chemical alterations of the antibody molecule introduced by the coupling procedure or because of steric hindrance. Magnetic field separation of free and bound hormone with ferric oxide conjugated to the first antibody may also be used; no centrifugation step is required, as the antibody-bound hormone is pelleted in a magnetic field and free hormone in the supernatant can simply be decanted. However, it is preferable to conduct this separation by conjugation of the second antibody to ferric oxide so that the same magnetic reagent can be used for several radioimmunoassays with different first antibodies. The techniques using complexed antibodies usually result in shorter incubation times and quicker assay results.

1.5 Calculations and presentation of results

Several methods are routinely used for the presentation of radioimmunoassay results (Fig. 2). The ratio between the counts bound at each unlabeled hormone concentration (B) to the counts bound at zero unlabeled hormone (B_0) is plotted against the mass of hormone, from which a non-linear plot is obtained (Fig. 2(a)). A linearization of this plot can be achieved by a logarithmic transformation of the hormone concentration (Fig. 2(b)) or by the log–logit plot (Fig. 2(c)). The two linear presentations are less accurate at the two ends of the curve. The log–logit plot tends to be steeper than the logarithmic plot at the low concentration of hormone, even when the assay is not sensitive in this range. Under such circumstances the log–logit plot will give erroneous results that an inexperienced worker may regard as accurate.

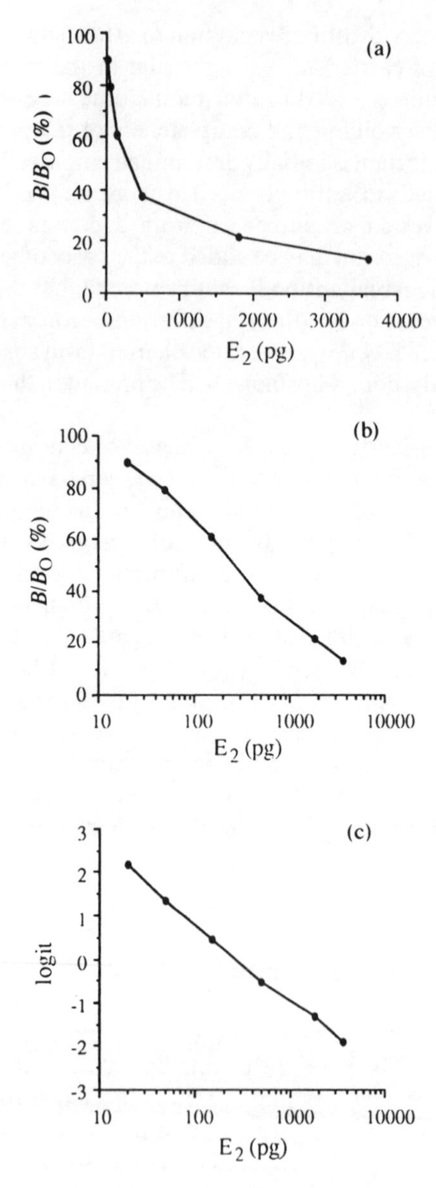

Fig. 2. Dose–response lines for 17β-estradiol (E_2) radioimmunoassay depicted wih three commonly used graphic presentations. Panel (a): B/B_0 versus linear dose of hormone; (b): Similar to (a) with a logarithmic scale for hormone level; (c) log–logit transformation of the data [logit = ln $(B/B_0)/(1 - B/B_0)$]; B = counts bound; B_0-maximum number of counts bound when only antibody and labeled hormone are incubated.

1.6 Validation

Radioimmunoassay differs from bioassay in that it is an immunochemical procedure that is not affected by biological variability of the test system or by the presence of substances that might inhibit or enhance biological action. The measurement depends only on the interaction of chemical agents in accordance with the law of mass action. However, non-specific factors in the biological fluids measured do interfere in this interaction, and cross-reacting prohormones, molecular fragments, and related antigens may alter the specificity of the immune reaction.

A necessary condition for proper validation of a radioimmunoassay is that the apparent content of an unknown sample be independent of the dilution at which it is assayed. Similarity of response between standard and clinical sample must be ascertained by constructing dose–response curves for each with graduated doses of both to ascertain whether parallelism exists between the lines. Parallelism should exist over a range of at least 100-fold between hormone standard and unknown sample for valid dose interpolation. Experimental errors often obscure the lack of superimposability when the concentration range is too small. A logarithmic dose–response plot is generally less sensitive to dissimilarity of standards and unknowns than is a linear plot.

Superimposability is a necessary but insufficient condition to ensure immunochemical identity of standards and unknowns. However, lack of superimposability means that the assay lacks quantitative validity even though it may be useful clinically if carefully interpreted [14]. Non-superimposability of dilution curves of standards and unknowns can arise from chemical interference with the antigen–antibody reaction, from degradation of labeled and unlabeled antigen and/or antibody, or from a variety of immunological factors, including the use of heterologous hormonal standards, the heterogeneity of immunologically related hormones of the same species, and the presence of precursors or metabolites in addition to the usual hormonal forms.

Another means of validating an assay is the measurement of pools of sera containing low, medium, and high concentrations of tested compound, in replicate on each assay, as a quality control check. Considerably less interassay variation is observed than with bioassay, so most laboratories do not assay clinical samples at multiple volumes in routine radioimmunoassays unless hormone heterogeneity is suspected.

Finally, an independent check of each new assay's validity should be made with both physiological and independent assay (preferably bioassay) correlations. In addition, changes in the concentration of the test compound should be introduced into the biological system by different methods and the expected physiological changes in concentration should be observed with the newly developed assay.

1.7 Assay sensitivity

Sensitivity, or lowest detectable dose, is defined as that concentration of ligand which induces a significant change in percentage binding of radioactively labeled hormone to the antibody when compared with a blank sample. The sensitivity of the assay is a function of the hormone's affinity for its specific antibody and of the specific radioactivity of the labeled antigen. After the antibody is harvested, studies are designed to assess the antiserum's sensitivity and specificity. The concentration of antibody within an antiserum is initially determined by checking, at serial dilutions, the binding of the antiserum to the radiolabeled antigen. This technique is referred to as titering. Thus, titer relates to the concentration of antibody molecules in the serum.The binding curve of the antigen to the antibody is hyperbolic. The optimal concentration of antibody to achieve the highest sensitivity is that amount which binds $\approx$40–50% of the labeled antigen. At antibody concentrations which bind more than 50% of the antigen, addition of unlabeled antigen will cause smaller changes in the bound radioactivity owing to the non-linearity of the curve in this range. At antibody concentrations which bind less than 50%, although the binding curve is linear, the amount of bound labeled antigen is smaller, which also impairs the sensitivity.

After the concentration of antibody which binds 40–50% of radiolabeled hormone at equilibrium has been determined, the sensitivity and specificity of the antibody are ascertained. Sera harvested from immunized animals for use in most clinical radioimmunoassays contain a population of antibodies with varying specificities and affinities. An antiserum is selected that has a population of antibodies with sufficiently high affinity and specificity for the measured metabolite or drug. Operationally, one can select out that population of high affinity antibody within a polyclonal antibody by using the antiserum in sufficiently high dilution, so that other lower affinity, and possibly non-specific, populations of antibody become relatively insignificant. If one chooses to use an antiserum at a significantly different dilution from that initially used to validate an assay, one must again determine the sensitivity and specificity of the antiserum at the new concentration, since other populations of antibody which may not be as sensitive and specific as when the antiserum was used at higher dilutions may now come to the fore. In contrast, monoclonal antibodies have unvarying sensitivity and specificity. The specificity of a monoclonal antibody is unaltered by antibody concentration, but the assay sensitivity decreases with higher antibody concentration as with a polyclonal antibody.

1.8 Assay specificity

Both the structure and the purity of a hormone used for immunization must be considered. The human glycoprotein hormones luteinizing hormone (LH), follicle stimulating hormone (FSH), thyroid stimulating hormone (TSH), and human chorionic gonadotropin (HCG) share a common quaternary structure of two dissimilar subunits, designated α and β [15]. The α subunits among these hormones are essentially identical in terms of their amino acid sequences. It is the β subunit that confers both immunological and biological specificity. Consequently, one must carefully select an antiserum generated towards one of the glycoprotein hormones and carefully check whether it is sufficiently specific for measuring hormone concentrations in the presence of high levels of the other glycoprotein hormones in the same sample. In some cases, cross-reactions resulting from common antigenic determinants can be eliminated by adsorbing the antibody with the common antigen. For example, if a TSH antiserum cross-reacts with FSH and LH and that cross-reaction is due to common α subunit determinants, the antiserum can be adsorbed with free α subunit. The resulting antiserum, ideally, should be specific for TSH with little, if any, loss of sensitivity for that hormone. If only a relatively crude hormonal preparation is available for immunization, then a hybridoma approach is preferable, since one can select antibody-producing clones which are appropriately sensitive and specific for assay requirements.

1.9 Problems

The wide use of radioimmunoassays in clinics, the commercialization of kits and the simplification of techniques may lead to erroneous results for an unexperienced user. For this reason it is important to always keep in mind the limitations of RIA methodology. Several of the most common problems will be discussed in some detail, among them: standardization of the assay incubation conditions, the effect of degradation of antigen or antibody, selecting the right reference material when a true standard is not available circulating endogenous antibody or binding proteins, and cross-reactivity caused by immunologically related ligands.

1.9.1 Effects of experimental conditions on the immune reaction

Changes in temperature, pH, ionic strength and chemical nature of buffering solutions, and the presence or absence of a variety of anticoagulants or protective agents can have profound effects on the reaction of antigen with antibody [12]. Experience with many RIA procedures shows that there is no general predictable effect of pH and ionic strength on the immune reaction. The effects depend on the particular ligand or hormone, the particular antiserum used and the buffer.

It is therefore desirable for standards and unknowns to have the same media and conditions during the incubation period. Thus, if acid–alcohol extracts are to be assayed, either the same volume of acid–alcohol must be added to standards as is used in the unknown sample, or it must be demonstrated that the volume of acid–alcohol used does not affect the standard curve. Similarly, incubation tubes of standards and unknowns should contain the same concentration of protective agents, anticoagulants, salts, proteins, and so forth.

The equilibrium constant for the reaction of antigen with antibody is generally higher at 4 °C than at room temperature [16]. Thus more sensitive assays are obtained by incubation at 4 °C. However, low temperature decreases the rates of association and dissociation. If a high sensitivity assay is not critical, incubation at room temperature or even in a water bath at 37 °C permits a more rapid assay. Dissociation of antigen–antibody complexes may occur if mixtures are incubated at 4 °C and the free and antibody-bound labeled antigens are then separated at room temperature.

1.9.2 Effect of degradation of labeled antigen and/or antibody

Many peptide hormones are subject to proteolytic damage by enzymes in blood and other biological fluids. Differential damage of the labeled antigen in standards and unknowns decreases the reliability of a radioimmunoassay procedure and, depending on the extent of damage, may even completely invalidate the results. If, during the incubation period, either labeled antigen or antibody is destroyed, immune precipitation is reduced, and such reduction may be interpreted erroneously as being due to high hormone content. For example, apparently high concentrations of oxytocin are erroneously detected in serum of pregnant women because of high specific proteolytic activity in such serum [17]. This activity degrades the labeled tracer, causing reduction in the bound hormone in the assay, which is normally interpreted as a high hormone level. In this case a specific inhibitor of the proteolytic enzyme may be included in the assay to prevent hormonal degradation.

Some general approaches to overcome this problem are available. If the peptide hormone in the unknown fluid can tolerate boiling, this method is perhaps the simplest for inactivating degratative enzymes. For small peptides, such as vasopressin, which are more susceptible to proteolytic degradation, acid extraction or acetone precipitation may be employed.

When a specific adsorbent method is used for separation of free from antibody-bound labeled antigen, a control mixture containing the labeled antigen and either the unknown sample or the diluent used for standards, but without antiserum, is used to evaluate the differential damage occurring during the incubation period. The incubation damage is evaluated by the failure of the damaged labeled antigen to be adsorbed by the specific adsorbent. It is important to determine whether the labeled degradation products are not adsorbed to the same matrix as the free hormone. If such an adsorption method is not available, any analytical separation method, such as HPLC, may be employed to detect changes in the labeled tracer.

The only certain method for ensuring integrity of the labeled antigen is demonstration of its ability to bind to antibody at the end of the incubation period. Excess antibody is usually used for this evaluation, even though this technique may fail to reveal subtle alterations in immunoreactivity. When the double antibody method is used for separation of the labeled immune complexes, damaged labeled antigen that does not bind to the first antibody is not precipitable by the second antibody and appears in the supernatant, along with free labeled antigen.

1.9.3 Heterologous hormone standards

The primary structures of many hormonal peptides and other proteins in biological fluids have diverged during the course of evolution. The species differences resulting from these mutations are most likely to be found in regions of the hormone not essential for its biological activity. However, because the immunogenicity of exogenous hormone from another species is likely to depend on the foreignness, i.e., the regions of difference, it is not unexpected that immunological and biological potencies of a hormone may differ among different species.

A heterologous radioimmunoassay is usually employed when the desired reference compound is not available. If a dilution curve of the unknown is not superimposable on a dilution curve of heterologous hormone standards, the assay of the unknown will certainly not have quantitative validity. Under those circumstances it is probably advisable to use crude tissue extracts or biological fluids containing as high a concentration of the material as the known samples. However, evaluation of the true absolute concentration does require preparation of a suitable reference standard from the appropriate species.

1.9.4 Immunologically related but different hormones

Consideration must be given to the possibility that plasma or tissue extracts may contain a biologically different substance that nevertheless cross-reacts immunologically. Early in the development of the assay for growth hormone, for example, it was noted that the hormone concentration decreased linearly with dilution in plasma obtained from umbilical cord blood, acromegalic patients, and stimulated control subjects, but not in plasma obtained from pregnant women [18]. The interfering substance in the pregnant women proved to be human placental lactogen (hPL), which is of placental origin and resembles growth hormone but is neither biologically nor immunologically identical with it. The synthesis by the placenta and other tissues of additional hormones (e.g., chorionic gonadotropin) that have biological and immunological properties similar to those of pituitary hormones may also pose problems of specificity in assays for the glycoproteins in the plasma of pregnant women (see Section 1.7).

1.9.5 Heterogeneity of the test compound

It is now clear, largely on the basis of studies involving RIA, that many biologically active peptides are found in more than one form in plasma and in tissue extracts. These forms may or may not have biological activity and may represent either precursor(s) or metabolic product(s) of the original biologically active compound. Such heterogeneity complicates the interpretation of the radioimmunoassay measurement. However, recognition of the problem has opened new vistas in our understanding of the paths of synthesis and metabolism of such peptides. For example, insulin, which is a 6 kDa peptide, is formed by cleavage of 1.5 kDa C-peptide from its precursor molecule, called proinsulin. It is indeed fortunate that the first RIA was described for insulin, since the predominant form of the hormone in the circulation of most subjects in the stimulated state is identical with the purified standard. Only in patients with insulinoma [19] or in those with a rare genetic abnormality that prevents cleavage of the C-peptide [20] does the prohormone appear to predominate.

1.9.6 Circulating endogenous antibody

If circulating endogenous antibodies are present in the patient's serum, spurious assay determinations may result. The radiolabeled hormone added to the test tube will bind both to the patient's antibody present within the serum sample and to the antibody added to the test tube. Since second antibody is generated to the IgG of the animal species used to generate the first antibody, the hormone bound to the human IgG will not be precipitated fully by that second antibody and much of it will remain as a soluble complex in the supernatant. Consequently, fewer counts representing antibody-bound counts will appear in the pellet. As the concentration of hormone in an assay tube is inversely proportional to the amount of radiolabeled hormone precipitated, a spuriously high hormone concentration will result. Other separation techniques which do not distinguish between the endogenous and exogenous antibodies will result in false concentrations as well, simply because more antibody is present in the reaction mixture. In this situation a falsely low concentration will be obtained because higher radioactivity will be present in the bound fraction. Circulating antibodies are commonly encountered among diabetic patients receiving exogenous insulin. If one suspects circulating antibody in a patient, assay tubes containing labeled ligand and the patient's serum or plasma and buffer are incubated without exogenous antibody. [^{125}I] Insulin–endogenous antibody complexes are then precipitated with appropriate concentrations of human immunoglobulin or ammonium sulfate.

Similar problems may arise in assays of ligands which are normally bound to specific binding protein in the body fluids. Good examples are insulin-like growth factors and steroid hormones. Steroid hormones usually circulate in serum bound to a variety of proteins. Since these serum proteins may have affinities similar to those of the antibodies used for RIA, the hormone usually must initially be separated from these serum proteins for valid assay. Such separation may be carried out by inexpensive methods such as silica adsorption or costly ones like affinity chromatography. More recently, octadecylsilyl cartridges (for example, C_{18} Sep-Pak cartridges) have been used to isolate insulin and insulin-like growth factors from plasma and tissue extracts. The peptide can readily be extracted from the cartridge in concentrated form, usually free from non-specific interfering substances such as proteins which bind insulin-like growth factors.

The binding proteins which interfere with radioimmunoassay were used initially for competitive protein binding assays. Cortisol-binding globulin (CBG) was used to assay cortisol and other related steroids, e.g., progesterone, which bind to that carrier protein. These assays are no longer extensively used , since each batch of serum from which the binding globulin is harvested must be characterized, and more extensive thin-layer or other chromatographic separations of steroids are required. The added laboratory time needed for these initial steps, as well as the wider interassay variation, has relegated that assay approach to a secondary place.

2 IMMUNORADIOMETRIC ASSAY

Immunoradiometric assay (IRMA) differs from RIA in that an antibody is labeled instead of the ligand or hormone. This technique provides a potential advantage in that immunoglobulins can be iodinated to a higher radioactivity than that of most small ligands, and the product is usually stable for a longer time than iodinated hormones. In the following example of an IRMA procedure, a mixture of iodinated antibodies and ferric oxide-conjugated antibodies which recognize different epitopes of the peptide is incubated until equilibrium is reached. The Fe-tagged antibody facilitates the separation of the entire complex by means of a magnetic field (Fig. 3). Increasing amounts of the measured compound will result in increased radioactivity in the separated complex and the ideal standard curve will be a linear ascending curve (Fig. 3). This is in contrast to RIA, where a non-linear descending curve is obtained. This by itself increases the sensitivity of the assay at the lower end of the concentration range, since the accuracy of measuring a small increase in radioactivity above the value obtained at zero hormone in IRMA is higher than measuring a small decrease in radioactivity in the same concentration range in RIA.

One of the important aspects of this procedure is the use of two antibodies which recognize different epitopes of the ligand [21]. The radioactive antibody will be detected in the complex separated by the magnet only if the two antibodies are both bound to the same ligand molecule. Thus, if the two epitopes are found at the two sides of a peptide hormone, only the full length hormone will be detected

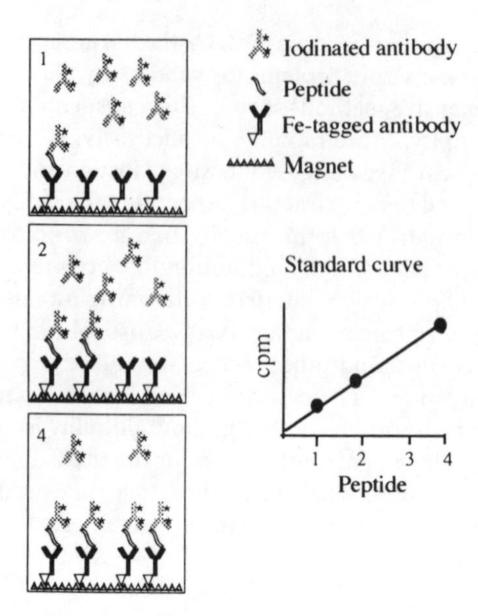

Fig. 3. Standard curve for immunoradiometric assay. The model is shown with a small number of antibody and peptide ligand molecules for simplicity. The relative amount of peptide is shown in the upper right corner of each box and in the standard curve. The Fe-tagged antibodies, separated with a magnet, are shown at the bottom of the boxes. The soluble antibodies and peptides are shown above. A similar situation will be achieved by other separation procedures which are specific for the unlabeled antibody.

but not its degradation products. Therefore this method may be more specific than RIA. However, such a high specificity may lead to erroneous results in cases where one of the fragments still has some of the biological activity of the native hormone.

A major disadvantage of IRMA is that biological samples must be analyzed in at least two different dilutions to ascertain that the concentration of the hormone is within the linear portion of the standard curve. When a high concentration of the hormone is present in the reaction mixture, a 'hook effect' is observed because there is a decrease in the counts of the labeled antibody owing to a limited amount of the labeled antibody or the Fe-tagged antibody, or both (Fig. 4). For those biological samples containing high hormone levels, a falsely low hormone concentration may be reported if only a single aliquot of undiluted serum is assayed because of the hook effect. Safety is the main reason that the amount of the iodinated antibody cannot be increased, since the amount in a regular IRMA assay is 5–10 times higher than that used in RIA. In addition, a higher amount of radioactive antibody will increase the blank values, thus reducing the sensitivity at the lower concentrations.

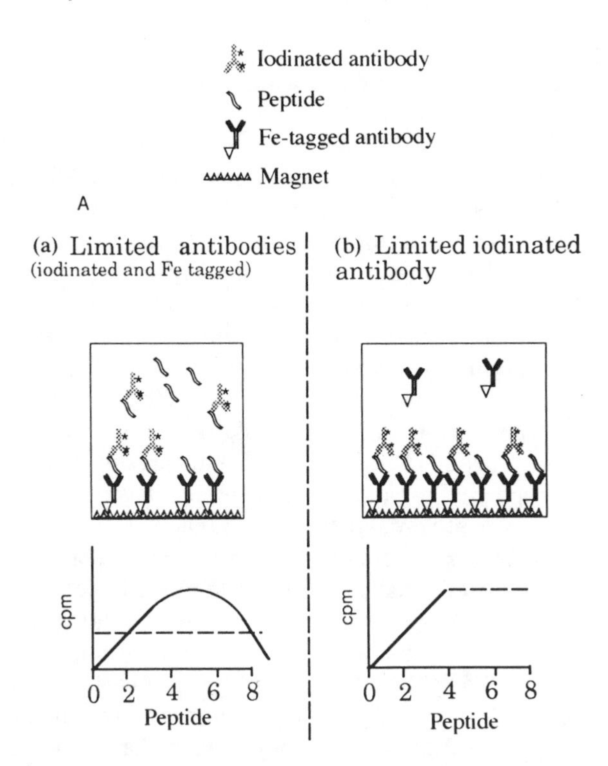

Fig. 4. Dose–response line for graduated volume of ligand sample added to an immunoradiometric assay. See Fig. 2 for explanation of the model. A 'hook effect' is observed at high concentration of ligand [in panel (a), a value of 8 gives the same result as a value of 2]. This may result from an insufficient amount of both antibodies (iodinated and Fe-tagged). If the concentration of hormone present in the sample were greater the hook effect would be greater, so that with increased volumes of the sample one would see the counts per minute bound approach the abscissa. If only the iodinated antibody is limited, the hook effect will not be seen, but this will limit the upper direction level of the ligand [values above 4 in panel (b)] In view of these relationships, it is imperative that all samples be analyzed in at least two different volumes to be certain that the level of hormone is within the linear dose–response range of the IRMA.

3 RADIORECEPTOR ASSAY

The principle of radioreceptor assay is essentially the same as that for RIA except that the hormone is bound to a specific hormone receptor in the plasma membrane, nuclei or cytosol instead of to an antibody. The specific receptors for most polypeptide hormones are on the external surface of plasma membranes, whereas biologically active steroids, as well as thyroxine and triiodothyronine, have specific intracellular receptors. Radioreceptor assay sensitivity is generally lower than that for RIA and for most in vitro bioassays. The hormone must have the appropriate conformation to interact with its receptor and consequently must be biologically active. A hormone may fail to bind to its specific receptor but yet be detected by an antibody in a RIA system. That disparity simply reflects the fact that radioimmunoassay and radioreceptor assays may detect different foci on the hormone molecule. The identity between the biological activities on the molecule and that which binds to the receptor is responsible for the major advantage of the radioreceptor assay in that non-active metabolites are not measured in this assay.

A variety of radioreceptor assays has been developed. In general, tissue is harvested from hormone-specific target organs and the receptor is isolated by the use of standard techniques. Isolated particulate plasma membrane receptor is relatively stable on storage at temperatures below −20 °C. However, solubilized polypeptide and steroid hormone receptors derived from either plasma membranes or subcellular fractions without bound ligand are unstable, as reflected by their significantly decreased capacity to bind specific hormones even when they are stored frozen for a relatively short period of time.

For radioreceptor assays, iodination of polypeptide hormones must be carried out by techniques which minimally alter the hormone's biological activity. For the preservation of biological activity, one iodine atom is introduced per molecule of hormone. Excessive iodination usually alters the hormone's biological activity much more markedly than its immunological activity. Moreover, some of the chemicals used in the iodination procedures are strong oxidizing or reducing agents, and consequently their concentration must be kept to a minimum. The variety of iodination techniques available is described above.

REFERENCES

1. S. A. Berson and R. S. Yalow, *Nature* 184, 1948 (1959).
2. I. R. Davies, E. A. Padlam and D. M. Segal, *Annu. Rev. Biochem.* **44**, 639 (1975).
3. T. L. Goodfriend, L. Levine and G. D. Fasman, *Science* **144**, 1344, (1964).
4. F. M. Richards and J. B. Knowles, *J. Mol. Biol.* **37**, 231 (1968).
5. P. Goldsmith, P. Gierschik, G. Milligan, C. G. Unson, R. Vinitsky, H. L. Malech and A. M. Spiegel, *J. Biol. Chem.* **262**, 14683 (1987).
6. J. Freund, *Annu. Rev. Microbiol.* **1**, 291 (1947).
7. A. G. Johnson and S. J. McDermott, *Lancet* **2**, 589, (1973).
8. G. Kohler and C. Milstein, *Nature* **256**, 495 (1975).
9. W. M. Hunter and F. C. Greenwood, *Nature* **194**, 495 (192).
10. A. E. Bolton and W. M. Hunter, *Biochem. J.* **133**, 529 (1973).
11. Y. Miyachi, L. Vaitukaitis, E. Nieschlag and M. B. Lipsett, *J. Clin. Endocrinol. Metab.* **34**, 23 (1972).
12. S. A. Berson and R. S. Yalow, General radioimmunoassay, in edited by S. A. Berson and R. S. Yalow, *Methods in Investigative and Diagnostic Endocrinology,* Part 1, North-Holland, Amsterdam (1973), pp. 84–120.

13. S. M. Glick and A. Kagan, Vasopressin, *Methods of Hormone Radioimmunoassay*, edited by B. M. Jaffe and H. R. Behrman, Academic Press, New York (1978), pp. 341–350.
14. R. Silverman and R. S. Yalow, *J. Clin. Invest.* **52**, 1958 (1973).
15. L. Vaitukaitis, G. T. Ross, G. D. Braunstein and P. L. Rayford, *Rec. Prog. Horm. Res.* **32**, 289 (1976).
16. S. A. Berson and R. S. Yalow, *J. Clin. Invest.* **38**, 1996 (1959).
17. A. Kagan and S. M. Glick, Oxytocin, in *Methods of Hormone Radioimmunoassay*, edited by B. M. Jaffe and H. R. Behrman, Academic Press, New York (1978), pp. 327–338.
18. S. M. Glick, J. Roth and R. S. Yalow, *Rec. Prog. Horm. Res.* **21**, 241 (1965).
19. S. J. Goldsmith, R. S. Yalow and S. A. Berson, *Diabetes* **18**, 834 (1969).
20. D. C. Robbins, P. M. Blix and A. H. Rubenstein, *Nature* **291**, 679 (1981).
21. J. W. Findling, W. C. Engeland and H. Raff, *Trends Endocrinol. Metab.* **1**, 283 (1990).

INDEX

absorption 145, 189
absorption coefficient 13
absorption edge 395
absorption of neutrons 203
ACAR 482, 487
accelerator 325
accuracy 133, 215
activate 79
activated sample 144
activation analysis 79, 121
activity 144, 249
acute effects 109
ADC 53, 247
aerosol 350
ALARA 111
ALARP 111
ALI 115
alloys 459
alpha particles 418
aluminium levels 368
Alzheimer's disease 363, 368
Am-Be 191
AMAC 197
americium-241 (^{241}Am) 396, 399, 407, 408, 411, 412
amorphous alloy 464
analogue to digital converter 53
angle of incidence 255
angular correlation of annihilation radiation 482, 487
angular distribution 311, 420
anisotropy coefficient 312
annihilation 18, 246
annihilation radiation 246, 248
annual limits of intake 115
anomalous muonium 318
anti-inhibition 481
anti-Compton 138
antibody 525
antigen 525
antiserum 529, 528
apparent mass attenuation coefficient 197
approximate methods 229
archaeological applications 355
archaeological samples 411
ash content 194
ashing 149, 151
assay sensitivity 536
assay specificity 536
asymmetric coincidence 254
asymmetry coefficient 311
asymmetry parameter 312
atherosclerosis 373
atopic dermatitis 372
attenuation 4
attenuation coefficient 6, 11–14, 423
attenuation gauge 195
Auger electrons 325, 329
austenitometers 463
average cross section 299
average stopping power 229
AVF cyclotron 72
azimuthally varing field cyclotron 72

B cells 378
backing 349
backscattering 194, 304, 421, 425, 429
BaF_2 detector 311
barn 80
beam charge 347
beam intensity 347
  monitoring 234
  reductions 222
Bechterew's disease 365
becquerel (Bq) 103
beryllium (^9Be) 63
beta particles 423
betatron 73
Bethge apparatus 150, 152

BGO 33, 170
biological applications 348
biological effects 108
biological fluids 525
biological half life 106
biological material 350
biological samples 412
biomedical application 361
bismuth germanate 33, 170
Bolton–Hunter reagent 531
bond-centered site 313, 319
borehole logging 206
boron 62, 93
boron content 204
Bragg additivity rule 258
Bragg peak 234
branching ratio 331
Breit–Rabi diagram 318
bremsstrahlung 9, 10, 137, 329, 332
  source 394
buildup factor 13
bulk measurement 163
bulk samples 295, 324
burnup 80

cadmium 60, 184, 203, 384, 412, 413
cadmium-109 (^{109}Cd) 396, 399, 407, 408, 410, 412
cadmium-lined 61
calibration 347
californium-252 (^{252}Cf) 63
carnallite 194
carriers 146, 147
cell damage 384
certified reference materials 346
channeling 308
channeltron electron detector 367
charged particle accelerator 61, 62, 81
charged particle activation 86
charged particle reactions 79, 80
charged particule detector 43
Chaudri method 227
chelating agents 160
chemical etch 238
chemical mass balance model 353
chemical quenching 480
chemical yield 155, 512
Cherenkov radiation 10
chloramine-T technique 531
chronic fatigue syndrome 380
chronic mononucleosis 383
circular streaker sampler 352
circulating antibody 540
cis-Pt chemotherapy 413
clinical trace element research 364
CMB model 353
cobalt-57 (^{57}Co) 156, 396, 407, 412
Cockroft Walton 62, 66
cofactor 364
cold neutrons 93
collimator 399, 400
collision kinematics 256
combustion 150
comparator 84, 126, 169
compound nucleus 259, 296
Compton background 129, 332
Compton continuum 55
Compton edge 55
Compton scattering (process) 12, 18, 54, 137, 397
Compton suppression 138, 181, 182
conduction band 33, 37
contamination 145
continuous background 332
conversion gain 53
cooling time 144
corrosion 456
Coster–Kronig transition 336
Coulomb barrier 218–221, 259, 294
Coulomb potential 258
Coulomb scattering 308
counting statistics 173
CPAA 89, 90, 215, 297
critical absorber 340
critical angle 303, 304
critical depth 401
cross-section 4, 5, 6, 60, 79, 85, 124, 125, 165, 221, 225
cryofixation 349
crystallization 465
curie (Ci) 103
cyclic accelerator 68
cyclotron 69, 83

DBS 489, 482
DC accelerator 66
DCA 247
dE/dx detector 44
dead time 27, 136, 225, 247, 249, 340
decay curve analysis 225, 249
decay time 144

decontamination factor 157
dees 69
defects 464, 478
delayed activation analysis 79
delayed analysis 77, 83, 96
delayed effects 108
delayed measurement 75
delayed neutrons 91
density maps 421
density measurement 201
dental amalgams 381
depletion layer 43
depletion region 41, 42
depth analysis 256
depth distribution 315
depth profiling 95, 280
depth resolution 300
depth dependence 294
dermatology 370
detection efficiency 243, 247
detection limits 128, 129, 137, 174, 215, 335, 339
detector efficiency 170
deuteron beam 62
DGNAA 94, 86, 91
DIDA 512
differential cross-section 258
digital filtering 344
dilution method 402
direct IDA 512
direct irradiation 397
direct reaction 78
dispersive recoil spectrometry 271
displacement substoichiometry 516
dissolution 148
distorted substitutional site 313
dopant 313
dopant atom 39
Doppler broadening spectroscopy 482, 489
dose 104
dose equivalent 104, 110
dose rate 104
double antibody 532
double escape peak 55
drugs 532
DRUM 352
dry ashing 149
dual energy gamma gauge 197
dynamic studies 364
dynodes 29, 33
EBS 254
ECPSSR model 329
EDE 105, 117
effective dose equivalent 105
effective half-life 106
elastic backscattering 253
elastic backscattering spectrometry 254
elastic collision 253
elastic recoil 253
elastic recoil coincidence spectrometry 271
elastic recoil detection 255
elastic recoil detection analysis 78
elastic recoil spectrometry 254, 255
elastic recoil spectrum 262
elastic scattering 20, 393
electric field gradient 311, 312
electron hole 48
electron microprobe 323
electron–hole pair 40, 54
electron–ion pair 40
electron–volt (eV) 105
electrostatic analyser 270
elemental analysis 323, 467
elemental mapping 379, 382, 384
endoergic reaction 77, 219
energy dispersive 325, 341, 362, 391
energy loss detector 271
energy loss rate 298
energy loss telescope 276, 282
energy reduction calculation 239
energy resolution 24, 170, 247
energy straggling 258, 276
energy–angle relationship 421
environmental samples 410
epitaxial 309
epithermal NAA 125, 138, 119
epithermal neutrons 58, 60, 81, 82, 85, 123, 124, 190
EPMA 323
epoxy polymer 492
Epstein–Barr virus (BEV) 383
ERCS 271
ERD 255
ERDA 78
ERS 255
essential trace elements 363, 383
etching 146, 238
EXB technique 269
excitation curve 222
excitation functions 60, 80, 222, 298

excitation source 394
exit angle 255
exoergic reaction 77, 219
exposure 103, 110, 112, 113
exposure rate 104, 115
external beam system (mode) 268, 324, 355
external irradiation 180
extrinsic semiconductor 39

F-factor 228, 250
FAST 353
fast neutron activation analysis 85
fast neutrons 58, 59, 60, 85, 123, 124, 190
fast positrons 482
fat 199, 202
field applications 164, 186
filter 347
filter paper 126
fire assay 354
fluorescence 391
fluorescence radiation 346
fluorescence yield 393, 336
flux distribution 306
FNAA 85, 86
forward recoil spectrometry 255
forward scattering 426
free radical generation 364
free volume compressibility 492
free volume holes 490
Frenkel pair 320
frequency distribution factor 312
FRS 255
funny filters 340
FWHM 48, 179

gallium arsenide 144
Gamgat gauges 197
gamma-globulin 528
gamma-ray spectra 129
gamma-ray spectrometer 51, 164, 168, 176, 179, 246
gamma–gamma coincidence 138, 256
Ge (Li) detector 46, 48, 367
Geiger counter 26, 27
Geiger tube 27
Geiger–Müller counter 26
Geiger–Müller region 28
gel filtration 532
genetic effects 108, 110
geological applications 353, 408
geological samples 149
glancing angle geometry 268
glancing node 268
graphite 348
gray 104
growth factor 540
Gy 104

hair 350
half-life 106
half-value thickness 106
HAP 145, 158
hapten 528
heavy ion Rutherford backscattering 254
heavy metals 379, 381
heavy water reactor 58
HEIR 255
heterologous radioimmuno assay 540
HgI_2 detector 399
high energy ion recoil 255
high purity germanium detector 47, 343
HIRBS 254
HIV 378
hole 37
homogeneity 402
hormone receptor 543
hormones 525
host compound 294
host lattice 305
Hostaphan 349
HPGe detector 47, 48, 170, 343
HRS 255
human leukocyte analysis 365
hydrated antimony pentoxide 145
hydrogen 185
hydrogen depth profiles 282
hydrogen profiling 254, 255, 280
hydrogen/recoil spectrometry 255
hyperfine field distribution 453

IDA 511
IgG 529
immobilization 533
immunogen 529
immunoradiometric assay 541
implants 363
impurity analysis 300
impurity content 254
in vivo 412
in vivo prompt gamma activation analysis 164

in vivo trace elements measurement 414
in-situ analysis 403, 422
in-situ prompt gamma activation analysis 164
INAA 122, 143, 356
indicator radionuclide 145
indium depth profile 280
induced activity 223
induced radioactivity 121
industrial application 403
inelastic collisions 418
inelastic scattering 21, 393
infinite thickness 401
infinitely thick 394
instrumental neutron activation analysis 122, 143
insulin 525, 528
interferences 130, 145, 224, 401
intermediate neutrons 190
internal contamination 114, 115
internal exposure 117
internal irradiation 117, 180
internal standard 230
internal standardization method 230
interstitial impurity 308
intrinsic concentration 37
intrinsic semiconductor 38
iodine-125 (^{125}I) 396, 398, 408, 412
ion backscattering 253, 274
ion backscattering spectrometer 253
ion chamber 25
ion channeling 265, 267, 293, 300
ion exchange 158
ion-induced X-ray 294
ionization chamber 24
ionization chamber region 27
ionization cross section 328, 332
ionization radiation 102, 108, 110
iridium (^{3}H)-labeled antigen 530
iridium-192 (^{192}Ir) 407
IRMA 541
IRN 145
irrigation management 191
isochronous cyclotron 72
isolated blood cell 383
isotope dilution alpha spectrometry 520
isotope dilution analysis 511
isotope ratio 511, 519
ISPGAA 164

junction-diode 41, 42

K-factor 105
K/Cl ratio 376
Kaposi's sarcoma 378
Kapton 349
kinematic factors 256
kinematic separation method 271
kinematics 421
k_0 method 85, 118
$K\alpha$ 331
$K\beta$ 331

lactoperoxide 531
late effects 109
lattice location 306, 300
lead 182, 384, 400, 412
least squares method 452
LEED 505
LEPD 505
LET 105
ligand 527
light-water reactor 50
limited proportionality 28
limits of detection 335
line shape 452
linear accelerator 69
linear attenuation coefficient 11
linear energy transfer 105
liniac 69, 84, 90
liquid levels 194
liquid scintillation counter 532
liquid scintillation counting 31
liquid scintillators 31
lithium drifted silicon detector 45
lithium drifting 45
lithium ion drifting 44
LOD 335
logging 192
logit 533
logs 192
low energy loss spectroscopy 505
LWR 58
lysosomal degranulation 360

MAC 196
magnetic spectrometer 268
magnetic transformation 463
mass absorption coefficient 337, 401
mass attenuation coefficient 12, 196, 197
matrix effect 400
matrix interference 224

MCA 53, 325
medical applications 348
Menke disease 365
mercury 194, 182, 381, 384
metal syndrome 381
metals 145
micro-PIXE 324
microanalysis 280, 320
microbeam 253
microprobe 323, 348
minimum yield 304
moderator 484
moisture content 191, 202
moisture gauges 194, 197
moisture measurement 200, 201, 202
monoclonal antibody 529, 336
Mössbauer
  absorber 444
  elements 445
  experiment, transmission 433, 434
  line intensity, angular dependence of 438–440
  line shape 452
  line shape, Lorentzian 452
  line shift, pressure dependence of 438, 439
  line width, thickness dependence of 440–441
  parameters 434–441, 435
  parameters, magnetic field dependence of 437, 438
  parameters, temperature dependence of 434–437
  pattern, elementary 447
  pattern, induced 447
  pattern, spontaneous 447
  pattern, superimposed 447
  pattern, transformed 448
  photometry 443
  source 444
  specra, analytical information from 447–455
  spectra, decomposition of 451–454
  spectra, least square fitting of 452
  spectra, magnetic splitting of 442
  spectra, measurement of 442–446
  spectra, stripping of 453
  spectrometer 443f
  spectrometer, velocity calibration of 444
  spectroscopy 433–475
  spectroscopy, analytical applications of 455–473
  spectroscopy, backscattering gamma ray 442
  spectroscopy, conversion electron 443
  spectroscopy, quantitative analysis by 454–455
  spectroscopy, source experiments 445
  spectroscopy, X ray 442
  technique, scattering 442
  technique, transmission 442
  thermal scan method 443
multi-element 294, 323, 362
multichannel analyser 53, 325
multiple scattering 259, 276
muon 315
muon depolarization 320, 321
muon spin rotation 315, 316
muonium 315, 316
Mylar 349

n, γ reaction 79
n-type semiconductor 39
Na ^{125}I 530
Na ^{131}I 530
Na/K ratio 371
NAA 84, 121, 124, 125, 294
NaI (Tl) 311
NaI (Tl) crystal 33
NaI (Tl) scintillation detector 399
NaI detector 33, 170
natural background radiation 116
NDP 95
NE213 195
near surface 337
negative muon 320
Neugat gauges 197, 200, 201
neutron activation analysis 84, 293
neutron capture 21, 79
neutron collimator 177
neutron depth profiling 95
neutron detection 25
neutron generator 61, 62, 82
neutorn reaction 81
neutron scattering 203
neutron transmission 195
neutron-induced fission 81
neutrophil granulocyte 382
nitrogen profiling 278
non-invasive 193, 202
non-destructive 323, 349

non-invasive analysis 193
non-invasive methods 193
non-peptide hormone 532
non-Rutherford scattering 258, 259
non-stochastic effects 108
normal muonium 318
NRA 293
nuclear fission 79
nuclear interference 224, 251
nuclear microscopy 361, 362
nuclear potential scattering 259
nuclear reaction 75, 77, 219
nuclear reaction analysis 293
nuclear reactor 57, 80, 81
nuclear resonance scattering 259
nuclear transmutation 80
nucleon evaporation 79

OLPGAA 164
on-line characterization 163
on-line determination 403, 407
on-line prompt gamma activation analysis 164
organic scintillator 30, 195, 200
ortho positronium 479
oxygen depth distribution 280
oxygen stoichiometry 278

p-type semiconductor 39
P/Ca ratio 368
PAA 90
PAES 500
pair production 12, 16, 18, 54
pair spectrometer 182
PAL 482, 485
para positronium 479
particle evaporation 83
particle-induced gamma ray emission 94
particle-induced prompt particle emission 94
particle-induced X-ray 323
PAS 477
peak area 343
peak to Compton ratio 247
Penning trap 501
perturbed angular correlation 314
perturbed angular distribution 311
PES 254
PESA 353
PGAA 163, 167, 176
PGEs 354
PGNAA 91, 92, 93
phantom 412
phase analysis 459
photoelectric absorption 14
photoelectric effect 14, 18, 391
photomultiplier tube 29, 33
photon activation analysis 90
photoneutrons 62
photopeak 55
physical half life 106
pick off annihilation 479
PIGA 353
PIGE 94
pile-up 52, 136, 225, 247, 268, 342, 343
PIPPE 94
PIXE 323
planar configuration 46
planar Ge detector 407
planar HPGe detector 399
platinum group elements 354
PMT 29, 33
pneumonia 378
polyclonal antibody 346
polycrystalline 312
polymers 490
polymethyl methacrylate 493
porosity logs 192
porosity measurement 191
porous media 497
positron annihilation 477
positron annihilation lifetime 482, 485
positron annihilation spectroscopy 477
positron annihilation-induced Auger electron spectroscopy 500
positron emitters 246
positron energy loss spectroscopy 505
positron ionization mass spectroscopy 501
positron microscopy 503
positron optics 484
positron polarimetry 505
positron sources 482
positron spectroscopy 484
positron–electron annihilation 477
positronium 477, 478
pre-concentration 354
precision 133, 215, 243, 346
projected range 218
prompt activation analysis 79
prompt analysis 77, 91, 96
prompt gamma detection 168

prompt gamma ray activation analysis 163
prompt gamma ray neutron activation analysis 91–93
prompt gamma rays 95, 165, 166
prompt measurement 75
proportional counter 24, 25, 399
proportional region 27
protein measurement 207
proteolytic damage 539
proton absorber 334
proton elastic scattering 254
Ps 478, 479
Ps inhibition 481
PSD 32, 200
psoriasis 371, 372
pulse shape discrimination 32, 200
pulser method 247

Q value 79, 219, 220, 294
QF 104
quality control 134, 139
quality factor 104
quenching 26
quenching of Ps 479

R 104
rad 104
radiation dose 115
radiation protection 101, 102
radiative capture 123, 124
radiative neutron capture 79
radioactive isotope 394
radioactive source 107, 394
radiochemical neutron activation analysis 143
radiochemical separations 144
radioimmunoassay 525
radioisotopes 107
radioisotopic sources 81, 83
radiolabeled antigen 536
radionuclides 223, 246
radionuclide source 403
radioreceptor assay 543
range 8, 11, 102, 124, 217, 240, 294, 297, 325, 418, 423
range straggling 294
rare earth elements 354
RBS 76, 254, 339, 362
reagent blank 151
recoil foil 241
reconstructed surface 309
REEs 354
reference material 127, 139, 244, 298
rem 106
resolution 48, 170, 341
resonance integral 82, 86, 124, 125
reverse IDA 512
REXRF 393, 403
RIA 525
Ricci methods 227
RIDA 512
RNAA 143
RNGA 356
Roentgen 103
Rutherford backscattering 253, 295, 339
Rutherford backscattering spectrometry 78, 254, 362
Rutherford energy region 253, 259, 274
Rutherford scattering 258, 259

safety regulation 101
sample chamber 346
sample handling 178
sample preparation 402
saturation factor 297
SCA 53
scanning transmission ion microscopy 370
scatter coefficient 13
scattering 189
scattering cross-section 256, 264
scattering modalities 418, 424
scattering of neutrons 194
scattering reaction 77, 78
scattering recoil coincidence spectrometry 272
Schöniger technique 149
scintillation counter 28
scintillator 29
screened Coulomb potential 303
sealed tube neutron generator 62
secondary electron bremsstrahlung 332
secondary fluorescence 343
secondary target irradiation 397
secondary X-ray emission 340
sector-focused cyclotron 72
self absorption 32, 122, 244, 245, 247
semiconductor detector 24, 35, 341
sensitivity 137, 536
separation factor 157
separation on columns 157